EXPONENTS AND RADICALS (2-1, 5-1, 5-2)

$a^n = a \cdot a \cdot a \cdot \cdots \cdot a$	n factors of a, $n \in N$
$a^0 = 1, \quad a \neq 0$	
$a^{-n} = \dfrac{1}{a^n}, \quad a \neq 0, n \in R$	
$b^{m/n} = \sqrt[n]{b^m}$	nth root of b^m

CARTESIAN COORDINATES (7-1, 7-2)

(x_1, y_1)	Coordinates of a point P_1
$d = \sqrt{(x_2 - x_1)^2 + (y_2 - y_1)^2}$	Distance between P_1 and P_2
$m = \dfrac{y_2 - y_1}{x_2 - x_1}, \quad x_1 \neq x_2$	Slope of a line through P_1 and P_2

EQUATIONS OF A LINE (7-2)

$x = C$	Vertical line
$y = C$	Horizontal line
$y = mx + C$	Slope–intercept form
$y - y_1 = m(x - x_1)$	Point–slope form

CONIC SECTIONS (7-5, 7-6)

$(x - h)^2 + (y - k)^2 = R^2$	Circle
$y^2 = 4ax, \quad x^2 = 4ay$	Parabola
$\dfrac{x^2}{a^2} + \dfrac{y^2}{b^2} = 1$	Ellipse
$\dfrac{x^2}{a^2} - \dfrac{y^2}{b^2} = 1, \quad \dfrac{y^2}{b^2} - \dfrac{x^2}{a^2} = 1$	Hyperbola

QUADRATIC FORMULA (6-2)

If $ax^2 + bx + c = 0$, $a \neq 0$, then

$$x = \frac{-b \pm \sqrt{b^2 - 4ac}}{2a}$$

FUNCTION NOTATION (9-1)

$f(x)$	Value of f at x
$f^{-1}(x)$	Value of inverse at x

$f^{-1}[f(x)] = f[f^{-1}(x)] = x$ if f is one-to-one

VARIATION (9-5)

$y = kx, \quad k \neq 0$	Direct
$y = \dfrac{k}{x}, \quad k \neq 0$	Inverse
$y = kxy, \quad k \neq 0$	Joint

EXPONENTIAL FUNCTIONS (10-1)

$f(x) = b^x, \quad b > 0, b \neq 1$

LOGARITHMIC FUNCTIONS (10-2, 10-4)

$f(x) = \log_b x, \quad b > 0, b \neq 1$

$\log_{10} x$	Common logarithm
$\log_e x = \ln x$	Natural logarithm

$y = \log_b x$ is equivalent to $x = b^y$

4, 6, 7

INTERMEDIATE ALGEBRA
Structure and Use

FOURTH EDITION

INTERMEDIATE ALGEBRA
Structure and Use

Raymond A. Barnett
Merritt College

Thomas J. Kearns
Northern Kentucky University

McGRAW-HILL PUBLISHING COMPANY

New York St. Louis San Francisco Auckland Bogotá Caracas Hamburg Lisbon London Madrid Mexico Milan Montreal New Delhi Oklahoma City Paris San Juan São Paulo Singapore Sydney Tokyo Toronto

This book was set in Times Roman
by York Graphic Services, Inc.
The editors were Robert A. Weinstein and
James W. Bradley;
the designer was Joan E. O'Connor;
the production supervisor was Leroy A. Young.
The cover was designed by John Hite.
New drawings were done by Fine Line Illustrations, Inc.
R. R. Donnelley & Sons Company was
printer and binder.

INTERMEDIATE ALGEBRA
Structure and Use

2 3 4 5 6 7 8 9 0 DOC/DOC 9 9 8 7 6 5 4 3 2 1 0

ISBN 0-07-003946-1

Library of Congress Cataloging-in-Publication Data

Barnett, Raymond A.
Intermediate algebra: structure and use / Raymond A. Barnett, Thomas J. Kearns. — 4th ed.
p. cm.
1. Algebra. I. Kearns, Thomas J. II. Title.
QA154.2.B38 1990
512.9—dc20 89-27007

ABOUT THE AUTHORS

Raymond A. Barnett is an experienced teacher and author. He received his B.A. in Mathematical Statistics from the University of California at Berkeley and his M.A. in Mathematics from the University of Southern California. He then went on to become a member of the Department of Mathematics at Merritt College and head of that department for four years. He is a member of the Mathematical Association of America, the National Council of Teachers of Mathematics, and the American Association for the Advancement of Science. He is the author or coauthor of 17 books in mathematics that are still in print—all with a reputation for extremely readable prose and high quality mathematics.

Thomas J. Kearns received his B.S. from the University of Santa Clara and his M.S. and Ph.D. from the University of Illinois. After several years of teaching at the University of Delaware, he was appointed to the faculty at Northern Kentucky University. He served as chairman of the Department of Mathematical Sciences there for 10 years. He is a member of the professional organizations listed above and has coauthored four texts with Raymond A. Barnett and has also coauthored texts in college algebra and elementary statistics.

CONTENTS

PREFACE

This is an algebra text written for students with only elementary algebra as background and for students who need a review of intermediate algebra before proceeding further. The improvements in this fourth edition evolved out of generous responses from users of the third edition. Most of the changes in this edition have been made with the aim of making the text even more accessible to students.

PRINCIPAL CHANGES FROM THE THIRD EDITION

1. Basic material on **solving equations** now appears earlier in the text (Section 1-6) and is used thereafter when appropriate.
2. **Word problems** are introduced earlier (Section 1-6) and are more evenly dispersed throughout the text, rather than concentrated in one or two chapters. **Mixture problems** are deferred until after systems of equations have been introduced, and both one- and two-variable methods are used to solve them (Sections 8-2, 8-3).

3. The method of **solving equations by factoring** has been moved to an earlier section of the text (Section 2-7) and is used thereafter when appropriate. **Exponent properties** not needed for polynomial algebra have been deferred to Chapter 5.
4. A new section on **graphing quadratic polynomials** (Section 7-4) has been added.
5. The material on **complex numbers**, **functions**, and **graphing polynomials** has been simplified. Functions are introduced without formal consideration of relations. Polynomials are graphed as usual, not using nested interval forms (Sections 5-8, 9-1, 9-3).
6. A new appendix is included to provide an alternative approach to **setting up word problems** (Appendix A).
7. The text has been **shortened** by eliminating several peripheral topics: nested intervals, relations, evaluation of logarithms by tables, interpolation, matrix operations.
8. Many **exercise sets** have been replaced and/or expanded. More problems involve fractions or decimals.
9. There are more **worked-out examples** with **matched problems**. There is more **boxed material** for emphasis and **schematics** have been added for clarity. **Applications** have been kept current.

10. Areas that are a source of **common student errors** are highlighted with a special "**caution**" symbol.

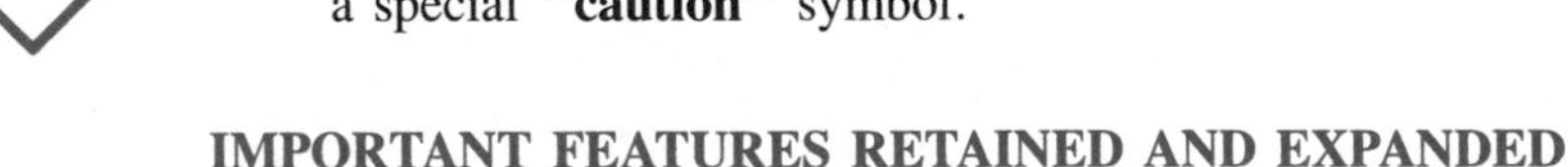

IMPORTANT FEATURES RETAINED AND EXPANDED FROM THE THIRD EDITION

1. The text is still **written for student comprehension**. Each concept is illustrated with an example, followed by a parallel problem with an answer (given at the end of the section) so that a student can immediately check his or her understanding of the concept. These follow-up problems also encourage active rather than passive reading of the text.
2. An **informal style** is used for exposition, definitions, and results. A **spiraling technique** is employed for many topics; that is, a topic is introduced in a relatively simple framework and then is returned to one or more times in successively more complex forms. For example:

   Solving equations: Sections 1-6, 2-7, 3-6, 4-1 to 4-5, 6-1 to 6-5, and 10-5
   Word problems: Sections 1-6, 2-7, 4-1 to 4-3, 6-3, 8-2, 8-3, 8-5, 9-5, and 10-5
   Graphing equations: Chapters 7 to 10
   Inequalities: Sections 4-4, 4-5, 6-5, and 8-5

   The use of this spiraling technique continues from the companion text *Elementary Algebra: Structure and Use (Fifth Edition).*
3. The text includes **more than 3,800 carefully selected and graded problems**. The exercises are divided into A, B, and C groupings: the A problems are easy and routine, the B problems are more challenging but still

emphasize mechanics, and the C problems are a mixture of theoretical and difficult mechanics. In short, the text is designed so that an average or below-average student will be able to experience success and a very capable student will be challenged.

4. The subject matter is related to the real world through many carefully selected **realistic applications** from the physical sciences, business and economics, life sciences, and social sciences. Thus, the text is equally suited for students interested in any of these areas.
5. As in the third edition, following each example is a **matched problem** to encourage active involvement in reading the text. The answers to these matched problems are at the end of each section just before the exercise set.
6. **Answers** to all chapter review exercises and to all odd-numbered problems from the other exercises are in the back of the book.
7. **Historical comments** are included for interest.
8. **Chapter review sections** include a summary of the chapter with all important terms and symbols, and a comprehensive review exercise. Answers to all review exercises are included in the back of the book and are keyed (with numbers in italics) to corresponding text sections.

ADDITIONAL STUDENT AIDS

1. **Think boxes** (dashed boxes) are used to enclose steps that are usually performed mentally (see Sections 1-4, 1-6, 2-1, 3-5).
2. **Annotation** of examples and developments is found throughout the text to help students through critical stages (see Sections 1-4, 1-6, 2-3, 3-4).
3. **Functional use of a second color** guides students through critical steps (see Sections 1-6, 2-1, 2-3, 3-5).
4. **Summaries** of formulas and symbols (keyed to the sections in which they are introduced) and the metric system are inside the front and back covers of the text for convenient reference.
5. A **solutions manual** is available at a nominal cost through a bookstore. The manual includes detailed solutions to all odd-numbered problems and all chapter review exercises.

INSTRUCTOR'S AIDS

This supplements package contains a wide and varied assortment of useful instructor's aids. They include:

1. A **student's solutions manual** contains detailed solutions to every chapter review exercise as well as all other odd-numbered problems. This supplement is available to students at a nominal fee.
2. An **answer manual** (which slips inside the back of the text) contains answers to the even-numbered problems not answered in the text. This supplement is available to adopters without charge.

3. An **instructor's resource manual** provides sample tests (chapter, mid-term, and final), transparency masters, and additional teaching suggestions and assistance.
4. A **computer testing system** is also available to adopters without cost. This system provides the instructor with numerous test questions from the text. Several test question types are available including multiple choice, open-ended, matching, true-false, and vocabulary. The testing system enables the instructor to find these questions by several different criteria. In addition, instructors may edit their own questions.
5. A printed and bound **test bank** is also available. This bank is a hard copy listing of the questions found in the computerized version.

ERROR CHECK

Because of the careful checking and proofing by a number of very competent people (acting independently), the authors and publisher believe this book to be substantially error-free. If any errors remain, the authors would be grateful if corrections were sent to: Mathematics Editor, College Division, 43rd floor, McGraw-Hill Book Company, 1221 Avenue of the Americas, New York, New York 10020.

ACKNOWLEDGMENTS

In addition to the authors, many others are involved in the publication of a book. The authors wish to thank the many users of the third edition for their kind remarks and helpful suggestions that were incorporated into this fourth edition. We particularly wish to thank the following for their detailed reviews: Margaret D. Dolgas, University of Delaware; Kendall Griggs, Hutchinson Community College; Martin Johnson, Montgomery Community College; Herbert F. Kramer, Longview Community College; Grace Malaney, Donnelly College; Peggy Miller, Kearney State College; Analy Scorsone, Lexington Community College; Mark Serebransky, Camden County College; Jane Sieberth, Franklin University; Deborah H. White, College of the Redwoods; Brad Wind, Marshall University; and Kelly Wyatt, Umpqua Community College.

We also wish to thank Fred Safier for his careful preparation of the Solutions Manual accompanying this book; Richard Morel, a developmental editor, for his detailed reading of the manuscript, his many helpful suggestions, and for his work on several of the text's supplements; Sr. Margaret Anne Kraemer, Northern Kentucky University, and Marie B. Jump, Erlanger-Elsmere (KY) School District, for their detailed checking of examples and exercises; Karen Hughes, for her editorial and coordinating work on many of the text's supplements; Laura Gurley, for her coordination of the reviews and manuscript; and Robert Weinstein, mathematics editor, for his support and useful ideas about the series as a whole.

Raymond A. Barnett
Thomas J. Kearns

TO THE STUDENT

The following suggestions will help you get the most out of this book and your efforts.

As you study the text we suggest a five-step process. For each section:

1. Read the mathematical development.
2. Work through the illustrative example.
3. Work the matched problem.
4. Review the main ideas in the section.
5. Work the assigned exercises at the end of the section.

(Steps 1–3: Repeat the 1-2-3 cycle until the section is finished.)

All of this should be done with plenty of paper, pencils, and a wastebasket at hand. In fact, no mathematics text should be read without pencil and paper in hand; mathematics is not a spectator sport. Just as you cannot learn to swim by watching someone else, you cannot learn mathematics simply by reading worked examples—you must work problems, lots of them.

If you have difficulty with the course, then, in addition to doing the regular assignments, spend more time on the examples and matched problems and work more A exercises, even if they are not assigned. If the A exercises continue to be difficult for you, see your instructor. If you find the course too easy, then work more C exercises, even if they are not assigned. If the C exercises are consistently easy for you, you are probably ready to start a higher level course. See your instructor.

Raymond A. Barnett
Thomas J. Kearns

INTERMEDIATE ALGEBRA
Structure and Use

1

PRELIMINARIES

Algebra is often referred to as "generalized arithmetic." In arithmetic we deal with basic arithmetic operations (addition, subtraction, multiplication, and division) on specific numbers. In algebra we continue to use all that we know in arithmetic, but, in addition, we reason and work with symbols that represent or are placeholders for one or more numbers. The rules for manipulating and reasoning with these symbols depend, in large part, on certain properties of numbers (since the symbols represent numbers). In this chapter we will review important number systems and some of their basic properties that determine how we can manipulate algebraic expressions. The final section of the chapter reviews the basic concepts involved in solving equations.

1-1 BASIC CONCEPTS

- Sets
- The Set of Real Numbers
- The Real Number Line

In this section, we will summarize the familiar number systems of arithmetic and beginning algebra, all contained in the set of real numbers identified with points on the number line. We begin with a brief introduction to the ideas and terminology of sets.

SETS

Our use of the word "set" will not differ appreciably from the way it is used in everyday language. Words such as "set," "collection," "bunch," and "flock" all convey the same idea. Thus, we think of a **set** as a collection of objects with the important property that given any object we can tell whether it is or is not in the set. Capital letters, such as A, B, and C, are often used to designate particular sets. For example,

$$A = \{3, 5, 7\} \qquad B = \{4, 5, 6\}$$

specify sets A and B consisting of the numbers enclosed in the braces $\{\ \}$.

Each object in a set is called a **member** or **element** of the set. Symbolically:

$a \in A$	means	"a is an element of set A"
$a \notin A$	means	"a is not an element of set A"

Referring to sets A and B above, we see that

$5 \in a$ 5 is an element of set A.
$3 \notin B$ 3 is not an element of set B.

A set is usually described in one of two ways:

1. By **listing** the elements between braces $\{\ \}$:

   $\{3, 5, 7\}$

   Note: The order in which elements are listed is irrelevant. Also, a given element is not listed more than once.
2. By enclosing a **rule** within braces $\{\ \}$ that determines the elements in the set:

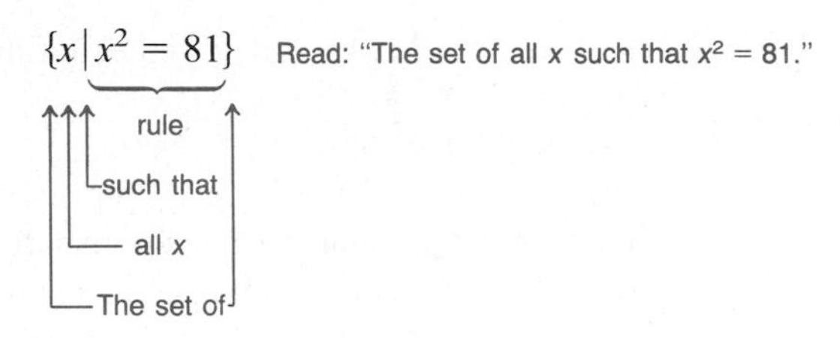

Example 1 Let A be the set of all numbers x such that $x^2 = 25$.

(A) Specify A by the listing method.
(B) Specify A by the rule method.
(C) Indicate true (T) or false (F): $5 \in A$ $\quad -5 \notin A$ $\quad 25 \in A$

Solution **(A)** *Listing method* $A = \{-5, 5\}$
(B) *Rule method* $A = \{x \mid x^2 = 25\}$ Read: "The set of all x such that $x^2 = 25$."
(C) $5 \in A$ is true since 5 does belong to the set A.
$-5 \notin A$ is false since -5 belongs to A.
$25 \in A$ is false since 25 is not one of the two numbers in set A.

Problem 1† Let B be the set of all x such that $x^2 = 49$.

(A) Specify B by the listing method.
(B) Specify B by the rule method.
(C) Indicate true (T) or false (F): $7 \in B$ $\quad 7 \notin B$ $\quad 49 \notin B$

The letter x introduced above is a variable. In general, a **variable** is a symbol used to represent unspecified elements from a set with two or more elements. (This set is called the **replacement set** for the variable.) A **constant**, on the other hand, is a symbol that names exactly one object. The symbol "8" is a constant, since it always names the number eight.

The introduction of variables into mathematics occurred about A.D. 1600. A French mathematician, François Vieta (1540–1603), is singled out as the one mainly responsible for this new idea. Many mark this point as the beginning of modern mathematics.

Some sets are **finite** (there is one counting number that indicates the total number of elements), some sets are **infinite** (in counting the elements, we never come to an end), and some sets are **empty** (the set contains no elements). Empty sets are also called **null** sets. Symbolically:

> $\varnothing$ represents "the empty set."

† Answers to matched problems following examples are located at the end of each section before the exercise set.

Example 2 Indicate whether the set is finite, infinite, or empty:

(A) $A = \{2, 4, 6\}$
(B) $N = \{1, 2, 3, 4, 5, \ldots\}$, the set of *natural,* or *counting numbers*. The three dots indicate that the pattern continues without end.
(C) $C = \{x \mid x \text{ is a counting number between 2 and 3}\}$

Solution **(A)** This set has three elements and is, therefore, finite.
(B) This set is infinite.
(C) There are no such numbers; this set is empty.

Problem 2 Indicate whether the set is finite, infinite, or empty:

(A) $U = \{x \mid x \text{ is a counting number less than } 0\}$
(B) $E = \{2, 4, 6, \ldots\}$
(C) $G = \{x \mid x^2 = 25\}$

If each element of set A is also an element of set B, we say that A is a **subset** of set B. The set of all women in a class is a subset of the whole class. (Note that the definition of a subset allows a set to be a subset of itself.) If two sets have exactly the same elements (the order of listing does not matter), the sets are said to be **equal**. Set A is equal to set B if and only if A is a subset of B and B is a subset of A.

Symbolically:

Subsets

$A \subset B$	means	"A is a subset of B"†	$\{3, 5\} \subset \{3, 5, 7\}$
$A = B$	means	"A is equal to B"	$\{4, 6\} = \{6, 4\}$

It is useful to note that:

$\varnothing$ is a subset of every set.

This is logically correct because it is true that every element of $\varnothing$ is an element of any given set, since $\varnothing$ has no elements. (Note that the symbol $\varnothing$ is not enclosed by braces.)

† Some texts use a variation of this notation where $A \subseteq B$ denotes that A is a subset of B and possibly equal to B, while $A \subset B$ denotes that A is a "proper" subset of B, that is, a subset not equal to B.

Do not confuse the two symbols $\subset$ and $\in$. The former is used only between two sets, and the latter is used only between an element of a set and a set. Note the following:

$4 \in \{2, 4, 6\}$	Correct usage of $\in$
~~$4 \subset \{2, 4, 6\}$~~	Incorrect usage of $\subset$
~~$\{4\} \in \{2, 4, 6\}$~~	Incorrect usage of $\in$
$\{4\} \subset \{2, 4, 6\}$	Correct usage of $\subset$

Example 3 Let $A = \{-3, 0, 5\}$, $B = \{0, 5, -3\}$, and $C = \{0, 5\}$. Indicate true (T) or false (F):

(A) $C \in A$ **(B)** $C \subset A$ **(C)** $A = B$
(D) $A \subset B$ **(E)** $A = C$ **(F)** $\varnothing \subset A$

Solution

(A) False: the set C is not one of the elements in the set A.
(B) True: each of the objects in C is also in A.
(C) True: A and B consist of the same three numbers.
(D) True: every object in A is also in B.
(E) False: A and C do not consist of the same numbers; -3 belongs to A but not to C.
(F) True: $\varnothing$ is a subset of every set.

Problem 3 Let $M = \{-4, 6\}$, $N = \{6, -4\}$, and $P = \{-4\}$. Indicate true (T) or false (F):

(A) $M \neq N$ **(B)** $P \subset N$ **(C)** $P \in N$
(D) $N \subset M$ **(E)** $\varnothing \subset P$ **(F)** $M \subset P$

THE SET OF REAL NUMBERS

The real number system is the number system you have used most of your life. In algebra we are interested in manipulating symbols in order to change or simplify algebraic expressions and to solve algebraic equations. Because many of these symbols represent real numbers, it is important to briefly review the set of real numbers and some of its important subsets (see Table 1 on page 6). Figure 1, also on page 6, illustrates how these sets of numbers are related to one another.

The set of integers contains all the natural numbers and something else (their negatives and 0). The set of rational numbers contains all the integers (for example, 5 can be expressed as the ratio of two integers, $\frac{5}{1}$) and something else (noninteger ratios of integers). And the set of real numbers contains all

TABLE 1 THE SET OF REAL NUMBERS

SYMBOL	NUMBER SYSTEM	DESCRIPTION	EXAMPLES
N	Natural numbers	Counting numbers (also called positive integers)	1, 2, 3, . . .
J	Integers	Set of natural numbers, their negatives, and 0	. . . , −2, −1, 0, 1, 2, . . .
Q	Rational numbers	Any number that can be represented as a/b, where a and b are integers and $b \neq 0$	$-4; \frac{-3}{5}; 0; 1; \frac{2}{3};$ 3.67
R	Real numbers	Set of all rational and irrational numbers (the irrational numbers are all the real numbers that are not rational)	$-4; \frac{-3}{5}; 0; 1; \frac{2}{3};$ $3.67; \sqrt{2}; \pi;$ $\sqrt[3]{5}$

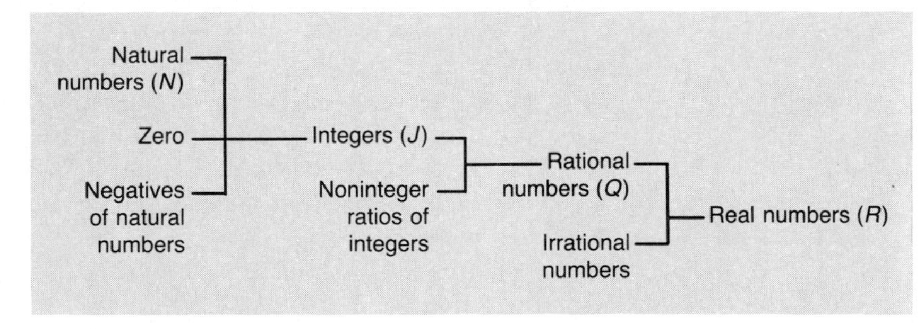

FIGURE 1 The real number system

the rational numbers and something else (irrational numbers). In short, N is a subset of J, J is a subset of Q, and Q is a subset of R. Symbolically,

$$N \subset J \subset Q \subset R$$

Rational numbers have repeating decimal representations, whereas irrational numbers have infinite nonrepeating decimal representations.† For example, the decimal representations of the rational numbers 2, $\frac{4}{3}$, and $\frac{5}{11}$ are, respectively,

$$2 = 2.0000\ldots \qquad \tfrac{4}{3}=1.333\ldots \qquad \tfrac{5}{11}=0.454545\ldots$$

whereas those of the irrational numbers $\sqrt{2}$ and π are, respectively,

$$\sqrt{2} = 1.41421356\ldots \qquad \pi = 3.14159265\ldots$$

† It is instructive to experiment with your hand calculator by dividing any integer by another nonzero integer. Also, see what happens when you take the square root of any number that is not a perfect square. More will be said about irrational numbers in Chapter 5.

THE REAL NUMBER LINE

A one-to-one correspondence exists between the set of real numbers and the set of points on a line; that is, each real number corresponds to exactly one point, and each point to exactly one real number. A line with a real number associated with each point, and vice versa, as in Figure 2, is called a **real number line**, or simply a **real line**. Each number associated with a point is called the **coordinate** of the point. The point with coordinate 0 is called the **origin**. The arrow on the number line indicates a positive direction; the coordinates of all points to the right of the origin are called **positive real numbers**, and those to the left of the origin are called **negative real numbers**.

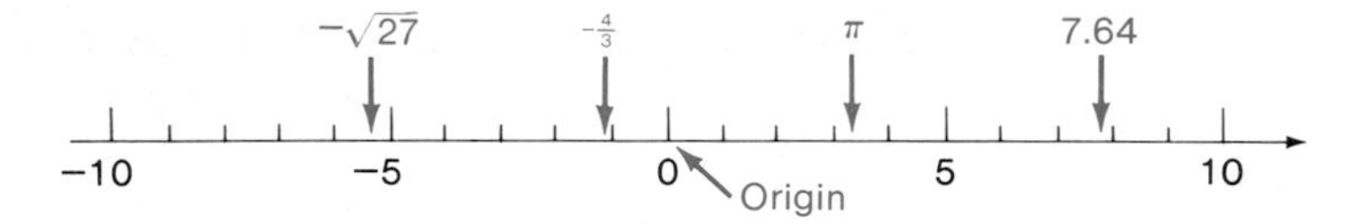

FIGURE 2 A real number line

ANSWERS TO MATCHED PROBLEMS

1. **(A)** $\{-7, 7\}$ **(B)** $\{x \mid x^2 = 49\}$ **(C)** T, F, T
2. **(A)** Empty **(B)** Infinite **(C)** Finite
3. **(A)** F **(B)** T **(C)** F **(D)** T **(E)** T **(F)** F

EXERCISE 1-1

A *In Problems 1–12, indicate true (T) or false (F).*

1. $3 \in \{2, 3, 4\}$

2. $4 \in \{3, 5, 7\}$

3. $5 \notin \{2, 3, 4\}$

4. $6 \notin \{2, 4, 6, 8\}$

5. $\{2, 3\} \subset \{1, 2, 3\}$

6. $\{2, 4\} \subset \{1, 2, 3\}$

7. $\{7, 3, 5\} = \{3, 5, 7\}$

8. $\{3, 1, 2\} \subset \{1, 2, 3\}$

9. $\varnothing \subset \{2, 5\}$

10. $\varnothing \subset \{1, 3\}$

11. $\{7\} \in \{3, 5, 7\}$

12. $\{7\} \subset \{3, 5, 7\}$

B **13.** Give an example of a negative integer, an integer that is neither positive nor negative, and a positive integer.

14. Give an example of a negative rational number, a rational number that is neither positive nor negative, and a positive rational number.

15. Give an example of a rational number that is not an integer.

16. Give an example of an integer that is not a natural number.

Write each set in Problems 17–28 using the listing method; that is, list the elements between braces.

17. $\{x \mid x$ is a counting number between 5 and 10$\}$

18. $\{x \mid x$ is a counting number between 4 and 8$\}$

19. $\{x \mid x \text{ is a letter in ``}\textit{status}\text{''}\}$

20. $\{x \mid x \text{ is a letter in ``}\textit{Illinois}\text{''}\}$

21. $\{x \mid x \text{ was a woman president of the United States}\}$

22. $\{x \mid x \text{ is a month starting with the letter } B\}$

23. $\{x \mid x - 5 = 0\}$ **24.** $\{x \mid x + 3 = 0\}$

25. $\{x \mid x + 9 = x + 1\}$ **26.** $\{x \mid x - 3 = x + 2\}$

27. $\{x \mid x^2 = 4\}$ **28.** $\{x \mid x^2 = 9\}$

29. Indicate which of the following are true:
(A) All natural numbers are integers.
(B) All real numbers are irrational.
(C) All rational numbers are real numbers.

30. Indicate which of the following are true:
(A) All integers are natural numbers.
(B) All rational numbers are real numbers.
(C) All natural numbers are rational numbers.

31. Each of the following real numbers lies between two successive integers on a real number line. Indicate which two:
(A) $\frac{15}{4}$ **(B)** $-\frac{4}{3}$ **(C)** $-\sqrt{7}$

32. Each of the following real numbers lies between two successive integers on a real number line. Indicate which two:
(A) $\frac{25}{7}$ **(B)** $-\frac{11}{4}$ **(C)** $-\sqrt{19}$

C **33.** If $A = \{1, 2, 3, 4\}$ and $B = \{2, 4, 6\}$, find:
(A) $\{x \mid x \in A \text{ or } x \in B\}$† **(B)** $\{x \mid x \in A \text{ and } x \in B\}$

34. If $M = \{-2, 0, 2\}$ and $N = \{-1, 0, 1, 2\}$, find:
(A) $\{x \mid x \in M \text{ or } x \in N\}$ **(B)** $\{x \mid x \in M \text{ and } x \in N\}$

35. Given the sets of numbers N (natural numbers), J (integers), Q (rational numbers), and R (real numbers), indicate to which sets each of the following numbers belongs:
(A) -3 **(B)** 3.14 **(C)** π **(D)** $\frac{2}{3}$

36. Given the sets of numbers N, J, Q, and R (see Problem 35), indicate to which sets each of the following numbers belongs:
(A) 8 **(B)** $\sqrt{2}$ **(C)** -1.414 **(D)** $\frac{-5}{2}$

† Unless otherwise stated, we will use ''or'' as it is usually used in mathematics—as an inclusive or. Thus, in this case, x is an element of the set if $x \in A$ or $x \in B$, or both.

37. If $c = 0.151515\ldots$, then $100c = 15.1515\ldots$ and

$$\begin{aligned} 100c - c &= (15.1515\ldots) - (0.151515\ldots) \\ 99c &= 15 \\ c &= \tfrac{15}{99} = \tfrac{5}{33} \end{aligned}$$

Proceeding similarly, convert the repeating decimal $0.090909\ldots$ into a fraction. (All repeating decimals are rational numbers, and all rational numbers have repeating decimal representations.)

38. Repeat Problem 37 for $0.181818\ldots$.

Express each number as a decimal fraction to the capacity of your calculator. Observe the repeating decimal representation of the rational numbers and the apparent nonrepeating decimal representation of the irrational numbers.

39. **(A)** $\frac{8}{9}$ **(B)** $\frac{3}{11}$ **(C)** $\sqrt{5}$ **(D)** $\frac{11}{8}$

40. **(A)** $\frac{13}{6}$ **(B)** $\sqrt{21}$ **(C)** $\frac{7}{16}$ **(D)** $\frac{29}{111}$

APPLICATIONS†

41. The executive committee of a student council consists of a president, vice president, secretary, and treasurer and is denoted by the set $\{P, V, S, T\}$. How many two-person subcommittees are possible; that is, how many two-element subsets can be formed?

42. How many three-person subcommittees are possible in Problem 41?

1-2 EQUALITY AND INEQUALITY

- Algebraic Expressions
- Equality Relation
- Inequality Relation

Basic algebraic terminology as well as equality and inequality relations are reviewed in this section.

ALGEBRAIC EXPRESSIONS

An **algebraic expression** is a meaningful symbolic form involving constants, variables, mathematical operations, and grouping symbols. For example,

$$\begin{array}{lll} 2 + 8 & 4 \cdot 3 - 7 & 16 - 3(7 - 4) \\ 5x - 3y & 7(x + 2y) & 4\{u - 3[u - 2(u + 1)]\} \end{array}$$

are all algebraic expressions.

† Applications will often be placed after the C-level exercises or calculator problems and may vary in level of difficulty from easy (no stars) and moderately difficult (marked ★) to difficult (marked ★★). In short, they are not necessarily C-level problems in difficulty.

Two or more algebraic expressions (each taken as a single entity) joined by plus or minus signs are called **terms**. (For reasons that will become clear in Chapter 2, a term includes the sign that precedes it.) Two or more algebraic expressions joined by multiplication are called **factors**. For example,

$$\underbrace{3(x - y)}_{\text{Term}} + \underbrace{(x + y)(x - y)}_{\text{Term}}$$

has two terms, $3(x - y)$ and $(x + y)(x - y)$, and each term has two factors. The first term has factors 3 and $(x - y)$, and the second term has factors $(x + y)$ and $(x - y)$. A term may contain several factors, and a factor may contain several terms.

EQUALITY RELATION

The use of an **equality sign (=)** between two expressions asserts that the two expressions are names or descriptions of exactly the same object. The symbol $\neq$ means **is not equal to**. Statements involving the use of an equality or inequality sign may be true or they may be false:

$15 - 3 = 4 \cdot 3$ True statement

$7 - 2 = \frac{8}{2}$ False statement

$7 - 2 \neq \frac{8}{2}$ True statement

It is interesting to note that the equality sign did not appear until rather late in history—the sixteenth century. It was introduced by the English mathematician Robert Recorde (1510–1558).

If two algebraic expressions involving at least one variable are joined with an equal sign, the resulting form is called an **algebraic equation**. The following are algebraic equations in one or more variables:

$$2x - 3 = 3(x - 5) \qquad a + b = b + a \qquad 3x + 5y = 7$$

Since a variable is a placeholder for constants from a given replacement set, an equation is neither true nor false as it stands; it does not become true or false until the variables have been replaced by constants. Formulating algebraic equations is an important first step in solving many practical problems using algebraic methods.

Example 4 Translate each statement into an algebraic equation using x as the only variable:

(A) 5 times a number is 3 more than twice the number.
(B) 4 times a number is 5 less than twice the number.

Solution **(A)** Let x = The unknown number; then the statement translates as follows:

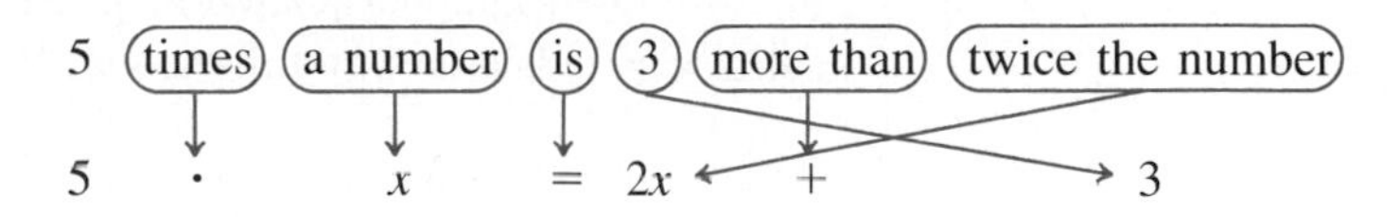

Thus,

$$5x = 2x + 3$$

(B) Let x = The unknown number; then the statement translates as follows:

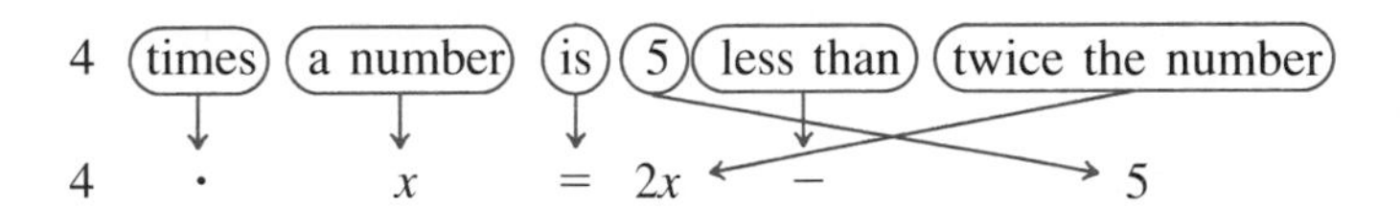

[*Note:* "Less than" in this context means "subtracted from."] Thus,

$$4x = 2x - 5 \quad \text{Not } 4x = 5 - 2x$$

Problem 4 Translate each statement into an algebraic equation using x as the only variable:

(A) 7 is 3 more than a certain number.
(B) 12 is 9 less than a certain number.
(C) 3 times a certain number is 6 less than twice that number.
(D) If 6 is subtracted from a certain number, the difference is twice a number that is 4 less than the original number.

Several important properties of the equality symbol (=) follow directly from its logical meaning. These properties must hold any time the symbol is used.

Basic Properties of Equality

If a, b, and c are names of objects, then:

1.	$a = a$.	REFLEXIVE PROPERTY
2.	If $a = b$, then $b = a$.	SYMMETRIC PROPERTY
3.	If $a = b$ and $b = c$, then $a = c$.	TRANSITIVE PROPERTY
4.	If $a = b$, then either may replace the other in any statement without changing the truth or falsity of the statement.	SUBSTITUTION PRINCIPLE

The properties of equality are used extensively throughout mathematics. For example, using the symmetric property, we may reverse the left and right sides of an equation any time we wish. That is,

If $A = P + Prt,$ then $P + Prt = A$

Using the transitive property, we find that if

$2x + 3x = (2 + 3)x$ and $(2 + 3)x = 5x$

then

$2x + 3x = 5x$

And, finally, if we know that

$C = \pi D$ and $D = 2R$

then, using the substitution principle, D in the first formula may be replaced by $2R$ from the second formula to obtain

$C = \pi(2R) = 2\pi R$

INEQUALITY RELATION

We now turn to the **inequality** or **order relation**. This relation has to do with "less than" and "greater than." If a and b are any real numbers and $a \neq b$, then a must be "less than" b or a must be "greater than" b.

Just as we use = to replace the words "is equal to," we will use the **inequality symbols** $<$ and $>$ to represent "is less than" and "is greater than," respectively. Thus, we can write the following symbolic forms and their corresponding equivalent verbal forms:

Inequality Symbols

$a < b$	a is less than b	$5 < 8$	
$a > b$	a is greater than b	$8 > 5$	
$a \leq b$	a is less than or equal to b	$5 \leq 8$	$5 \leq 5$
$a \geq b$	a is greater than or equal to b	$8 \geq 5$	$5 \geq 5$

It no doubt seems obvious to you that

$5 < 8$

is true, but it may not seem equally obvious that

$$-8 < -5 \qquad \tfrac{3}{7} < \tfrac{17}{38} \qquad \sqrt{5} < \tfrac{43}{19}$$

To make the inequality relation precise so that we can interpret it relative to *all* real numbers, we need a precise definition of the concept.

Definition of $a < b$ and $b > a$

For a and b real numbers, we say that **a is less than b** or **b is greater than a** and write

$$a < b \qquad \text{or} \qquad b > a$$

if there exists a positive real number p such that $a + p = b$ (or equivalently $b - a = p$).

We would expect that if a positive number is added to *any* real number, the sum will be larger than the original. That is essentially what the definition states. When we write

$$a \leq b$$

we mean **a is less than or equal to b**, and when we write

$$a \geq b$$

we mean that **a is greater than or equal to b**.

The inequality symbols $<$ and $>$ have a very clear geometric interpretation on the real number line. If $a < b$, then a is to the left of b; if $c > d$, then c is to the right of d (Figure 3).

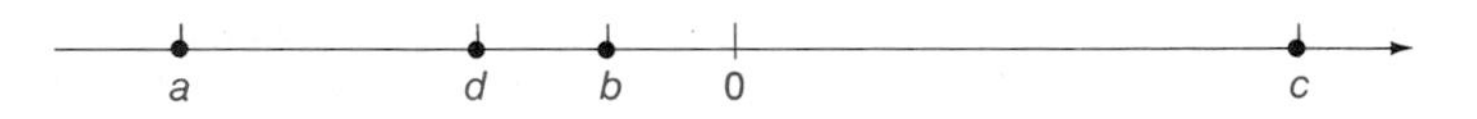

FIGURE 3 $a < b$, $c > d$

Example 5 Indicate true (T) or false (F). Letters a, b, c, and d refer to Figure 3.

(A) $-3{,}000 > 0$ **(B)** $-10 \leq 2$ **(C)** $-5 \geq -5$
(D) $0 < -25$ **(E)** $a < c$ **(F)** $d \geq b$

Solution **(A)** F **(B)** T **(C)** T **(D)** F **(E)** T **(F)** F

Problem 5 Indicate true (T) or false (F). Letters a, b, c, and d refer to Figure 3.

(A) $0 < 25$ **(B)** $-35 \geq 3$ **(C)** $-5 \leq -5$
(D) $-25 > 0$ **(E)** $b \leq c$ **(F)** $d < b$

We assume that the following two important inequality properties hold for all real numbers:

Basic Inequality Properties

For any real numbers a, b, and c:

1. Either $a < b$, $a = b$, or $a > b$. TRICHOTOMY PROPERTY
2. If $a < b$ and $b < c$, then $a < c$. TRANSITIVE PROPERTY

The **double inequality** $a < x \leq b$ means that $a < x$ and $x \leq b$; that is, x is between a and b, including b but not including a. Similar interpretations are given to forms such as $a \leq x < b$, $a \leq x \leq b$, and $a < x < b$.

Let us now turn to simple **inequality statements** (inequality forms involving at least one variable) of the form

$$x > 2 \qquad -2 < x \leq 3$$
$$x \leq -3 \qquad 0 \leq x \leq 5$$

We are interested in graphing such statements on a real number line. In general, to **graph an inequality statement** in one variable on a real number line is to graph the set of all real number replacements of the variable that make the statement true. This set is called the **solution set** of the inequality statement.

Example 6 Graph on a real number line: **(A)** $x > 2$ **(B)** $-2 < x \leq 3$

Solution **(A)** The solution set for $x > 2$ is the set of *all* real numbers greater than 2. Graphically, this set includes *all* the points to the right of 2:

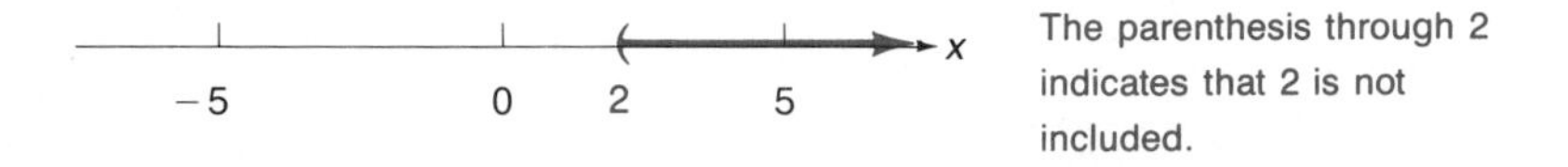

(B) The solution set for $-2 < x \leq 3$ is the set of *all* real numbers between -2 and 3, including 3 but not -2. Graphically:

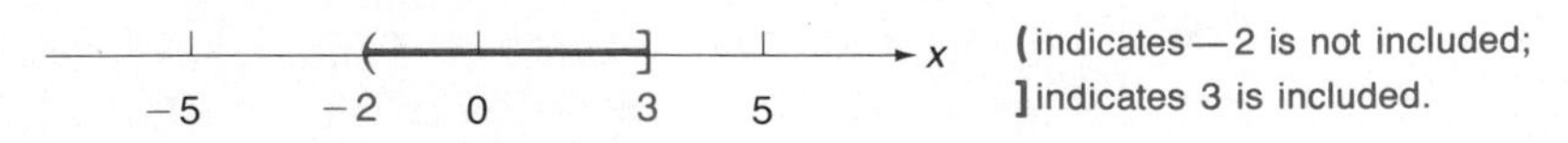

Problem 6 Graph on a real number line: **(A)** $x \leq -3$ **(B)** $0 < x < 5$

More will be said about solving equations and inequalities of a more complicated nature in Chapter 4. Just as formulating algebraic equations is an important step in solving many practical problems, formulating appropriate algebraic inequality statements is an important step in solving other types of practical problems.

Example 7 Translate each statement into an algebraic inequality statement using x as the only variable:

(A) 8 times a number is greater than or equal to 10 more than the number.
(B) 4 less than twice a number is less than 6 times the number.

Solution **(A)** Let $x =$ The unknown number(s); then the statement translates as follows:

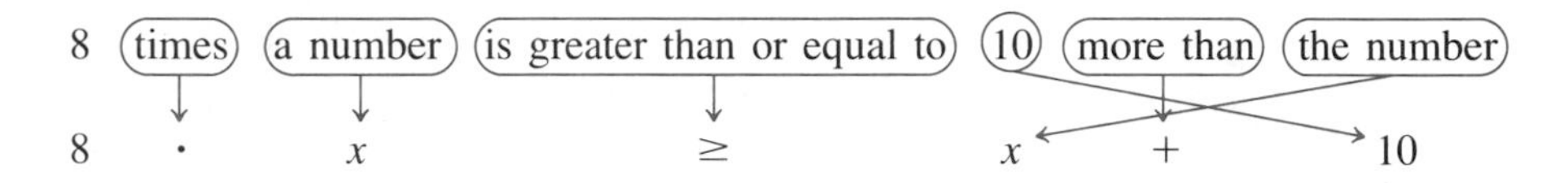

Or, more compactly,

$$8x \geq x + 10$$

(B) Let $x =$ The unknown number(s); then the statement translates as follows:

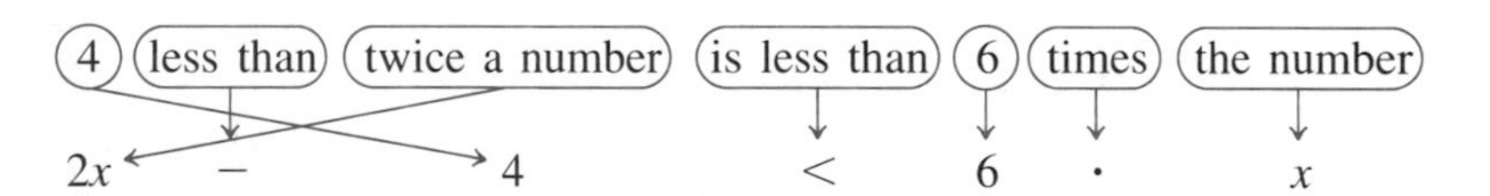

Or, more compactly,

$$2x - 4 < 6x$$

Note: "Less than" in the first part of the statement means "subtract from"; "is less than" in the middle of the statement means $<$.

Problem 7 Translate each statement into an algebraic inequality statement using x as the only variable:

(A) 7 less than twice a number is greater than 5 times the number.
(B) 3 more than a number is less than or equal to 5 less than twice the number.

ANSWERS TO MATCHED PROBLEMS

4. **(A)** $7 = 3 + x$ **(B)** $12 = x - 9$ **(C)** $3x = 2x - 6$
(D) $x - 6 = 2(x - 4)$

5. **(A)** T **(B)** F **(C)** T **(D)** F **(E)** T **(F)** T

6. **(A)** x: −5 −3 0 5

(B) x: −5 0 5

7. **(A)** $2x - 7 > 5x$ **(B)** $3 + x \le 2x - 5$

EXERCISE 1-2

A *Write in symbolic form.*

1. $11 - 5$ is equal to $\frac{12}{2}$.
2. $\frac{36}{2}$ is equal to $3 \cdot 6$.
3. 4 is greater than -18.
4. 3 is greater than -7.
5. -12 is less than -3.
6. -20 is less than 1.
7. x is greater than or equal to -8.
8. x is less than or equal to 12.
9. x is between -2 and 2.
10. x is between 0 and 10.

Replace equality or inequality symbols with appropriate verbal forms.

11. $\frac{48}{4} = 15 - 3$ **12.** $24 + 3 = 9 \cdot 3$ **13.** $12 > 11$

14. $5 > -2$ **15.** $11 < 12$ **16.** $-2 < 5$

17. $x \le 4$ **18.** $x \le -3$ **19.** $6 - 4 \ne \frac{9}{3}$

20. $\frac{15}{3} \ne \frac{10}{5}$ **21.** $2x - 1 \ge 3$ **22.** $3(x + 4) \ge 25$

Replace each question mark with =, <, or > to form a true statement.

23. $-3 ? 7$ **24.** $1 ? -6$ **25.** $-2 ? -5$

26. $0 ? -100$ **27.** $-1.35 ? -1$ **28.** $-6.33 ? -6$

29. $\frac{55}{5} ? \frac{22}{2}$ **30.** $10 + 3 ? \frac{39}{3}$ **31.** $6 - 4 ? \frac{24}{6}$

32. $6 \cdot 2 ? 4 + 6$ **33.** $\frac{3}{2} ? \frac{3}{4}$ **34.** $-\frac{3}{2} ? -\frac{3}{4}$

35. $-10 ? -2 ? -1$ **36.** $-2 ? 0 ? 2$

Referring to the number line below, replace each question mark in Problems 37–42 with either < or > to form a true statement.

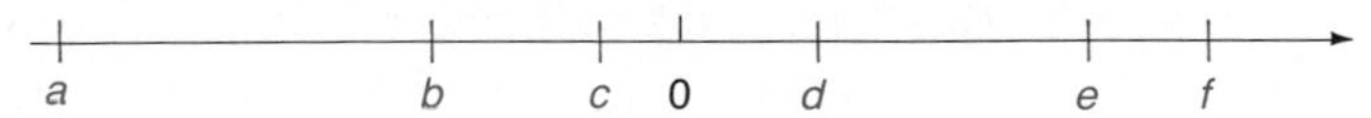

37. $e ? a$ **38.** $a ? d$ **39.** $c ? b$

40. $e ? f$ **41.** $0 ? d$ **42.** $0 ? a$

B **43.** If we add a positive real number to any real number, will the sum be greater than or less than the original number?

44. If we add a positive real number to any real number, will the sum be to the right or left of the original number on a number line?

Graph on a real number line for x a real number.

45. $x \leq 3$ **46.** $x \geq -2$ **47.** $-5 < x \leq -1$

48. $-4 \leq x < 1$ **49.** $x > -4$ **50.** $x > 1$

51. $-1 < x < 3$ **52.** $-2 \leq x \leq 2$

Translate each statement into an algebraic equation or inequality statement using x as the only variable.

53. $x - 8$ is positive. **54.** $2x - 8$ is negative.

55. $x + 4$ is not negative. **56.** $4x + 1$ is not positive.

57. 80 is 3 more than twice a certain number.

58. 18 is 3 times a certain number.

59. x is greater than or equal to -3 and less than 4.

60. x is greater than 0 and less than 10.

61. 26 is 12 less than a certain number.

62. 32 is 5 less than a certain number.

63. x is less than 6 less than twice x.

64. x is greater than 9 more than 4 times x.

65. 6 times a number is 4 more than 3 times the number.

66. 7 times a number is 12 less than 4 times the number.

67. 6 less than a certain number is 5 times the number that is 7 more than the certain number.

68. 5 more than a certain number is 3 times the number that is 4 less than the certain number.

69. $\frac{9}{5}C + 32$ is between 63 and 72, inclusive.

70. $\frac{5}{9}(F - 32)$ is between 20 and 25, inclusive.

71. The sum of three consecutive natural numbers is 186.

72. The sum of three consecutive natural numbers is 372.

C *Replace each question mark with an appropriate symbol to make the statement an illustration of the given property of equality or inequality.*

73. ***Symmetry property*** If $-5 = t$, then ?.

74. ***Transitive property*** If $x < 3$ and $3 < y$, then ?.

75. ***Transitive property*** If $5x + 7x = (5 + 7)x$ and $(5 + 7)x = 12x$, then ?.

76. ***Trichotomy property*** If $x \neq 3$, then either $x < 3$ or ?.

77. ***Reflexive property*** $3 - x = ?$

78. ***Substitution principle*** If $2x + 3y = 5$ and $y = x - 3$, then $2x + 3(?) = 5$.

79. What is wrong with the following argument? Four is an even number and 8 is an even number; hence we can write "4 = Even number" and "8 = Even number." By the symmetric law for equality, we can write "Even number = 8," and we can conclude, using the transitive law for equality (since 4 = Even number and Even number = 8), that 4 = 8.

80. What is wrong with the following argument? Rod is a human and Jan is a human; hence we can write "Rod = Human" and "Jan = Human." By the symmetric law for equality, we can write "Human = Jan," and we can conclude, using the transitive law for equality (since Rod = Human and Human = Jan), that Rod = Jan.

APPLICATIONS

81. In a rectangle with area 90 square meters, the length is 3 meters less than twice the width. Write an equation relating the area with the length and width, using x as the only variable. [*Note:* Area = (Length)(Width).]

82. In a rectangle with area 500 square inches, the length is 5 inches longer than the width. Write an equation relating the area with the length and the width, using x as the only variable.

★83. In a rectangle with perimeter 210 centimeters, the length is 10 centimeters less than 3 times the width. Write an equation relating the perimeter with the length and width, using x as the only variable.

★84. In a rectangle with perimeter 186 feet, the width is 7 feet less than the length. Write an equation relating the perimeter with the length and the width, using x as the only variable.

★★85. A rectangular lot is to be fenced with 400 feet of wire fencing.

(A) If the lot is x feet wide, write a formula for the area of the lot in terms of x.

(B) What real number values can x assume? Write the answer in terms of a double inequality statement that also allows the area to be 0.

86. A flat sheet of cardboard in the shape of an 18 by 12 inch rectangle is to be used to make an open-topped box by cutting an x by x inch square out of each corner and folding the remaining part appropriately.

(A) Write an equation for the volume of the box in terms of x.

(B) What real number values can x assume? Formulate the answer in terms of a double inequality statement that also allows the volume to be 0.

[*Note:* Volume = (Length)(Width)(Height).]

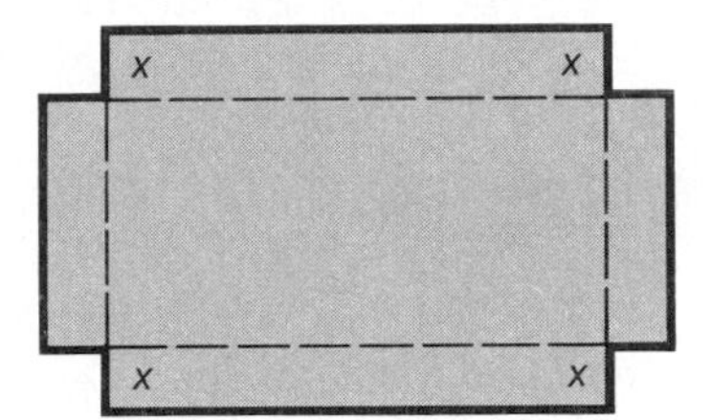

1-3 REAL NUMBER PROPERTIES

- Commutative Properties
- Associative Properties
- Distributive Properties
- Simplifying Algebraic Expressions
- Identity Properties

In the last section we discussed algebraic expressions, equations, and inequality statements. We now take a closer look at some of the basic properties of the set of real numbers that enable us to convert algebraic expressions into "equivalent forms." These assumed properties become operational rules in the algebra of real numbers.

Basic Properties of the Set of Real Numbers

Let R be the set of real numbers and x, y, and z arbitrary elements of R.

ADDITION PROPERTIES

CLOSURE:	$x + y$ is a unique element in R, that is, the sum of two real numbers is also a real number.
ASSOCIATIVE:	$(x + y) + z = x + (y + z)$
COMMUTATIVE:	$x + y = y + x$
IDENTITY:	0 is the additive identity; that is, $0 + x = x + 0 = x$ for all $x \in R$, and 0 is the only element in R with this property.
INVERSE:	For each $x \in R$, $-x$ is its unique additive inverse; that is, $x + (-x) = (-x) + x = 0$, and $-x$ is the only element in R relative to x with this property.

Basic Properties of the Set of Real Numbers *(Continued)*

MULTIPLICATION PROPERTIES

CLOSURE:	xy is a unique element in R; that is, the product of two real numbers is also a real number.
ASSOCIATIVE:	$(xy)z = x(yz)$
COMMUTATIVE:	$xy = yx$
IDENTITY:	1 is the multiplicative identity; that is, for $x \in R$, $1x = x1 = x$, and 1 is the only element in R with this property.
INVERSE:	For each $x \in R$, $x \neq 0$, $1/x$ is its unique multiplicative inverse; that is, $x(1/x) = (1/x)x = 1$, and $1/x$ is the only element in R relative to x with this property.

COMBINED PROPERTIES

DISTRIBUTIVE:	$x(y + z) = xy + xz$ $(x + y)z = xz + yz$

The names of the properties may appear bothersome, but it is useful to be able to refer to them. Most of the ideas presented here are quite simple. In fact, you have been using many of these properties in arithmetic for a long time. In this section we will informally consider several of the properties and will discuss others as needed in other parts of the book.

COMMUTATIVE PROPERTIES

You are already familiar with the commutative properties for addition and multiplication. They simply indicate that the order in which addition or multiplication is performed doesn't matter: $2 + 3 = 3 + 2$ and $2 \cdot 3 = 3 \cdot 2$.

Example 8 Use an appropriate commutative property to replace each question mark with an appropriate symbol:

(A) $x + 9 = 9 + ?$ **(B)** $7y = ?7$
(C) $2x + 3y = ?+2x$ **(D)** $3 + yx = 3 + ?y$

Solution **(A)** $x + 9 = 9 + x$ **(B)** $7y = y7$
(C) $2x + 3y = 3y + 2x$ **(D)** $3 + yx = 3 + xy$

Problem 8 Use the appropriate commutative property to replace each question mark with an appropriate symbol:

(A) $x + 3 = 3 + ?$ **(B)** $3x + 5y = 5y + ?$
(C) $5(y + x) = 5(? + y)$ **(D)** $y + x4 = y + 4?$

Does the commutative property hold relative to subtraction and division? That is, do $x - y = y - x$ and $x \div y = y \div x$ for all real numbers x and y, division by 0 excluded? The answer is no, since, for example, $5 - 3 \neq 3 - 5$ and $8 \div 4 \neq 4 \div 8$.

ASSOCIATIVE PROPERTIES

When computing

$$4 + 3 + 5 \qquad \text{or} \qquad 4 \cdot 3 \cdot 5$$

why don't we need parentheses to show us which two numbers are to be added or multiplied first? The answer is to be found in the associative properties. These properties allow us to write

$$(4 + 3) + 5 = 4 + (3 + 5) \qquad \text{and} \qquad (4 \cdot 3) \cdot 5 = 4 \cdot (3 \cdot 5)$$

so it doesn't matter how we group relative to either operation.

Example 9 Use an appropriate associative property to replace each question mark with an appropriate symbol:

(A) $(x + 7) + 2 = x + (?)$ **(B)** $3(5y) = (?)y$
(C) $x + (x + 3) = (?) + 3$ **(D)** $(2y)y = 2(?)$

Solution
(A) $(x + 7) + 2 = x + (7 + 2) = x + 9$
(B) $3(5y) = (3 \cdot 5)y = 15y$
(C) $x + (x + 3) = (x + x) + 3 = 2x + 3$
(D) $(2y)y = 2(yy) = 2y^2$ Recall that $y^2 = y \cdot y$, $y^3 = y \cdot y \cdot y$, etc.

The associative property tells us how parentheses can be moved relative to addition and multiplication.

Problem 9 Using the appropriate associative property, replace each question mark with an appropriate symbol:

(A) $(x + 3) + 9 = x + (?)$ **(B)** $6(3y) = (?)y$
(C) $(5 + z) + z = 5 + (?)$ **(D)** $(3z)z = 3(?)$

Does the associative property hold for subtraction and division? The answer is no, since, for example, $(8 - 4) - 2 \neq 8 - (4 - 2)$ and $(8 \div 4) \div 2 \neq 8 \div (4 \div 2)$. Evaluate each side of each equation to see why.

Conclusion

Relative to addition, commutativity and associativity permit us to change the order of addition at will and insert or remove parentheses as we please. The same is true for multiplication, but not for subtraction and division.

DISTRIBUTIVE PROPERTIES

The distributive properties allow us to convert a product $x(y + z)$ to a sum $xy + xz$, and vice versa. The process of rewriting $x(y + z)$ as $xy + xz$ is usually called **multiplying out**, or just **multiplying**. The process of rewriting the sum $xy + xz$ as the product $x(y + z)$ is usually called **factoring out the common factor**; in this case the common factor is x.

Example 10 Multiply, using a distributive property:

(A) $3(x + y)$ **(B)** $5(2x^2 + 3)$
(C) $2x(x + 1)$ **(D)** $(3 + x)y$

Solution **(A)** $3(x + y) = 3x + 3y$ **(B)** $5(2x^2 + 3) = 10x^2 + 15$
(C) $2x(x + 1) = 2x^2 + 2x$ **(D)** $(3 + x)y = 3y + xy$

Problem 10 Multiply, using a distributive property:

(A) $5(m + n)$ **(B)** $4(3y^3 + 5)$ **(C)** $3y(2 + y)$ **(D)** $(a + 4)b$

Example 11 Take out factors common to both terms:

(A) $6x + 6y$ **(B)** $7x^2 + 14$ **(C)** $ax + ay$

Solution **(A)** $6x + 6y = \mathbf{6}x + \mathbf{6}y = \mathbf{6}(x + y)$
(B) $7x^2 + 14 = \mathbf{7}x^2 + \mathbf{7} \cdot 2 = \mathbf{7}(x^2 + 2)$
(C) $ax + ay = \mathbf{a}x + \mathbf{a}y = \mathbf{a}(x + y)$

Problem 11 Take out factors common to both terms:

(A) $9m + 9n$ **(B)** $6y^3 + 12$ **(C)** $du + dy$

Several other useful distributive forms follow from the basic versions given above and from other properties discussed in this chapter:

Additional Distributive Properties

For real numbers $a, b, c, \ldots$:

$$(ab - ac) = a(b - c) = (b - c)a$$

$$a(b + c + d + \cdots + f) = ab + ac + ad + \cdots + af$$

$$ab + ac + ad + \cdots + af = a(b + c + d + \cdots + f) = (b + c + d + \cdots + f)a$$

SIMPLIFYING ALGEBRAIC EXPRESSIONS

The commutative, associative, and distributive properties provide us with our first tools for transforming algebraic expressions into equivalent forms. Two algebraic expressions are **equivalent** over given replacement sets for the variables if both yield equal numbers for each replacement of the variables by numbers from their respective replacement sets. Let us use the associative and commutative properties to formally transform

$$(x + 8) + (y + 2)$$

into a simpler equivalent form. By formally transforming, we mean that each step will be justified by a basic property or an earlier stated property. We will then indicate how most of these steps are performed mentally.

$(x + 8) + (y + 2) = [(x + 8) + y] + 2$	Associative property for +
$= [x + (8 + y)] + 2$	Associative property for +
$= [x + (y + 8)] + 2$	Commutative property for +
$= x + [(y + 8) + 2]$	Associative property for +
$= x + [y + (8 + 2)]$	Associative property for +
$= x + (y + 10)$	Substitution property for =
$= (x + y) + 10$	Associative property for +

Normally, we do most of these steps mentally and simply write

$$(x + 8) + (y + 2) = x + 8 + y + 2$$
$$= x + y + 10$$

Even though we did not write each step as in the formal treatment, you should not lose sight of the fact that the associative and commutative properties are behind the mental steps taken in the simpler version.

Example 12 Simplify mentally, using commutative and associative properties:

(A) $(3x + 5) + (2y + 7)$ **(B)** $(6x)(3y)$

Solution **(A)** $(3x + 5) + (2y + 7) = 3x + 5 + 2y + 7$
$= 3x + 2y + 12$

(B) $(6x)(3y) = 6x3y$
$= 6 \cdot 3xy$
$= 18xy$

Problem 12 Repeat Example 12 for

(A) $(2a + 3) + (3b + 11)$ **(B)** $(4m)(6n)$

IDENTITY PROPERTIES

The identity property for addition states that 0 is the unique real number having the property that when it is added to any real number we get that number back again. Thus,

$$5 + 0 = 5 \qquad 0 + 3x = 3x \qquad (x + y) + 0 = x + y$$

Similarly, the number 1 plays the same role relative to multiplication. That is, when any number is multiplied by 1, we get that number back again. Thus, we can write

$$1 \cdot 8 = 8 \qquad 1x = x \qquad 1xy = xy$$

The other properties will be discussed as needed.

ANSWERS TO MATCHED PROBLEMS

8. **(A)** x **(B)** $3x$ **(C)** x **(D)** x
9. **(A)** $3 + 9$ **(B)** $6 \cdot 3$ **(C)** $z + z$ **(D)** zz
10. **(A)** $5m + 5n$ **(B)** $12y^3 + 20$ **(C)** $6y + 3y^2$ **(D)** $ab + 4b$
11. **(A)** $9(m + n)$ **(B)** $6(y^3 + 2)$ **(C)** $d(u + y)$
12. **(A)** $2a + 3b + 14$ **(B)** $24mn$

EXERCISE 1-3

All variables represent real numbers.

A *Replace each question mark with an appropriate expression that will illustrate the use of the indicated real number property.*

1. ***Commutative property*** **(+)** $x + 3 = ?$
2. ***Commutative property*** **(+)** $m + n = ?$

3. ***Associative property*** **(·)** $5(7z) = ?$

4. ***Associative property*** **(·)** $(uv)w = ?$

5. ***Commutative property*** **(·)** $nm = ?$

6. ***Commutative property*** **(·)** $dc = ?$

7. ***Associative property*** **(+)** $9 + (11 + M) = ?$

8. ***Associative property*** **(+)** $(x + 7) + 5 = ?$

9. ***Distributive property*** $3(x + 1) = ?$

10. ***Distributive property*** $x(x + 1) = ?$

11. ***Distributive property*** $(2 + x)x = ?$

12. ***Distributive property*** $(4 + x)3 = ?$

13. ***Identity property*** **(+)** $7x + 0 = ?$

14. ***Identity property (+)*** $0 + (x + z) = ?$

15. ***Identity property*** **(·)** $1(x + y) = ?$

16. ***Identity property*** **(·)** $1(uv) = ?$

State the justifying property for each statement.

17. $12 + w = w + 12$
18. $2x + 3 = 3 + 2x$
19. $m + (n + 3) = (m + n) + 3$
20. $(3x + y) + 5 = 3x + (y + 5)$
21. $20x = x20$
22. $MN = NM$
23. $2(x + 5) = 2x + 10$
24. $x(y + 1) = xy + x \cdot 1$
25. $4(8y) = (4 \cdot 8)y$
26. $(12u)v = 12(uv)$
27. $3x + 0 = 3x$
28. $0 + (2x + 3) = 2x + 3$
29. $1m = m$
30. $uv = 1uv$
31. $4 + 2x = 2(2 + x)$
32. $3x + 9 = 3(x + 3)$

Remove parentheses and simplify:

33. $(x + 7) + 2$
34. $3 + (5 + m)$
35. $4(5y)$
36. $6(8n)$
37. $12 + (u + 3)$
38. $(4 + x) + 13$
39. $(3x)7$
40. $4(y3)$
41. $0 + 1x$
42. $(1y + 3) + 0$

B *State the justifying property for each statement.*

43. $2 + (y + 3) = 2 + (3 + y)$

44. $(3m)n = n(3m)$

45. $7(y4) = 7(4y)$

46. $5 + (y + 2) = (y + 2) + 5$

47. $3x + 2y = 2y + 3x$

48. $3x + (2x + 5y) = (3x + 2x) + 5y$

49. $(2x)(x + 3) = 2[x(x + 3)]$

50. $(x + 3) + (2 + y) = x + [3 + (2 + y)]$

Remove parentheses and simplify:

51. $(x + 7) + (y + 4) + (z + 1)$

52. $(7 + m) + (8 + n) + (3 + p)$

53. $(3x + 5) + (4y + 6)$

54. $(3a + 7) + (5b + 2)$

55. $0 + (1x + 3) + (y + 2)$

56. $1(x + 3) + 0 + (y + 2)$

57. $(12m)(3n)(1p)$

58. $(8x)(4y)(2z)$

Use the distributive property to rewrite each sum as a product and each product as a sum.

59. $(x + 5)x$

60. $(x + 2) \cdot 3$

61. $2x + 18$

62. $7x + 21$

63. $(x + 2)8$

64. $8x + 2$

65. $6x + 12$

66. $6(x + 12)$

C **67.** Indicate whether true (T) or false (F), and for each false statement find real number replacements for a and b that will illustrate its falseness. For all real numbers a and b:

(A) $a + b = b + a$ **(B)** $a - b = b - a$
(C) $ab = ba$ **(D)** $a \div b = b \div a$

68. Indicate whether true (T) or false (F), and for each false statement find real number replacements for a, b, and c that will illustrate its falseness. For all real numbers a, b, and c:

(A) $(a + b) + c = a + (b + c)$ **(B)** $(a - b) - c = a - (b - c)$
(C) $a(bc) = (ab)c$ **(D)** $(a \div b) \div c = a \div (b \div c)$

69. Supply a reason for each step.

STATEMENT	REASON
1. $(x + 3) + (y + 4) = (x + 3) + (4 + y)$	**1.**
2. $= x + [3 + (4 + y)]$	**2.**
3. $= x + [(3 + 4) + y]$	**3.**

STATEMENT		REASON
4.	$= x + (7 + y)$	4.
5.	$= x + (y + 7)$	5.
6.	$= (x + y) + 7$	6.

70. Supply a reason for each step.

STATEMENT		REASON
1.	$(5x)(2y) = (x5)(2y)$	1.
2.	$= x[5(2y)]$	2.
3.	$= x[(5 \cdot 2)y]$	3.
4.	$= x(10y)$	4.
5.	$= (x10)y$	5.
6.	$= (10x)y$	6.
7.	$= 10(xy)$	7.

1-4 ADDITION AND SUBTRACTION

- The Negative of a Number
- The Absolute Value of a Number
- Addition of Real Numbers
- Subtraction of Real Numbers
- Combined Operations

In this section, we review addition and subtraction of real numbers. Before doing so, we need to say a few words about the operations "the negative of" and "the absolute value of." These operations are useful in describing operations related to addition, subtraction, multiplication, and division of real numbers.

THE NEGATIVE OF A NUMBER

For each real number x, we denoted its additive inverse by

$$-x \quad \begin{cases} \text{Additive inverse of } x \\ \text{Opposite of } x \\ \text{Negative of } x \end{cases}$$

All the names on the right describe the same thing and are used interchangeably. You will recall that the **opposite of** or **negative of a number x** is obtained from x by changing its sign. The opposite of or negative of 0 is 0.

Example 13 Find:

(A) $-(+5)$ **(B)** $-(-8)$ **(C)** $-(0)$ **(D)** $-[-(-4)]$

Solution **(A)** $-(+5) = -5$ **(B)** $-(-8) = +8$ or 8
(C) $-(0) = 0$ **(D)** $-[-(-4)] = -(+4) = -4$

Examples 13(A) and (B) illustrate the fact that:

$-x$ is not necessarily a negative number.

That is, $-x$ represents a negative number if x is positive and a positive number if x is negative.

It can be shown that:

Double Negative Property

For a any real number:

$$-(-a) = a$$

Problem 13 Find:

(A) $-(+11)$ **(B)** $-(-12)$ **(C)** $-(0)$ **(D)** $-[-(+6)]$

It is now important to note the three distinct uses of the minus sign.

Multiple Uses of the Minus Sign

1. As the operation "subtract": $9 \overset{\downarrow}{-} 3 = 6$
2. As the operation "the negative or opposite of": $\overset{\downarrow}{-}(-8) = 8$
3. As part of a number symbol: $\overset{\downarrow}{-}4$

THE ABSOLUTE VALUE OF A NUMBER

The **absolute value** of a number x is an operation on x, denoted by the symbol

$$|x|$$

(not square brackets), which gives the magnitude of x without its sign. The absolute value of a number can also be thought of geometrically as the distance of the number from 0 on the real number line expressed as a positive number or 0. For example, both 5 and -5 are 5 units from 0 (see Figure 4).

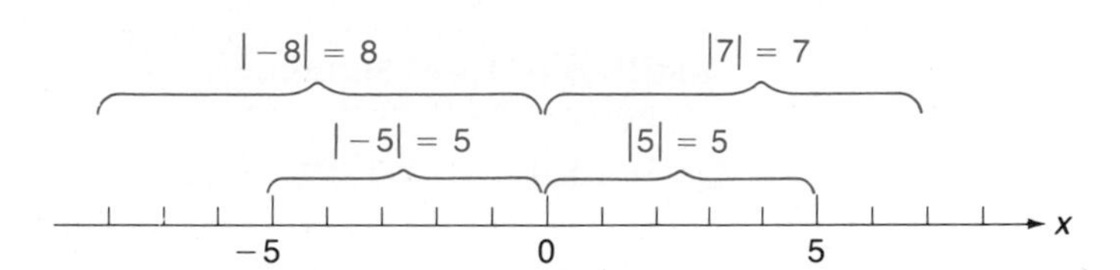

FIGURE 4 Absolute value

Thus, we can write $|5| = 5$ and $|-5| = 5$. Figure 4 also illustrates the fact that $|-8| = 8$ and $|7| = 7$.

Symbolically, and more formally, we define absolute value as follows:

Absolute Value

$$|x| = \begin{cases} x & \text{if } x \text{ is positive} \\ 0 & \text{if } x \text{ is } 0 \\ -x & \text{if } x \text{ is negative} \end{cases}$$

Note: $-x$ is positive if x is negative.

It is important to remember that:

The absolute value of a number is never negative.

Example 14 Evaluate:

(A) $|24|$ **(B)** $|-7|$ **(C)** $|0|$ **(D)** $-(|-8| + |-3|)$

Solution **(A)** $|24| = 24$ **(B)** $|-7| = 7$ **(C)** $|0| = 0$
(D) $-(|-8| + |-3|) = -(8 + 3) = -(11) = -11$

Note: $-(11)$ represents the opposite or negative of the positive number 11, while -11 represents a negative number.

Problem 14 Evaluate:

(A) $|-13|$ **(B)** $|43|$ **(C)** $-|-4|$ **(D)** $-(|-6| - |+2|)$

ADDITION OF REAL NUMBERS

We are now ready to review addition of real numbers.

Addition of Real Numbers: $a + b$

1. If a is 0: $0 + b = b$ $0 + 8 = 8$
   If b is 0: $a + 0 = a$ $(-13) + 0 = -13$
   This is the identity property for addition.
2. If a and b are both positive: add as you have learned in arithmetic.
3. If a and b are both negative: take the opposite of the sum of the absolute values. $(-3) + (-8) = -(3 + 8) = -11$
4. If a and b have opposite signs: subtract the smaller absolute value from the larger; then attach the sign of the original number with the larger absolute value. $3 + (-8) = -(8 - 3) = -5$

Part 3 can be rephrased as follows: mentally block out the signs of the two numbers (take absolute values), add as in arithmetic, and then attach a minus sign to the result. For example,

$$(-5) + (-3) \boxed{= -(5 + 3)}^{\dagger} = -8$$

Similarly, part 4 can be restated: mentally block out the signs of the two numbers (take their absolute values), subtract the smaller from the larger, and then attach the sign of the number with the largest absolute value. For example,

$$(-3) + (+9) \boxed{= +(9 - 3)} = 6$$

$$(+3) + (-9) \boxed{= -(9 - 3)} = -6$$

To add three or more numbers, add all of the positive numbers together, add all of the negative numbers together (the commutative and associative properties justify this procedure), and then add the two resulting sums as above.

Example 15 Add: $3 + (-6) + 8 + (-4) + (-5)$

Solution

$$\begin{aligned} 3 + (-6) + 8 + (-4) + (-5) &= (3 + 8) + [(-6) + (-4) + (-5)] \\ &= 11 + (-15) \boxed{= -(15 - 11)} = -4 \end{aligned}$$

Problem 15 Add: $6 + (-8) + (-4) + 10 + (-3) + 1$

† Throughout the book dashed boxes are used to represent steps that are usually done mentally.

SUBTRACTION OF REAL NUMBERS

We define subtraction in terms of addition. We can perform numerical computation as always, but this definition will make algebraic manipulation easier.

Subtraction

For a and b any real numbers:

$$a - b = a + (-b)$$

To subtract b from a, add the opposite of b to a.

Recall that the opposite of b is the same as the negative of b and the additive inverse of b:

Opposite of −9

$$(-3) - (-9) = (-3) + 9 = 6$$

Change to addition

You should get to the point where you can perform this type of subtraction mentally, and simply write down the answer.

Example 16 Subtract:

(A) $8 - (-5)$ **(B)** $(-8) - 5$
(C) $(-8) - (-5)$ **(D)** $0 - 5$

Solution **(A)** $8 - (-5) = 8 + 5 = 13$ Change subtraction to addition and replace −5 by its opposite.

(B) $(-8) - 5 = (-8) + (-5) = -13$

(C) $(-8) - (-5) = (-8) + 5 = -3$

(D) $0 - 5 = 0 + (-5) = -5$

Problem 16 Subtract:

(A) $4 - 7$ **(B)** $7 - (-4)$ **(C)** $(-7) - 4$
(D) $(-7) - (-4)$ **(E)** $(-4) - (-7)$ **(F)** $0 - (-4)$

COMBINED OPERATIONS

When three or more terms are combined by addition and subtraction and symbols of grouping are omitted, we convert (mentally) any subtraction to addition and add. Thus,

$$8 - 5 + 3 \;\boxed{= 8 + (-5) + 3}\; = 6$$

Think

Example 17 Evaluate:

(A) $2 - 3 - 7 + 4$ **(B)** $-4 - 8 + 2 + 9$

Solution **(A)** $2 - 3 - 7 + 4 \;\boxed{= 2 + (-3) + (-7) + 4}\; = -4$

Think

(B) $-4 - 8 + 2 + 9 \;\boxed{= (-4) + (-8) + 2 + 9}\; = -1$

Think

Problem 17 Evaluate: **(A)** $5 - 8 + 2 - 6$ **(B)** $-6 + 12 - 2 - 1$

Recall that to evaluate an algebraic expression for particular values of the variables means to replace each variable by a given value and calculate the resulting arithmetic value.

Example 18 Evaluate each for $x = 2$, $y = -3$, and $z = -9$:

(A) $x + y$ **(B)** $y - z$ **(C)** $y - (z - x)$ **(D)** $|(-y) - |z||$

Solution **(A)** $x + y$

$\boxed{(\quad) + (\quad)}$ Use of parentheses as indicated prevents many sign errors.

$(2) + (-3) = -1$

(B) $y - z$

$\boxed{(\quad) - (\quad)}$

$(-3) - (-9)$

$\boxed{= (-3) + (9)}\; = 6$

(C) $y - (z - x)$ — Replace parentheses with brackets (another form of parentheses), substitute values, and then evaluate starting inside the square brackets.

$$(-3) - [(-9) - 2]$$
$$= (-3) - (-11)$$
$$\boxed{= (-3) + (11)} = 8$$

(D) $\left|(-y) - |z|\right|$

$\left|[-(-3)] - |-9|\right|$ — Evaluate $-(-3)$ and $|-9|$ first.

$= |(3) - (9)|$ — Subtract inside absolute value signs.

$= |-6| = 6$ — Take the absolute value.

Problem 18 Evaluate for $x = -4$, $y = 5$, and $z = -11$:

(A) $y + z$ **(B)** $x - y$ **(C)** $(z - x) + y$ **(D)** $\left|(x) - |z|\right|$

Example 19 Using a calculator, evaluate each for $x = -504.394$, $y = 829.077$, and $z = -1{,}023.998$:

(A) $y - x$ **(B)** $x - (y - z)$

Solution There are a number of ways to proceed, depending on your calculator.

(A) This problem can be entered directly to obtain

$$829.077 - (-504.394) = 1{,}333.471$$

(B) Evaluate $y - z = 829.077 - (-1{,}023.998)$
$= 1{,}853.075$

Then evaluate $x - 1{,}853.075$ to obtain $-2{,}357.469$. [Rearranging the expression $x - (y - z)$ algebraically, as will be done in Section 1-5, or making use of available memory in your calculator can simplify the solution.]

Problem 19 Use the values of x, y, and z in Example 19 to evaluate:

(A) $x - y$ **(B)** $x - (z - y)$

ANSWERS TO MATCHED PROBLEMS

13. **(A)** -11 **(B)** 12 **(C)** 0 **(D)** 6
14. **(A)** 13 **(B)** 43 **(C)** -4 **(D)** -4
15. 2

16. **(A)** -3 **(B)** 11 **(C)** -11 **(D)** -3 **(E)** 3 **(F)** 4

17. **(A)** -7 **(B)** 3

18. **(A)** -6 **(B)** -9 **(C)** -2 **(D)** 15

19. **(A)** $-1{,}333.471$ **(B)** $1{,}348.681$

EXERCISE 1-4

A *Evaluate.*

1. $-(+7)$ **2.** $-(+12)$ **3.** $-(-6)$

4. $-(-8)$ **5.** $|+2|$ **6.** $|+9|$

7. $|-27|$ **8.** $|-32|$ **9.** $|0|$

10. $-(0)$ **11.** $(-7) + (-3)$ **12.** $(-7) + (+3)$

13. $(+7) + (-3)$ **14.** $(-12) + (+8)$ **15.** $(+3) - (+9)$

16. $(+3) - (-9)$ **17.** $(+9) - (-3)$ **18.** $(-9) - (-3)$

19. The negative of a number is (*always, sometimes, never*) a negative number.

20. The absolute value of a number is (*always, sometimes, never*) a positive number.

B *Evaluate.*

21. $-[-(-3)]$ **22.** $-[-(+6)]$

23. $-|-(+2)|$ **24.** $-|-(-3)|$

25. $-(|-9| - |-3|)$ **26.** $-(|-14| - |-8|)$

27. $(-2) + (-6) + 3$ **28.** $(-2) + (-8) + 5$

29. $5 - 7 - 3$ **30.** $3 - 2 + 4$

31. $-7 + 6 - 4$ **32.** $-4 + 7 - 6$

33. $-2 - 3 + 6 - 2$ **34.** $-4 + 7 - 3 - 2$

35. $6 - [3 - (-9)]$ **36.** $(-10) - [(-6) + 3]$

37. $[6 - (-8)] - [(-8) - 6]$ **38.** $[3 - 5] + [(-5) - (-2)]$

Evaluate each for $x = 23.417$, $y = -52.608$, *and* $z = -13.012$.

39. $y + x$ **40.** $x + z$ **41.** $x - y$

42. $z - y$ **43.** $(x - z) + y$ **44.** $(y + z) - z$

Replace each question mark with an appropriate real number.

45. $-(?) = 5$

46. $-(?) = -8$

47. $|?| = 7$

48. $|?| = -4$

49. $(-3) + ? = -8$

50. $? + 5 = -6$

51. $(-3) - ? = -8$

52. $?-(-2) = -4$

Evaluate for $x = 3$, $y = -8$, and $z = -2$.

53. $x + y$

54. $y + z$

55. $y - x$

56. $y - z$

57. $(x - z) + y$

58. $y - (z - x)$

59. $\left|(-z) - |y|\right|$

60. $\left||-y| - 12\right|$

61. $-\left||y| - |x|\right|$

62. $-\left||-10| - |x|\right|$

C *Which of the following hold for all integers a, b, and c? Illustrate each false statement with an example showing that it is false.*

63. $a + b = b + a$

64. $a + (-a) = 0$

65. $a - b = b - a$

66. $a - b = a + (-b)$

67. $(a + b) + c = a + (b + c)$

68. $(a - b) - c = a - (b - c)$

69. $|a + b| = |a| + |b|$

70. $|a - b| = |a| - |b|$

71. Supply the reasons for each of the following steps:

STATEMENT	REASON
1. $b + [a + (-b)] = b + [(-b) + a]$	**1.**
2. $= [b + (-b)] + a$	**2.**
3. $= 0 + a$	**3.**
4. $= a$	**4.**

72. Supply the reasons for each step:

STATEMENT	REASON
1. $(a + b) + [(-a) + (-b)] = (b + a) + [(-a) + (-b)]$	**1.**
2. $= [(b + a) + (-a)] + (-b)$	**2.**
3. $= \{b + [a + (-a)]\} + (-b)$	**3.**
4. $= (b + 0) + (-b)$	**4.**
5. $= b + (-b)$	**5.**
6. $= 0$	**6.**
7. Therefore, $-(a + b) = (-a) + (-b)$.	

APPLICATIONS

73. You own a stock that is traded on a stock exchange. On Monday it closed at \$23.50 per share; it fell \$3.25 on Tuesday and another \$6.75 on Wednesday; it rose \$2.50 on Thursday; and it finished strongly on Friday by rising \$7.75. Use addition of signed numbers to determine the closing price of the stock on Friday.

74. Find, using subtraction of signed numbers, the difference in the height between the highest point in the United States, Mount McKinley (20,270 feet) and the lowest point in the United States, Death Valley (-280 feet).

1-5 MULTIPLICATION, DIVISION, AND ORDER OF OPERATIONS

- Multiplication of Real Numbers
- Sign Properties for Multiplication
- Division of Real Numbers
- Order of Operations

Having discussed addition and subtraction, we now turn to multiplication and division. We will then be in a position to consider problems involving all four arithmetic operations ($+$, $-$, $\cdot$, $\div$).

MULTIPLICATION OF REAL NUMBERS

Multiplication of real numbers is defined as follows:

Multiplication of Real Numbers: *ab*

1. If a is 0: $0 \cdot b = 0$ $\quad 0 \cdot 7 = 0$
   If b is 0: $a \cdot 0 = 0$ $\quad -3 \cdot 0 = 0$
2. If a and b are both positive: multiply as you have learned in arithmetic.
3. If a and b are both negative: take the product of the absolute values.
   $(-3)(-7) = 3 \cdot 7 = 21$
4. If a and b have opposite signs: take the opposite of the product of the absolute values. $(-3) \cdot 7 = -(3 \cdot 7) = -21$

Part 1 is a result that can be proved on the basis of the material in the preceding sections (see Problem 83 in Exercise 1-5). To see why numbers with opposite signs are multiplied as in part 4, consider the product

$$(+2)(-7)$$

We start with something we know is true and then proceed through a sequence of logical steps to a conclusion that must also be true:

$(+7) + (-7) = 0$	Inverse property for addition
$(+2)[(+7) + (-7)] = (+2)0$	Property of equality (if $a = b$, then $ca = cb$) that follows from properties of equality in Section 1-2
$(+2)0 = 0$	Part 4 of definition
$(+2)(+7) + (+2)(-7) = 0$	Distributive property and transitive property for equality
$(+14) + (+2)(-7) = 0$	Substitution principle for equality
$(+2)(-7) = -14$	Additive inverse property [$(+2)(-7)$ must be the additive inverse of $(+14)$]

Thus, we see that if the properties of the real numbers stated in Section 1-3 hold, then the product of $(+2)$ and (-7) must be -14. There is no other choice! Similar arguments are used to justify parts 3 and 4 of the definition. From parts 2, 3, and 4, we see that:

The product of two numbers with unlike signs is negative.
The product of two numbers with like signs is positive.

Example 20 Evaluate:

(A) $2(-7)$ **(B)** $(-2)(-7)$ **(C)** $0(-7)$

Solution **(A)** $2(-7) \boxed{= -(2 \cdot 7)} = -14$

(B) $(-2)(-7) \boxed{= 2 \cdot 7} = 14$

(C) $0(-7) = 0$

Problem 20 Evaluate: **(A)** $4(-3)$ **(B)** $(-4)3$ **(C)** $(-4)(-3)$ **(D)** $0(-3)$

SIGN PROPERTIES FOR MULTIPLICATION

Several important sign properties for multiplication are summarized in the following result.

Sign Properties for Multiplication

For a and b any real numbers:

(A) $(-1)a = -a$ **(B)** $(-a)b = -(ab)$ **(C)** $(-a)(-b) = ab$

Example 21 Evaluate $(-a)b$ and $-(ab)$ for $a = -5$ and $b = 4$.

Solution

$$(-a)b = [-(-5)]4 = 5 \cdot 4 = 20$$

$$-(ab) = -[(-5)4] = -(-20) = 20$$

Problem 21 Evaluate $(-a)(-b)$ and ab for $a = -5$ and $b = 4$.

Expressions of the form

$$-ab$$

occur frequently and at first glance may seem confusing. If you were asked to evaluate $-ab$ for $a = -3$ and $b = +2$, how would you proceed? Would you take the negative of a and then multiply it by b, or multiply a and b first and then take the negative of the product? Actually it does not matter! Because of the sign properties we get the same result either way since $(-a)b = -(ab)$. If we consider other material in this section, we find that

$$-ab = \begin{cases} (-a)b \\ a(-b) \\ -(ab) \\ (-1)ab \end{cases}$$

For example,
$-3(4) = (-3)(4) = 3(-4) = -(3)(4) = (-1)(3)(4)$

and we are at liberty to replace any one of these five forms with another from the same group.

Example 22 Evaluate $-ab$ for $a = -7$ and $b = 4$.

Solution Most people would proceed in one of the two following ways:

$$-(-7)(4) = \begin{cases} -[(-7)(4)] = -(-28) = 28 \\ [-(-7)](4) = (7)(4) = 28 \end{cases}$$

Both are correct.

Problem 22 Evaluate $-ab$ for $a = 6$ and $b = -3$ two different ways.

DIVISION OF REAL NUMBERS

You will recall in arithmetic that to check the division problem

$$9\overline{)36}^{\;4}$$

we multiply 9 by 4 to obtain 36. We will use this checking requirement to transform division into multiplication. Instead of asking "What is 9 divided into 36?", we ask "What number times 9 is 36?" Both questions have the same answer. The latter way of looking at division is the more useful of the two because of its generalization to other number systems.

Definition of Division

We write

$$\left.\begin{array}{ll} & a \div b = Q \\ & \quad Q \\ \text{or} & b\overline{)a} \end{array}\right\} \quad \text{if and only if} \quad Qb = a \text{ and } Q \text{ is unique}$$

The quotient Q is the number that must be multiplied times b to produce a.

Let us use the definition to find

$$12 \div (-3) = ? \qquad \text{or} \qquad \begin{array}{r} ? \\ -3\overline{)12} \end{array}$$

We ask "What number must (-3) be multiplied by to produce 12?" From our discussion of multiplication, we know the answer to be -4. Thus, we write

$$12 \div (-3) = -4 \qquad \text{or} \qquad \begin{array}{r} -4 \\ -3\overline{)12} \end{array}$$

since $(-4)(-3) = 12$.

What about division involving 0?

$$\begin{array}{r} ? \\ 5\overline{)0} \end{array} \quad ? \cdot 5 = 0$$

$$\begin{array}{r} ? \\ 0\overline{)5} \end{array} \quad ? \cdot 0 = 5$$

$$\begin{array}{r} ? \\ 0\overline{)0} \end{array} \quad ? \cdot 0 = 0$$

In the first case the quotient is 0, since $0 \cdot 5 = 0$. In the second case we find that no real number times 0 can produce 5; hence, this quotient is not defined.

In the third case we have the other extreme—any real number will produce 0 when multiplied times 0; hence, the quotient is not unique. We conclude that:

Zero cannot be used as a divisor—ever!
Division by zero is undefined.

The two division symbols $\div$ and $\overline{)\quad}$ from arithmetic are not used a great deal in algebra and higher mathematics. The horizontal bar (—) and slash mark (/) are the symbols most frequently used. Thus

$$a/b \qquad \frac{a}{b} \qquad a \div b \qquad \text{and} \qquad b\overline{)a}$$

In each case b is the divisor.

all name the same number (assuming the quotient is defined), and we can write

$$a/b = \frac{a}{b} = a \div b = b\overline{)a}$$

Now to the mechanics of division:

Division of Real Numbers: $\frac{a}{b}$

1. If $a = 0$ and $b \neq 0$: $\frac{a}{b} = 0$ $\quad \frac{0}{3} = 0 \quad \frac{0}{-15} = 0$

   If $b = 0$: $\frac{a}{b}$ is not defined $\quad \frac{3}{0}, \frac{-15}{0}, \frac{0}{0}$ are not defined.
2. If a and b are both positive: divide as you learned in arithmetic.
3. If a and b are both negative: take the quotient of the absolute values.

   $\frac{-15}{-3} = \frac{15}{3} = 5$
4. If a and b have opposite signs: take the opposite of the quotient of the absolute values.

   $\frac{-15}{3} = -\left(\frac{15}{3}\right) = -5 \qquad \frac{15}{-3} = -\left(\frac{15}{3}\right) = -5$

From parts 2, 3 and 4, we see that:

The quotient of two numbers with unlike signs is negative.
The quotient of two numbers with like signs is positive.

Example 23 Evaluate:

(A) $\dfrac{-22}{11}$ (B) $\dfrac{-36}{-12}$ (C) $\dfrac{48}{-16}$

Solution (A) $\dfrac{-22}{11} = -2$ (B) $\dfrac{-36}{-12} = 3$ (C) $\dfrac{48}{-16} = -3$

Problem 23 Evaluate: (A) $\dfrac{-36}{-9}$ (B) $\dfrac{24}{-8}$ (C) $\dfrac{-72}{12}$

Several important sign properties for division are summarized in the following result.

Sign Properties for Division

For all real numbers a and b, $b \neq 0$:

(A) $\dfrac{-a}{-b} = \dfrac{a}{b} = -\dfrac{-a}{b} = -\dfrac{a}{-b}$ $\qquad \dfrac{-3}{-4} = \dfrac{3}{4} = -\dfrac{-3}{4} = -\dfrac{3}{-4}$

(B) $\dfrac{-a}{b} = \dfrac{a}{-b} = -\dfrac{a}{b}$ $\qquad \dfrac{-3}{4} = \dfrac{3}{-4} = -\dfrac{3}{4}$

Example 24 Evaluate $\dfrac{-a}{b}$, $\dfrac{a}{-b}$, and $-\dfrac{a}{b}$ for $a = -6$ and $b = 2$.

Solution

$$\frac{-a}{b} = \frac{-(-6)}{2} = \frac{6}{2} = 3 \qquad \frac{a}{-b} = \frac{(-6)}{-(2)} = \frac{-6}{-2} = 3 \qquad -\frac{a}{b} = -\frac{-6}{2} = -(-3) = 3$$

Problem 24 Evaluate $\dfrac{-a}{-b}$, $\dfrac{a}{b}$, and $-\dfrac{-a}{b}$ for $a = -6$ and $b = 2$.

ORDER OF OPERATIONS

We now consider problems involving various combinations of the **arithmetic operations** $+$, $-$, $\cdot$, and $\div$ as well as **grouping symbols** such as **parentheses** ( ), **brackets** [ ], **braces** { }, and **fraction bars** ——.

To start, suppose several people were asked to evaluate

$$6 - 4(-3) + \frac{-6}{2}$$

To get the same result from all (a reasonable request), we need an agreement indicating the order in which the operations should be performed.

Order of Operations

(A) IF NO GROUPING SYMBOLS ARE PRESENT:

1. Perform any multiplication and division first, proceeding from left to right.
2. Then perform any addition and subtraction, proceeding from left to right.

(B) IF SYMBOLS OF GROUPING ARE PRESENT:

1. Simplify above and below any fraction bars following the steps in (A).
2. Simplify within other symbols of grouping, generally starting with the innermost and working outward, following the steps in (A).

Thus,

$$6 - 4(-3) + \frac{-6}{2} = 6 - (-12) + (-3) = 6 + 12 + (-3) = 15$$

Example 25 Evaluate for $x = -24$, $y = 2$, and $z = -3$:

(A) $2x - 3yz + \frac{x}{z}$ **(B)** $\frac{x}{y} - \frac{16z + xy}{y + z}$ **(C)** $-x - y(x - 5yz)$

Solution **(A)**

$$2x - 3yz + \frac{x}{z}$$

$$2(\quad) - 3(\quad)(\quad) + \frac{(\quad)}{(\quad)}$$

Using parentheses as indicated will help to reduce sign errors.

$$2(-24) - 3(2)(-3) + \frac{(-24)}{(-3)}$$

Multiplication and division precede addition and subtraction.

$$= (-48) - (-18) + 8$$
$$= (-48) + 18 + 8 = -22$$

(B)

$$\frac{x}{y} - \frac{16z + xy}{y + z}$$

$$\frac{(-24)}{2} - \frac{16(-3) + (-24)(2)}{2 + (-3)}$$

$$= (-12) - \frac{(-48) + (-48)}{-1}$$

$$= (-12) - \frac{-96}{-1} = (-12) - 96 = -108$$

(C)

$$-x - y(x - 5yz)$$

$$-(-24) - 2[(-24) - 5(2)(-3)]$$

$$= 24 - 2[(-24) - (-30)]$$

$$= 24 - 2(6)$$

$$= 24 - 12$$

$$= 12$$

Notice how brackets and parentheses are used. Brackets are just another kind of parentheses; each may replace the other as desired.

Problem 25 Evaluate for $u = 36$, $v = -4$, and $w = -3$:

(A) $3vw - \dfrac{u}{3w} + 4v$ **(B)** $\dfrac{9w - 8v}{v - w} - \dfrac{u}{v}$

(C) $u - [7 - 2(u - 4vw)]$

ANSWERS TO MATCHED PROBLEMS

20. **(A)** -12 **(B)** -12 **(C)** 12 **(D)** 0
21. Both are -20.
22. $-[(6)(-3)] = -(-18) = 18$ and $(-6)(-3) = 18$
23. **(A)** 4 **(B)** -3 **(C)** -6
24. All three are -3.
25. **(A)** 24 **(B)** 4 **(C)** 5

EXERCISE 1-5 A

Evaluate, performing the indicated operations.

1. $(-3)(-5)$ **2.** $(-7)(-4)$ **3.** $(-18) \div (-6)$

4. $(-20) \div (-4)$ **5.** $(-2)(+9)$ **6.** $(+6)(-3)$

7. $\dfrac{-9}{+3}$ **8.** $\dfrac{+12}{-4}$ **9.** $0(-7)$

10. $(-6)0$ **11.** $0/5$ **12.** $0/(-2)$

13. $3/0$ **14.** $-2/0$ **15.** $0 \div 0$

16. $\frac{0}{0}$ **17.** $\frac{-21}{3}$ **18.** $\frac{-36}{-4}$

19. $(-4)(-2) + (-9)$

20. $(-7) + (-3)(+2)$

21. $(+5) - (-2)(+3)$

22. $(-7) - (-3)(-4)$

23. $5 - \frac{-8}{2}$

24. $7 - \frac{-16}{-2}$

25. $(-1)(-8)$ and $-(-8)$

26. $(-1)(+3)$ and $-(+3)$

27. $-12 + \frac{-14}{-7}$

28. $\frac{-10}{5} + (-7)$

29. $\frac{6(-4)}{-8}$

30. $\frac{5(-3)}{3}$

31. $\frac{22}{-11} - (-4)(-3)$

32. $3(-2) - \frac{-10}{-5}$

33. $\frac{-16}{2} - \frac{3}{-1}$

34. $\frac{27}{-9} - \frac{-21}{-7}$

35. $(+5)(-7)(+2)$

36. $(-6)(-3)(+4)$

37. $(-22)(+36)(0)$

38. $(+19)(0)(-35)$

B *Evaluate, performing the indicated operations.*

39. $[(+2) + (-7)][(+8) - (+10)]$

40. $[(-3) - (+8)][(+4) + (-2)]$

41. $12 - 7[(-4)(5) - 2(-8)]$

42. $9 - 5[(-2) - 3]$

43. $\frac{9}{-3} - \frac{3 + 9(-2)}{-2 - (-3)}$

44. $\frac{4(-2) - (-5)}{(-9) - (-6)} - \frac{-24}{-8}$

45. $\{[8/(-2)] - [21 + 5(-3)]\} - (-2)(-4)$

46. $7 - \{9 - [5 - 2(-3)] - (8/2)\}$

47. $[8 - (9 - 7)]/6 - 3$

48. $[8 - (9 - 7)]/(6 - 3)$

49. $[8 - (9 - 7)]/(6 - 4) + (4 + 2)/2$

50. $[8 - (9 - 7)]/([(6 - 5) + (4 - 3)]/2)$

51. $8 - [(9 - 3)/2 - 2 + 4 - 4/2]$

52. $(8 - 1) - 7/(6 - 5) + 3 - 4/2$

53. $\dfrac{(8 - 1) - 8/(6 - 2)}{4 + 2/2}$

54. $\dfrac{8 - (9 - 6/(3 - 2))}{(4 - 2)/2}$

Evaluate Problems 55–76 for $w = 2$, $x = -3$, $y = 0$, *and* $z = -24$.

55. z/w **56.** z/x **57.** w/y **58.** y/x

59. $\dfrac{z}{x} - wz$ **60.** $wx - \dfrac{z}{w}$ **61.** $\dfrac{xy}{w} - xyz$ **62.** $wxy - \dfrac{y}{z}$

63. $-|w||x|$ **64.** $(|x||z|)$ **65.** $\dfrac{|z|}{|x|}$ **66.** $-\dfrac{|z|}{|w|}$

67. $(wx - z)(z - 8x)$ **68.** $(5x - z)(wx - 3w)$

69. $wx + \dfrac{z}{wx} + wz$ **70.** $xyz + \dfrac{y}{z} + x$

71. $\dfrac{8x}{z} - \dfrac{z - 6x}{wx}$ **72.** $\dfrac{w - x}{w + x} - \dfrac{z}{2x}$

73. $\dfrac{24}{3w - 2x} - \dfrac{24}{3w + 2x}$ **74.** $\dfrac{48}{z + 8x} - \dfrac{48}{z - 8x}$

75. $\dfrac{z}{wx} - 2[z + 3(2x - w)]$ **76.** $\dfrac{8wx}{-z} - x[5 + 2(z + 9w)]$

77. Any integer divided by 0 is (*always, sometimes, never*) 0.

78. Zero divided by *any* integer is (*always, sometimes, never*) 0.

79. A product made up of an odd number of negative factors is (*sometimes, always, never*) negative.

80. A product made up of an even number of negative factors is (*sometimes, always, never*) negative.

C **81.** If the quotient $\dfrac{x}{y}$ exists, and neither x nor y is 0, when is it equal to $\dfrac{-|x|}{|y|}$?

82. If the quotient $\dfrac{x}{y}$ exists, and neither x nor y is 0, when is it equal to $\dfrac{|x|}{|y|}$?

83. Provide the reasons for each step in the proof that $a0 = 0$ for all real numbers a (part 1, Multiplication of Real Numbers).

STATEMENT	REASON
1. $a0 = a(0 + 0)$	**1.**
2. $a0 = a0 + a0$	**2.**
3. $a0 + [-(a0)] = (a0 + a0) + [-(a0)]$	**3.**
4. $0 = a0 + \{a0 + [-(a0)]\}$	**4.**
5. $0 = a0 + 0$	**5.**
6. $0 = a0$	**6.**
7. $a0 = 0$	**7.**

84. Provide the reasons for each step in the proof that $(-1)a = -a$ [part (A), Sign Properties for Multiplication].

STATEMENT	REASON
1. $a + (-1)a = 1a + (-1)a$	**1.**
2. $= a[1 + (-1)]$	**2.**
3. $= a \cdot 0$	**3.**
4. $= 0$	**4.**
5. Therefore, $(-1)a = -a$.	**5.**

1-6 SOLVING EQUATIONS

- Combining Like Terms
- Basic Concepts
- Equations with Integer Coefficients
- Literal Equations
- A Strategy for Solving Word Problems
- Number Problems

In this section we review basic methods of solving equations such as

$$2(3x - 5) - 2 = 5 - (3x + 2) \tag{1}$$

As a first step we recall the process of combining like terms. We then consider what it means to solve an equation and describe the basic strategy for doing so. The strategy is applied to equations of the form (1), to literal equations of the form (2),

$$2(3x - y) - 2 = y - (3x + 2) \tag{2}$$

and to word problems.

COMBINING LIKE TERMS

A constant that is present as a factor in a term is called the **numerical coefficient** (or simply the **coefficient**) of the term. For example, in the term $4x^2$, 4 is the coefficient of x^2. If no constant appears in the term, then the coefficient is understood to be 1. Two terms are called **like terms** if they have exactly the same variable factors to the same powers. The numerical coefficients may or may not be the same. Since constant terms involve no variables, all constant terms are like terms. If an algebraic expression contains two or more like terms, these terms can be combined into a single term by making use of the distributive law.

Example 26 Combine like terms:

(A) $3x + 7x$ **(B)** $6m - 9m$
(C) $3z + 5 - z + 2$ **(D)** $5x^3y - 2xy - x^3y - 2x^3y$

Solution

(A) $3x + 7x = (3 + 7)x = 10x$

(B) $6m - 9m = (6 - 9)m = -3m$

(C)
$$\begin{aligned} 3z + 5 - z + 2 &= 3z - z + 5 + 2 \\ &= (3 - 1)z + (5 + 2) \\ &= 2z + 7 \end{aligned}$$

(D)
$$\begin{aligned} 5x^3y - 2xy - x^3y - 2x^3y &= 5x^3y - x^3y - 2x^3y - 2xy \\ &= (5 - 1 - 2)x^3y - 2xy \\ &= 2x^3y - 2xy \end{aligned}$$

Problem 26 Combine like terms:

(A) $5y + 4y$ **(B)** $2u - 6u$ **(C)** $4x - 1 + 3x + 2$
(D) $6mn^2 - m^2n - 3mn^2 - mn^2$

It should be clear that free use has been made of the properties discussed in this chapter. Most of the steps illustrated in the dashed boxes are done mentally. The process is quickly mechanized as follows:

> Like terms are combined by adding their numerical coefficients.

Example 27 Combine like terms mentally:

(A) $3x - 5y + 6x + 2y$
(B) $x^3y^2 - 2x^2y^3 + 5x^2y^2 - 4x^2y^3 - x^3y^2 - 5x^2y^2$

Solution **(A)** $3x - 5y + 6x + 2y \boxed{= 3x + 6x - 5y + 2y} = 9x - 3y$

(B) $x^3y^2 - 2x^2y^3 + 5x^2y^2 - 4x^2y^3 - x^3y^2 - 5x^2y^2$

$$\boxed{= x^3y^2 - x^3y^2 - 2x^2y^3 - 4x^2y^3 + 5x^2y^2 - 5x^2y^2}$$

$$= -6x^2y^3$$

Problem 27 Combine like terms mentally:

(A) $7m + 8n - 5m - 10n$
(B) $2u^4v^2 - 3uv^3 - u^4v^2 + 6u^4v^2 + 2uv^3 - 6u^4v^2$

BASIC CONCEPTS

A **solution** or **root** of an equation involving a single variable is a replacement of the variable by a constant that makes the left side of the equation equal to the right side. For example, $x = 4$ is a solution of

$$2x - 1 = x + 3$$

since

$$2(\mathbf{4}) - 1 = \mathbf{4} + 3 \quad \text{That is, } 7 = 7.$$

The set of all solutions is called the **solution set**. To **solve an equation** is to find its solution set.

Knowing what we mean by a solution set for an equation is one thing, but finding it is another. Our objective is to develop a systematic method of solving equations that is free of guesswork. We start by introducing the idea of equivalent equations. We say that two equations are **equivalent** if they both have the same solution set—any solution of one is a solution of the other.

The basic idea in solving equations is to perform operations on equations that produce *simpler* equivalent equations and to continue the process until we reach an equation whose solution is obvious—generally an equation such as

$$x = -5 \qquad \text{or} \qquad x = \tfrac{1}{2} \qquad \text{or} \qquad x = 7$$

The following properties of equality produce equivalent equations when applied.

Equality Properties

For a, b, and c any real numbers:

1. If $a = b$, then $a + c = b + c$. ADDITION PROPERTY
   If $x - 2 = 3$, then $(x - 2) + 2 = 3 + 2$.
2. If $a = b$, then $a - c = b - c$. SUBTRACTION PROPERTY
   If $x + 4 = 5$, then $(x + 4) - 4 = 5 - 4$.
3. If $a = b$, then $ca = cb$, $c \neq 0$. MULTIPLICATION PROPERTY
   If $\frac{x}{2} = 3$, then $2 \cdot \frac{x}{2} = 2 \cdot 3$.
4. If $a = b$, then $\frac{a}{c} = \frac{b}{c}$, $c \neq 0$. DIVISION PROPERTY
   If $5x = 10$, then $\frac{5x}{5} = \frac{10}{5}$.

These properties of equality follow directly from the basic equality properties discussed in Section 1-2.

We can think of the process of solving an equation as a game. The objective of the game is to isolate the variable (with a coefficient of 1) on one side of the equation (usually the left), leaving a constant on the other side. We are now ready to solve equations. The following strategy might prove helpful.

Equation-Solving Strategy

1. Use the multiplication property to remove fractions if present.†
2. Simplify the left and right sides of the equation by removing grouping symbols and combining like terms.
3. Use the equality properties to get all variable terms on one side (usually the left) and all constant terms on the other side (usually the right). Combine like terms in the process.
4. Isolate the variable (with a coefficient of 1), using the division or multiplication property of equality.

EQUATIONS WITH INTEGER COEFFICIENTS

Example 28 Solve $3x - 2(2x - 5) = 2(x + 3) - 8$ and check.

† Our initial examples will not include fractions. This step is included here for completeness, and will be necessary for later examples where fractions will be present.

LITERAL EQUATIONS

An equation involving more than one variable can be solved for a specific variable using the strategy developed in this section.

Example 29 Solve for y in terms of x:

$$2(3x - y) - 2 = y - (3x + 2)$$

Solution

$$2(3x - y) - 2 = y - (3x + 2)$$

$$6x - 2y - 2 = y - 3x - 2$$ Clear parentheses. Treat $-(3x + 2)$ as $(-1)(3x + 2)$ so that removing parentheses gives $-3x - 2$.

$$6x - 2y - 2 + 2 = y - 3x - 2 + 2$$ Addition property

$$6x - 2y = y - 3x$$

$$6x - 2y - y = y - 3x - y$$ Subtraction property

$$6x - 3y = -3x$$

$$6x - 3y - 6x = -3x - 6x$$ Subtraction property

$$-3y = -9x$$

$$\frac{-3y}{-3} = \frac{-9x}{-3}$$ Division property

$$y = 3x$$

Because our solution still involves other variables (letters), such equations are often called **literal equations**.

Problem 29 Solve for y in terms of x:

$$3(x + y) + 1 = 3(3x + 1) + y$$

A STRATEGY FOR SOLVING WORD PROBLEMS

A great many practical problems can be solved using algebraic techniques—so many, in fact, there is no one method of attack that will work for all. However, we can formulate a strategy that may help you organize your approach.

A Strategy for Solving Word Problems

1. Read the problem carefully—several times if necessary—until you understand the problem, know what is to be found, and know what is given.
2. If appropriate, draw figures or diagrams and label known and unknown parts. Look for formulas connecting the known quantities with the unknown quantities.
3. Let one of the unknown quantities be represented by a variable, say x, and try to represent all other unknown quantities in terms of x. This is an important step and must be done carefully. Be sure you clearly understand what you are letting x represent.
4. Form an equation relating the unknown quantities with the known quantities. This step may involve the translation of an English sentence into an algebraic sentence, the use of relationships in a geometric figure, the use of certain formulas, and so on.
5. Solve the equation and write answers to *all* parts of the problem requested.
6. Check all solutions in the original problem.

Step 4 in the strategy is often difficult, and there is no systematic procedure that will work for all problems. If you have trouble with this step, Appendix A suggests an approach that may be helpful.

NUMBER PROBLEMS

In earlier sections you had experience in translating verbal forms into symbolic forms. We now take advantage of that experience to solve a variety of number problems.

Example 30 Find a number such that 16 more than twice the number is 1 more than 5 times the number.

Solution Let

$$x = \text{The number}$$

We symbolize each part of the problem as follows:

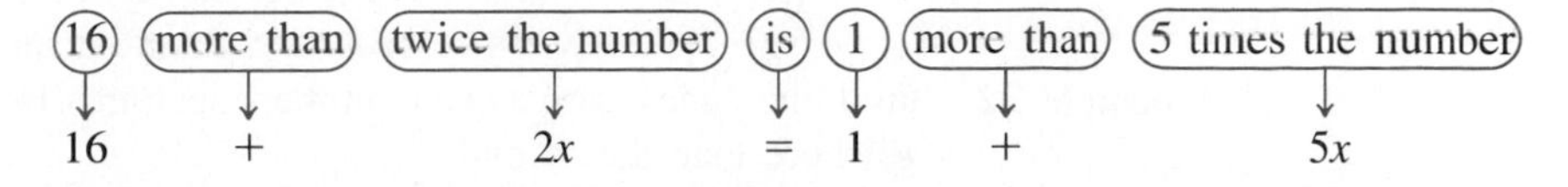

We now solve the equation:

$$16 + 2x = 1 + 5x$$
$$15 = 3x$$
$$5 = x \qquad \text{that is, } x = 5$$

Check 16 more than twice the number: $16 + 2(5) = 26$
1 more than 5 times the number: $1 + 5(5) = 26$

Problem 30 Find a number such that 12 more than 3 times the number is 5 times the number.

Example 31 Find a number such that 2 less than twice the number is 5 times the quantity that is 2 more than the number.

Solution Let

$x =$ The number

Symbolize each part:

2 less than twice the number: $2x - 2$ Not $2 - 2x$

is: $=$

5 times the quantity
that is 2 more than the number: $5(x + 2)$ Why would $5x + 2$ be incorrect?

Write an equation and solve:

$$2x - 2 = 5(x + 2)$$
$$2x - 2 = 5x + 10$$
$$-3x = 12$$
$$x = -4$$

Checking is left to you.

Problem 31 Find a number such that 4 times the quantity that is 2 less than the number is 1 more than 3 times the number. Write an equation and solve.

Example 32 Find three consecutive even numbers such that twice the first plus the third is 10 more than the second.

Solution Let

$$x = \text{First of three consecutive even numbers}$$

$$x + 2 = \text{Second consecutive even number}$$

$$x + 4 = \text{Third consecutive even number}$$

The difference between any two consecutive even numbers is 2.

Form an equation and solve:

(twice the first) (plus) (the third) (is) (10) (more than) (the second)

$$2x \quad + \quad (x + 4) \quad = \quad 10 \quad + \quad (x + 2)$$

$$2x + x + 4 = 10 + x + 2$$

$$3x + 4 = x + 12$$

$$2x = 8$$

$$x = 4$$

$$x + 2 = 6$$

$$x + 4 = 8$$

Three consecutive even numbers

Thus, the three consecutive even numbers are 4, 6, and 8.

Check Twice the first plus the third: $2 \cdot 4 + 8 = 8 + 8 = 16$

10 more than the second: $10 + 6 = 16$

Problem 32 Find three consecutive odd numbers such that 4 times the first minus the third is the same as the second. Write an equation and solve.

ANSWERS TO MATCHED PROBLEMS

26. **(A)** $9y$ **(B)** $-4u$ **(C)** $7x + 1$ **(D)** $2mn^2 - m^2n$
27. **(A)** $2m - 2n$ **(B)** $u^4v^2 - uv^3$
28. $x = -9$ **29.** $y = 3x + 1$ **30.** $x = 6$
31. $4(x - 2) = 1 + 3x$; $x = 9$
32. $4x - (x + 4) = x + 2$; 3, 5, 7

EXERCISE 1-6

A *Simplify by removing parentheses, if any, and combining like terms.*

1. $9x + 8x$ **2.** $7x + 3x$

3. $9x - 8x$ **4.** $7x - 3x$

5. $5x + x + 2x$ **6.** $3x + 4x + x$

7. $4t - 8t - 9t$ **8.** $2x - 5x + x$

9. $4y + 3x + y$

10. $2x + 3y + 5x$

11. $8 + 4x - 4$

12. $-3 - x + 5$

13. $5m + 3n - m - 9n$

14. $2x + 8y - 7x - 5y$

15. $3(u - 2v) + 2(3u + v)$

16. $2(m + 3n) + 4(m - 2n)$

17. $4(m - 3n) - 3(2m + 4n)$

18. $2(x - y) - 3(3x - 2y)$

19. $(2u - v) + (3u - 5v)$

20. $(x + 3y) + (2x - 5y)$

21. $(3x + 2) + (x - 5)$

22. $(y + 7) - (4 - 2y)$

23. $2(x + 5) - (x - 1)$

24. $3(z - 2) - 2(z - 3)$

Solve and check.

25. $3(x + 2) = 5(x - 6)$

26. $5x + 10(x - 2) = 40$

27. $4(x - 2) = 4x - 8$

28. $3y + 6 = 3(y + 2)$

29. $5 + 4(t - 2) = 2(t + 7) + 1$

30. $7x - (8x - 4) - 10 = 5 - (4x + 2)$

31. $3x - (x + 2) = 5x - 3(x - 1)$

32. $x - 2(x - 4) = 3x - 2(2x + 1)$

33. $10x + 25(x - 3) = 275$

34. $x + (x + 2) + (x + 4) = 54$

35. $5x - (7x - 4) - 2 = 5 - (3x + 2)$

36. $-3(4 - t) = 5 - (t + 1)$

37. $2(3x + 1) - 8x = 2(1 - x)$

38. $3(4x - 2) - 8x = 6x - 2(x + 3)$

39. $x(x - 1) + 5 = x^2 + x - 3$

40. $x(x + 2) = x(x + 4) - 12$

41. $x(x - 4) - 2 = x^2 - 4(x + 3)$

42. $t(t - 6) + 8 = t^2 - 6t - 3$

B *Solve for the indicated variable in terms of the other variables.*

43. $10x + 5y = 150$; solve for x

44. $10x + 5y = 150$; solve for y

45. $3x - 4y + 5 = x - 2y + 3$; solve for x

46. $3x - 4y + 5 = x - 2y + 3$; solve for y

47. $2 - 3(x + 3y) = x - 5(y + 6)$; solve for y

48. $2 - 3(x + 3y) = x - 5(y + 6)$; solve for x

49. $3(x + y) - 1 = 17 - 3y$; solve for x

50. $3(y - 2x) = 9(x - 1)$; solve for y

If x represents a number, write an algebraic expression for each of the following numbers:

51. Twice x

52. 3 times x

53. 3 less than x

54. 2 more than x

55. 2 more than twice x

56. 5 less than twice x

57. 4 less than 3 times x

58. 5 more than 6 times x

59. 1 more than the negative of $2x$

60. 1 less than the negative of $3x$

C *Find numbers meeting each of the indicated conditions. Write an equation using x and solve.*

61. 7 times a number is 12 less than 4 times the number.

62. 6 times a number is 24 more than 3 times the number.

63. 3 more than twice the number is 12 less than 3 times the number.

64. 1 more than 5 times the number is 6 times 1 more than the number.

65. Three consecutive integers whose sum is 96

66. Three consecutive integers whose sum is 78

67. Three consecutive even numbers whose sum is 42

68. Three consecutive even numbers whose sum is 54

1-7 CHAPTER REVIEW

A **set** is a collection of objects called **members** or **elements** of the set. Sets are usually described by **listing** {list of elements} or by a **rule** $\{x|$rule that determines that x is a member$\}$. That an object a belongs to set A is denoted $a \in A$; that it does not is denoted by $a \notin A$. The set containing no elements is called the **empty set** or the **null set** and is denoted $\varnothing$. Sets are **finite** if the elements can be counted (and there is an end); they are **infinite** otherwise. If each element of set A is also in set B, we say A is a **subset** of B and write $A \subset B$. A **variable** is a symbol that represents unspecified elements from a **replacement set**; a **constant** is a symbol for one object in a set. *(1-1)*

The **real number** system consists of

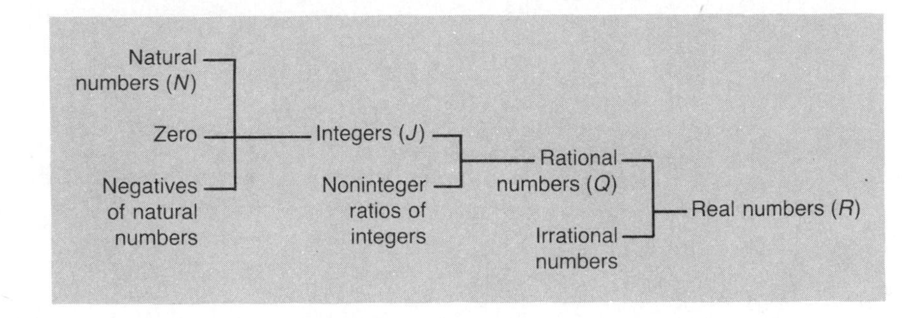

The real numbers can be represented as points on a **real number line** (**real line**) where each point is associated with a single real number called the **coordinate** of the point. The point with coordinate 0 is called the **origin**, with **positive real numbers** to the right and **negative real numbers** to the left. *(1-1)*

An **algebraic expression** is made up from variables, constants, mathematical operation signs, and grouping symbols. Expressions joined by plus or minus signs are called **terms**; those joined by multiplication are **factors**. Algebraic expressions joined by an equal sign are called **algebraic equations**. Equality satisfies these **equality properties**:

Reflexive property:	$a = a$
Symmetric property:	If $a = b$, then $b = a$.
Transitive property:	If $a = b$ and $b = c$, then $a = c$.
Substitution property:	If $a = b$, then either may be substituted for the other.

The **inequality symbols** $<$, $>$, $\leq$, $\geq$ denote **less than**, **greater than**, **less than or equal to**, and **greater than or equal to**, respectively; $a < b$ and $b > a$ mean $a + p = b$ for some positive number p. For an **inequality statement**—that is, an inequality involving a variable—the numbers that make it true are called the **solution set**. The solution set can be represented by a **graph of the inequality statement**. *(1-2)*

The real numbers under addition and multiplication satisfy several basic **properties** that are listed and named in Section 1-3. Among these are the **commutative properties** ($a + b = b + a$, $a \cdot b = b \cdot a$), the **associative properties** [$(a + b) + c = a + (b + c)$ and $(ab)c = a(bc)$], the **distributive**

properties $[a(b + c) = ab + ac,\ (a + b)c = ac + bc]$, and the **identity properties** $(a + 0 = 0 + a = a,\ 1 \cdot a = a \cdot 1 = a)$. These properties allow us to manipulate algebraic expressions. Two algebraic expressions are **equivalent** if they yield the same number for all possible values of the variables. *(1-3)*

For any real number x, $-x$ denotes its **additive inverse**, the number we add to x to get 0; this is also called the **negative of the number**. For any real number x, its **absolute value** $|x|$ is the distance from x to the origin, a nonnegative quantity. Rules for addition of real numbers are given in Section 1-4. Subtraction is accomplished in terms of addition by $a - b = a + (-b)$. *(1-4)*

Rules for multiplication of real numbers are given in Section 1-5. Division is defined in terms of multiplication: $a \div b$ is equal to a **quotient** Q if $bQ = a$ and Q is unique. Division by 0 is not defined. Multiplication and division satisfy these **sign properties**:

$$(-a)b = -(ab) = a(-b) \qquad (-a)(-b) = ab$$

$$\frac{(-a)}{b} = \frac{a}{-b} = -\frac{a}{b} \qquad \frac{-a}{-b} = \frac{a}{b} = -\frac{-a}{b} = -\frac{a}{-b}$$

The **order of operations** for arithmetic operations, unless **grouping symbols** indicate otherwise, is first to do multiplications and divisions left to right and then to do additions and subtractions left to right. *(1-5)*

The distributive properties are used to **combine like terms**—that is, terms with identical variables to the same power—a common step in solving equations. A **solution** or **root** of an equation in one variable x is a replacement value for x that makes the equality true. To **solve** an equation means to find all solutions—the **solution set**. Two equations are **equivalent** if they have the same solution set. Equations are generally solved by performing operations that produce simpler equivalent equations. The following **properties of equality** yield equivalent equations:

If $a = b$, then $a + c = b + c$ and $a - c = b - c$.

If $a = b$ and $c \neq 0$, then $a \cdot c = b \cdot c$ and $a/c = b/c$.

An **equation-solving strategy** is to

1. Clear fractions.
2. Simplify both sides.
3. Get all variable terms to one side, constants to the other, and combine like terms.
4. Isolate the variable.

A **linear equation** $ax + b = 0$ has the unique solution $x = -b/a$ if $a \neq 0$ and either no solution or infinitely many if $a = 0$. The same strategy can be applied to solving certain **literal equations**, that is, equations involving more than one variable that are to be solved for one of the variables. A **general strategy for solving word problems** is, briefly:

1. Read the problem very carefully.
2. Write down the important facts and relationships.
3. Identify the unknown quantities in terms of one variable.
4. Find the equation.
5. Solve the equation. Write down all solutions asked for.
6. Check the solutions. *(1-6)*

REVIEW EXERCISE 1-7

Work through all the problems in this chapter review and check answers in the back of the book. (Answers to all problems are there, and following each answer is a number in italics indicating the section in which that type of problem is discussed.) Where weaknesses show up, review appropriate sections in the text.

A

1. For $A = \{1, 2, 3, 4, 5\}$, $B = \{1, 3, 5\}$, and $C = \{5, 1, 3\}$, indicate true (T) or false (F):
(A) $4 \in A$ **(B)** $4 \notin C$ **(C)** $B \in A$ **(D)** $A \subset B$
(E) $C \subset A$ **(F)** $B \subset C$ **(G)** $\varnothing \subset B$ **(H)** $B \neq C$

2. 3.127127 . . . represents a (*rational, irrational*) number?

Evaluate.

3. $3 \cdot 7 - 4$ **4.** $7 + 2 \cdot 3$ **5.** $(-8) + 3$

6. $(-9) + (-4)$ **7.** $(-3) - (-9)$ **8.** $4 - 7$

9. $0 - (-3)$ **10.** $(-12) - 0$ **11.** $(-7)(-4)$

12. $3(-6)$ **13.** $(-16)/4$ **14.** $(-12)/(-2)$

15. $(-6)/0$ **16.** $0/(-3)$ **17.** $10 - 3(6 - 4)$

18. $(-8) - (-2)(-3)$ **19.** $(-9) - [(-12)/3]$

20. $(4 - 8) + 4(-2)$ **21.** $|-8|$

22. $-(-5)$ **23.** $-[-(-3)]$

24. $-|-(-2)|$ **25.** $-(|-3| + |-2|)$

26. $-(|-8| - |3|)$

Remove parentheses and simplify using commutative, associative, and distributive properties.

27. $7 + (x + 3)$

28. $(3x)5$

29. $(2x)(4y)$

30. $(y + 7) + (x + 2) + (z + 3)$

31. $0 + (1x + 2)$

32. $(x + 0) + 0y$

33. $3(x - 5)$

34. $(a + b)7$

35. $2 + 3(x + 4)$

36. $(x - 1)2 + 3$

Replace each question mark with < or > to form a true statement.

37. $7 ? 2$

38. $-7 ? -2$

39. $-12 ? 0$

40. $-342 ? -3$

41. $0 ? -45$

42. $-50 ? 20$

B **43.** For N = The set of natural numbers, J = The set of integers, Q = The set of rational numbers, R = The set of real numbers, indicate true (T) or false (F):
(A) $-5 \in N$ **(B)** $-5 \in R$ **(C)** $Q \subset R$ **(D)** $Q \subset J$
(E) $1.43 \in Q$ **(F)** $-\frac{2}{3} \in R$ **(G)** $\sqrt{2} \in Q$ **(H)** $\pi \in R$

44. Graph on a real number line: **(A)** $x < -1$ **(B)** $-4 \le x < 3$

Evaluate.

45. $2[9 - 3(3 - 1)]$

46. $6 - 2 - 3 - 4 + 5$

47. $[(-3) - (-3)] - (-4)$

48. $[-(-4)] + (-|-3|)$

49. $[(-16)/2] - (-3)(4)$

50. $(-2)(-4)(-3) - \dfrac{-36}{(-2)(9)}$

51. $2\{9 - 2[(3 + 1) - (1 + 1)]\}$

52. $(-3) - 2\{5 - 3[2 - 2(3 - 6)]\}$

53. $\dfrac{12 - (-4)(-5)}{4 + (-2)} - \dfrac{-14}{7}$

54. $\dfrac{24}{(-4) + 4} - \dfrac{24}{(-4) + 4}$

55. $3[14 - x(x + 1)]$ for $x = 3$

56. $-(-x)$ for $x = -2$

57. $-(|x| - |w|)$ for $x = -2$ and $w = -10$

58. $(x + y) - z$ for $x = 6$, $y = -8$, $z = 4$

59. $\left(2x - \frac{z}{x}\right) - \frac{w}{x}$ for $w = -10$, $x = -2$, $z = 0$

60. $\frac{(xyz + xz) - z}{z}$ for $x = -6$, $y = 0$, $z = -3$

61. $\frac{x - 3y}{z - x} - \frac{z}{xy}$ for $x = -3$, $y = 2$, and $z = -12$

Solve for x.

62. $2x - 3 = 4x + 5$

63. $x - y = 2x + 3y + 4$

Translate each statement into an algebraic equation or inequality statement using x as the only variable. Do not solve the equation.

64. $x - 1$ is positive.

65. $2x + 3$ is not negative.

66. 50 is 10 less than twice a certain number.

67. x is less than 12 less than twice x.

68. x is greater than or equal to -5 and less than 5.

69. 8 more than a certain number is 5 times the number that is 6 less than the certain number.

70. In a rectangle with area 1,200 square centimeters, the width is 10 centimeters less than the length x. Write an equation relating the area with the length and width, using x as the only variable.

71. If the length of a rectangle is 5 meters longer than its width x and the perimeter is 43 meters, write an algebraic equation relating the sides and the perimeter.

Replace each question mark with an appropriate symbol to form a true statement.

72. $-a + ? = 0$

73. $a + (-a) = ?$

74. $a + ? = a$

75. $0 + ? = a$

76. $? \cdot a = a$

77. $a \cdot ? = 1$, $a \neq 0$

78. If $a + p = b$ for some positive number p, then a is (*greater than, less than*) b.

79. If $a - p = b$ for some positive number p, then a is (*greater than, less than*) b.

Replace each question mark with an appropriate symbol to make the statement an illustration of the stated property. (All variables represent real numbers.)

80. ***Symmetry property*** If $P + Prt = A$, then ?.

81. ***Transitive property*** If $x + y < 5$ and $5 < z$, then ?.

82. ***Commutative property*** $P + Q = ?$

83. ***Distributive property*** $AB + B = (?)B$

84. ***Inverse property*** $x + ? = 0$

85. ***Substitution principle*** If $3u - 2v = 5$ and $v = u + 4$, then $3u - 2(?) = 5$.

86. ***Identity property*** $(?)x = x$

87. ***Associative property*** $(x + 3) + 5 = ?$

88. ***Trichotomy property*** If $y \neq 5$, then either $y > 5$ or ?.

State the real number property that justifies each statement.

89. $5 + (x + 3) = 5 + (3 + x)$

90. $5 + (3 + x) = (5 + 3) + x$

91. $5(x3) = 5(3x)$

92. $5(3x) = (5 \cdot 3)x$

93. $(x + y) + 0 = x + y$

94. $(ab) + [-(ab)] = 0$

95. $3(x - 4) = 3x - 12$.

C

96. Evaluate $uv - 3\{x - 2[(x + y) - (x - y)] + u\}$ for $u = -2$, $v = 3$, $x = 2$, and $y = -3$.

97. Evaluate $\dfrac{5w}{x - 7} - \dfrac{wx - 4}{x - w}$ for $w = -4$ and $x = 2$.

98. Replace the question marks with appropriate symbols:

$$\{3, 4, 5, 6\} = \{x \in N | ? \leq x \; ? \; 7\}$$

99. Write each set by the listing method:
(A) $\{x \in J \mid -2 < x \leq 2\}$ **(B)** $\{x \in N \mid 5 < x < 6\}$

100. If $M = \{3, 4, 5, 7\}$ and $N = \{4, 5, 6\}$, find:
(A) $\{x \mid x \in M \text{ or } x \in N\}$ **(B)** $\{x \mid x \in M \text{ and } x \in N\}$

101. Describe the elements in each set:
(A) $\{x \in R \mid |x| = x\}$ **(B)** $\{x \in R \mid |x| = -x\}$

102. Find a number such that 3 times the quantity that is 4 more than the number is 5 times the original number.

103. Three consecutive natural numbers have the property that twice the sum of the first two is 11 more than 3 times the third. Find the numbers.

104. Two consecutive natural numbers have the property that their product is 7 less than the product of the smaller number times the successor of the larger. Find the numbers.

2

POLYNOMIALS AND FACTORING

In this chapter, we will consider the operations of addition, subtraction, and multiplication on polynomials. We will then look at the process of factoring a polynomial, that is, writing the polynomial as a product, and finally we will use factoring to solve equations.

2-1 ADDITION AND SUBTRACTION OF POLYNOMIALS

- Natural Number Exponents
- Polynomials; Degree
- Removing Symbols of Grouping
- Addition and Subtraction

Algebraic expressions such as

$$2x + 3 \qquad 4x^2 - x + 5 \qquad x^2 + 2xy + y^2$$

are called polynomials.

In order to consider basic operations on polynomials, we need first to recall a basic property of natural number exponents. It will also be helpful to introduce some terminology for describing polynomials.

NATURAL NUMBER EXPONENTS

Recall that

$$a^5 = a \cdot a \cdot a \cdot a \cdot a \quad \text{Five factors of } a$$

and, in general:

Natural Number Exponent

For n a natural number and a a real number:

$$a^n = a \cdot a \cdot \cdots \cdot a \quad n \text{ factors of } a$$

(n: Exponent; a: Base)

$2^3 = 2 \cdot 2 \cdot 2 = 8$

Exponent forms are encountered so frequently in algebra that it is essential for you to become completely familiar with their basic properties and uses. The first of these properties is given below; additional properties will be considered in detail in Chapter 5.

Consider:

$$a^3a^4 = \overbrace{(a \cdot a \cdot a)}^{3 \text{ factors}}\overbrace{(a \cdot a \cdot a \cdot a)}^{4 \text{ factors}} = \overbrace{(a \cdot a \cdot a \cdot a \cdot a \cdot a \cdot a)}^{3+4 \text{ factors}} = a^{3+4} = a^7$$

which suggests that for any real number a and any positive-integer exponents m and n:

Property 1

$a^m a^n = a^{m+n}$

$a^5a^2 = a^{5+2} = a^7 \qquad 2^3 \cdot 2^4 = 2^7$

Example 1 Simplify by rewriting the expression so that the variable occurs only once:

(A) $x^{12}x^{13}$ **(B)** $(3x^4)(5x^3)$

Solution **(A)** $x^{12}x^{13} = x^{12+13} = x^{25}$

(B) $(3x^4)(5x^3) = (3 \cdot 5)(x^4x^3)$
$= 15x^{4+3}$
$= 15x^7$

Problem 1 Simplify by rewriting the expression so that the variable occurs only once:

(A) a^7a^3 **(B)** $(2y^2)(6y^3)$

Recall that in the order of operations agreed to in Section 1-5, multiplications and divisions are to be done before additions and subtractions unless grouping symbols indicate otherwise. This scheme is extended to include the taking of powers (and later roots) by agreeing that:

Powers take precedence over multiplication and division.

Thus, for example,

$$\frac{2^3}{5} = \frac{8}{5} = 1.6$$

The exponent applies only to the 2.

and

$$5 \cdot 2^3 = 5 \cdot 8 = 40$$

but

$$\left(\frac{2}{5}\right)^3 = (0.4)^3 = 0.064$$

Here the parentheses indicate that the exponent applies to the expression within.

and

$$(5 \cdot 2)^3 = 10^3 = 1,000$$

Particular care is required when applying exponents to expressions involving negative numbers:

$$-4^2 \neq (-4)^2 \qquad \text{because} \qquad \begin{cases} -4^2 = -(4 \cdot 4) = -16 \\ (-4)^2 = (-4)(-4) = 16 \end{cases}$$

In evaluating -4^2, the absence of parentheses indicates that the square applies only to the 4 and does not include the negative sign.

POLYNOMIALS; DEGREE

An algebraic expression involving only the operations of addition, subtraction, and multiplication on variables and constants is called a **polynomial**. Phrased differently, a polynomial is a sum of terms, each of which is a product of constants and variables raised to whole-number powers. Here are some examples:

POLYNOMIALS

$$3x - 1 \qquad x \qquad 2x^2 - 3x + 2 \qquad 5$$
$$x^3 - 3x^2y - 4y^2 \qquad 0 \qquad x^2 - \tfrac{2}{3}xy + 2y^2 \qquad (x + y)^2$$

In a polynomial, a variable cannot appear in a denominator, as an exponent, within a radical, or within absolute-value bars. The following expressions are, therefore, not polynomials:

NONPOLYNOMIALS

$$\frac{2x + 1}{3x^2 - 5x + 7} \qquad 3^x \qquad x^3 - 2\sqrt{x} + \frac{1}{x^3} \qquad |2x^3 - 5| \qquad \frac{1}{x} \qquad \sqrt{x}$$

We see that a polynomial in one variable x is constructed by adding or subtracting constants and terms of the form ax^n, where a is a real number and n is a natural number. A polynomial in two variables x and y is constructed by adding or subtracting constants and terms of the form ax^my^n, where again a is a real number and m and n are natural numbers.

It is convenient to identify certain types of polynomials for more efficient study. The concept of degree is used for this purpose. The **degree of a term** in a polynomial is the sum of the powers of the variables in the term. Thus, if a term has only one variable, the degree of the term is the power of the variable. A nonzero constant term—that is, a term without any variables—is assigned degree 0. The **degree of a polynomial** is the degree of its term with the highest degree. The constant 0, either as a term or as a polynomial, is not assigned a degree.

Example 2 What is the degree of each term in the following polynomials? What is the degree of the polynomial?

(A) $4x^3$ **(B)** $3x^3y^2$ **(C)** $3x^5 - 2x^4 + x^2 - 3$
(D) $x^2 - 2xy + y^2 + 2x - 3y + 2$

Solution

(A) This polynomial and its only term are of degree 3.
(B) This polynomial and its only term are of degree 5.
(C) The degrees of the four terms, in order, are 5, 4, 2, and 0; the degree of the polynomial is 5.
(D) The degrees of the six terms, in order, are 2, 2, 2, 1, 1, and 0; the degree of the polynomial is 2.

Problem 2 What is the degree of each term in the following polynomials? What is the degree of the polynomial?

(A) $7x^5$ **(B)** $3x^3y^4$ **(C)** $5x^3 - 7x^2 + x - 9$
(D) $4x^2y - xy^2 + xy + y^2 + x + 6$

We also call a one-term polynomial a **monomial**, a two-term polynomial a **binomial**, and a three-term polynomial a **trinomial**. For example,

$4x^3 - 3x + 7$	$5x - 2y$	$6x^4y^3$	7
Trinomial degree 3	Binomial degree 1	Monomial degree 7	Monomial degree 0

Recall that the constant factor in a term in a polynomial is called the coefficient. The constant term in a polynomial is sometimes also referred to as a coefficient. The coefficient of a term in a polynomial includes the sign that precedes it, so that a term like $-3x^2$ should be thought of as $(-3)x^2$.

Example 3 What is the coefficient of each term in the polynomial

$$3x^4 - 2x^3 + x^2 - x + 3$$

Solution

Coefficient of x^3 Coefficient of x

$$3x^4 - 2x^3 + x^2 - x + 3 = 3x^4 + (-2)x^3 + 1 \cdot x^2 + (-1)x + 3$$

Coefficient of x^4 Coefficient of x^2 Constant term or coefficient

The coefficient of x^4 is 3, that of x^3 is -2, that of x^2 is 1, and that of x is -1. The constant term is 3.

Problem 3 What is the coefficient of each term in the polynomial

$$5x^4 - x^3 - 3x^2 + x - 7$$

REMOVING SYMBOLS OF GROUPING

How can we simplify expressions such as

$$2(3x - 5y) - 2(x + 3y)$$

You no doubt would guess that we could rewrite this expression, using the various forms of the distributive property, as

$$6x - 10y - 2x - 6y$$

and combine like terms to obtain

$$4x - 16y$$

and your guess would be correct.

Example 4 Remove parentheses and simplify:

(A) $2(3x^2 - 2x + 5) + (x^2 + 3x - 7)$

$$= 2(3x^2 - 2x + 5) + 1(x^2 + 3x - 7)$$

Think

$$= 6x^2 - 4x + 10 + x^2 + 3x - 7$$
$$= 7x^2 - x + 3$$

(B) $(x^3 - 2x - 6) - (2x^3 - x^2 + 2x - 3)$

$$= 1(x^3 - 2x - 6) + (-1)(2x^3 - x^2 + 2x - 3)$$

Think

Be careful with the sign here.

$$= x^3 - 2x - 6 - 2x^3 + x^2 - 2x + 3$$
$$= -x^3 + x^2 - 4x - 3$$

(C) $[3x^2 - (2x + 1)] - (x^2 - 1)$

$$= [3x^2 - 2x - 1] - (x^2 - 1)$$

Remove inner parentheses first.

$$= 3x^2 - 2x - 1 - x^2 + 1$$
$$= 2x^2 - 2x$$

(D) $y - \{x - [2y - (x + y)]\}$

$= y - \{x - [2y - x - y]\}$ Remove innermost parentheses first.

$= y - \{x - [y - x]\}$ Simplify within brackets.

$= y - \{x - y + x\}$ Remove inner brackets.

$= y - x + y - x$

$= 2y - 2x$

Problem 4 Remove parentheses and simplify:

(A) $3(u^2 - 2v^2) + (u^2 + 5v^2)$

(B) $(m^3 - 3m^2 + m - 1) - (2m^3 - m + 3)$

(C) $(x^3 - 2) - [2x^3 - (3x + 4)]$

(D) $\{x - [y - (z + x)] + x\} - [y - (x + z)]$

ADDITION AND SUBTRACTION

Addition and subtraction of polynomials can be thought of in terms of removing parentheses and combining like terms, as illustrated in Example 4. Horizontal and vertical arrangements are illustrated in the next two examples. You should be able to work either way, letting the situation dictate the choice.

Example 5 Add:

$$x^4 - 3x^3 + x^2 \qquad -x^3 - 2x^2 + 3x \qquad \text{and} \qquad 3x^2 - 4x - 5$$

Solution Add horizontally:

$$\begin{aligned}(x^4 - 3x^3 + x^2) &+ (-x^3 - 2x^2 + 3x) + (3x^2 - 4x - 5)\\ &= x^4 - 3x^3 + x^2 - x^3 - 2x^2 + 3x + 3x^2 - 4x - 5\\ &= x^4 - 4x^3 + 2x^2 - x - 5\end{aligned}$$

or vertically by lining up like terms and adding their coefficients:

$$\begin{array}{rrrrr} x^4 & - 3x^3 & + x^2 & & \\ & - x^3 & - 2x^2 & + 3x & \\ & & 3x^2 & - 4x & - 5 \\ \hline x^4 & - 4x^3 & + 2x^2 & - x & - 5 \end{array}$$

Problem 5 Add horizontally and vertically:

$$3x^4 - 2x^3 - 4x^2 \qquad x^3 - 2x^2 - 5x \qquad \text{and} \qquad x^2 + 7x - 2$$

Example 6 Subtract: $4x^2 - 3x + 5$ from $x^2 - 8$

Solution $(x^2 - 8) - (4x^2 - 3x + 5)$
$= x^2 - 8 - 4x^2 + 3x - 5$
$= -3x^2 + 3x - 13$

or

$$\begin{array}{r} x^2 - 8 \\ -4x^2 + 3x - 5 \\ \hline -3x^2 + 3x - 13 \end{array}$$

Change signs and add.

Problem 6 Subtract: $2x^2 - 5x + 4$ from $5x^2 - 6$

ANSWERS TO MATCHED PROBLEMS

1. **(A)** a^{10} **(B)** $12y^5$
2. **(A)** 5 **(B)** 7
(C) Degrees of terms are 3, 2, 1, and 0; degree of polynomial is 3.
(D) Degrees of terms are 3, 3, 2, 2, 1, and 0; degree of polynomial is 3.
3. The coefficients in order are 5, −1, −3, 1, and −7.
4. **(A)** $4u^2 - v^2$ **(B)** $-m^3 - 3m^2 + 2m - 4$
(C) $-x^3 + 3x + 2$ **(D)** $4x - 2y + 2z$
5. $3x^4 - x^3 - 5x^2 + 2x - 2$ **6.** $3x^2 + 5x - 10$

EXERCISE 2-1

A *Replace the question marks with appropriate symbols.*

1. $y^2y^7 = y^?$ **2.** $x^7x^5 = x^?$ **3.** $y^8 = y^3y^?$
4. $x^{10} = x^?x^6$

Indicate the degree of each monomial.

5. $3x^5$ **6.** $-7x^3$ **7.** $-2xy^3$
8. $5x^2y$ **9.** $21x^2yz^3$ **10.** $-13xy^3z^4$

Identify each of the following as a monomial, binomial, or trinomial and give its degree.

11. $3x^2 + 7$ **12.** $1 + 2y + 3y^2$
13. $-u^5 + u^4v^2 - u^3v$ **14.** $-7a^2 + 2b^5$
15. $-t^3 + 8$ **16.** $13x^3y^2z$
17. $5p^2r^3st^2$ **18.** $xyz + x^2z^2 + yz^2$

Given the polynomial $7x^4 - 3x^3 - x^2 + x - 3$, indicate the following:

19. The coefficient of the second term
20. The coefficient of the third term
21. The exponent of the variable in the second term

22. The exponent of the variable in the fourth term

23. The coefficient of the fourth term

24. The coefficient of the first term

Simplify by rewriting the expression so that the variable occurs only once.

25. $(5x^2)(2x^9)$

26. $(2x^3)(3x^7)$

27. $(4y^3)(3y)(y^6)$

28. $(2x^2)(3x^3)(x^4)$

Simplify.

29. $(5 \times 10^8)(y \times 10^9)$

30. $(2 \times 10^3)(3 \times 10^{12})$

Add.

31. $6x + 5$ and $3x - 8$

32. $3x - 5$ and $2x + 3$

33. $7x - 5$, $-x + 3$, and $-8x - 2$

34. $2x + 3$, $-4x - 2$, and $7x - 4$

35. $5x^2 + 2x - 7$, $2x^2 + 3$, and $-3x - 8$

36. $2x^2 - 3x + 1$, $2x - 3$, and $4x^2 + 5$

Subtract.

37. $3x - 8$ from $2x - 7$

38. $4x - 9$ from $2x + 3$

39. $2y^2 - 6y + 1$ from $y^2 - 6y - 1$

40. $x^2 - 3x - 5$ from $2x^2 - 6x - 5$

B *Simplify by removing symbols of grouping, if any, and combining like terms.*

41. $-x^2y + 3x^2y - 5x^2y$

42. $-4r^3t^3 - 7r^3t^3 - 7r^3t^3 + 9r^3t^3$

43. $y^3 + 4y^2 - 10 + 2y^3 - y + 7$

44. $3x^2 - 2x + 5 - x^2 + 4x - 8$

45. $a^2 - 3ab + b^2 + 2a^2 + 3ab - 2b^2$

46. $2x^2y + 2xy^2 - 5xy + 2xy^2 - xy - 4x^2y$

47. $x - 3y - 4(2x - 3y)$

48. $a + b - 2(a - b)$

49. $y - 2(x - y) - 3x$

50. $x - 3(x + 2y) + 5y$

51. $-2(-3x + 1) - (2x + 4)$

52. $-3(-t + 7) - (t - 1)$

53. $2(x - 1) - 3(2x - 3) - (4x - 5)$

54. $-2(y - 7) - 3(2y + 1) - (-5y + 7)$

55. $4t - 3[4 - 2(t - 1)]$

56. $3x - 2[2x - (x - 7)]$

57. $3[x - 2(x + 1)] - 4(2 - x)$

58. $2(3x + y) - 2[y - (3x + 2)]$

Replace each question mark with an appropriate algebraic expression.

59. $5 + m - 2n = 5 - (?)$

60. $2 + 3x - y = 2 - (?)$

61. $w^2 - y - z = w^2 - (?)$

62. $x - y - z + 5 = x - (?)$

Add.

63. $2x^4 - x^2 - 7$, $3x^3 + 7x^2 + 2x$, and $x^2 - 3x - 1$

64. $3x^3 - 2x^2 + 5$, $3x^2 - x - 3$, and $2x + 4$

Subtract.

65. $5x^3 - 3x + 1$ from $2x^3 + x^2 - 1$

66. $3x^3 - 2x^2 - 5$ from $2x^3 - 3x + 2$

67. Subtract the sum of the first two polynomials from the sum of the last two: $3m^3 - 2m + 5$, $4m^2 - m$, $3m^2 - 3m - 2$, and $m^3 + m^2 + 2$

68. Subtract the sum of the last two polynomials from the sum of the first two: $2x^2 - 4xy + y^2$, $3xy - y^2$, $x^2 - 2xy - y^2$, and $-x^2 + 3xy - 2y^2$

C *Remove symbols of grouping and combine like terms.*

69. $2t - 3\{t + 2[t - (t + 5)] + 1\}$

70. $x - \{x - [x - (x - 1)]\}$

71. $w - \{x - [z - (w - x) - z] - (x - w)\} + x$

72. $3x^2 - 2\{x - x[x + 4(x - 3)] - 5\}$

73. $\{[(x - 1) - 1] - x\} - 1$

74. $1 - \{1 - [1 - (1 - x)]\}$

75. $x - \{1 - [x - (1 - x)]\}$

76. $x - \{[1 - (x - 1)] - x\}$

APPLICATIONS

77. The width of a rectangle is 5 meters less than its length. If x is the length of the rectangle, write an algebraic expression that represents the perimeter P of the rectangle and simplify the expression.

78. Repeat Problem 77 if the length of the rectangle is 3 meters more than twice its width.

79. A pile of coins consists of nickels, dimes, and quarters. There are 5 fewer dimes than nickels and 2 more quarters than dimes. If x equals the number of nickels, write an algebraic expression that represents the value of the pile in cents. Simplify the expression. [*Hint:* If x represents the number of nickels, then what do $x - 5$ and $(x - 5) + 2$ represent?]

80. A parking meter contains dimes and quarters only. There are 4 fewer quarters than dimes. If x represents the number of dimes, write an algebraic expression that represents the total value of all coins in the meter in cents. Simplify the expression.

★81. A board is to be cut into four pieces. The largest piece is to be 3 times as long as the smallest; the other two pieces are each to be twice as long as the smallest. If x represents the length of the smallest piece, write an algebraic expression for the total length of the board. Simplify the expression.

★82. A wire is to be cut into three pieces so that the second piece is 3 feet longer than the first and the third is 8 feet longer than the sum of the other two. Let x represent the length of the smallest piece. Write an algebraic expression for the total length of the wire in feet. Simplify the expression.

★83. A jogger runs for some time at a rate of 8 kilometers per hour and then runs twice as long at a rate of 12 kilometers per hour. Let t be the time, in hours, run at the slower pace. Write an algebraic expression for the total distance run. Simplify the expression. (Recall that Distance = Rate × Time).

★84. A racer drives for 2 hours at one speed and then for another hour and a half at a speed 12 miles per hour slower. Let s be her initial speed. Write an algebraic expression for the total distance driven. Simplify the expression.

2-2 MULTIPLICATION OF POLYNOMIALS

- Multiplication of Monomials
- Multiplication of Polynomials
- Mental Multiplication of Binomials
- Squaring Binomials

To develop multiplication of polynomials, we first multiply monomials using Property 1 of exponents. We can then multiply any polynomial by making repeated use of distributive properties. Multiplication and squaring of binomials will be required so often that it will be helpful to be able to do these mentally; this section will show you how.

MULTIPLICATION OF MONOMIALS

Monomials were multiplied in Example 1 using a property of exponents. We review the process.

Example 7 Multiply: **(A)** x^3x^5 **(B)** $(3m^{12})(5m^{23})$ **(C)** $(-3x^3y^4)(2x^2y^3)$

Solution **(A)** $x^3x^5 \boxed{= x^{3+5}} = x^8$ $x^3x^5 \neq x^{3 \cdot 5}$

(B) $(3m^{12})(5m^{23}) \boxed{= 3 \cdot 5m^{12+23}} = 15m^{35}$

(C) $(-3x^3y^4)(2x^2y^3) \boxed{= (-3)(2)x^{3+2}y^{4+3}} = -6x^5y^7$

Problem 7 Multiply: **(A)** y^4y^7 **(B)** $(9x^4)(3x^2)$ **(C)** $(4u^3v^2)(-3uv^3)$

MULTIPLICATIONS OF POLYNOMIALS

How do we multiply polynomials with more than one term? The distributive property plays a central role in the process and leads directly to the following mechanical rule:

> **Mechanics of Multiplying Two Polynomials**
>
> To multiply two polynomials, multiply each term of one by each term of the other. Then add like terms.

Example 8 Multiply: **(A)** $3x^2(2x^2 - 3x + 4)$ **(B)** $(2x - 3)(3x^2 - 2x + 3)$

Solution **(A)** $3x^2(2x^2 - 3x + 4) = 6x^4 - 9x^3 + 12x^2$

(B) $(2x - 3)(3x^2 - 2x + 3)$

$$\boxed{= 2x(3x^2 - 2x + 3) - 3(3x^2 - 2x + 3)}$$
$$= 6x^3 - 4x^2 + 6x - 9x^2 + 6x - 9$$
$$= 6x^3 - 13x^2 + 12x - 9$$

or

$$\begin{array}{rrrr} 3x^2 & - \;\; 2x & + \;\; 3 & \\ 2x & - \;\; 3 & & \\ \hline 6x^3 & - \;\; 4x^2 & + \;\; 6x & \\ & - \;\; 9x^2 & + \;\; 6x & - \;\; 9 \\ \hline 6x^3 & - \;\; 13x^2 & + \;\; 12x & - \;\; 9 \end{array}$$

Note that either way, each term in $3x^2 - 2x + 3$ is multiplied by each term in $2x - 3$. In the vertical arrangement, by multiplying by $2x$ first the like terms line up more conveniently. Students usually prefer a vertical arrangment for this type of problem.

Problem 8 Multiply:

(A) $2m^3(3m^2 - 4m - 3)$ **(B)** $\begin{array}{r} 2x^2 + 3x - 1 \\ \underline{3x \quad - 4} \end{array}$ **(C)** $\begin{array}{r} 2x^2 + 3x - 2 \\ \underline{3x^2 - 2x + 1} \end{array}$

MENTAL MULTIPLICATION OF BINOMIALS

For reasons that will become clear shortly, it is essential that you learn to multiply first-degree polynomials of the type $(3x + 2)(2x - 1)$ and $(3x - y)(x + 2y)$ mentally. We will use a horizontal arrangement and try to discover a method that will enable us to do this. We start by multiplying each term in the first binomial times each term in the second binomial:

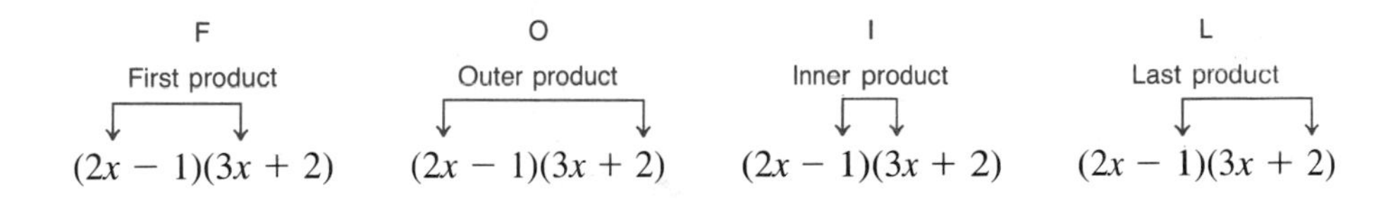

Performing these four operations on one line, we obtain

	F First product	O Outer product	I Inner product	L Last product
	↓	↓	↓	↓
$(2x - 1)(3x + 2) =$	$6x^2$	$+ 4x$	$- 3x$	$- 2$

The inner and outer products are like terms and hence combine into one term. Thus,

$$(2x - 1)(3x + 2) = 6x^2 + x - 2$$

To speed up the process we combine the inner and outer products mentally. The method just described is called the **FOIL method**. We note again that the product of two first-degree polynomial is a second-degree polynomial. A simple three-step process for carrying out the FOIL method is illustrated in Example 9.

Example 9 Multiply by the FOIL method:

(A) $(3x + 1)(2x - 5)$ **(B)** $(x - 4y)(x + 4y)$
(C) $(x + a)(x + a)$

Solution

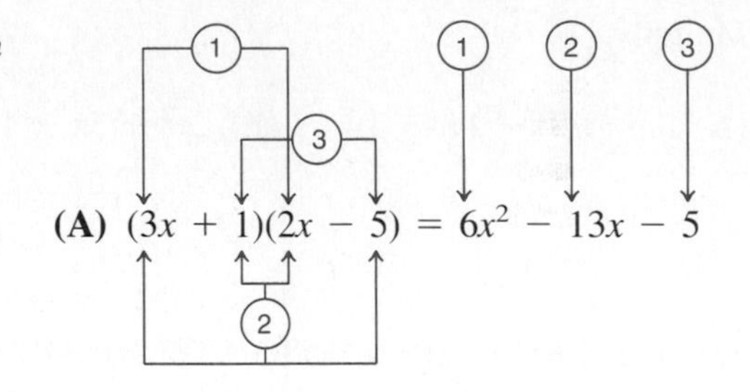

The like terms are obtained in step 2 by multiplying the inner and outer products, and they are combined mentally.

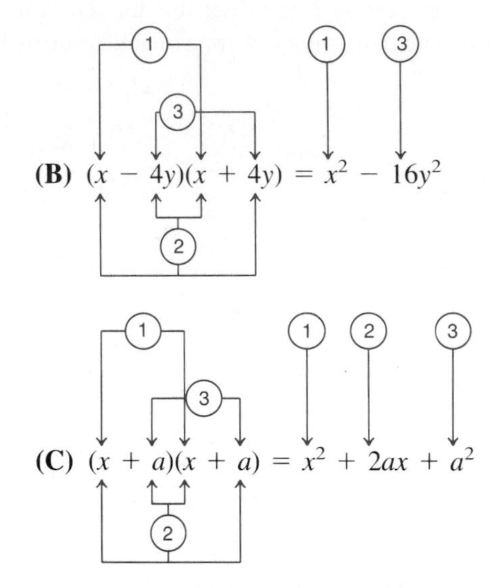

Notice in part (B) that there is no middle term because its coefficient is 0. In part (C), we used $xa = ax$ to get $xa + ax = 2ax$.

Problem 9 Multiply by the FOIL method:

(A) $(2x - 1)(x + 4)$ **(B)** $(3x - y)(x + y)$
(C) $(a - b)(a - b)$ **(D)** $(3x + 4)(3x - 4)$

In the next section, we will try to reverse this process. Given a polynomial such as $6x^2 - 13x - 5$, we will want to find first-degree factors that produce this polynomial as a product. To test and check possible factors, it is necessary to be able to multiply factors such as $3x + 1$ and $2x - 5$ quickly and accurately.

The FOIL method works equally well on any binomials, not just those of degree 1. For arbitrary binomials, however, the outer and inner products will not necessarily be like terms.

Example 10 Multiply by the FOIL method:

$$(a^2 + 3b^2)(a - 2b^3)$$

Solution

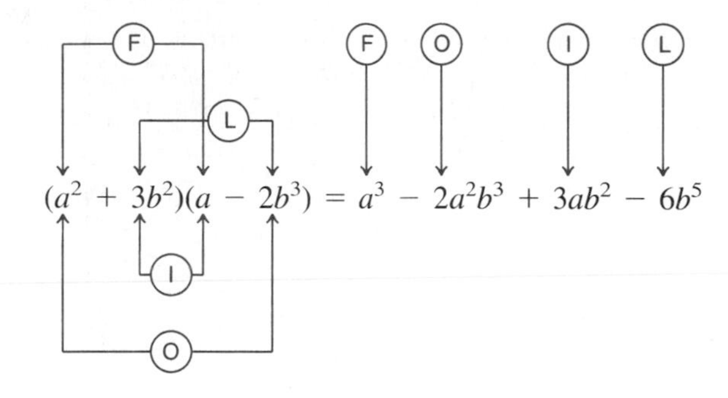

Problem 10 Multiply by the FOIL method:

$$(x^3 - 1)(x^2 + 3y)$$

SQUARING BINOMIALS

Since

$$(A + B)^2 = (A + B)(A + B) = A^2 + 2AB + B^2$$
$$(A - B)^2 = (A - B)(A - B) = A^2 - 2AB + B^2$$

we can formulate a simple mechanical rule for squaring any binomial directly.

Mechanical Rule for Squaring Binomials

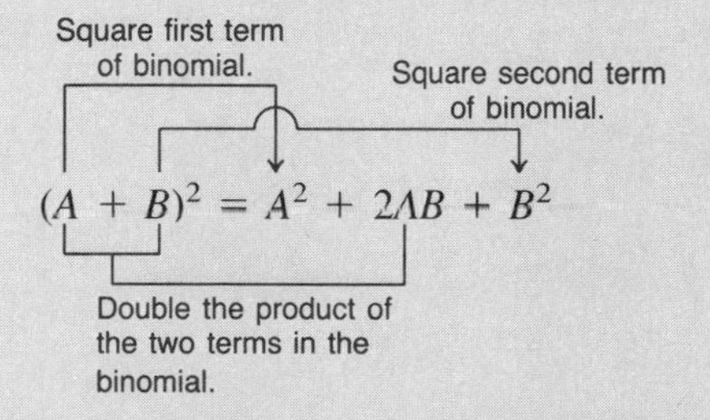

We do the same thing for $(A - B)^2$, except that the sign of the middle term on the right becomes negative:

$$(A - B)^2 = A^2 - 2AB + B^2$$

Example 11 Square each binomial:

(A) $(x + h)^2$ **(B)** $(2x + 1)^2$ **(C)** $(2x - 3y)^2$

Solution

(A) $(x + h)^2 = x^2 + 2(x)(h) + h^2 = x^2 + 2hx + h^2$

(B) $(2x + 1)^2 = (2x)^2 + 2(2x)(1) + 1^2 = 4x^2 + 4x + 1$

(C) $(2x - 3y)^2 = (2x)^2 + 2(2x)(-3y) + (-3y)^2$

$= 4x^2 - 12xy + 9y^2$

Problem 11 Square each binomial:

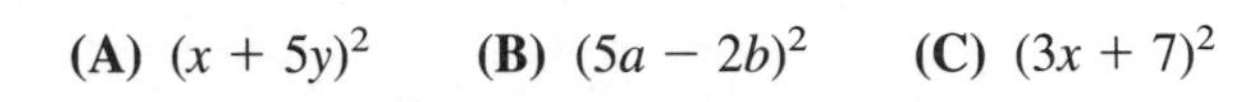

(A) $(x + 5y)^2$ **(B)** $(5a - 2b)^2$ **(C)** $(3x + 7)^2$

Note that, in general,

$$(a + b)^2 \neq a^2 + b^2$$
$$(a - b)^2 \neq a^2 - b^2$$

ANSWERS TO MATCHED PROBLEMS

7. (A) y^{11} (B) $27x^6$ (C) $-12u^4v^5$
8. (A) $6m^5 - 8m^4 - 6m^3$ (B) $6x^3 + x^2 - 15x + 4$ (C) $6x^4 + 5x^3 - 10x^2 + 7x - 2$
9. (A) $2x^2 + 7x - 4$ (B) $3x^2 + 2xy - y^2$ (C) $a^2 - 2ab + b^2$ (D) $9x^2 - 16$
10. $x^5 + 3x^3y - x^2 - 3y$
11. (A) $x^2 + 10xy + 25y^2$ (B) $25a^2 - 20ab + 4b^2$ (C) $9x^2 + 42x + 49$

EXERCISE 2-2

Multiply.

A

1. y^2y^3
2. x^3x^2
3. $(5y^4)(2y)$
4. $(2x)(3x^4)$
5. $(8x^{11})(-3x^9)$
6. $(-7u^9)(5u^7)$
7. $(-3u^4)(2u^5)(-u^7)$
8. $(2x^3)(-3x)(-4x^5)$
9. $(cd^2)(c^2d^2)$
10. $(a^2b)(ab^2)$
11. $(-3xy^2z^3)(-5xyz^2)$
12. $(-2xy^3z)(2x^3yz)$
13. $y(y + 7)$
14. $x(1 + x)$
15. $5y(2y - 7)$
16. $3x(2x - 5)$
17. $3a^2(a^3 + 2a^2)$
18. $2m^2(m^2 + 3m)$

19. $2y(y^2 + 2y - 3)$

20. $3x(2x^2 - 3x + 1)$

21. $7m^3(m^3 - 2m^2 - m + 4)$

22. $3x^2(2x^3 + 3x^2 - x - 2)$

23. $5uv^2(2u^3v - 3uv^2)$

24. $4m^2n^3(2m^3n - mn^2)$

25. $2cd^3(c^2d - 2cd + 4c^3d^2)$

26. $3x^2y(2xy^3 + 4x - y^2)$

B **27.** $(3y + 2)(2y^2 + 5y - 3)$

28. $(2x - 1)(x^2 - 3x + 5)$

29. $(m + 2n)(m^2 - 4mn - n^2)$

30. $(x - 3y)(x^2 - 3xy + y^2)$

31. $(2m^2 + 2m - 1)(3m^2 - 2m + 1)$

32. $(x^2 - 3x + 5)(2x^2 + x - 2)$

33. $(a + b)(a^2 - ab + b^2)$

34. $(a - b)(a^2 + ab + b^2)$

35. $(2x^2 - 3xy + y^2)(x^2 + 2xy - y^2)$

36. $(a^2 - 2ab + b^2)(a^2 + 2ab + b^2)$

Multiply mentally.

37. $(x + 3)(x + 2)$

38. $(m - 2)(m - 3)$

39. $(a + 8)(a - 4)$

40. $(m - 12)(m + 5)$

41. $(t + 4)(t - 4)$

42. $(u - 3)(u + 3)$

43. $(m - n)(m + n)$

44. $(a + b)(a - b)$

45. $(4t - 3)(t - 2)$

46. $(3x - 5)(2x + 1)$

47. $(3x + 2y)(x - 3y)$

48. $(2x - 3y)(x + 2y)$

49. $(2m - 7)(2m + 7)$

50. $(3y + 2)(3y - 2)$

51. $(6x - 4y)(5x + 3y)$

52. $(3m + 7n)(2m - 5n)$

53. $(2x - 3t)(3s - t)$

54. $(2x - 3y)(3x - 2y)$

55. $(x^3 + y^3)(x + y)$

56. $(x^2 + y^3)(x^3 + y)$

57. $(2x \quad y^2)(x^2 + 3y)$

58. $(x^3 - y^2)(2x + y^2)$

Square each binomial, using the mechanical rule.

59. $(3x + 2)^2$

60. $(4x + 3y)^2$

61. $(2x - 5y)^2$

62. $(2x - 7)^2$

63. $(6u + 5v)^2$

64. $(7p + 2q)^2$

65. $(2m - 5n)^2$

66. $(4x - 1)^2$

C *Simplify.*

67. $(x + 2y)^3$

68. $(2m - n)^3$

69. $(3x - 1)(x + 2) - (2x - 3)^2$

70. $(2x + 3)(x - 5) - (3x - 1)^2$

71. $2(x - 2)^3 - (x - 2)^2 - 3(x - 2) - 4$

72. $(2x - 1)^3 - 2(2x - 1)^2 + 3(2x - 1) + 7$

73. $-3x\{x[x - x(2 - x)] - (x + 2)(x^2 - 3)\}$

74. $2\{(x - 3)(x^2 - 2x + 1) - x[3 - x(x - 2)]\}$

75. $(2x - 1)(2x + 1)(3x^3 - 4x + 3)$

76. $(x - 1)(x - 2)(2x^3 - 3x^2 - 2x - 1)$

77. $[(3x - 2) + y]^2$

78. $[y + (x - 1)]^2$

79. $[(x + y) - 2][(x + y) + 1]$

80. $[(x - 1) + y][(x + 1) - y]$

81. If you are given two polynomials, one of degree m and another of degree n, and $m > n$, then what is the degree of their product?

82. What is the degree of the sum of the two polynomials in Problem 81?

APPLICATIONS

83. The length of a rectangle is 8 meters more than its width. If y is the length of the rectangle, write an algebraic expression that represents its area. Change the expression to a form without parentheses.

84. Repeat Problem 83 if the length of the rectangle is 3 meters less than twice the width.

2-3 FACTORING OUT COMMON FACTORS; FACTORING BY GROUPING

- Factoring Out Common Factors
- Factoring by Grouping

We can view the distributive property

$$a(b + c) = ab + ac$$

from left to right as multiplying out a product. If we view it from right to left, we are rewriting a sum as a product, that is, factoring the right-hand side. In this case we are taking out the common factor a. This is the basis for the factoring methods considered in this section.

FACTORING OUT COMMON FACTORS

The simplest common factors to recognize are monomials.

Example 12 Factor out factors common to all terms:

(A) $8x^2 - 14x$ **(B)** $3a^2b - 6ab + 15a^3b^2$

Solution **(A)** $8x^2 - 14x = \mathbf{2x} \cdot 4x - \mathbf{2x} \cdot 7$

$= \mathbf{2x}(4x - 7)$

(B) $3a^2b - 6ab + 15a^3b^2 = \mathbf{3ab} \cdot a - \mathbf{3ab} \cdot 2 + \mathbf{3ab} \cdot 5a^2b$

$= \mathbf{3ab}(a - 2 + 5a^2b)$

Problem 12 Factor out factors common to all terms:

(A) $20x^2 + 4x$ **(B)** $4u^3v^2 - 10u^2v^4 + 18u^2v$

Common factors do not have to be monomials. Compare the following:

$\boldsymbol{a}x + \boldsymbol{a}y = \boldsymbol{a}(x + y)$
$(\boldsymbol{w} + \mathbf{1})x + (\boldsymbol{w} + \mathbf{1})y = (\boldsymbol{w} + \mathbf{1})(x + y)$

The factoring is in essence the same. We must be ready to recognize that $w + 1$ is a single number, just as a is in the first example. Similarly, the following two examples are also essentially the same:

$3x\boldsymbol{z} + 4y\boldsymbol{z} = (3x + 4y)\boldsymbol{z}$
$3x(\mathbf{2x + y}) + 4y(\mathbf{2x + y}) = (3x + 4y)(\mathbf{2x + y})$

Common factors can be taken out on either side, left or right, because multiplication is commutative.

Example 13 Factor out factors common to all terms:

(A) $3y(y + 2) - 5(y + 2)$ **(B)** $x^2(x - 1) + 2(x - 1)$

Solution **(A)** $3y(y + 2) - 5(y + 2) = 3y(\mathbf{y + 2}) - 5(\mathbf{y + 2})$

$= (3y - 5)(\mathbf{y + 2})$

(B) $x^2(x - 1) + 2(x - 1) = (x^2 + 2)(x - 1)$

Problem 13 Factor out factors common to all terms:

(A) $x(x^2 + y^2) - y(x^2 + y^2)$ **(B)** $2x(x - 5) - 3(x - 5)$

FACTORING BY GROUPING

In some situations we may be able to take a polynomial with no apparent common factor and find one when the terms are properly grouped. In the next example, grouping the first two and last two terms will yield a common factor in two steps.

Example 14 Factor by grouping:

(A) $x^2 + x + 3x + 3$ **(B)** $3y^2 + 6y - 5y - 10$
(C) $2x^2 - 2xy - xy^2 + y^3$

Solution **(A)** $x^2 + x + 3x + 3$ Group the first two and last two terms.

$= (x^2 + x) + (3x + 3)$ Remove common factors from each group.

$= x(\mathbf{x + 1}) + 3(\mathbf{x + 1})$ Take out the common factor $x + 1$ to complete the factoring.

$= (x + 3)(\mathbf{x + 1})$

(B) $3y^2 + 6y - 5y - 10$ Group the first two and last two terms.

$= (3y^2 + 6y) + (-5y - 10)$ Remove common factors from each group.

$= 3y(\mathbf{y + 2}) + (-5)(\mathbf{y + 2})$ Take out the common factor $y + 2$.

$= (3y - 5)(\mathbf{y + 2})$

(C) $2x^2 - 2xy - xy^2 + y^3$

$= (2x^2 - 2xy) + (-xy^2 + y^3)$

$= 2x(x - y) + y^2(-x + y)$ Recognize $-x + y = (-1)(x - y)$

$= 2x(\mathbf{x - y}) + (y^2)(-1)(\mathbf{x - y})$

$= (2x - y^2)(\mathbf{x - y})$

Problem 14 Factor by grouping:

(A) $x^3 + 3x^2 + x + 3$ **(B)** $2x^2 + 4x - 3x - 6$
(C) $3ab^2 + b^3 + 3a^2 + ab$

Not all polynomials can be regrouped for factoring as easily as those in Example 14. You may have to rearrange the terms in order to regroup them successfully, and even this may not always work—some polynomials with integer coefficients cannot be factored at all using integer coefficients.

Example 15 Factor $2x^2 + y^3 - xy^2 - 2xy$ by grouping.

Solution

$$2x^2 + y^3 - xy^2 - 2xy$$

$$= (2x^2 + y^3) + (-xy^2 - 2xy)$$

If we group the first two and last two terms, no common factor appears.

$$= (2x^2 + y^3) + (-xy)(y - 2)$$

$$2x^2 + y^3 - xy^2 - 2xy$$

Rearrange the terms and try again.

$$= 2x^2 - xy^2 + y^3 - 2xy$$
$$= (2x^2 - xy^2) + (y^3 - 2xy)$$
$$= x(2x - y^2) + y(y^2 - 2x)$$
$$= x(\mathbf{2x - y^2}) + y(-1)(\mathbf{2x - y^2}) \qquad y^2 - 2x = (-1)(2x - y^2)$$
$$= (x - y)(\mathbf{2x - y^2})$$

Still another rearrangement of the terms

$$2x^2 + y^3 - xy^2 - 2xy = 2x^2 - 2xy - xy^2 + y^3$$

yields the polynomial as it appeared in Example 14(C), where grouping also worked.

Problem 15 Factor $y^3 + 3x^2 + xy + 3xy^2$ by grouping.

ANSWERS TO MATCHED PROBLEMS

12. **(A)** $4x(5x + 1)$ **(B)** $2u^2v(2uv - 5v^3 + 9)$
13. **(A)** $(x - y)(x^2 + y^2)$ **(B)** $(2x - 3)(x - 5)$
14. **(A)** $(x^2 + 1)(x + 3)$ **(B)** $(2x - 3)(x + 2)$
(C) $(b^2 + a)(3a + b)$
15. $(y + 3x)(x + y^2)$

EXERCISE 2-3

Factor out factors common to all terms.

A

1. $3xz - 6x$
2. $2ab + 4$
3. $8x + 2y$
4. $5xy - y$
5. $6x^2 + 9x$
6. $8xy - 20y^2$
7. $14x^2y - 7xy^2$
8. $2x^2y^3 + 3x^2y^2$
9. $2x(x - 3) + z(x - 3)$
10. $a(c + d) + b(c + d)$
11. $(a + b)c - (a + b)d$
12. $3u(v + 1) - u(v + 1)$
13. $x(x - y) - y(x - y)$
14. $a(x + h) + 3(x + h)$
15. $ab(c + d) - (c + d)$
16. $x(pq - r) + y(pq - r)$

17. $x^5 + x^4 + x^3$

18. $y^2 - y^4 - y^8$

19. $a^2b - a^3b^2 - a^4b$

20. $rst - rs + rt$

21. $abc + bcd + cde$

22. $2xy + 8xz - 4yz$

23. $xy^2z^3 - x^2y^3z + x^3yz^2$

24. $2xy^2 + 4x^2y + 6x^2y^2$

B **25.** $a(a - 2) + 3(2 - a)$

26. $(2x - 3) + x(3 - 2x)$

27. $x(x - 1) - (1 - x)$

28. $x(x - y) - y(y - x)$

29. $(x - 2) - 5(2 - x)$

30. $2(a - b) - 8(b - a)$

31. $3(x - 1) + 6(1 - x)$

32. $4(x^2 - 2) + 10(2 - x^2)$

Factor out common factors from each group and, if possible, complete the factoring.

33. $(2x^3 + x^2) + (6x + 3)$

34. $(x^3 - 5x) + (20x^2 - 100)$

35. $(abc - a^2b^2) - (c^3 - abc^2)$

36. $(4x^2 - 8x) - (6x - 12)$

37. $(3xy + 6x) - (y^2 + 2y)$

38. $(2x + 4y) - (xy + y^2)$

Factor by grouping if possible. (Compare these problems to Problems 33–38.)

39. $2x^3 + x^2 + 6x + 3$

40. $x^3 - 5x + 20x^2 - 100$

41. $abc - a^2b^2 - c^3 + abc^2$

42. $4x^2 - 8x - 6x + 12$

43. $3xy + 6x - y^2 - 2y$

44. $3x + 4y - xy - y^2$

C *Factor by grouping if possible.*

45. $2x^4 - 3x^2 + 6x^2 - 9$

46. $x^3 - x^2 + 3x - 3$

47. $x^3y^2 - x^2y^3 + 3x - 3y$

48. $xy^3 + x^3y + x^2 + y^2$

49. $4ab - b^2 - 4a^2 + ab$

50. $2x^2 - x^3 - 2y^2 + xy^2$

51. $xy^2z - xyz^2 - xy + xz$

52. $4b^2 - ab + a^2 - 4ab$

53. $2x - 6y - 9yz + 3xz$

54. $x - xy - z + zy$

55. $3a^2 + bc + ab + 3ac$

56. $x^2 + 6y + x^2y + 6$

57. $12 + xy^2 - 4x - 3y^2$

58. $5xy - z^2 + xyz - 5z$

59. $ac + bd + ab + cd$

60. $ab + cd - bd - ac$

61. $yz + 2xy - 2xz - y^2$

62. $2xy + 9 - 6y - 3x$

2-4 FACTORING SECOND-DEGREE POLYNOMIALS

- Recognizing Perfect Squares
- Factoring Second-Degree Polynomials

Products of the form

$$(ax + b)(cx + d) = acx^2 + (ad + bc)x + bd$$

or

$$(ax + by)(cx + dy) = acx^2 + (ad + bc)xy + bdy^2$$

are easily obtained by multiplying out. We would like to be able to reverse the process and factor expressions of the form

$$Ax^2 + Bx + C$$
$$Ax^2 + Bxy + Cy^2$$

For example, can we factor

$$2x^2 - x - 6$$

into two first-degree polynomials with integer coefficients:

$$2x^2 - x - 6 = (ax + b)(cx + d)$$

We can easily check that

$$2x^2 - x - 6 = (2x + 3)(x - 2)$$

but if the factors were not supplied, how could we find them?

We begin this section by looking at a special kind of factoring—recognizing those second-degree polynomials that are perfect squares—and then return to the problem of factoring second-degree polynomials in general.

RECOGNIZING PERFECT SQUARES

Consider the product

$$(x + b)^2 = (x + b)(x + b) = x^2 + 2bx + b^2$$

The role played by b identifies the expression $x^2 + 2bx + b^2$ as a perfect square:

$$x^2 + 2bx + b^2 \quad = (x + b)^2$$

The last term is the square of one-half the coefficient of x.

One-half the coefficient of x

Example 16 Identify the perfect squares among these second-degree polynomials; factor those that are.

(A) $x^2 + 10x + 25$ **(B)** $x^2 + 4x + 16$
(C) $x^2 - 6x + 9$ **(D)** $4x^2 + 20x + 25$

Solution **(A)** $x^2 + 10x + 25$ is a perfect square, since the constant term is the square of one-half the coefficient of x: $25 = (\frac{1}{2} \cdot 10)^2$ Here $x^2 + 10x + 25 = (x + 5)^2$.

(B) $x^2 + 4x + 16$ is not a perfect square: $16 \neq (\frac{1}{2} \cdot 4)^2$.

(C) $x^2 - 6x + 9$ is a perfect square: $9 = [\frac{1}{2}(-6)]^2$. To factor, recognize that here $b = \frac{1}{2}(-6) = -3$, so

$$x^2 - 6x + 9 = [x + (-3)]^2 = (x - 3)^2$$

(D) In $4x^2 + 20x + 25$, notice that the first term, $4x^2$, is $(2x)^2$. Rewrite the first-degree term, $20x$, as $10(2x)$ and take one-half the coefficient of $2x$ rather than of x, that is, 5. The constant is 5^2, so this is a perfect square:

$$4x^2 + 20x + 25 = (2x + 5)^2$$

A substitution may make this process easier to see. Let $y = 2x$. Then

$$\begin{aligned} 4x^2 + 20x + 25 &= y^2 + 10y + 25 \\ &= (y + 5)^2 \\ &= (2x + 5)^2 \end{aligned}$$

Problem 16 Identify the perfect squares among these second-degree polynomials; factor those that are.

(A) $x^2 + 8x + 8$ **(B)** $x^2 + 4x + 4$
(C) $x^2 - 12x + 36$ **(D)** $9x^2 + 6x + 1$

FACTORING SECOND-DEGREE POLYNOMIALS

We now develop a trial-and-error method for factoring any second-degree polynomial. Consider the polynomial $x^2 - 5x + 6$. We would like to factor

$$x^2 - 5x + 6 = (x + a)(x + b)$$

If we multiply out $(x + a)(x + b)$, we obtain

$$x^2 \underbrace{-5}x + \underbrace{6} = x^2 + \underbrace{(a + b)}x + \underbrace{ab}$$

so we must have $ab = 6$ and $a + b = -5$. At this stage, we might try all possible a and b that multiply to 6 and find that $a = -2$ and $b = -3$ work; that is,

$$x^2 - 5x + 6 = (x - 2)(x - 3)$$

Some observations, however, will limit the number of possible a and b to try and thus shorten our work:

(i) a and b must have the same sign since the product is positive.
(ii) The sign must be negative to make the sum negative.
(iii) The order does not matter since

$$(x + a)(x + b) = (x + b)(x + a)$$

that is, we would not have to check both $a = -1$, $b = -6$ and also $a = -6$, $b = -1$.

We, therefore, look for two negative numbers that multiply to 6 and add to -5. The only pairs that multiply to 6 are $-2, -3$ and $-1, -6$. Of these, only the first pair, $-2, -3$, add to -5.

The trial-and-error method is somewhat more complicated when the coefficient of x^2 is not 1. Consider again the polynomial $2x^2 - x + 6$. We begin by comparing the polynomial and the multiplied-out form of $(ax + b)(cx + d)$:

$$2x - x - 6 = acx^2 + (ad + bc)x + bd$$

Thus we need to look for coefficients a, b, c, and d such that

$$\begin{aligned} ac &= 2 \\ bd &= -6 \\ ad + bc &= -1 \end{aligned}$$

There are only a limited number of integer pairs that meet the first two conditions:

$ac = 2$	$bd = -6$
$a \cdot c$	$b \cdot d$
$1 \cdot 2$	$2(-3)$
$2 \cdot 1$	$(-2)3$
$(-1)(-2)$	$3(-2)$
$(-2)(-1)$	$(-3)2$

We need only try the first combination here for a and c since the other three will be related to it in this way:

$$\left.\begin{array}{l}(1 \cdot x + b)(2 \cdot x + d) \\ (2 \cdot x + d)(1 \cdot x + b) \\ (-x - b)(-2x - d) \\ (-2x - d)(-x - b)\end{array}\right\} \text{All multiply to } 2x^2 + (2b + d)x + bd.$$

(If there were essentially different factorizations of $ac = 2$, we would, however, possibly have to try other pairs for ac.)

Next we try each pair for bd until we either find one that works or exhaust all the possibilities. In this case, $b = -2$ and $d = 3$ give the correct middle term. Thus

$$\begin{aligned} 2x^2 - x - 6 &= (ax + b)(cx + d) \\ &= (x - 2)(2x + 3) \end{aligned}$$

If checking all possibilities does not yield a factorization, we will say that the polynomial is not factorable using integers.

Example 17 Factor each polynomial, if possible, using integer coefficients:

(A) $x^2 - x - 12$ **(B)** $2x^2 - 5xy + 2y^2$ **(C)** $x^2 + x + 4$
(D) $12x^2 + x - 6$

Solution **(A)** $x^2 - x - 12$ Put in what we know.

$= (x - \quad)(x + \quad)$ The signs must be opposite.

Now test all possible factors of 12 to see if any produce the correct middle term.

12
$1 \cdot 12$
$12 \cdot 1$
$2 \cdot 6$
$6 \cdot 2$
$3 \cdot 4$
$4 \cdot 3$

The last choice, $4 \cdot 3$, yields the desired results, so

$$x^2 - x - 12 = (x - 4)(x + 3)$$

(B) $2x^2 - 5xy + 2y^2$

$= (2x - \quad)(x - \quad)$ Both signs must be negative to give a positive coefficient to y^2 and a negative in the middle term.

Try all possible factors for 2.

2
$1 \cdot 2$
$2 \cdot 1$

The first works:

$$(2x - y)(x - 2y) = 2x^2 - 5xy + 2y^2$$

(C) $x^2 + x + 4 = (x + \quad)(x + \quad)$

We try all ways of factoring 4 with positive factors:

$1 \cdot 4$
$4 \cdot 1$
$2 \cdot 2$

None of these provides the correct middle term, so $x^2 + x + 4$ is not factorable using integer coefficients.

(D) $12x^2 + x - 6 = (\underset{\uparrow\,?}{\quad} x + \underset{\uparrow\,?}{\quad})(\underset{\uparrow\,?}{\quad} x - \underset{\uparrow\,?}{\quad})$

The signs must be opposite to produce the negative constant and positive coefficient of x.

We now consider all positive factors for 12 and 6:

12	6
$1 \cdot 12$	$1 \cdot 6$
$2 \cdot 6$	$2 \cdot 3$
$3 \cdot 4$	$3 \cdot 2$
$4 \cdot 3$	$6 \cdot 1$
$6 \cdot 2$	
$12 \cdot 1$	

We check each choice for 12 with each possibility for 6—there are 24 such combinations—until we find one that produces the correct middle term. Eventually, we will obtain

$$(4x + 3)(3x - 2) = 12x^2 + x - 6$$

If none of the combinations had worked, we would have concluded the polynomial is not factorable using integer coefficients.

In problems like those found in Example 17(D), if the coefficients of the first and last terms get larger and larger with more and more factors, the number of combinations that need to be checked increases very rapidly. And it is quite possible in most practical situations that none of the combinations will work. It is important, however, that you understand the approach presented here, since it will work for most of the simpler factoring problems you will encounter. The next (optional) section introduces a systematic approach to the problem of factoring that will reduce the amount of trial and error substantially and even tell you whether a polynomial can be factored before you proceed too far.

In conclusion, we point out that if a, b, and c are selected at random out of the integers, the probability that

$$ax^2 + bx + c$$

is not factorable in the integers is much greater than the probability that it is. But even being able to factor only some second-degree polynomials leads to marked simplification of some algebraic expressions and an easy way to solve some second-degree equations, as will be seen later.

Problem 17 Factor each polynomial, if possible, using integer coefficients.

(A) $x^2 + 7x + 12$ **(B)** $x^2 + x - 5$
(C) $x^2 - xy - 6y^2$ **(D)** $4x^2 + 11x - 3$

ANSWERS TO MATCHED PROBLEMS

16. **(A)** Not a perfect square **(B)** $(x + 2)^2$ **(C)** $(x - 6)^2$ **(D)** $(3x + 1)^2$

17. **(A)** $(x + 4)(x + 3)$ **(B)** Not factorable using integers **(C)** $(x - 3y)(x + 2y)$ **(D)** $(4x - 1)(x + 3)$

EXERCISE 2-4

Factor in the integers, if possible. If the expression is not factorable, say so.

A

1. $x^2 + 3x + 2$ **2.** $x^2 + 8x + 15$ **3.** $x^2 + 7x + 10$

4. $x^2 + 4x + 3$ **5.** $x^2 - 2x - 3$ **6.** $x^2 - 2x - 8$

7. $x^2 + 3x - 4$
8. $x^2 + 5x - 6$
9. $x^2 - 9x + 20$
10. $x^2 - 9x + 18$
11. $x^2 - 6x + 5$
12. $x^2 - 6x + 8$
13. $x^2 - 6x + 9$
14. $x^2 + 8x + 16$
15. $x^2 + 14x + 49$
16. $x^2 - 4x - 4$
17. $x^2 - 6xy + 9y^2$
18. $x^2 + 8xy + 16y^2$
19. $2x^2 - x - 1$
20. $2x^2 + 7x + 6$
21. $2x^2 + 7x + 5$
22. $2x^2 + x - 3$
23. $3x^2 + 7x - 4$
24. $3x^2 - 11x + 10$

B
25. $2x^2 - xy + y^2$
26. $2x^2 + xy - 6y^2$
27. $2x^2 - 9xy + 10y^2$
28. $2x^2 - 7xy + 3y^2$
29. $4x^2 + 4x - 3$
30. $4x^2 + 11x - 3$
31. $4x^2 + 3x - 1$
32. $4x^2 + 4x - 15$
33. $2x^2 + 4x + 3$
34. $4x^2 + 4x + 1$
35. $4x^2 + 12x + 9$
36. $2x^2 - 4x - 5$
37. $x^2 - x - 2$
38. $9x^2 - 6x + 1$
39. $9x^2 + 12x + 4$
40. $x^2 + x - 1$
41. $6x^2 - 7x + 2$
42. $6x^2 - 5x - 4$
43. $6x^2 - xy - y^2$
44. $6x^2 - xy - 2y^2$

C
45. $9x^2 + 6xy + y^2$
46. $9x^2 + 12xy - 4y^2$
47. $4x^2 - 4xy - y^2$
48. $4x^2 - 12xy + 9y^2$
49. $8x^2 + 14x + 7$
50. $16x^2 + 7x + 6$
51. $8x^2 - 14x - 15$
52. $16x^2 + 2x - 3$
53. $6x^2 + 10xy + 5y^2$
54. $9x^2 + 11xy + 4y^2$
55. $6x^2 + 21xy - 12y^2$
56. $9x^2 + 26xy - 3y^2$

A polynomial such as $x^4 - x^2 - 12$ is not second-degree but can still be factored using the techniques of this section. We can think of x^2 as the variable and rewrite the polynomial as $(x^2)^2 - (x^2) - 12$, and factor it as $(x^2 - 4)(x^2 + 3) = (x - 2)(x + 2)(x^2 + 3)$ as in Example 17(A). Use this method to factor the following polynomials:

57. $x^4 + 4x^2 + 3$
58. $x^4 - x^2 - 2$
59. $2x^4 - x^2 - 3$
60. $6x^4 - 7x^2 - 3$
61. $x^6 + x^3 - 6$
62. $2x^6 - 7x^3 - 4$

2-5 *ac* TEST AND FACTORING (OPTIONAL)

In Example 17(D), the polynomial $12x^2 + x - 6$ was factored by trial and error as

$$12x^2 + x - 6 = (4x + 3)(3x - 2)$$

The following process leads to the same result:

$$12x^2 + x - 6$$ Rewrite x as $9x - 8x$; we'll see in a moment what leads to this choice.

$$= 12x^2 + 9x - 8x - 6$$ Group the first two and last two terms.

$$= (12x^2 + 9x) - (8x + 6)$$ Factor out common factors.

$$= 3x(4x + 3) - 2(4x + 3)$$ Factor out common factors again.

$$= (3x - 2)(4x + 3)$$

The first step in the process, rewriting x as $9x - 8x$, is what makes it work. To see why this choice works, consider the process again:

$$12x^2 + x - 6$$ Rewrite x as $px + qx$; note that $p + q$ must be 1.

$$= 12x^2 + px + qx - 6$$
$$= (12x^2 + px) + (qx - 6)$$
$$= x(12x + p) + (qx - 6)$$

For there to be a common factor, $12x + p$ must be a multiple of $qx - 6$. This means that the ratio of 12 to q must be the same as p to -6; that is,

$$\frac{12}{q} = \frac{p}{-6}$$

so $pq = -72$. Therefore, we need two integers p and q with $p + q = 1$ and $pq = -72$; 9 and -8 work.

This process can be generalized and applied to second-degree polynomials of the following types:

$$ax^2 + bx + c \quad \text{and} \quad ax^2 + bxy + cy^2 \qquad (1)$$

The process provides a test, called the ***ac* test for factorability**, that not only tells us if the polynomials of type (1) can be factored using integer coefficients, but, in addition, leads to a direct way of factoring those that are factorable.

ac Test for Factorability

If in polynomials of type (1) the product ac has two integer factors p and q whose sum is the coefficient of the middle term b; that is, if integers p and q exist so that

$$pq = ac \qquad \text{and} \qquad p + q = b \tag{2}$$

then polynomials of type (1) have first-degree factors with integer coefficients. If no integers p and q exist that satisfy Equations (2), then polynomials of type (1) will not have first-degree factors with integer coefficients.

Once we find integers p and q in the ac test, if they exist, our work is almost finished, since we can then write polynomials of type (1), splitting the middle term, in the forms

$$ax^2 + px + qx + c \qquad \text{and} \qquad ax^2 + pxy + qxy + cy^2 \tag{3}$$

Then the factoring can be completed in a couple of steps using factoring by grouping discussed at the end of Section 2-3.

Let us make the discussion concrete through several examples.

Example 18 Factor, if possible, using integer coefficients.

(A) $2x^2 + 7x - 4$ **(B)** $4x^2 + 5x - 6$ **(C)** $8x^2 + 12xy + 9y^2$

Solution **(A)** *Step 1* Test $2x^2 + 7x - 4$ for factorability using the ac test:

$$ac = 2(-4) = -8$$

We need two integers p and q such that $pq = -8$ and $p + q = 7$. We write (or think of) all integer pairs that multiply to -8:

$pq = -8$
$2(-4)$
$(-2)4$
$1(-8)$
$(-1)8$

and test to see if any of these add to 7, the coefficient of the middle term. The last pair works; that is,

$$\overset{p}{(-1)}\overset{q}{(8)} = \overset{ac}{-8} \qquad \text{and} \qquad \overset{p}{(-1)} + \overset{q}{(8)} = \overset{b}{7}$$

We can conclude that the polynomial can be factored.

Step 2 We now split the middle term $7x$ of the original equation using $p = -1$ and $q = 8$.

$$2x^2 + \overset{b}{7x} - 4 = 2x^2 - \overset{p}{1x} + \overset{q}{8x} - 4$$

Step 3 Factor the resulting polynomial by grouping. This will always work if we can get to this step, and it won't matter which order of p and q is used—the process will work if they are reversed.

$2x^2 + 7x - 4$

$= 2x^2 - x + 8x - 4$ Group the first two and last two terms.

$= (2x^2 - x) + (8x - 4)$ Factor out common factors.

$= x(2x - 1) + 4(2x - 1)$

$= (x + 4)(2x - 1)$

This process can be reduced to a few key operational steps when all of the commentary is eliminated and some of the process is done mentally. The only trial and error occurs in step 1, and with a little practice that step will go fairly fast.

(B) Compute ac for $4x^2 + 5x - 6$:

$$ac = 4(-6) = -24$$

We need two integers, p and q, whose product is -24 and whose sum is 5, the coefficient of the middle term.

$pq = -24$
$1(-24)$
$(-1)24$
$12(-2)$
$(-12)2$
$8(-3)$
$(-8)3$
$6(-4)$
$(-6)4$

We find 8 and -3 will work, so we split the middle term $5x = 8x - 3x$ and factor by grouping:

$$\begin{aligned} 4x^2 + 5x - 6 &= 4x^2 + 8x - 3x - 6 \\ &= (4x^2 + 8x) - (3x + 6) \\ &= 4x(x + 2) - 3(x + 2) \\ &= (4x + 3)(x + 2) \end{aligned}$$

(C) Compute ac for $8x^2 + 12xy + 9y^2$:

$$ac = 8 \cdot 9 = 72$$

Thus, we need p and q with $pq = 72$ and $p + q = 12$. In this case, both p and q must be positive, so we need only look at these possibilities:

$pq = 72$
$1 \cdot 72$
$2 \cdot 36$
$3 \cdot 24$
$4 \cdot 18$
$6 \cdot 12$
$8 \cdot 9$

None of these add up to 12, so we can conclude that $8x^2 + 12xy + 9y^2$ is not factorable using integer coefficients.

Problem 18 Factor, if possible, using integer coefficients.

(A) $3x^2 - 2xy - y^2$ **(B)** $6x^2 - 7x - 5$ **(C)** $4x^2 + 10x + 9$

ANSWER TO MATCHED PROBLEM

18. **(A)** $(3x + y)(x - y)$ **(B)** $(3x - 5)(2x + 1)$
(C) Not factorable

EXERCISE 2-5

Factor Problems 1–42, if possible, using integer coefficients. Use the ac test and proceed as in Example 18.

A

1. $4x^2 + 11x - 3$ **2.** $4x^2 - x - 3$ **3.** $4x^2 + 4x - 3$

4. $4x^2 - 8x - 5$ **5.** $4x^2 + 11x + 8$ **6.** $4x^2 + 21x + 5$

7. $4x^2 - 25x + 6$ **8.** $6x^2 + 12x + 9$ **9.** $6x^2 - 25x + 4$

10. $6x^2 - 29x - 5$ **11.** $4x^2 + 4x - 15$ **12.** $4x^2 - 4x - 3$

B

13. $4x^2 + 4x + 3$ **14.** $6x^2 - 7x - 10$ **15.** $6x^2 - 29x + 20$

16. $8x^2 - 12x + 5$ **17.** $6x^2 + 7x - 3$ **18.** $6x^2 + 5x - 4$

19. $6x^2 + 7x - 20$ **20.** $6x^2 + 14x + 9$ **21.** $6x^2 + x - 15$

22. $6x^2 - 43x + 7$ **23.** $6x^2 + 20x + 5$ **24.** $6x^2 + 37x + 6$

25. $4x^2 - 13xy + 10y^2$ **26.** $4x^2 - xy - 14y^2$

27. $4x^2 - 9xy - 5y^2$ **28.** $6x^2 - xy - 2y^2$

29. $6x^2 + 14xy - 12y^2$ **30.** $10x^2 + 24xy + 25y^2$

C

31. $12x^2 + 19xy + 5y^2$ **32.** $6x^2 - 16xy + 3y^2$

33. $6x^2 - 20xy - 25y^2$ **34.** $9x^2 - 14xy + 4y^2$

35. $4x^2 - 17xy - 15y^2$ **36.** $4x^2 - 25xy + 12y^2$

37. $6x^2 + 28x - 5$ **38.** $6x^2 - 28x + 15$

39. $18x^2 + 37x - 20$ **40.** $18x^2 + 5x - 2$

41. $18x^2 + 9xy - 20y^2$ **42.** $18x^2 + 3xy - 28y^2$

43. Find all integers b such that $x^2 + bx - 18$ can be factored using integer coefficients.

44. Find all integers b such that $x^2 + bx + 20$ can be factored using integer coefficients.

45. Find all integers c between 0 and 10 such that $x^2 + 5x + c$ can be factored using integer coefficients.

46. Find all integers c between 0 and 10 such that $x^2 - 6x + c$ can be factored using integer coefficients.

Refer to the approach introduced in Exercise 2-4 (Problems 57–62) for factoring certain polynomials of degree higher than 2. Apply the ac test and factor, if posible, using integer coefficients.

47. $6x^2 + 7x^2 + 2$ **48.** $5x^4 - 11x^2 + 2$

49. $2x^4 + 5x^2y^2 + 3y^4$

50. $3x^4 + 5x^2y^2 - 2y^4$

51. $2x^6 - x^3 - 3$

52. $3x^6 - 10x^3 - 8$

2-6 MORE FACTORING

- Sum and Difference of Two Squares
- Sum and Difference of Two Cubes
- Combined Forms
- Higher-Degree Polynomials
- General Strategy

In this section we consider several additional factoring forms and suggest a general strategy for applying the various factoring processes.

SUM AND DIFFERENCE OF TWO SQUARES

If we multiply $(A - B)$ and $(A + B)$, we obtain

$$(A - B)(A + B) = A^2 - B^2$$

which is a difference of two squares. Writing this result in reverse order, we obtain a very useful factoring formula. If we try to factor the sum of two squares, $A^2 + B^2$, we find that it cannot be factored using integer coefficients unless A and B have common factors. (Try it to see why.)

Sum and Difference of Two Squares

$A^2 + B^2$ cannot be factored using integer coefficients unless A and B have common factors. (1)

$A^2 - B^2 = (A - B)(A + B)$ (2)

Example 19 Factor, if possible, using integer coefficients:

(A) $x^2 - 16$ **(B)** $4x^2 - y^2$ **(C)** $x^2 + 9y^2$ **(D)** $8x^2 - 50y^2$

Solution **(A)** $x^2 - 16 \boxed{= x^2 - 4^2} = (x - 4)(x + 4)$

(B) $4x^2 - y^2 \boxed{= (2x)^2 - y^2} = (2x - y)(2x + y)$

(C) $x^2 + 9y^2$ is the sum of two squares and, therefore, not factorable using integer coefficients.

(D) $8x^2 - 50y^2 = 2(4x^2 - 25y^2) = 2[(2x)^2 - (5y)^2]$

$$= 2(2x - 5y)(2x + 5y)$$

Problem 19 Factor, if possible, using integer coefficients.

(A) $x^2 + 16$ **(B)** $x^2 - 9y^2$ **(C)** $3x^2 - 48$ **(D)** $9x^2 - 49y^2$

SUM AND DIFFERENCE OF TWO CUBES

We can verify, by direct multiplication of the right sides, the following factoring formulas for the sum and difference of two cubes:

Sum and Difference of Two Cubes

$$A^3 + B^3 = (A + B)(A^2 - AB + B^2) \quad (3)$$
$$A^3 - B^3 = (A - B)(A^2 + AB + B^2) \quad (4)$$

You should memorize these formulas. Observing the pattern of signs in them may help you to better remember them:

$$A^3 + B^3 = (A + B)(A^2 - AB + B^2)$$

Same sign | Opposite sign

$$A^3 - B^3 = (A - B)(A^2 + AB + B^2)$$

Same sign | Opposite sign

Neither $A^2 - AB + B^2$ nor $A^2 + AB + B^2$ can be factored using integer coefficients. (See Problems 83 and 84, Exercise 2-6.)

To factor

$$y^3 - 27$$

we first note that it can be written in the form

$$y^3 - 3^3$$

Thus, we are dealing with the difference of two cubes. If in the factoring formula (4) we let $A = y$ and $B = 3$, we obtain

$$y^3 - 27 = y^3 - 3^3 = (y - 3)(y^2 + 3y + 9)$$

Example 20 Factor as far as possible using integer coefficients:

(A) $x^3 + 64$ **(B)** $8x^3 - 1$ **(C)** $16x^3 + 2y^3$ **(D)** $8x^3 - 27y^3$

Solution **(A)** $x^3 + 64 = x^3 + 4^3 = (x + 4)(x^2 - 4x + 16)$

(B) $8x^3 - 1 = (2x)^3 - 1^3 = (2x - 1)[(2x)^2 + 1 \cdot 2x + 1^2]$

$= (2x - 1)(4x^2 + 2x + 1)$

(C) $16x^3 + 2y^3 = 2[8x^3 + y^3] = 2[(2x)^3 + y^3]$

$= 2(2x + y)[(2x)^2 - 2xy + y^2]$
$= 2(2x + y)(4x^2 - 2xy + y^2)$

(D) $8x^2 - 27y^3 = (2x)^3 - (3y)^3$

$= (2x - 3y)[(2x)^2 + (2x)(3y) + (3y)^2]$
$= (2x - 3y)(4x^2 + 6xy + 9y^2)$

Problem 20 Factor as far as possible using integer coefficients:

(A) $27x^3 + 1$ **(B)** $x^3 - 64$ **(C)** $81x^3 - 3y^3$ **(D)** $8x^3 + 125y^3$

COMBINED FORMS

In Examples 19(D) and 20(C), we first removed common factors before proceeding. In general, factoring will be simpler if we do this.

General Factoring Principle

Remove common factors first before proceeding further.

Example 21 Factor as far as possible using integer coefficients:

(A) $3x^2 + 12x + 12$ **(B)** $4x^2 - 100$
(C) $5x^2 - 5x - 60$ **(D)** $11x^3 - 297y^3$

Solution **(A)** $3x^2 + 12x + 12$ Factor out the common factors.

$= 3(x^2 + 4x + 4)$ Recognize the perfect square.

$= 3(x + 2)^2$

(B) $4x^2 - 100$ — Factor out the common factor 4.

$= 4(x^2 - 25)$ — Factor the difference of two squares.

$= 4(x - 5)(x + 5)$

Since both $4x^2$ and 100 are perfect squares, this problem could also have been treated as the difference of two squares initially:

$$\begin{aligned} 4x^2 - 100 &= (2x)^2 - 10^2 = (2x - 10)(2x + 10) \\ &= 2(x - 5)2(x + 5) \\ &= 4(x - 5)(x + 5) \end{aligned}$$

(C) $5x^2 - 5x - 60$ — Factor out the common factor 5.

$= 5(x^2 - x - 12)$ — Factor $x^2 - x - 12$ by trial and error.

$= 5(x - 4)(x + 3)$

(D) $11x^3 - 297y^3$ — Factor out the common factor 11.

$= 11(x^3 - 27y^3)$ — Factor the difference of two cubes.

$= 11(x - 3y)(x^2 + 3xy + 9y^2)$

Problem 21 Factor as far as possible using integer coefficients:

(A) $2x^3 + 250$ **(B)** $6x^2 + 18x - 24$
(C) $12x^2 - 108$ **(D)** $4x^2 + 40x + 100$

Occasionally, polynomial forms of a more general nature than we considered in Section 2-3 can be factored by appropriate grouping of terms. The following example illustrates the process.

Example 22 Factor by grouping terms:

(A) $x^2 + 3xy - 2x - 6y$ **(B)** $x^3 + 3x^2 + 4x + 12$

Solution **(A)** $x^2 + 3xy - 2x - 6y$ — Group the first two and last two terms.

$= (x^2 + 3xy) + (-2x - 6y)$ — Remove common factors.

$= x(x + 3y) - 2(x + 3y)$
$= (x - 2)(x + 3y)$

(B) $x^3 + 3x^2 + 4x + 12$ — Group the first two and last two terms.

$= (x^3 + 3x^2) + (4x + 12)$ — Remove the common factors.

$= x^2(x + 3) + 4(x + 3)$ — Remove the common factor $x + 3$.

$= (x^2 + 4)(x + 3)$

Problem 22 Factor by grouping terms:

(A) $x^2 - xy + 3x - 3y$ (B) $2x^3 - x^2 + 4x - 2$

HIGHER-DEGREE POLYNOMIALS

The techniques and examples of factoring introduced so far have dealt mainly with polynomials of degree 2 or 3. Some higher-degree polynomials can be factored in the same ways.

Example 23 Factor: **(A)** $x^4 - 81$ **(B)** $x^4 + 6x^2 + 8$ **(C)** $x^6 + y^3$

Solution **(A)** $x^4 - 81$ — Think of x^4 as $(x^2)^2$.

$= (x^2)^2 - 9^2$ — Factor as the difference of two squares.

$= (x^2 - 9)(x^2 + 9)$
$= (x - 3)(x + 3)(x^2 + 9)$

(B) $x^4 + 6x^2 + 8$ — Think of x^4 as $(x^2)^2$.

$= (x^2)^2 + 6(x^2) + 8$ — Factor as a second-degree polynomial in x^2.

$= (x^2 + 4)(x^2 + 2)$

(C) $x^6 + y^3$ — Think of x^6 as $(x^2)^3$.

$= (x^2)^3 + y^3$ — Factor as the sum of two cubes.

$= (x^2 + y)[(x^2)^2 - (x^2)y + y^2]$ — Simplify the second factor.

$= (x^2 + y)(x^4 - x^2y + y^2)$

It may be easier to see the structure of the polynomial with a substitution. In part (B), we are to think of x^4 as $(x^2)^2$. If we write $u = x^2$, then $u^2 = x^4$ and the factoring process becomes

$x^4 + 6x^2 + 8$ — Replace x^4 by u^2, x^2 by u.

$= u^2 + 6u + 8$ — Factor.

$= (u + 4)(u + 2)$ — Replace u by x^2.

$= (x^2 + 4)(x^2 + 2)$

Problem 23 Factor: **(A)** $x^4 - y^4$ **(B)** $x^4 - 3x^2 + 2$ **(C)** $8x^6 - 27$

GENERAL STRATEGY

There is no standard procedure (**algorithm**) for factoring a polynomial. However, the following strategy may be helpful.

General Strategy for Factoring

1. Remove common factors (Section 2-3).
2. If the polynomial has two terms, look for a difference of two squares or a sum or difference of two cubes (Section 2-6).
3. If the polynomial has three terms:
   (A) See if it is a perfect square (Section 2-4).
   (B) Try trial and error (Section 2-4).
   (C) Or use the *ac* test (Section 2-5).
4. If the polynomial has more than three terms, try grouping (Sections 2-3 and 2-6).

Factoring requires skill, creativity, and perseverance. Often the appropriate technique is not immediately apparent, and practice is necessary to develop your recognition of what might work on a given problem. Exercises 2-6 and 2-8 contain numerous problems for this purpose.

ANSWERS TO MATCHED PROBLEMS

19. **(A)** Not factorable using integer coefficients
(B) $(x - 3y)(x + 3y)$ **(C)** $3(x - 4)(x + 4)$
(D) $(3x - 7y)(3x + 7y)$

20. **(A)** $(3x + 1)(9x^2 - 3x + 1)$ **(B)** $(x - 4)(x^2 + 4x + 16)$
(C) $3(3x - y)(9x^2 + 3xy + y^2)$
(D) $(2x + 5y)(4x^2 - 10xy + 25y^2)$

21. **(A)** $2(x + 5)(x^2 - 5x + 25)$ **(B)** $6(x + 4)(x - 1)$
(C) $12(x - 3)(x + 3)$ **(D)** $4(x + 5)^2$

22. **(A)** $(x + 3)(x - y)$ **(B)** $(x^2 + 2)(2x - 1)$

23. **(A)** $(x^2 + y^2)(x - y)(x + y)$ **(B)** $(x^2 - 2)(x - 1)(x + 1)$
(C) $(2x^2 - 3)(4x^4 + 6x^2 + 9)$

EXERCISE 2-6

Factor Problems 1–84 as far as possible using integer coefficients.

A

1. $v^2 - 25$ **2.** $x^2 - 81$ **3.** $9x^2 - 4$

4. $4m^2 - 1$ **5.** $x^2 + 49$ **6.** $y^2 + 64$

7. $9x^2 - 16y^2$ **8.** $25u^2 - 4v^2$ **9.** $x^3 + 1$

10. $y^3 - 1$
11. $m^3 - n^3$
12. $p^3 + q^3$
13. $8x^3 + 27$
14. $u^3 - 8v^3$
15. $6u^2v^2 - 3uv^3$
16. $2x^3y - 6x^2y^3$
17. $2x^2 - 8$
18. $3y^2 - 27$
19. $2x^3 + 8x$
20. $3x^4 + 27x^2$
21. $12x^3 - 3xy^2$
22. $2u^3v - 2uv^3$
23. $2x^4 + 2x$
24. $xy^3 + x^4$
25. $6x^2 + 36x + 48$
26. $4x^2 - 4x - 24$
27. $3x^3 - 6x^2 + 15x$
28. $2x^3 - 2x^2 + 8x$

B

29. $x^2y^2 - 16$
30. $m^2n^2 - 36$
31. $a^3b^3 + 8$
32. $27 - x^3y^3$
33. $4x^3y + 14x^2y^2 + 6xy^3$
34. $3x^3y - 15x^2y^2 + 18xy^3$
35. $4u^3 + 32v^3$
36. $54x^3 - 2y^3$
37. $60x^2y^2 - 200xy^3 - 35y^4$
38. $60x^4 + 68x^3y - 16x^2y^2$
39. $xy + 2x + y^2 + 2y$
40. $x^2 + 3x + xy + 3y$
41. $x^2 - 5x + xy - 5y$
42. $x^2 - 3x - xy + 3y$
43. $ax - 2bx - ay + 2by$
44. $mx + my - 2nx - 2ny$
45. $15ac - 20ad + 3bc - 4bd$
46. $2am - 3an + 2bm - 3bn$
47. $x^3 - 2x^2 - x + 2$
48. $x^3 - 2x^2 + x - 2$
49. $(y - x)^2 - y + x$
50. $x^2(x - 1) - x + 1$
51. $x^2y^2 - xy - 6$
52. $a^2b^2 - 7ab + 12$
53. $z^4 - z^2 - 6$
54. $x^4 + 4x^2 + 4$

C

55. $x^8 - 4$
56. $a^6 + 8a^3 - 20$
57. $r^4 - s^4$
58. $16a^4 - b^4$
59. $x^4 - 3x^2 - 4$
60. $x^4 - 7x^2 - 18$
61. $(x - 3)^2 - 16y^2$
62. $(x + 2)^2 - 9y^2$
63. $(a - b)^2 - 4(c - d)^2$
64. $(x^2 - x)^2 - 9(y^2 - y)^2$
65. $25(4x^2 - 12xy + 9y^2) - 9a^2b^2$
66. $18a^3 - 8a(x^2 + 8x + 16)$
67. $x^6 - 1$
68. $a^6 - 64b^6$
69. $2x^3 - x^2 - 8x + 4$
70. $4y^3 - 12y^2 - 9y + 27$
71. $25 - a^2 - 2ab - b^2$
72. $x^2 - 2xy + y^2 - 9$

73. $16x^4 - x^2 + 6xy - 9y^2$

74. $x^4 - x^2 + 4x - 4$

75. $x^3 - 2x^2 + 3x - 6$

76. $x^3 + 2x^2 - 5x - 10$

77. $x^5 - x^4 + x - 1$

78. $x^5 + x^3 + 2x^2 + 2$

79. $3x^3 - x^2 + 12x - 4$

80. $2x^3 + x^2 + 4x + 2$

81. $x^2 + 4x - y^2 + 4$

82. $x^2 + y^2 - z^2 + 2xy$

83. Use the *ac* test to show that $a^2 - ab + b^2$ cannot be factored using integer coefficients.

84. Use the *ac* test to show that $a^2 + ab + b^2$ cannot be factored using integer coefficients.

2-7 SOLVING EQUATIONS BY FACTORING

- Zero Factor Property
- Solving Equations by Factoring

An equation in which a polynomial is equated to 0 can be solved readily if the polynomial can be completely factored. We solve the original equation by setting each factor equal to 0. A property of real numbers, called the zero factor property, justifies this process.

ZERO FACTOR PROPERTY

When nonzero real numbers are multiplied, the product will never be 0. Looked at differently, if a product is 0, then at least one of the factors must be 0. This is the **zero factor property**.

Zero Factor Property

For real numbers a and b:

$a \cdot b = 0$ if and only if $a = 0$ or $b = 0$ (or both)

We apply this property to solving equations of the form $a \cdot b = 0$. The property allows us to solve the simpler equations $a = 0$ or $b = 0$.

SOLVING EQUATIONS BY FACTORING

Example 24 Solve by factoring, if possible:

(A) $x^2 + 2x - 15 = 0$ **(B)** $4x^2 = 6x$ **(C)** $2x^2 - 8x + 3 = 0$

Solution **(A)** $x^2 + 2x - 15 = 0$

$(x - 3)(x + 5) = 0$ — $(x - 3)(x + 5) = 0$ if and only if $(x - 3) = 0$ or $(x + 5) = 0$.

$x - 3 = 0$ or $x + 5 = 0$

$x = 3$ $\quad$ $x = -5$

(B) $4x^2 = 6x$ — If both sides are divided by x, we lose one solution ($x = 0$). But we can simplify the equation by dividing both sides by 2, a common factor of both coefficients.

$2x^2 = 3x$

$2x^2 - 3x = 0$ — Rewrite the equation as a polynomial equal to 0.

$x(2x - 3) = 0$ — $x(2x - 3) = 0$ if and only if $x = 0$ or $2x - 3 = 0$.

$x = 0$ or $2x - 3 = 0$

$x = 0$ $\quad$ $x = \frac{3}{2}$

(C) The polynomial cannot be factored using integer coefficients; hence, another method must be used to find the solution. This will be discussed later.

Problem 24 Solve by factoring, if possible:

(A) $x^2 - 2x - 8 = 0$ $\quad$ **(B)** $9t^2 = 6t$ $\quad$ **(C)** $x^2 - 3x = 3$

Example 25 The length of a rectangle is 1 inch more than twice its width. If the area is 21 square inches, find its dimensions.

Solution Draw a figure and label the sides, as shown.

$$x(2x + 1) = 21$$
$$2x^2 + x - 21 = 0$$
$$(2x + 7)(x - 3) = 0$$

$2x + 7 = 0$ or $x - 3 = 0$

$2x = -7$ $\quad$ $x = 3$ inches — Width

~~$x = -\frac{7}{2}$~~ $\quad$ $2x + 1 = 7$ inches — Length

A negative length is not possible, so must be discarded.

Note: In practical problems involving second-degree equations, one of two solutions must often be discarded because it will not make sense in the problem.

Problem 25 The base of a triangle is 6 meters longer than its height. If the area is 20 square meters, find its dimensions ($A = \frac{1}{2}bh$).

ANSWERS TO MATCHED PROBLEMS

24. **(A)** $x = 4, -2$ **(B)** $t = 0, \frac{2}{3}$
(C) Cannot be factored using integer coefficients

25. Height 4 meters, base 10 meters

EXERCISE 2-7

Solve by factoring.

A

1. $u^2 + 5u = 0$
2. $v^2 - 3v = 0$
3. $3A^2 = -12A$
4. $4u^2 = 8u$
5. $x^2 - 11x - 12 = 0$
6. $y^2 - 6y + 5 = 0$
7. $x^2 + 4x - 5 = 0$
8. $x^2 - 4x - 12 = 0$
9. $3Q^2 - 10Q - 8 = 0$
10. $2d^2 + 15d - 8 = 0$

B

11. $u^2 = 2u + 3$
12. $m^2 + 2m = 15$
13. $3x^2 = x + 2$
14. $2x^2 = 3 - 5x$
15. $y^2 = 5y - 2$
16. $3 = t^2 + 7t$
17. $2x(x - 1) = 3(x + 1)$
18. $3x(x - 2) = 2(x - 2)$
19. $t^2 = 4$
20. $y^2 = 9$
21. $m^2 + 4m = 12$
22. $A^2 = 2A + 8$
23. $2y^2 = 2 + 3y$
24. $L^2 - 2L = 15$
25. $2x^2 + 2 = 5x$
26. $x^2 - 3x = 10$
27. $x^2 + x = 6$
28. $x^2 + x = 2$

29. The width of a rectangle is 8 inches less than its length. If its area is 33 square inches, find its dimensions.

30. Find the base and height of a triangle with area 2 square feet if its base is 3 feet longer than its height ($A = \frac{1}{2}bh$).

C *Solve by factoring.*

31. $x^4 - 81 = 0$
32. $2x^4 - 32 = 0$
33. $(x^3 - 1)(x^2 - 4) = 0$
34. $(x^4 - 1)(x^3 - 8) = 0$
35. $x^4 - 18x^2 + 81 = 0$
36. $x^4 - 8x^2 + 16 = 0$
37. $x^6 + 7x^3 - 8 = 0$
38. $x^6 - 26x^3 - 27 = 0$
39. $x^2 + 3x = 1 + 3x$
40. $x(x + 1) - (x + 1) = x^2$
41. $(x + 1)(x + 2) - 6x = (2x - 1)(x - 1)$
42. $2x(x - 1) = (x - 2) + x$

2-8 CHAPTER REVIEW

For a natural number n, the **base** a raised to the **exponent** n is $a^n = a \cdot a \cdot \cdots \cdot a$ (n factors). Natural number exponents satisfy this **property of exponents**: $a^m a^n = a^{m+n}$. In combined operations, powers take precedence over multiplication and division.

A **polynomial** is an algebraic expression involving only the operations of addition, subtraction, and multiplication on variables and constants. The constant factor in a term is called the **coefficient**. The **degree of a term** in a polynomial is the sum of the powers of the variables; constants are assigned degree 0. The **degree of a polynomial** is the highest degree of its terms. Polynomials with one, two, and three terms are called **monomials**, **binomials**, and **trinomials**, respectively. Addition and subtraction of polynomials are accomplished by **grouping** and combining like terms. *(2-1)*

Monomials are multiplied by using the above property of exponents. Polynomials are multiplied by multiplying each term of one by each term of the other and adding. Binomials can be multiplied mentally by the FOIL method

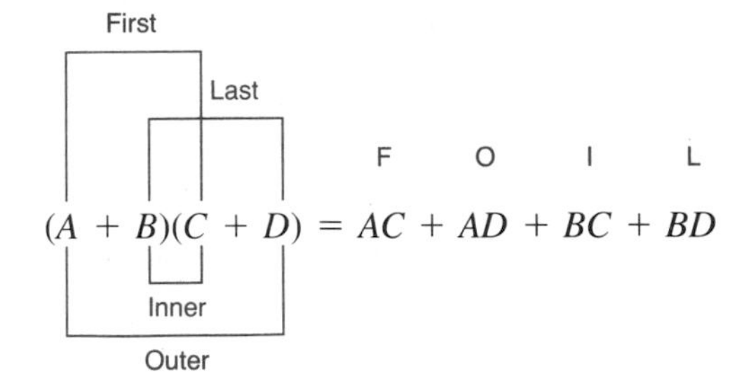

and squared by

$$(A + B)^2 = A^2 + 2AB + B^2 \qquad (A - B)^2 = A^2 - 2AB + B^2 \quad (2\text{-}2)$$

To **factor** a polynomial means to find factors whose product is the given polynomial. Factors common to all terms can be factored out by the distributive law. **Grouping** may lead to factoring out common factors. *(2-3)*

Second-degree polynomials can be factored by trial and error *(2-4)* or by the ***ac* test**. The *ac* test to factor $ax^2 + bx + c$ or $ax^2 + bxy + cy^2$ involves finding two integers p and q such that $pq = ac$, $p + q = b$, rewriting the middle term as $px + qx$ or $pxy + qxy$, and grouping. *(2-5)*

The sum $A^2 + B^2$ of two squares cannot be factored unless there are common factors. The difference of two squares is factored

$$A^2 - B^2 = (A - B)(A + B)$$

The sum and difference of two cubes are factored

$$A^3 + B^3 = (A + B)(A^2 - AB + B^2)$$
$$A^3 - B^3 = (A - B)(A^2 + AB + B^2)$$

A **general strategy for factoring** is

1. Remove common factors.
2. If the polynomial has two terms, look for a difference of two squares or a sum or difference of two cubes.
3. If the polynomial has three terms:
   (A) See if it is a perfect square.
   (B) Try trial and error.
   (C) Or use the *ac* test.
4. If a polynomial has four or more terms, try grouping. *(2-6)*

Equations where a polynomial is set equal to 0 can be solved by factoring the polynomial and using the **zero factor property**: $ab = 0$ if and only if $a = 0$ or $b = 0$. *(2-7)*

REVIEW EXERCISE 2-8

Work through all the problems in this chapter review and check answers in the back of the book. (Answers to all problems are there, and following each answer is a number in italics indicating the section in which that type of problem is discussed.) Where weaknesses show up, review appropriate sections in the text.

A *Simplify, using natural number exponents only.*

1. x^3x^8

2. $2x^3 \cdot 5x^4$

3. Given the polynomial $3x^5 - 2x^3 + 7x^2 - x + 2$:
(A) What is its degree?
(B) What is the degree of the second term?

4. Consider the product $(3x^3y^2)(-2xy^2z^3)$ of two monomials.
(A) What is the degree of the first factor?
(B) What is the degree of the product?

Add, subtract, or multiply as indicated.

5. $(2x + 5) + (x^2 - 4)$

6. $(2x + 5) - (x^2 - 4)$

7. $(2x + 5)(x^2 - 4)$

8. $(3x^2 + 2x + 1) - (2x^2 - 3x + 4)$

9. $(3x^2 + 2x + 1) + (2x^2 - 3x + 4)$

10. $(3x^2 + 2x + 1)(2x^2 - 3x + 4)$

11. $[x^2 - (1 - x - x^2)] - (x + 1)$

12. $-2x\{(x^2 + 2)(x - 3) - x[x - x(3 - x)]\}$

Factor as far as possible in the integers.

13. $3x^4 + 9x^3$

14. $2x^2y^3 + x^3y$

B

15. $2x^2 - 10x + x - 5$

16. $4x^2 + 11x - 3$

17. $4x^2 - 25$

18. $x^2 + 3xy + 2y^2$

19. $a^2 + 9$

20. $x^2 + 2x + 4 + 2x$

21. $2x^2 - 5x - 3$

22. $x^3 - 3x + 9 - 3x$

23. $3x^2 + 2x - 3x - 2$

24. $a^3 - 8$

25. $x^2 + 6x + 8$

26. $2x^2 - 5xy + 2y^2$

27. $x^3 + 64$

28. $a^2 - 9b^2$

Solve by factoring.

29. $x^2 + 7x + 12 = 0$

30. $x^4 - 16 = 0$

C

Factor as far as possible in the integers.

31. $x^6 + 1$

32. $2x^4 + 6x^3 + 3x^2 + x^3$

33. $x^2a^3 - 4xa^3 + 4a^3$

34. $3x^3y - 3xy^3$

35. $x^3 - 3x^2 + x - 3$

36. $x^6 + x^5 + x^2 + x$

37. $2x^3 - 2x^2 - 4x$

38. $x^2 + 2xy + y^2 - 1$

39. $x^4 - 1$

40. $8x^3 - 125$

41. $3x^3 + 2x^2 - 15x - 10$

42. $x^4 + 2x^3 - 3x^2$

43. $-x^5y^3 - 2x^4y^2 - x^3y$

44. $2a^4 + 2a$

Solve by factoring.

45. $2x^2 - x = 3$

46. $x^2 - x = 20$

3

ALGEBRAIC FRACTIONS

Polynomials are built up from constants and variables using the operations of addition, subtraction, and multiplication. If we allow division as well, we obtain algebraic forms called *rational expressions*. In this chapter we consider these expressions: how to change the form; how to add, subtract, multiply, and divide them; and, finally, how to solve equations involving them.

3-1 RATIONAL EXPRESSIONS

- Fundamental Principle of Fractions
- Reducing to Lowest Terms
- Raising to Higher Terms

Fractional forms in which the numerator and denominator are polynomials are called **rational expressions**. For example,

$$\frac{1}{x} \qquad \frac{-6}{y-3} \qquad \frac{x-2}{x^2-2x+5} \qquad \frac{x^2-3xy+y^2}{3x^3y^4} \qquad \frac{3x^2-8x-4}{5}$$

are all rational expressions. (Recall that a nonzero constant is a polynomial of degree 0.) More generally, a rational expression is an algebraic expression involving only the operations of addition, subtraction, multiplication, and division on variables and constants.

Each rational expression names a real number for real-number replacements of the variables, division by 0 excluded. Hence, all properties of the real numbers apply to those expressions. In particular, the basic properties of fractions apply. In this section we recall the fundamental principle of fractions that allows us to rewrite a fraction in higher or lower terms, and use this principle to change the form of rational expressions.

FUNDAMENTAL PRINCIPLE OF FRACTIONS

You will recall from arithmetic that

$$\frac{8}{12} = \boxed{\frac{8 \div 4}{12 \div 4}} = \frac{2}{3} \qquad \text{and} \qquad \frac{3}{5} = \boxed{\frac{2 \cdot 3}{2 \cdot 5}} = \frac{6}{10}$$

The first example illustrates reducing a fraction to lowest terms, while the second example illustrates raising a fraction to higher terms. Both illustrate the use of the **fundamental principle of fractions**.

The Fundamental Principle of Fractions

For all polynomials P, Q, and K with $Q, K \neq 0$:

$$\frac{PK}{QK} = \frac{P}{Q}$$ We may divide out a common factor K from both the numerator and denominator. This is called **reducing to lower terms**.

$$\frac{P}{Q} = \frac{PK}{QK}$$ We may multiply the numerator and denominator by the same nonzero factor K. This is called **raising to higher terms**.

REDUCING TO LOWEST TERMS

To reduce a fraction to **lowest terms** is to divide out *all* common factors from the numerator and denominator.

It is important to keep in mind when reducing fractions to lowest terms that it is only common *factors* in products that can be divided out: common *terms* in sums or differences cannot be divided out. For example:

Common factors may be divided out: $\frac{2y}{3y} = \frac{2\cancel{y}}{3\cancel{y}} = \frac{2}{3}$

Common terms may not be divided out: $\frac{2+y}{3+y} \neq \frac{2+\cancel{y}}{3+\cancel{y}} \neq \frac{2}{3}$

Example 1 Reduce to lowest terms:

(A) $\frac{8x^2y}{12xy^2}$ **(B)** $\frac{6x^2 - 3x}{3x}$ **(C)** $\frac{x^2y - xy^2}{x^2 - xy}$ **(D)** $\frac{3x^3 - 2x^2 - 8x}{x^4 - 8x}$

Solution **(A)** $\frac{8x^2y}{12xy^2} = \frac{\cancel{(4xy)}(2x)}{\cancel{(4xy)}(3y)} = \frac{2x}{3y}$ Divide out common factors.

(B) $\frac{6x^2 - 3x}{3x} = \frac{\cancel{3x}(2x-1)}{\cancel{3x}} = 2x - 1$ Factor the top, then divide out common factors.

Note: $\frac{6x^2 - \cancel{3x}}{\cancel{3x}}$ is wrong. Only common *factors* can be divided out.

(C) $\frac{x^2y - xy^2}{x^2 - xy} = \frac{\cancel{x}y\cancel{(x-y)}}{\cancel{x}\cancel{(x-y)}} = y$ Factor the top and bottom, then divide out common factors.

(D) $\frac{3x^3 - 2x^2 - 8x}{x^4 - 8x} = \frac{\cancel{x}(3x^2 - 2x - 8)}{\cancel{x}(x^3 - 8)}$

$= \frac{(3x+4)\cancel{(x-2)}}{\cancel{(x-2)}(x^2+2x+4)}$

$= \frac{3x+4}{x^2+2x+4}$

Problem 1 Reduce to lowest terms:

(A) $\dfrac{16u^5v^3}{24u^3v^6}$ (B) $\dfrac{4x}{8x^2 - 4x}$ (C) $\dfrac{x^2 - 3x}{x^2y - 3xy}$ (D) $\dfrac{2x^3 - 8x}{2x^4 + 16x}$

RAISING TO HIGHER TERMS

The following example shows how to use the fundamental principle of fractions to raise fractions to higher terms. We will need to use this technique later when we add and subtract rational expressions.

Example 2 Complete the raising-to-higher-terms process by replacing the question marks with appropriate expressions

(A) $\dfrac{3}{2x} = \dfrac{?}{16xy^2}$ (B) $\dfrac{2x}{3y} = \dfrac{?}{3xy - 3y^2}$

(C) $\dfrac{x - 2y}{2x + y} = \dfrac{2x^2 - 5xy + 2y^2}{?}$

Solution (A) $\dfrac{3}{2x} = \dfrac{(\mathbf{8y^2})(3)}{(\mathbf{8y^2})(2x)} = \dfrac{24y^2}{16xy^2}$ (B) $\dfrac{2x}{3y} = \dfrac{(\mathbf{x - y})2x}{(\mathbf{x - y})3y} = \dfrac{2x^2 - 2xy}{3xy - 3y^2}$

(C) $\dfrac{x - 2y}{2x + y} = \dfrac{(\mathbf{2x - y})(x - 2y)}{(\mathbf{2x - y})(2x + y)} = \dfrac{2x^2 - 5xy + 2y^2}{4x^2 - y^2}$

Problem 2 Complete the raising-to-higher-terms process by replacing the question marks with appropriate expressions:

(A) $\dfrac{3u^2}{4v} = \dfrac{?}{20u^2v^2}$ (B) $\dfrac{3m}{2n} = \dfrac{3m^3 - 3m}{?}$

(C) $\dfrac{2x - 3}{x - 4} = \dfrac{?}{3x^2 - 14x + 8}$

ANSWERS TO MATCHED PROBLEMS

1. (A) $\dfrac{2u^2}{3v^3}$ (B) $\dfrac{1}{2x - 1}$ (C) $\dfrac{1}{y}$ (D) $\dfrac{x - 2}{x^2 - 2x + 4}$
2. (A) $15u^4v$ (B) $2m^2n - 2n$ (C) $6x^2 - 13x + 6$

EXERCISE 3-1

A *Reduce to lowest terms.*

1. $\dfrac{3x^3}{6x^5}$ 2. $\dfrac{9u^5}{3u^6}$ 3. $\dfrac{14x^3y}{21xy^2}$

4. $\frac{20m^4n^6}{15m^5n^2}$ 5. $\frac{15y^3(x-9)^3}{5y^4(x-9)^2}$ 6. $\frac{2x^2(x+7)}{6x(x+7)}$

7. $\frac{(2x-1)(2x+1)}{3x(2x+1)}$ 8. $\frac{(x+3)(2x+5)}{2x^2(2x+5)}$

9. $\frac{x^2-2x}{2x-4}$ 10. $\frac{2x^2-10x}{4x-20}$

11. $\frac{m^2-mn}{m^2n-mn^2}$ 12. $\frac{a^2b+ab^2}{ab+b^2}$

Complete the raising-to-higher-terms process by replacing the question marks with appropriate expressions.

13. $\frac{3}{2x}=\frac{?}{8x^2y}$ 14. $\frac{5x}{3}=\frac{10x^3y^2}{?}$

15. $\frac{7}{3y}=\frac{?}{6x^3y^2}$ 16. $\frac{5u}{4v^2}=\frac{20u^3v}{?}$

B *Reduce to lowest terms.*

17. $\frac{x^2+6x+8}{3x^2+12x}$ 18. $\frac{x^2+5x+6}{2x^2+6x}$ 19. $\frac{x^2-9}{x^2+6x+9}$

20. $\frac{x^2-4}{x^2+4x+4}$ 21. $\frac{4x^2-9y^2}{4x^2y+6xy^2}$ 22. $\frac{a^2-16b^2}{4ab-16b^2}$

23. $\frac{x^2-xy+2x-2y}{x^2-y^2}$ 24. $\frac{u^2+uv-2u-2v}{u^2+2uv+v^2}$

25. $\frac{6x^3+28x^2-10x}{12x^3-4x^2}$ 26. $\frac{12x^3-78x^2-42x}{16x^4+8x^3}$

27. $\frac{x^3-8}{x^2-4}$ 28. $\frac{y^3+27}{2y^3-6y^2+18y}$

29. $\frac{x^2+10x+21}{x^2+5x-14}$ 30. $\frac{x^2-4x+3}{x^2+2x-15}$

31. $\frac{6x^2z+4xyz}{3xyz+2y^2z}$ 32. $\frac{2x^2+x-3}{4x^2+x-5}$

33. $\frac{x^3-3x^2+2x-6}{x^2-5x+6}$ 34. $\frac{x^3-x^2+x-1}{x^2-1}$

Complete the raising-to-higher-terms process by replacing the question marks with appropriate expressions.

35. $\dfrac{3x}{4y} = \dfrac{?}{4xy + 4y^2}$

36. $\dfrac{4m}{5n} = \dfrac{4m^2 - 4mn}{?}$

37. $\dfrac{x - 2y}{x + y} = \dfrac{x^2 - 3xy + 2y^2}{?}$

38. $\dfrac{2x + 5}{x - 3} = \dfrac{?}{3x^2 - 8x - 3}$

C *Reduce to lowest terms.*

39. $\dfrac{x^3 - y^3}{3x^3 + 3x^2y + 3xy^2}$

40. $\dfrac{2u^3v - 2u^2v^2 + 2uv^3}{u^3 + v^3}$

41. $\dfrac{ux + vx - uy - vy}{2ux + 2vx + uy + vy}$

42. $\dfrac{mx - 2my + nx - 2ny}{mx - 2my - nx + 2ny}$

43. $\dfrac{x^4 - y^4}{(x^2 - y^2)(x + y)^2}$

44. $\dfrac{x^4 - 2x^2y^2 + y^4}{x^4 - y^4}$

45. $\dfrac{x^4 + x}{x^3 - x^2 + x}$

46. $\dfrac{x^5 + x^3 + x}{x^4 + x^2 + 1}$

47. $\dfrac{x^2 + y^2}{x^3 + y^3}$

48. $\dfrac{z^3 - z}{z^3 - z^2 - z + 1}$

49. $\dfrac{x^2u + uxy + vxy + vy^2}{xu^2 + uvy + uvx + yv^2}$

50. $\dfrac{x^2y^2z - xyz^2 - xy + z}{x^2yz - x - xyz^2 + yz}$

3-2 MULTIPLICATION AND DIVISION

- Multiplication of Rational Expressions
- Division of Rational Expressions

Multiplication and division of rational expressions are based upon the corresponding properties of real fractions.

MULTIPLICATION OF RATIONAL EXPRESSIONS

We start with the process of multiplying rational forms.

Multiplication

If P, Q, R, and S are polynomials ($Q, S \neq 0$), then

$$\frac{P}{Q} \cdot \frac{R}{S} = \frac{P \cdot R}{Q \cdot S}$$

Example 3 Multiply and reduce to lowest terms:

(A) $\dfrac{3a^2b}{4c^2d} \cdot \dfrac{8c^2d^3}{9ab^2}$ **(B)** $(x^2 - 4) \cdot \dfrac{2x - 3}{x + 2}$

(C) $\dfrac{4a^2 - 9b^2}{4a^2 + 12ab + 9b^2} \cdot \dfrac{6a^2b}{8a^2b^2 - 12ab^3}$

Solution **(A)** $$\frac{3a^2b}{4c^2d} \cdot \frac{8c^2d^3}{9ab^2} = \frac{(3a^2b) \cdot (8c^2d^3)}{(4c^2d) \cdot (9ab^2)} = \frac{24a^2bc^2d^3}{36ab^2c^2d}$$
$$= \frac{(2ad^2)\overset{1}{\cancel{(12abc^2d)}}}{(3b)\underset{1}{\cancel{(12abc^2d)}}} = \frac{2ad^2}{3b}$$

This process is easily shortened to the following when it is realized that, in effect, any factor in either numerator may be divided out with any like factor in either denominator. Thus,

$$\frac{\overset{1\cdot a\cdot 1}{\cancel{3}\cancel{a^2}\cancel{b}}}{\underset{1\cdot 1\cdot 1}{\cancel{4}\cancel{c^2}d}} \cdot \frac{\overset{2\cdot 1\cdot d^2}{\cancel{8}\cancel{c^2}\cancel{d^3}}}{\underset{3\cdot 1\cdot b}{\cancel{9}\cancel{a}\cancel{b^2}}} = \frac{2ad^2}{3b}$$

(B) $$(x^2 - 4) \cdot \frac{2x - 3}{x + 2} = \frac{\overset{1}{\cancel{(x + 2)}}(x - 2)}{1} \cdot \frac{(2x - 3)}{\underset{1}{\cancel{(x + 2)}}}$$
$$= (x - 2)(2x - 3)$$

Factor where possible, divide out common factors, then multiply and write the answer.

(C) $$\frac{4a^2 - 9b^2}{4a^2 + 12ab + 9b} \cdot \frac{6a^2b}{8a^2b^2 - 12ab^3}$$
$$= \frac{\overset{1}{\cancel{(2a - 3b)}}\overset{1}{\cancel{(2a + 3b)}}}{\underset{(2a + 3b)}{\cancel{(2a + 3b)^2}}} \cdot \frac{\overset{3a}{\cancel{6a^2b}}}{\underset{2b}{\cancel{4ab^2}}\underset{1}{\cancel{(2a - 3b)}}}$$
$$= \frac{3a}{2b(2a + 3b)}$$

Problem 3 Multiply and reduce to lowest terms:

(A) $\dfrac{4x^2y^3}{9w^2z} \cdot \dfrac{3wz^2}{2xy^4}$ **(B)** $\dfrac{x + 5}{x^2 - 9} \cdot (x + 3)$

(C) $\dfrac{x^2 - 9y^2}{x^2 - 6xy + 9y^2} \cdot \dfrac{6x^2y}{2x^2 + 6xy}$

DIVISION OF RATIONAL EXPRESSIONS

The following result follows from the definition of division: $A \div B = Q$ if and only if $QB = A$ and Q is unique.

Division

If P, Q, R, and S are polynomials (Q, R, $S \neq 0$), then

Divisor is R/S — Reciprocal of divisor is S/R

$$\frac{P}{Q} \div \frac{R}{S} = \frac{P}{Q} \cdot \frac{S}{R}$$

That is, to divide one fraction by another, multiply by the reciprocal of the divisor.

To prove that the indicated process is a valid procedure, one has only to show that the product of R/S and $(P/Q) \cdot (S/R)$ is P/Q, a problem left to the exercises.

Example 4 Divide and reduce to lowest terms:

(A) $\dfrac{6a^2b^3}{5cd} \div \dfrac{3a^2c}{10bd}$ **(B)** $(4 - x) \div \dfrac{2x^2 - 32}{6xy}$

(C) $\dfrac{10x^3y}{3xy + 9y} \div \dfrac{4x^2 - 12x}{x^2 - 9}$

Solution **(A)** $\dfrac{6a^2b^3}{5cd} \div \dfrac{3a^2c}{10bd} = \dfrac{\overset{2 \cdot 1}{\cancel{6}\cancel{a^2}b^3}}{\underset{1 \cdot 1}{\cancel{5}c\cancel{d}}} \cdot \dfrac{\overset{2 \cdot 1}{\cancel{10}b\cancel{d}}}{\underset{1 \cdot 1}{\cancel{3}\cancel{a^2}c}} = \dfrac{4b^4}{c^2}$

(B) $(4 - x) \div \dfrac{2x^2 - 32}{6xy} = (4 - x) \cdot \dfrac{6xy}{2(x - 4)(x + 4)}$

$$= (-1)\overset{1}{\cancel{(x - 4)}} \cdot \frac{\overset{3}{\cancel{6}}xy}{\underset{1}{\cancel{2}}\underset{1}{\cancel{(x - 4)}}(x + 4)}$$

$$= \frac{-3xy}{x + 4}$$

Note here that we rewrote

$$4 - x = (-1)(x - 4)$$

More generally, recall

$$a - b = (-1)(b - a) = -(b - a)$$

Thus,

$$\frac{a - b}{b - a} = -1$$

(C) $\dfrac{10x^3y}{3xy + 9y} \div \dfrac{4x^2 - 12x}{x^2 - 9} = \dfrac{\overset{5 \cdot x^2 \cdot 1}{\cancel{10x^3y}}}{\underset{1 \cdot 1}{\cancel{3y(x + 3)}}} \cdot \dfrac{\overset{1}{\cancel{(x + 3)}}\overset{1}{\cancel{(x - 3)}}}{\underset{2 \cdot 1 \cdot 1}{\cancel{4x(x - 3)}}} = \dfrac{5x^2}{6}$

Problem 4 Divide and reduce to lowest terms:

(A) $\dfrac{8w^2z^2}{9x^2y} \div \dfrac{4wz}{6xy^2}$ **(B)** $\dfrac{2x^2 - 8}{4x} \div (x + 2)$

(C) $\dfrac{x^2 - 4x + 4}{4x^2y - 8xy} \div \dfrac{x^2 + x - 6}{6x^2 + 18x}$

ANSWERS TO MATCHED PROBLEMS

3. **(A)** $\dfrac{2xz}{3wy}$ **(B)** $\dfrac{x + 5}{x - 3}$ **(C)** $\dfrac{3xy}{x - 3y}$

4. **(A)** $\dfrac{4wzy}{3x}$ **(B)** $\dfrac{x - 2}{2x}$ **(C)** $\dfrac{3}{2y}$

EXERCISE 3-2

Do not change improper fractions in your answers to mixed fractions; that is, write $\frac{7}{2}$, not $3\frac{1}{2}$.

A *Multiply and reduce to lowest terms.*

1. $\dfrac{10}{9} \cdot \dfrac{12}{15}$ **2.** $\dfrac{3}{7} \cdot \dfrac{14}{9}$ **3.** $\dfrac{2a}{3bc} \cdot \dfrac{9c}{a}$

4. $\dfrac{2x}{3yz} \cdot \dfrac{6y}{4x}$ **5.** $\dfrac{3x^2}{4} \cdot \dfrac{16y}{12x^3}$ **6.** $\dfrac{2x^2}{3y^2} \cdot \dfrac{9y}{4x}$

Divide and reduce to lowest terms.

7. $\dfrac{9m}{8n} \div \dfrac{3m}{4n}$ **8.** $\dfrac{6x}{5y} \div \dfrac{3x}{10y}$ **9.** $\dfrac{a}{4c} \div \dfrac{a^2}{12c^2}$

10. $\dfrac{2x}{3y} \div \dfrac{4x}{6y^2}$ **11.** $\dfrac{x}{3y} \div 3y$ **12.** $2xy \div \dfrac{x}{y}$

Perform the indicated operations and reduce to lowest terms.

13. $\dfrac{8x^2}{3xy} \cdot \dfrac{12y^3}{6y}$ **14.** $\dfrac{6a^2}{7c} \cdot \dfrac{21cd}{12ac}$

15. $\dfrac{21x^2y^2}{12cd} \div \dfrac{14xy}{9d}$ **16.** $\dfrac{3uv^2}{5w} \div \dfrac{6u^2v}{15w}$

17. $\dfrac{9u^4}{4v^3} \div \dfrac{-12u^2}{15v}$ **18.** $\dfrac{-6x^3}{5y^2} \div \dfrac{18x}{10y}$

19. $\dfrac{3c^2d}{a^3b^3} \div \dfrac{3a^3b^3}{cd}$

20. $\dfrac{uvw}{5xyz} \div \dfrac{5vy}{uwxz}$

B **21.** $\dfrac{3x^2y}{x-y} \cdot \dfrac{x-y}{6xy}$

22. $\dfrac{x+3}{2x^2} \cdot \dfrac{4x}{x+3}$

23. $\dfrac{x+3}{x^3+3x^2} \cdot \dfrac{x^3}{x-3}$

24. $\dfrac{a^2-a}{a-1} \cdot \dfrac{a+1}{a}$

25. $\dfrac{x-2}{4y} \div \dfrac{x^2+x-6}{12y^2}$

26. $\dfrac{4x}{x-4} \div \dfrac{8x^2}{x^2-6x+8}$

27. $\dfrac{6x^2}{4x^2y-12xy} \cdot \dfrac{x^2+x-12}{3x^2+12x}$

28. $\dfrac{2x^2+4x}{12x^2y} \cdot \dfrac{6x}{x^2+6x+8}$

29. $(t^2-t-12) \div \dfrac{t^2-9}{t^2-3t}$

30. $\dfrac{2y^2+7y+3}{4y^2-1} \div (y+3)$

31. $\dfrac{m+n}{m^2-n^2} \div \dfrac{m^2-mn}{m^2-2mn+n^2}$

32. $\dfrac{x^2-6x+9}{x^2-x-6} \div \dfrac{x^2+2x-15}{x^2+2x}$

33. $-(x^2-3x) \cdot \dfrac{x-2}{x-3}$

34. $-(x^2-4) \cdot \dfrac{3}{x+2}$

35. $\left(\dfrac{d^5}{3a} \div \dfrac{d^2}{6a^2}\right) \cdot \dfrac{a}{4d^3}$

36. $\dfrac{d^5}{3a} \div \left(\dfrac{d^2}{6a^2} \cdot \dfrac{a}{4d^3}\right)$

37. $\dfrac{2x^2}{3y^3} \cdot \dfrac{-6yz}{2x} \cdot \dfrac{y}{-xz}$

38. $\dfrac{-a}{-b} \cdot \dfrac{12b^2c}{15ac} \cdot \dfrac{-10}{4b}$

C **39.** $\dfrac{9-x^2}{x^2+5x+6} \cdot \dfrac{x+2}{x-3}$

40. $\dfrac{2-m}{2m+m^2} \cdot \dfrac{m^2+4m+4}{m^2-4}$

41. $\dfrac{x^2-xy}{xy+y^2} \div \left(\dfrac{x^2-y^2}{x^2+2xy+y^2} \div \dfrac{x^2-2xy+y^2}{x^2y+xy^2}\right)$

42. $\left(\dfrac{x^2-xy}{xy+y^2} \div \dfrac{x^2-y^2}{x^2+2xy+y^2}\right) \div \dfrac{x^2-2xy+y^2}{x^2y+xy^2}$

43. $(x^2-x-6)/(x-3) = x+2$, except for what values of x?

44. $(x^2-1)/(x-1)$ and $x+1$ name the same real numbers for (*all, all but one, no*) replacements of x by real numbers.

45. Prove that $\dfrac{P}{Q} \div \dfrac{R}{S} = \dfrac{P}{Q} \cdot \dfrac{S}{R}$ by using the definition of division.

3-3 ADDITION AND SUBTRACTION

Addition and subtraction of rational expressions are based on the corresponding properties of real fractions. Thus:

Addition and Subtraction

If P, D, and Q are polynomials ($D \neq 0$), then

$$\frac{P}{D} + \frac{Q}{D} = \frac{P+Q}{D} \qquad (1)$$

$$\frac{P}{D} - \frac{Q}{D} = \frac{P-Q}{D} \qquad (2)$$

In words: if the denominators of two rational expressions are the same, we may either add or subtract the expressions by adding or subtracting the numerator and placing the result over the common denominator. If the denominators are not the same, we use the fundamental principle of fractions to change the form of each fraction so they have a common denominator, then use either Property 1 or 2.

Even though any common denominator will do, the problem will generally become less involved if the least common denominator (LCD) is used. Recall that the least common denominator is the least common multiple (LCM) of all the denominators; that is, it is the ''smallest'' quantity exactly divisible by each denominator.

If the LCD is not obvious (often it is), then it is found as follows:

Finding the Least Common Denominator (LCD)

Step 1 Factor each denominator completely using integer coefficients.

Step 2 The LCD must contain each *different* factor that occurs in any of the denominators to the highest power it occurs in any one denominator.

Example 5 Find the least common multiple for

(A) $18x^3$, $15x$, $10x^2$ **(B)** $6(x-3)$, x^2-9, $4x^2+24x+36$

Solution **(A)** Write each expression in completely factored form, including coefficients:

$$18x^3 = 2 \cdot 3^2x^3 \qquad 15x = 3 \cdot 5x \qquad 10x^2 = 2 \cdot 5x^2$$

The LCM must contain each different factor (2, 3, 5, and x) to the highest power it occurs in any one expression. Thus,

$$\text{LCM} = 2 \cdot 3^2 \cdot 5x^3 = 90x^3$$

(B) Factor each expression completely:

$$6(x - 3) = 2 \cdot 3(x - 3) \qquad x^2 - 9 = (x - 3)(x + 3)$$
$$4x^2 + 24x + 36 = 4(x^2 + 6x + 9) = 2^2(x + 3)^2$$

The different factors of these expressions and the highest power to which each occurs are

2	2^2 in the third expression
3	3 in the first expression
$x - 3$	$x - 3$ in the first and second expressions
$x + 3$	$(x + 3)^2$ in the third expression

Thus,

$$\text{LCM} = 2^2 \cdot 3(x - 3)(x + 3)^2 = 12(x - 3)(x + 3)^2$$

Problem 5 Find the LCM for

(A) $15y^2, 12y, 9y^4$ **(B)** $3x^2 - 12, x^2 - 4x + 4, 12(x + 2)$

Example 6 Combine into a single fraction and simplify:

(A) $\dfrac{x + 1}{x - 2} + \dfrac{3x - 2}{x - 2}$ **(B)** $\dfrac{1}{x - 3} - \dfrac{x - 2}{x - 3}$

Solution **(A)** $\dfrac{(x + 1)}{x - 2} + \dfrac{(3x - 2)}{x - 2}$ When a numerator has more than one term, place terms in parentheses before proceeding. Since denominators are the same, use Property 1 to add.

$$= \frac{(x + 1) + (3x - 2)}{x - 2}$$ Simplify the numerator.

$$= \frac{x + 1 + 3x - 2}{x - 2}$$

$$= \frac{4x - 1}{x - 2}$$

(B) $\dfrac{1}{x-3} - \dfrac{(x-2)}{x-3}$ Use Property 2 to subtract.

$= \dfrac{1-(x-2)}{x-3}$ Simplify the numerator. Watch signs.

$= \dfrac{1-x\overset{\downarrow}{+}2}{x-3}$ Sign errors are frequently made where the arrow points.

$= \dfrac{3-x}{x-3}$ $3-x \neq x-3;\ 3-x = -(x-3)$

$= \dfrac{-(x-3)}{(x-3)}$ Reduce to lowest terms.

$= -1$

Problem 6 Combine into a single fraction and simplify:

$$\frac{x+3}{2x-5} - \frac{3x-2}{2x-5}$$

Example 7 Combine into a single fraction and simplify:

$$\frac{3}{2y} - \frac{1}{3y^2} + 1$$

Solution $\dfrac{3}{2y} - \dfrac{1}{3y^2} + 1$ LCD $= 6y^2$

$= \dfrac{\mathbf{3y} \cdot 3}{\mathbf{3y} \cdot 2y} - \dfrac{\mathbf{2} \cdot 1}{\mathbf{2} \cdot 3y^2} + \dfrac{\mathbf{6y^2}}{\mathbf{6y^2}}$ Use the fundamental principle of fractions to make each denominator $6y^2$.

$= \dfrac{9y}{6y^2} - \dfrac{2}{6y^2} + \dfrac{6y^2}{6y^2}$

$= \dfrac{9y-2+6y^2}{6y^2}$ Arrange the numerator in descending powers of y.

$= \dfrac{6y^2+9y-2}{6y^2}$

Problem 7 Combine into a single fraction and simplify:

$$\frac{5}{4x^3} - \frac{1}{3x} + 2$$

Example 8 Combine into a single fraction and simplify:

$$\frac{4}{3x^2 - 27} - \frac{x - 1}{4x^2 + 24x + 36}$$

Solution

$$\frac{4}{3x^2 - 27} - \frac{x - 1}{4x^2 + 24x + 36}$$

Factor each denominator completely.

$$= \frac{4}{3(x - 3)(x + 3)} - \frac{(x - 1)}{2^2(x + 3)^2}$$

LCD = $12(x - 3)(x + 3)^2$

$$= \frac{\mathbf{4(x + 3)} \cdot 4}{\mathbf{4(x + 3)} \cdot 3(x - 3)(x + 3)} - \frac{\mathbf{3(x - 3)} \cdot (x - 1)}{\mathbf{3(x - 3)} \cdot 2^2(x + 3)^2}$$

Use the fundamental principle of fractions to make each denominator $12(x - 3)(x + 3)^2$.

$$= \frac{16(x + 3)}{12(x - 3)(x + 3)^2} - \frac{3(x - 3)(x - 1)}{12(x - 3)(x + 3)^2}$$

$$= \frac{16(x + 3) - 3(x - 3)(x - 1)}{12(x - 3)(x + 3)^2}$$

Note that neither $x + 3$ nor $x - 3$ can be removed as a common factor here; the whole numerator must be written as a product before any factors can be divided out.

$$= \frac{16x + 48 - 3(x^2 - 4x + 3)}{12(x - 3)(x + 3)^2}$$

$$= \frac{16x + 48 - 3x^2 + 12x - 9}{12(x - 3)(x + 3)^2}$$

$$= \frac{-3x^2 + 28x + 39}{12(x - 3)(x + 3)^2}$$

Problem 8 Combine into a single fraction and simplify:

$$\frac{3}{2x^2 - 8x + 8} - \frac{x + 1}{3x^2 - 12}$$

ANSWERS TO MATCHED PROBLEMS

5. **(A)** $180y^4$ **(B)** $12(x - 2)^2(x + 2)$

6. -1 **7.** $\dfrac{24x^3 - 4x^2 + 15}{12x^3}$

8. $\dfrac{-2x^2 + 11x + 22}{6(x - 2)^2(x + 2)}$

EXERCISE 3-3

A *Find the least common multiple (LCM) for each group of expressions.*

1. $3, x$ **2.** $4, y$ **3.** $x, 1$

4. $y, 1$ **5.** v^2, v, v^3 **6.** x, x, x^2

7. $3x,\ 6x^2,\ 4$ **8.** $8u^3,\ 6u,\ 4u^2$ **9.** $x + 1,\ x - 2$

10. $x - 2,\ x + 3$ **11.** $y + 3,\ 3y$ **12.** $x - 2,\ 2x$

Combine into single fractions and simplify.

13. $\frac{7x}{5x^2} + \frac{2}{5x^2}$

14. $\frac{3m}{2m^2} + \frac{1}{2m^2}$

15. $\frac{4x}{2x - 1} - \frac{2}{2x - 1}$

16. $\frac{5a}{a - 1} - \frac{5}{a - 1}$

17. $\frac{y}{y^2 - 9} - \frac{3}{y^2 - 9}$

18. $\frac{2x}{4x^2 - 9} + \frac{3}{4x^2 - 9}$

19. $\frac{5}{3k} - \frac{6x - 4}{3k}$

20. $\frac{1}{2a^2} - \frac{2b - 1}{2a^2}$

21. $\frac{3x}{y} + \frac{1}{4}$

22. $\frac{2}{x} - \frac{1}{3}$

23. $\frac{2}{y} + 1$

24. $x + \frac{1}{x}$

25. $\frac{u}{v^2} - \frac{1}{v} + \frac{u^3}{v^3}$

26. $\frac{1}{x} - \frac{y}{x^2} + \frac{y^2}{x^3}$

27. $\frac{2}{3x} - \frac{1}{6x^2} + \frac{3}{4}$

28. $\frac{1}{8u^3} + \frac{5}{6u} - \frac{3}{4u^2}$

29. $\frac{2}{x + 1} + \frac{3}{x - 2}$

30. $\frac{1}{x - 2} + \frac{1}{x + 3}$

31. $\frac{3}{y + 3} - \frac{2}{3y}$

32. $\frac{2}{x - 2} - \frac{3}{2x}$

B *Find the LCM for each group of expressions.*

33. $12x^3,\ 8x^2y^2,\ 3xy^2$

34. $9u^3v^2,\ 6uv,\ 12v^3$

35. $15x^2y,\ 25xy,\ 5y^2$

36. $18m^4n^2,\ 12m^2n^4,\ 9mn$

37. $6(x - 1),\ 9(x - 1)^2$

38. $8(y - 3)^2,\ 6(y - 3)$

39. $6(x - 7)(x + 7),\ 8(x + 7)^2$

40. $3(x - 5)^2,\ 4(x + 5)(x - 5)$

41. $x^2 - 4,\ x^2 + 4x + 4$

42. $x^2 - 6x + 9,\ x^2 - 9$

43. $3x^2 + 3x,\ 4x^2,\ 3x^2 + 6x + 3$

44. $3m^2 - 3m,\ m^2 - 2m + 1,\ 5m^2$

Combine into a single fraction and simplify.

45. $\dfrac{2}{9u^3v^2} - \dfrac{1}{6uv} + \dfrac{1}{12v^3}$

46. $\dfrac{1}{12x^3} + \dfrac{3}{8x^2y^2} - \dfrac{2}{3xy^2}$

47. $\dfrac{4t - 3}{18t^3} + \dfrac{3}{4t} - \dfrac{2t - 1}{6t^2}$

48. $\dfrac{3y + 8}{4y^2} - \dfrac{2y - 1}{y^3} - \dfrac{5}{8y}$

49. $\dfrac{t + 1}{t - 1} - 1$

50. $2 + \dfrac{x + 1}{x - 3}$

51. $5 + \dfrac{a}{a + 1} - \dfrac{a}{a - 1}$

52. $\dfrac{1}{y + 2} + 3 - \dfrac{2}{y - 2}$

53. $\dfrac{2}{3(x - 5)^2} - \dfrac{1}{4(x + 5)(x - 5)}$

54. $\dfrac{1}{6(x - 7)(x + 7)} + \dfrac{3}{8(x + 7)^2}$

55. $\dfrac{5}{6(x - 1)} + \dfrac{2}{9(x - 1)^2}$

56. $\dfrac{3}{8(y - 3)^2} - \dfrac{1}{6(y - 3)}$

57. $\dfrac{3}{x + 3} - \dfrac{3x + 1}{(x - 1)(x + 3)}$

58. $\dfrac{4}{2x - 3} - \dfrac{2x + 1}{(2x - 3)(x + 2)}$

59. $\dfrac{3s}{3s^2 - 12} + \dfrac{1}{2s^2 + 4s}$

60. $\dfrac{2t}{3t^2 - 48} + \dfrac{t}{4t + t^2}$

61. $\dfrac{3}{x^2 - 4} - \dfrac{1}{x^2 + 4x + 4}$

62. $\dfrac{2}{x^2 - 6x + 9} - \dfrac{1}{x^2 - 9}$

63. $\dfrac{2}{x + 3} - \dfrac{1}{x - 3} + \dfrac{2x}{x^2 - 9}$

64. $\dfrac{2x}{x^2 - y^2} + \dfrac{1}{x + y} - \dfrac{1}{x - y}$

For Problems 65–70 note that $b - a = -(a - b)$*; thus,* $3 - y = -(y - 3)$, $1 - x = -(x - 1)$, *and so on.*

65. $\dfrac{5}{y - 3} - \dfrac{2}{3 - y}$

66. $\dfrac{3}{x - 1} + \dfrac{2}{1 - x}$

67. $\dfrac{3}{x - 3} + \dfrac{x}{3 - x}$

68. $\dfrac{-2}{2 - y} - \dfrac{y}{y - 2}$

69. $\dfrac{1}{5x - 5} - \dfrac{1}{3x - 3} + \dfrac{1}{1 - x}$

70. $\dfrac{x + 7}{ax - bx} + \dfrac{y + 9}{by - ay}$

C **71.** $\dfrac{x}{x^2 - x - 2} - \dfrac{1}{x^2 + 5x - 14} - \dfrac{2}{x^2 + 8x + 7}$

72. $\dfrac{m^2}{m^2 + 2m + 1} + \dfrac{1}{3m + 3} - \dfrac{1}{6}$

73. $\dfrac{1}{3x^2 + 3x} + \dfrac{1}{4x^2} - \dfrac{1}{3x^2 + 6x + 3}$

74. $\dfrac{1}{3m(m - 1)} + \dfrac{1}{m^2 - 2m + 1} - \dfrac{1}{5m^2}$

75. $\dfrac{xy^2}{x^3 - y^3} - \dfrac{y}{x^2 + xy + y^2}$

76. $\dfrac{x}{x^2 - xy + y^2} - \dfrac{xy}{x^3 + y^3}$

3-4 QUOTIENTS OF POLYNOMIALS

- Algebraic Long Division
- Synthetic Division

There are times when it is useful to find the quotients of polynomials by a long-division process similar to that used in arithmetic. For some division problems the process can also be shortened notationally by a technique called *synthetic division*.

ALGEBRAIC LONG DIVISION

If we divide 33 by 7, we obtain a quotient of 4 with remainder 5. This means that $33 = 4 \cdot 7 + 5$. Similarly, if we divide a polynomial A by a polynomial B and obtain a quotient of Q with remainder R, it means $A = QB + R$. The actual division can be accomplished by algebraic long division that is analogous to the same process in arithmetic. Several examples will illustrate the process.

Example 9 Divide using the long-division process and check:

(A) $(3x^2 + 11x - 20)/(x + 5)$
(B) $(x^3 - 27)/(x - 3)$
(C) $(x^3 + 2x^2 - 3x + 4)/(x - 2)$

Solution **(A)** $x + 5\overline{)3x^2 + 11x - 20}$

Arrange both polynomials in descending powers of the variable if this is not already done.

$$\begin{array}{r} 3x \phantom{{}+ 11x - 20} \\ x + 5\overline{)3x^2 + 11x - 20} \end{array}$$

Divide the first term of the divisor into the first term of the dividend. That is, what must x be multiplied by so that the product is exactly $3x^2$? Answer: $3x$.

$$\begin{array}{r} 3x \phantom{{}+ 11x - 20} \\ x + 5\overline{)3x^2 + 11x - 20} \\ \underline{3x^2 + 15x \phantom{{}-20}} \\ -\ 4x - 20 \end{array}$$

Multiply the divisor by $3x$, line up like terms as indicated, subtract, and bring down -20 from above.

$$\begin{array}{r} 3x \quad - \quad 4 \\ x + 5\overline{)3x^2 + 11x - 20} \\ \underline{3x^2 + 15x} \quad\quad \\ -\ 4x - 20 \\ \underline{-\ 4x - 20} \\ 0 \end{array}$$

Repeat the process until the degree of the remainder is less than that of the divisor, or the remainder is 0.

0 Remainder.

To check,

$$(x + 5)(3x - 4) = 3x^2 + 11x - 20$$

(B)

$$\begin{array}{r} x^2 + 3x \ + 9 \\ x - 3\overline{)x^3 + 0x^2 + 0x - 27} \\ \underline{x^3 - 3x^2} \quad\quad\quad\quad\quad \\ 3x^2 + 0x \quad\quad \\ \underline{3x^2 - 9x} \quad\quad \\ 9x - 27 \\ \underline{9x - 27} \\ 0 \end{array}$$

Note that the terms $0x^2$ and $0x$ need to be included. Now proceed as in part (A).

0 Remainder.

To check, note that as the difference of two cubes,

$$x^3 - 27 = (x - 3)(x^2 + 3x + 9)$$

(C)

$$\begin{array}{r} x^2 + 4x \ + 5 \\ x - 2\overline{)x^3 + 2x^2 - 3x + \ 4} \\ \underline{x^3 - 2x^2} \quad\quad\quad\quad\quad \\ 4x^2 - 3x \quad\quad \\ \underline{4x^2 - 8x} \quad\quad \\ 5x + \ 4 \\ \underline{5x - 10} \\ 14 \end{array}$$

Proceed as above until the degree of the remainder is less than the degree of the divisor.

14 Remainder.

We interpret the remainder 14 just as we would in an arithmetic problem. If we divide 19 by 5, we obtain a quotient of 3 with remainder 4 and write $19 = 5 \cdot 3 + 4$, or $\frac{19}{5} = 3 + \frac{4}{5}$.

For the polynomial division in part (C), we have

$$x^3 + 2x^2 - 3x + 4 = (x - 2)(x^2 + 4x + 5) + 14$$

or

$$\frac{x^3 + 2x^2 - 3x + 4}{x - 2} = x^2 + 4x + 5 + \frac{14}{x - 2}$$

We check the problem by verifying either of these equalities; the first is easier.

Problem 9 Divide using the long-division process and check:

(A) $(4x^3 - 12x^2 - x + 3)/(x - 3)$
(B) $(x^3 + 125)/(x + 5)$
(C) $(4x^3 + 3x^2 - 2x + 1)/(x - 1)$

Each of the quotients in Example 9 involved division by a linear polynomial. The same process works for divisors of degrees larger than 1. When the divisor is linear in the form $x - a$, however, there is a relationship between the remainder and the value of the polynomial at a. Consider the polynomials in Example 9.

POLYNOMIAL	DIVISOR	a	REMAINDER	VALUE OF POLYNOMIAL AT $x = a$
$3x^2 + 11x - 20$	$x + 5$	-5	0	0
$x^3 - 27$	$x - 3$	3	0	0
$x^3 + 2x^2 - 3x + 4$	$x - 2$	2	14	14

If we write $P = (x - a)Q + R$, we can see that R is the value of P when $x = a$ because $(x - a)Q$ becomes $(0)Q = 0$.

Example 10 Divide $x^5 - 3x^3 + x^2 - 8$ by $x - 2$, and compare the remainder with the value of $x^5 - 3x^3 + x^2 - 8$ at $x = 2$.

$$
\begin{array}{r}
x^4 + 2x^3 + x^2 + 3x + 6 \\
x - 2\overline{)x^5 - 0x^4 - 3x^3 + x^2 - 0x - 8} \\
\underline{x^5 - 2x^4} \\
2x^4 - 3x^3 \\
\underline{2x^4 - 4x^3} \\
x^3 + x^2 \\
\underline{x^3 - 2x^2} \\
3x^2 - 0x \\
\underline{3x^2 - 6x} \\
6x - 8 \\
\underline{6x - 12} \\
4 = R
\end{array}
$$

The value of the polynomial at $x = 2$ is

$$2^5 - 3 \cdot 2^3 + 2^2 - 8 = 32 - 24 + 4 - 8 = 4 = R$$

as claimed.

Problem 10 Repeat Example 10, dividing the polynomial by $x + 1$ and comparing the remainder with the value of the polynomial of $x = -1$.

SYNTHETIC DIVISION

Any polynomial can be divided by a first-degree polynomial of the form $x - r$ using the algebraic long-division process described earlier in this section. In some circumstances such divisions have to be done repeatedly and a quicker, more concise method, called **synthetic division**, is useful. The method is most easily understood through an example. Let us start by dividing $P = 2x^4 + 3x^3 - x - 5$ by $x + 2$, using ordinary long division. The critical parts of the process are indicated in color.

$$
\begin{array}{lrl}
 & 2x^3 - 1x^2 + 2x - 5 & \text{Quotient} \\
\text{Divisor}\quad x + 2 & \overline{)\,2x^4 + 3x^3 + 0x^2 - 1x - 5} & \text{Dividend} \\
 & 2x^4 + 4x^3 \qquad\qquad\qquad\quad & \\
 & \overline{-1x^3 + 0x^2} \qquad\qquad\quad & \\
 & -1x^3 - 2x^2 \qquad\qquad\quad & \\
 & \overline{2x^2 - 1x} \qquad\quad & \\
 & 2x^2 + 4x \qquad\quad & \\
 & \overline{-5x - 5} & \\
 & -5x - 10 & \\
 & \overline{5} & \text{Remainder}
\end{array}
$$

The numerals printed in color, which represent the essential part of the division process, are arranged more conveniently as

$$
\begin{array}{c|ccccc}
 & \multicolumn{5}{c}{\text{Dividend coefficients}} \\
 & 2 & 3 & 0 & -1 & -5 \\
 & & 4 & -2 & 4 & -10 \\
\hline
2 & 2 & -1 & 2 & -5 & 5 \\
 & \multicolumn{4}{c}{\text{Quotient coefficients}} & \text{Remainder}
\end{array}
$$

We see that the second and third rows of numerals are generated as follows. The first coefficient 2 of the dividend is brought down and multiplied by 2 from the divisor, and the product 4 is placed under the second dividend coefficient 3 and subtracted. The difference -1 is again multiplied by the 2 from the divisor, and the product is placed under the third coefficient from the dividend and subtracted. This process is repeated until the remainder is reached. The process can be made a little faster, and less prone to sign errors, by changing $+2$ from the divisor to -2 and adding instead of subtracting.

Thus,

	Dividend coefficients			
	2	3 0	−1	−5
		−4 2	−4	10
−2	2	−1 2	−5	5
	Quotient coefficients			Remainder

Key Steps in the Synthetic Division Process: $P \div (x - r)$

1. Arrange the coefficients of P in order of descending powers of x. (Write 0 as the coefficient for each missing power.)
2. After writing the divisor in the form $x - r$, use r to generate the second and third row of numbers as follows. Bring down the first coefficient of the dividend and multiply it by r; then add the product to the second coefficient of the dividend. Multiply this sum by r, and add the product to the third coefficient of the dividend. Repeat the process until a product is added to the constant term of P. [*Note:* This process is well suited to hand calculator use. Store r; then proceed from left to right recalling r and using it as indicated.]
3. The last number in the third row of numbers is the remainder; the other numbers in the third row are the coefficients of the quotient, which is of degree 1 less than P.

Example 11 Use synthetic division to find the quotient and remainder resulting from dividing $P = 4x^5 - 30x^3 - 50x - 2$ by $x + 3$.

Solution $x + 3 = x - (-3)$; therefore, $r = -3$.

	4	0	−30	0	−50	−2
		−12	36	−18	54	−12
−3	4	−12	6	−18	4	−14

The quotient is $4x^4 - 12x^3 + 6x^2 - 18x + 4$ with a remainder of -14.

Problem 11 Repeat Example 11 with $P = 3x^4 - 11x^3 - 18x + 8$ and divisor $x - 4$.

ANSWERS TO MATCHED PROBLEMS

9. **(A)** $4x^2 - 1$ **(B)** $x^2 - 5x + 25$ **(C)** $4x^2 + 7x + 5$, $R = 6$

10. $R = -5$ **11.** Quotient $3x^3 + x^2 + 4x - 2$; remainder 0

EXERCISE 3-4

A *Divide using the long-division process. Check the answers.*

1. $(3x^2 - 5x - 2)/(x - 2)$

2. $(2x^2 + x - 6)/(x + 2)$

3. $(2y^3 + 5y^2 - y - 6)/(y + 2)$

4. $(x^3 - 5x^2 + x + 10)/(x - 2)$

5. $(3x^2 - 11x - 1)/(x - 4)$

6. $(2x^2 - 3x - 4)/(x - 3)$

7. $(8x^2 - 14x + 13)/(2x - 3)$

8. $(6x^2 + 5x - 6)/(3x - 2)$

9. $(6x^2 + x - 13)/(2x + 3)$

10. $(6x^2 + 11x - 12)/(3x - 2)$

11. $(x^2 - 4)/(x - 2)$

12. $(y^2 - 9)/(y + 3)$

Divide using synthetic division. Write the quotient and indicate the remainder.

13. $(x^3 + 2x^2 + 3x + 4) \div (x - 2)$

14. $(x^3 + 2x^2 + 3x + 4) \div (x - 1)$

15. $(x^3 + 2x^2 + 3x + 4) \div (x + 1)$

16. $(x^3 + 2x^2 + 3x + 4) \div (x + 2)$

17. $(2x^3 - x^2 + x - 2) \div (x - 3)$

18. $(2x^3 - x^2 + x - 2) \div (x - 1)$

19. $(2x^3 - x^2 + x - 2) \div (x + 1)$

20. $(2x^3 - x^2 + x - 2) \div (x + 3)$

21. $(x^4 + x^3 + 3x^2 + 3x + 5) \div (x - 4)$

22. $(x^4 + x^3 + 3x^2 + 3x + 5) \div (x - 2)$

23. $(x^4 + x^3 + 3x^2 + 3x + 5) \div (x + 2)$

24. $(x^4 + x^3 + 3x^2 + 3x + 5) \div (x + 4)$

B *Divide using the long-division process. Check the answers.*

25. $(12x^2 + 11x - 2)/(3x + 2)$

26. $(8x^2 - 6x + 6)/(2x - 1)$

27. $(8x^2 + 7)/(2x - 3)$

28. $(9x^2 - 8)/(3x - 2)$

29. $(-7x + 2x^2 - 1)/(2x + 1)$

30. $(13x - 12 + 3x^2)/(3x - 2)$

31. $(x^3 - 1)/(x - 1)$

32. $(a^3 + 27)/(a + 3)$

33. $(x^4 - 81)/(x - 3)$

34. $(x^4 - 16)/(x + 2)$

35. $(4a^2 - 22 - 7a)/(a - 3)$

36. $(8c + 4 + 5c^2)/(c + 2)$

37. $(x + 5x^2 - 10 + x^3)/(x + 2)$

38. $(5y^2 - y + 2y^3 - 6)/(y + 2)$

39. $(3 + x^3 - x)/(x - 3)$

40. $(3y - y^2 + 2y^3 - 1)/(y + 2)$

Divide using synthetic division. Write the quotient and indicate the remainder.

41. $(x^3 - 2x + 4) \div (x - 2)$ **42.** $(x^3 - 2x + 4) \div (x - 1)$

43. $(x^3 - 2x + 4) \div (x + 1)$ **44.** $(x^3 - 2x + 4) \div (x + 2)$

45. $(x^4 - 3x + 5) \div (x - 4)$ **46.** $(x^4 - 3x + 5) \div (x - 2)$

47. $(x^4 - 3x + 5) \div (x + 2)$ **48.** $(x^4 - 3x + 5) \div (x + 4)$

C *Divide using the long-division process. Check the answers.*

49. $(9x^4 - 2 - 6x - x^2)/(3x - 1)$

50. $(4x^4 - 10x - 9x^2 - 10)/(2x + 3)$

51. $(8x^2 - 7 - 13x + 24x^4)/(3x + 5 + 6x^2)$

52. $(16x - 5x^3 - 8 + 6x^4 - 8x^2)/(2x - 4 + 3x^2)$

53. $(9x^3 - x + 2x^5 + 9x^3 - 2 - x)/(2 + x^2 - 3x)$

54. $(12x^2 - 19x^3 - 4x - 3 + 12x^5)/(4x^2 - 1)$

55. Given polynomials $P = x^3 - 6x^2 + 12x - 4$ and $D = x^2 - 3x + 2$, find polynomials Q and R such that $P = DQ + R$ and the degree of R is less than the degree of D, or $R = 0$.

56. Repeat the preceding problem for $P = x^4 - 4x^2 + 7x + 2$ and $D = x^2 - x + 1$.

Evaluate the polynomial $P = x^5 - 3x^3 + x^2 - 1$ *for the following values of* x *by using long division.*

57. 3 **58.** 2 **59.** 1 **60.** -1 **61.** -2 **62.** -3

Synthetic division can be used with any values for r, *not just integer values. Use synthetic division with a hand calculator to evaluate to six decimal places the polynomial* $P = x^3 - 2x^2 - 3x + 4$ *for the following values of* x:

63. 1.1 **64.** 1.01 **65.** 1.35 **66.** 2.11

67. -3.3 **68.** 1.001 **69.** 3.102 **70.** -3.141

3-5 COMPLEX FRACTIONS

A fractional form with fractions in its numerator or denominator, such as

$$\frac{x+1}{\frac{2}{3}} \quad \text{or} \quad \frac{1+\frac{1}{x}}{\frac{x}{2}}$$

is called a **complex fraction**. It is often necessary to represent a complex fraction as a **simple fraction**—that is (in all cases we will consider), as the quotient of two polynomials. The process does not involve any new concepts. It is a matter of applying old concepts in the right way. In particular, we will find the fundamental principle of fractions,

$$\frac{PK}{QK} = \frac{P}{Q} \qquad Q, K \neq 0 \tag{1}$$

of considerable use. Several examples should clarify the process.

Example 12 Express as simple fractions: **(A)** $\dfrac{\frac{3}{5}}{\frac{9}{10}}$ **(B)** $\dfrac{1\frac{1}{2}}{3\frac{2}{3}}$

Solution **(A)** The problem is easily solved by treating the complex fraction as the quotient of two fractions and dividing:

$$\frac{\frac{3}{5}}{\frac{9}{10}} = \frac{3}{5} \div \frac{9}{10} = \frac{\overset{1}{\cancel{3}}}{\underset{1}{\cancel{5}}} \cdot \frac{\overset{2}{\cancel{10}}}{\underset{3}{\cancel{9}}} = \frac{2}{3}$$

We can also simplify the complex fraction by multiplying numerator and denominator by a number that will clear both of fractions; the LCD of the numerator and denominator, in this case 10, will do this:

$$\frac{\frac{3}{5}}{\frac{9}{10}} = \frac{\mathbf{10} \cdot \frac{3}{5}}{\mathbf{10} \cdot \frac{9}{10}} = \frac{2 \cdot 3}{1 \cdot 9} = \frac{6}{9} = \frac{2}{3}$$

Although the first approach appears to be easier on the surface, the second can be very helpful when dealing with complex algebraic fractions.

(B) Recall that $1\frac{1}{2}$ and $3\frac{2}{3}$ represent sums and not products; that is, $1\frac{1}{2} = 1 + \frac{1}{2}$ and $3\frac{2}{3} = 3 + \frac{2}{3}$. Thus,

$$\frac{1\frac{1}{2}}{3\frac{2}{3}} = \frac{1 + \frac{1}{2}}{3 + \frac{2}{3}}$$ Write mixed fractions as sums.

$$= \frac{\mathbf{6}(1 + \frac{1}{2})}{\mathbf{6}(3 + \frac{2}{3})}$$ Multiply top and bottom by 6, the LCD of all fractions within the main fraction.

$$= \frac{\mathbf{6} \cdot 1 + \mathbf{6} \cdot \frac{1}{2}}{\mathbf{6} \cdot 3 + \mathbf{6} \cdot \frac{2}{3}}$$ The denominators 2 and 3 divide out.

$$= \frac{6 + 3}{18 + 4} = \frac{9}{22}$$ A simple fraction.

Problem 12 Express as simple fractions: **(A)** $\dfrac{\frac{3}{5}}{\frac{1}{4}}$ **(B)** $\dfrac{2\frac{3}{4}}{4\frac{1}{3}}$

Example 13 Express as simple fractions:

(A) $\dfrac{1 - \dfrac{1}{x^2}}{1 + \dfrac{1}{x}}$ **(B)** $\dfrac{\dfrac{a}{b} - \dfrac{b}{a}}{\dfrac{a}{b} + 2 + \dfrac{b}{a}}$

Solution **(A)** $\dfrac{1 - \dfrac{1}{x^2}}{1 + \dfrac{1}{x}}$

Multiply top and bottom by x^2, the LCD of the numerator and denominator.

$$= \frac{x^2\left(1 - \dfrac{1}{x^2}\right)}{x^2\left(1 + \dfrac{1}{x}\right)}$$

$$= \frac{x^2 \cdot 1 - x^2 \cdot \dfrac{1}{x^2}}{x^2 \cdot 1 + x^2 \cdot \dfrac{1}{x}}$$

$$= \frac{x^2 - 1}{x^2 + x}$$

Factor top and bottom to reduce to lowest terms.

$$= \frac{(x - 1)\overset{1}{\cancel{(x + 1)}}}{x\underset{1}{\cancel{(x + 1)}}}$$

$$= \frac{x - 1}{x}$$

(B) $\dfrac{\dfrac{a}{b} - \dfrac{b}{a}}{\dfrac{a}{b} + 2 + \dfrac{b}{a}}$

LCD $= ab$

$$= \frac{ab\left(\frac{a}{b} - \frac{b}{a}\right)}{ab\left(\frac{a}{b} + 2 + \frac{b}{a}\right)}$$

Use the fundamental principle of fractions; that is, multiply top and bottom by *ab* to clear fractions in the numerator and denominator.

$$= \frac{ab \cdot \frac{a}{b} - ab \cdot \frac{b}{a}}{ab \cdot \frac{a}{b} + ab \cdot 2 + ab \cdot \frac{b}{a}}$$

$$= \frac{a^2 - b^2}{a^2 + 2ab + b^2}$$

Reduce to lowest terms.

$$= \frac{(a - b)(a + b)}{(a + b)^2}$$

$$= \frac{a - b}{a + b}$$

A simple fraction.

Problem 13 Express as simple fractions:

(A) $\dfrac{1 - \frac{1}{3x}}{1 - \frac{1}{9x^2}}$ **(B)** $\dfrac{\frac{x}{y} + 1 - \frac{2y}{x}}{\frac{x}{y} - \frac{y}{x}}$

ANSWERS TO MATCHED PROBLEMS

12. **(A)** $\frac{12}{5}$ **(B)** $\frac{33}{52}$

13. **(A)** $\dfrac{3x}{3x + 1}$ **(B)** $\dfrac{x + 2y}{x + y}$

EXERCISE 3-5

In Problems 1–30, rewrite the expression as a simple fraction reduced to lowest terms.

A

1. $\dfrac{\frac{1}{2}}{\frac{2}{3}}$ **2.** $\dfrac{\frac{1}{4}}{\frac{2}{3}}$ **3.** $\dfrac{\frac{3}{8}}{\frac{5}{12}}$ **4.** $\dfrac{\frac{4}{15}}{\frac{5}{6}}$

5. $\dfrac{1\frac{1}{3}}{2\frac{1}{6}}$ **6.** $\dfrac{3\frac{1}{10}}{2\frac{1}{5}}$ **7.** $\dfrac{1\frac{2}{9}}{2\frac{5}{6}}$ **8.** $\dfrac{2\frac{4}{15}}{1\frac{7}{10}}$

9. $\dfrac{\frac{x}{y}}{\frac{1}{y^2}}$ **10.** $\dfrac{\frac{1}{b^2}}{\frac{a}{b}}$ **11.** $\dfrac{\frac{y}{2x}}{\frac{1}{3x^2}}$ **12.** $\dfrac{\frac{2x}{5y}}{\frac{1}{3x}}$

B

13. $\dfrac{1 + \dfrac{3}{x}}{x - \dfrac{9}{x}}$

14. $\dfrac{1 - \dfrac{2}{x}}{x - \dfrac{4}{x}}$

15. $\dfrac{1 - \dfrac{y^2}{x^2}}{1 - \dfrac{y}{x}}$

16. $\dfrac{\dfrac{a^2}{b^2} - 1}{\dfrac{a}{b} - 1}$

17. $\dfrac{\dfrac{1}{x} + \dfrac{1}{y}}{\dfrac{y}{x} - \dfrac{x}{y}}$

18. $\dfrac{b - \dfrac{a^2}{b}}{\dfrac{1}{a} - \dfrac{1}{b}}$

19. $\dfrac{\dfrac{x}{y} - 2 + \dfrac{y}{x}}{\dfrac{x}{y} - \dfrac{y}{x}}$

20. $\dfrac{1 + \dfrac{2}{x} - \dfrac{15}{x^2}}{1 + \dfrac{4}{x} - \dfrac{5}{x^2}}$

21. $\dfrac{\dfrac{a^2}{a-b} - a}{\dfrac{b^2}{a-b} + b}$

22. $\dfrac{n - \dfrac{n^2}{n-m}}{1 + \dfrac{m^2}{n^2 - m^2}}$

23. $\dfrac{\dfrac{m}{m+2} - \dfrac{m}{m-2}}{\dfrac{m+2}{m-2} - \dfrac{m-2}{m+2}}$

24. $\dfrac{\dfrac{y}{x+y} - \dfrac{x}{x-y}}{\dfrac{x}{x+y} + \dfrac{y}{x-y}}$

C

25. $1 - \dfrac{1}{1 - \dfrac{1}{x}}$

26. $2 - \dfrac{1}{1 - \dfrac{2}{x+2}}$

27. $1 - \dfrac{x - \dfrac{1}{x}}{1 - \dfrac{1}{x}}$

28. $\dfrac{t - \dfrac{1}{1 + \dfrac{1}{t}}}{t + \dfrac{1}{t - \dfrac{1}{t}}}$

29. $1 + \dfrac{1}{1 + \dfrac{1}{1 + \dfrac{1}{1+x}}}$

30. $1 - \dfrac{1}{1 - \dfrac{1}{1 - \dfrac{1}{1-x}}}$

APPLICATIONS

31. A formula for the average rate r for a round-trip between two points, where the rate going is r_G and the rate returning is r_R, is given by the complex fraction

$$r = \frac{2}{\dfrac{1}{r_G} + \dfrac{1}{r_R}}$$

Express r as a simple fraction.

32. The airspeed indicator on a jet aircraft registers 500 miles per hour. If the plane is traveling with an airstream moving at 100 miles per hour, then the plane's ground speed would be 600 miles per hour—or would it? According to Einstein, velocities must be added according to the following formula:

$$v = \frac{v_1 + v_2}{1 + \dfrac{v_1 v_2}{c^2}}$$

where v is the resultant velocity, c is the speed of light, and v_1 and v_2 are the two velocities to be added. Convert the right side of the equation into a simple fraction.

3-6 SOLVING EQUATIONS

- Equations Involving Fractions: No Variables in Denominators
- Equations Involving Fractions: Variables in Some Denominators

Recall the strategy for solving equations given in Section 1-6:

Equation-Solving Strategy

1. Use multiplication to remove fractions if present.
2. Simplify the left and right sides of the equation by removing grouping symbols and combining like terms.
3. Use equality properties to get all variable terms on one side (usually the left) and all constant terms on the other side (usually the right). Combine like terms in the process.
4. Isolate the variable (with a coefficient of 1), using the division or multiplication property of equality.

In this section we apply the strategy to equations involving fractions.

EQUATIONS INVOLVING FRACTIONS: NO VARIABLES IN DENOMINATORS

To solve equations involving fractions with no variables in denominators, we can start by using the multiplication property of equality to clear the fractions and then proceed as we did for equations with integer coefficients. What do we multiply both sides by to clear the fractions? We use any common multiple, preferably the LCM, of all denominators present in the equation.

Example 14 Solve: $\frac{x}{3} - \frac{1}{2} = \frac{5}{6}$

Solution

$$\frac{x}{3} - \frac{1}{2} = \frac{5}{6}$$

Clear fractions by multiplying both sides by 6, the LCM of all the denominators.

$$\mathbf{6} \cdot \left(\frac{x}{3} - \frac{1}{2}\right) = \mathbf{6} \cdot \frac{5}{6}$$

Clear parentheses.

$$\mathbf{6} \cdot \frac{x}{3} - \mathbf{6} \cdot \frac{1}{2} = \mathbf{6} \cdot \frac{5}{6}$$

$$2x - 3 = 5$$

The equation is now free of fractions.

$$2x = 8$$
$$x = 4$$

Problem 14 Solve: $\frac{1}{4}x - \frac{2}{3} = \frac{5}{12}x$ *Note:* $\frac{1}{4}x = \frac{x}{4}$ and $\frac{5}{12}x = \frac{5x}{12}$

Example 15 Solve: $0.2x + 0.3(x - 5) = 13$

Solution Some equations involving decimal-fraction coefficients are readily solved by first clearing decimals:

$$0.2x + 0.3(x - 5) = 13$$

Multiply by 10 to clear decimals.

$$\mathbf{10}(0.2x) + \mathbf{10}[0.3(x - 5)] = \mathbf{10} \cdot 13$$

$$2x + 3(x - 5) = 130$$
$$2x + 3x - 15 = 130$$
$$5x = 145$$
$$x = 29$$

Problem 15 Solve: $0.3(x + 2) + 0.5x = 3$

Example 16 Solve: $5 - \frac{2x - 1}{4} = \frac{x + 2}{3}$

Solution Before multiplying both sides by 12, the LCM of the denominators, enclose any numerator with more than one term in parentheses:

$$5 - \frac{(2x - 1)}{4} = \frac{(x + 2)}{3}$$

Multiply both sides by 12.

$$12 \cdot 5 - \overset{3}{\cancel{12}} \cdot \frac{(2x - 1)}{\underset{1}{\cancel{4}}} = \overset{4}{\cancel{12}} \cdot \frac{(x + 2)}{\underset{1}{\cancel{3}}}$$

12 is exactly divisible by each denominator.

$$
\begin{aligned}
60 - 3(2x - 1) &= 4(x + 2) \\
60 - 6x + 3 &= 4x + 8 \\
-6x + 63 &= 4x + 8 \\
-10x &= -55 \\
x &= \tfrac{11}{2} \quad \text{or} \quad 5.5
\end{aligned}
$$

Problem 16 Solve: $\dfrac{x+3}{4} - \dfrac{x-4}{2} = \dfrac{3}{8}$

A very common error occurs about now—students tend to confuse *algebraic expressions* involving fractions with *algebraic equations* involving fractions. Consider the two problems:

(A) Solve: $\dfrac{x}{2} + \dfrac{x}{3} = 10$ **(B)** Add: $\dfrac{x}{2} + \dfrac{x}{3} + 10$

The problems look very much alike but are actually very different. To solve the equation in (A) we multiply both sides by 6 (the LCM of 2 and 3) to clear the fractions. This works so well for equations, students want to do the same thing for problems like (B). The only catch is that (B) is not an equation and the multiplication property of equality does not apply. If we multiply (B) by 6, we obtain an expression 6 times as large as the original. To add in (B) we find the LCD and proceed as in Section 3-3.

Compare the following:

(A) $\dfrac{x}{2} + \dfrac{x}{3} = 10$

$$\boxed{\mathbf{6} \cdot \frac{x}{2} + \mathbf{6} \cdot \frac{x}{3} = \mathbf{6} \cdot 10}$$

$$
\begin{aligned}
3x + 2x &= 60 \\
5x &= 60 \\
x &= 12
\end{aligned}
$$

(B) $\dfrac{x}{2} + \dfrac{x}{3} + 10$

$$\boxed{= \frac{\mathbf{3} \cdot x}{\mathbf{3} \cdot 2} + \frac{\mathbf{2} \cdot x}{\mathbf{2} \cdot 3} + \frac{\mathbf{6} \cdot 10}{\mathbf{6} \cdot 1}}$$

$$= \frac{3x}{6} + \frac{2x}{6} + \frac{60}{6}$$

$$= \frac{5x + 60}{6}$$

EQUATIONS INVOLVING FRACTIONS: VARIABLES IN SOME DENOMINATORS

If an equation involves a variable in one or more denominators, such as

$$\frac{2}{3} - \frac{2}{x} = \frac{4}{x}$$

we may proceed in essentially the same way as above,

but we must avoid any value of x that makes a denominator 0.

Example 17 Solve: $\frac{2}{3} - \frac{2}{x} = \frac{4}{x}$

Solution

$$\frac{2}{3} - \frac{2}{x} = \frac{4}{x} \qquad x \neq 0$$

We note that $x \neq 0$, then multiply both sides by $3x$, the LCM of the denominators. If 0 turns up later as a "solution," it must be discarded.

$$\mathbf{3x} \cdot \frac{2}{3} - \mathbf{3x} \cdot \frac{2}{x} = \mathbf{3x} \cdot \frac{4}{x}$$

$3x$ is exactly divisible by each denominator.

$$2x - 6 = 12$$
$$2x = 18$$
$$x = 9$$

Problem 17 Solve: $\frac{3}{x} - \frac{1}{2} = \frac{4}{x}$

Example 18 Solve: $\frac{3x}{x-2} - 4 = \frac{14 - 4x}{x - 2}$

Solution

$$\frac{3x}{x-2} - 4 = \frac{14 - 4x}{x - 2} \qquad x \neq 2$$

If 2 turns up as a "solution," it must be discarded.

$$\mathbf{(x-2)}\frac{3x}{(x-2)} - 4\mathbf{(x-2)} = \mathbf{(x-2)}\frac{(14-4x)}{(x-2)}$$

Multiply by $(x - 2)$, the LCM of the denominators. Also place all binomial numerators and denominators in parentheses.

$$3x - 4(x - 2) = 14 - 4x$$
$$3x - 4x + 8 = 14 - 4x$$
$$-x + 8 = 14 - 4x$$
$$3x = 6$$
$$x = 2$$

2 cannot be a solution to the original equation (see comments above).

Since $x = 2$ cannot be a solution, the equation has no solution. (Hence, the solution set is empty.)

Problem 18 Solve: $\frac{2x}{x-1} - 3 = \frac{7 - 3x}{x - 1}$

Example 19 Solve: $2 - \frac{3x}{1-x} = \frac{8}{x-1}$

Solution

$$2 - \frac{3x}{1-x} = \frac{8}{x-1} \qquad \textit{Recall: } 1 - x = -(x-1)$$

$$2 - \frac{3x}{-(x-1)} = \frac{8}{x-1} \qquad \textit{Recall: } -\frac{a}{-b} = \frac{a}{b}$$

$$2 + \frac{3x}{x-1} = \frac{8}{x-1}$$

Multiply both sides by $(x - 1)$, keeping in mind that $x \neq 1$.

$$(x-1)(2) + (x-1)\left(\frac{3x}{x-1}\right) = (x-1)\left(\frac{8}{x-1}\right)$$

$$2x - 2 + 3x = 8$$
$$5x = 10$$
$$x = 2$$

Problem 19 Solve: $\frac{5x}{x-2} + \frac{10}{2-x} = 3$

ANSWERS TO MATCHED PROBLEMS

14. $x = -4$
15. $x = 3$ **16.** $x = \frac{19}{2}$ or 9.5
17. $x = -2$ **18.** $x = 2$
19. No solution

EXERCISE 3-6 *Solve.*

A

1. $\frac{x}{5} - 2 = \frac{3}{5}$ **2.** $\frac{x}{7} - 1 = \frac{1}{7}$ **3.** $\frac{x}{3} + \frac{x}{6} = 4$

4. $\frac{y}{4} + \frac{y}{2} = 9$ **5.** $\frac{m}{4} - \frac{m}{3} = \frac{1}{2}$ **6.** $\frac{n}{5} - \frac{n}{6} = \frac{6}{5}$

7. $\frac{5}{12} - \frac{m}{3} = \frac{4}{9}$ **8.** $\frac{2}{3} - \frac{x}{8} = \frac{5}{6}$ **9.** $0.7x = 21$

10. $0.9x = 540$ **11.** $0.7x + 0.9x = 32$

12. $0.3x + 0.5x = 24$ **13.** $\frac{1}{2} - \frac{2}{x} = \frac{3}{x}$

14. $\frac{2}{x} - \frac{1}{3} = \frac{5}{x}$ **15.** $\frac{1}{m} - \frac{1}{9} = \frac{4}{9} - \frac{2}{3m}$

16. $\frac{1}{2t} + \frac{1}{8} - \frac{2}{t} - \frac{1}{4}$ **17.** $\frac{x-2}{3} + 1 = \frac{x}{7}$

18. $\frac{x+3}{2} - \frac{x}{3} = 4$

19. $\frac{2x-3}{9} - \frac{x+5}{6} = \frac{3-x}{2} - 1$

20. $\frac{3x+4}{3} - \frac{x-2}{5} = \frac{2-x}{15} - 1$

21. $0.1(x-7) + 0.05x = 0.8$

22. $0.4(x+5) - 0.3x = 17$

23. $0.02x - 0.5(x-2) = 5.32$

24. $0.3x - 0.04(x+1) = 2.04$

B **25.** $\frac{7}{y-2} - \frac{1}{2} = 3$

26. $\frac{9}{A+1} - 1 = \frac{12}{A+1}$

27. $\frac{3}{2x-1} + 4 = \frac{6x}{2x-1}$

28. $\frac{5x}{x+5} = 2 - \frac{25}{x+5}$

29. $\frac{2E}{E-1} = 2 + \frac{5}{2E}$

30. $\frac{3N}{N-2} - \frac{9}{4N} = 3$

31. $\frac{n-5}{6n-6} = \frac{1}{9} - \frac{n-3}{4n-4}$

32. $\frac{1}{3} - \frac{s-2}{2s+4} = \frac{s+2}{3s+6}$

33. $5 + \frac{2x}{x-3} = \frac{6}{x-3}$

34. $\frac{6}{x-2} = 3 + \frac{3x}{x-2}$

35. $\frac{x^2+2}{x^2-4} = \frac{x}{x-2}$

36. $\frac{5}{x-3} = \frac{33-x}{x^2-6x+9}$

C **37.** $\frac{3x}{24} - \frac{2-x}{10} = \frac{5+x}{40} - \frac{1}{15}$

38. $\frac{2x}{10} - \frac{3-x}{14} = \frac{2+x}{5} - \frac{1}{2}$

39. $\frac{5t-22}{t^2-6t+9} - \frac{11}{t^2-3t} - \frac{5}{t} = 0$

40. $\frac{x-33}{x^2-6x+9} + \frac{5}{x-3} = 0$

41. $5 - \frac{2x}{3-x} = \frac{6}{x-3}$

42. $\frac{3x}{2-x} + \frac{6}{x-2} = 3$

43. $\frac{1}{c^2-c-2} - \frac{3}{c^2-2c-3} = \frac{1}{c^2-5c+6}$

44. $\frac{5t-22}{t^2-6t+9} - \frac{11}{t^2-3t} = \frac{5}{t}$

45. $\frac{x+3}{(x-1)(x-2)} + \frac{x+1}{(x-2)(x-3)} = \frac{2}{(x-1)(x-3)}$

46. $\frac{3}{x^2+x} - \frac{2}{x^2-x} = \frac{1}{x^2-1}$

47. $\dfrac{1}{x} - \dfrac{1}{x^2} = \dfrac{12}{x^3}$

48. $\dfrac{3}{x^2 + 2x + 1} - \dfrac{5}{x + 1} = 2$

49. $\dfrac{1}{(x - 1)^2} + \dfrac{1}{(x + 1)^2} = \dfrac{2}{x^2 - 1}$

50. $\dfrac{1}{x^2 + 3x + 2} + \dfrac{1}{x^2 - 3x + 2} = \dfrac{2}{x^2 - x - 2}$

3-7 CHAPTER REVIEW

A **rational expression** is an algebraic expression involving only the operations of addition, subtraction, multiplication, and division on variables and constants. A rational expression A/B can be **reduced to lowest terms** or **raised to higher terms** by using the **fundamental principle of fractions**:

$$\frac{A}{B} = \frac{AK}{BK} \quad \textit{(3-1)}$$

Rational expressions are multiplied and divided as follows:

$$\frac{A}{B} \cdot \frac{C}{D} = \frac{A \cdot C}{B \cdot D} \qquad \frac{A}{B} \div \frac{C}{D} = \frac{A}{B} \cdot \frac{D}{C} = \frac{A \cdot D}{B \cdot C} \quad \textit{(3-2)}$$

Rational expressions with a common denominator are added and subtracted as follows:

$$\frac{A}{D} + \frac{B}{D} = \frac{A + B}{D} \qquad \frac{A}{D} - \frac{B}{D} = \frac{A - B}{D}$$

Rational expressions with different denominators are converted to ones with a common denominator and then added or subtracted. *(3-3)*

A polynomial P can be divided by a polynomial **divisor** D by using **algebraic long division** to yield a **quotient** Q and **remainder** R so that $P = D \cdot Q + R$ and either R is 0 or the degree of R is smaller than that of D. When the divisor is of the form $x - r$, the remainder is the value of the polynomial P at $x = r$. In this case, the process can be done quickly by **synthetic division**. *(3-4)*

A **complex fraction** is a fraction in which the numerator or denominator contains a fraction. The fundamental principle of fractions and the operations of addition, subtraction, multiplication, and division are used to convert complex fractions to **simple fractions**—that is, to quotients of two polynomials. *(3-5)*

Equations involving fractions are solved by clearing fractions and then proceeding as usual on the resulting equation with integer coefficients. *(3-6)*

REVIEW EXERCISE 3-7

Work through all the problems in this chapter review and check answers in the back of the book. (Answers to all problems are there, and following each answer is a number in italics indicating the section in which that type of problem is discussed.) Where weaknesses show up, review appropriate sections in the text.

A *Perform the indicated operation and simplify. Express each answer as a simple fraction in lowest terms.*

1. $\dfrac{18x^3y^2(z+3)}{12xy^2(z+3)^2}$ **2.** $\dfrac{x^2+2x+1}{x^2-1}$ **3.** $1+\dfrac{2}{3x}$

4. $\dfrac{2}{x}-\dfrac{1}{6x}+\dfrac{1}{3}$ **5.** $\dfrac{1}{6x^3}-\dfrac{3}{4x}-\dfrac{2}{3}$ **6.** $\dfrac{4x^2y^3}{3a^2b^2}\div\dfrac{2xy^2}{3ab}$

7. $\dfrac{6x^2}{3(x-1)}-\dfrac{6}{3(x-1)}$ **8.** $1-\dfrac{m-1}{m+1}$

9. $\dfrac{3}{x-2}-\dfrac{2}{x+1}$ **10.** $(d-2)^2\div\dfrac{d^2-4}{d-2}$

11. $\dfrac{x+1}{x+2}-\dfrac{x+2}{x+3}$ **12.** $\dfrac{\frac{1}{4}}{\frac{2}{3}}$

13. $\dfrac{2\frac{3}{4}}{1\frac{1}{2}}$ **14.** $\dfrac{1-\dfrac{2}{y}}{1+\dfrac{1}{y}}$

Solve.

15. $0.4x+0.3x=6.3$ **16.** $-\frac{3}{5}y=\frac{2}{3}$

17. $\dfrac{x}{4}-3=\dfrac{x}{5}$

B **18.** $\dfrac{x}{4}-\dfrac{x-3}{3}=2$ **19.** $\dfrac{2}{3m}-\dfrac{1}{4m}=\dfrac{1}{12}$

20. $\dfrac{3x}{x-5}-8=\dfrac{15}{x-5}$ **21.** $0.05n+0.1(n-3)=1.35$

22. $\dfrac{5}{2x+3} - 5 = \dfrac{-5x}{2x+3}$

23. $\dfrac{3}{x} - \dfrac{2}{x+1} = \dfrac{1}{2x}$

24. $\dfrac{11}{9x} - \dfrac{1}{6x^2} = \dfrac{3}{2x}$

25. $\dfrac{u-3}{2u-2} = \dfrac{1}{6} - \dfrac{1-u}{3u-3}$

26. $\dfrac{x}{x^2-6x+9} - \dfrac{1}{x^2-9} = \dfrac{1}{x+3}$

Divide to find the quotient and remainder.

27. $(x^3 - 3x^2 + x - 3) \div (x - 1)$

28. $(x^3 + x) \div (x^2 + 1)$

29. $(x^4 + 2x^3 + 3x^2 + 4x + 5) \div (x^2 + 2)$

30. $(x^4 + x^2 - 1) \div (x + 2)$

31. $(x^4 - 1) \div (x - 1)$

32. $(x^4 + x + x^3) \div (1 + x + x^2)$

Perform the indicated operation and simplify. Express each answer as a simple fraction in lowest terms.

33. $\dfrac{2}{5b} - \dfrac{4}{3b^3} - \dfrac{1}{6a^2b^2}$

34. $\dfrac{2}{2x-3} - 1$

35. $\dfrac{4x^2y}{3ab^2} \div \left(\dfrac{2a^2x^2}{b^2y} \cdot \dfrac{6a}{2y^2}\right)$

36. $\dfrac{x}{x^2+4x} + \dfrac{2x}{3x^2-48}$

37. $\dfrac{x^3-x}{x^2-x} \div \dfrac{x^2+2x+1}{x}$

38. $\dfrac{\dfrac{x}{y} - \dfrac{y}{x}}{\dfrac{x}{y} + 1}$

39. $\dfrac{x}{x^3-y^3} - \dfrac{1}{x^2+xy+y^2}$

40. $\dfrac{\dfrac{y^2}{x^2-y^2} + 1}{\dfrac{x^2}{x-y} - x}$

41. $\dfrac{x^3-1}{x^2+x+1} \div \dfrac{x^2-1}{x^2+2x+1}$

42. $\dfrac{1}{3x^2-27} - \dfrac{x-1}{4x^3+24x^2+36x}$

C **43.** $\dfrac{4}{s^2 - 4} + \dfrac{1}{2 - s}$

44. $\dfrac{y^2 - y - 6}{(y + 2)^2} \cdot \dfrac{2 + y}{3 - y}$

45. $\dfrac{y}{x^2} \div \left(\dfrac{x^2 + 3x}{2x^2 + 5x - 3} \div \dfrac{x^3y - x^2y}{2x^2 - 3x + 1}\right)$

46. $\dfrac{1 - \dfrac{1}{1 + \dfrac{x}{y}}}{1 - \dfrac{1}{1 - \dfrac{x}{y}}}$

47. $\left(x - \dfrac{1}{1 - \dfrac{1}{x}}\right) \div \left(\dfrac{x}{x + 1} - \dfrac{x}{1 - x}\right)$

Solve.

48. $\dfrac{x - 3}{12} - \dfrac{x + 2}{9} = \dfrac{1 - x}{6} - 1$

49. $\dfrac{7}{2 - x} = \dfrac{10 - 4x}{x^2 + 3x - 10}$

50. $\dfrac{1}{x^2} + \dfrac{1}{(x + 1)^2} = \dfrac{2}{x^2 + x}$

51. $\dfrac{1}{x^2 + x} + \dfrac{1}{x^2 + 2x + 1} = \dfrac{2x + 1}{x(x + 1)^2}$

4

LINEAR EQUATIONS AND INEQUALITIES IN ONE VARIABLE

We have learned how to solve linear equations in Chapter 1, equations with a factorable polynomial equal to 0 in Chapter 2, and equations involving fractions in Chapter 3. We have also used these techniques to solve applied problems. In this chapter, we will deal with a wider variety of applications that require equation solving, and then develop techniques for solving linear inequalities similar to those already developed for linear equations.

4-1 SOLVING EQUATIONS AND APPLICATIONS

- Geometric Problems
- Ratio and Proportion Problems

Recall the strategy we have developed for solving equations:

Equation-Solving Strategy

1. Clear fractions.
2. Simplify both sides.
3. Get all variable terms to one side, constants to the other, and combine like terms.
4. Isolate the variable.

Recall also a general strategy for solving word problems:

A Strategy for Solving Word Problems

1. Read the problem carefully—several times if necessary—until you understand the problem, know what is to be found, and know what is given.
2. If appropriate, draw figures or diagrams and label known and unknown parts. Look for formulas connecting the known quantities with the unknown quantities.
3. Let one of the unknown quantities be represented by a variable, say x, and try to represent all other unknown quantities in terms of x. This is an important step and must be done carefully. Be sure you clearly understand what you are letting x represent.
4. Form an equation relating the unknown quantities with the known quantities. This step may involve the translation of an English sentence into an algebraic sentence, the use of relationships in a geometric figure, the use of certain formulas, and so on.
5. Solve the equation and write answers to *all* parts of the problem requested.
6. Check all solutions in the original problem.

In this section we will use these strategies to solve geometric problems and problems involving ratio and proportion.

GEOMETRIC PROBLEMS

Recall that the **perimeter of a triangle or rectangle** is the distance around the figure. Symbolically:

TRIANGLE

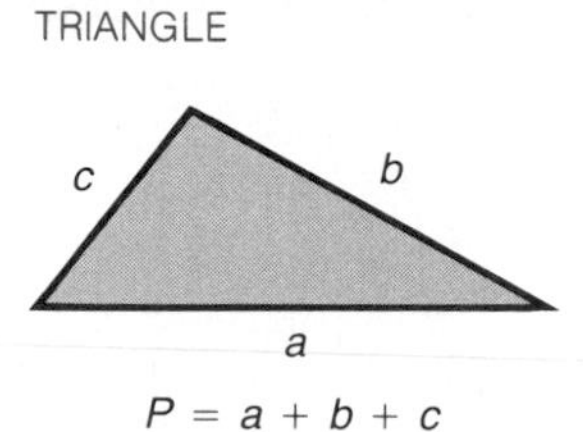

$P = a + b + c$

RECTANGLE

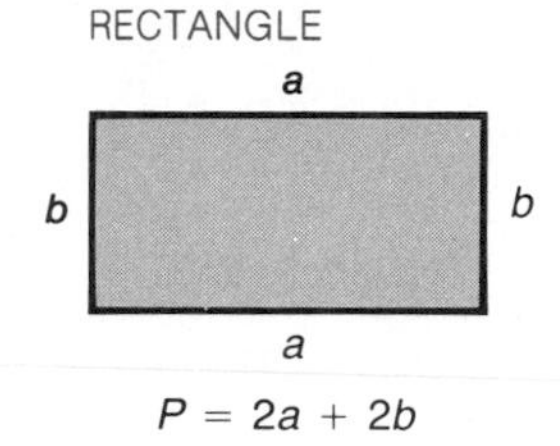

$P = 2a + 2b$

Example 1 If one side of a triangle is one-fourth the perimeter, the second side is 7 meters, and the third side is two-fifths the perimeter, what is the perimeter?

Solution Let

P = Perimeter

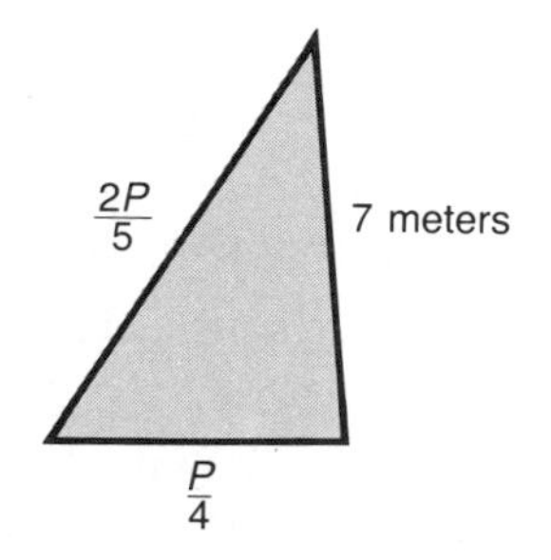

Draw a triangle and label sides, as shown. Thus,

$$P = a + b + c$$

$$P = \frac{P}{4} + 7 + \frac{2P}{5}$$

$$\mathbf{20} \cdot P = \mathbf{20} \cdot \frac{P}{4} + \mathbf{20} \cdot 7 + \mathbf{20} \cdot \frac{2P}{5} \quad \text{Clear fractions.}$$

$$20P = 5P + 140 + 8P$$
$$7P = 140$$
$$P = 20 \text{ meters}$$

Check

$$\text{Side 1} = \frac{P}{4} = \frac{20}{4} = 5 \text{ meters}$$

$$\text{Side 2} = 7 \text{ meters}$$

$$\text{Side 3} = \frac{2P}{5} = \frac{2 \cdot 20}{5} = 8 \text{ meters}$$

20 meters Perimeter

Problem 1 If one side of a triangle is one-third the perimeter, the second side is 7 centimeters, and the third side is one-fifth the perimeter, what is the perimeter of the triangle? Set up an equation and solve.

Example 2 Find the dimensions of a rectangle with perimeter 84 centimeters if its width is two-fifths the length.

Solution Draw a rectangle and label sides. If x = Length, then $\frac{2}{5}x$ = Width.
Begin with the formula for the perimeter of a rectangle:

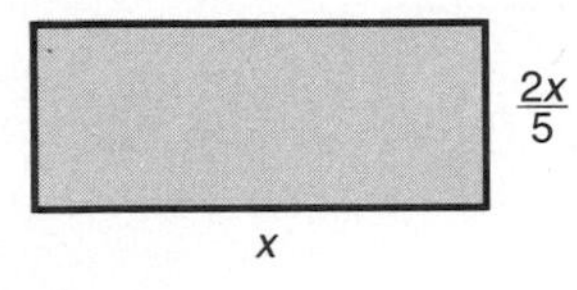

$$2a + 2b = P$$

$$2x + 2 \cdot \frac{2x}{5} = 84$$

$$2x + \frac{4x}{5} = 84$$

$$\mathbf{5} \cdot 2x + \mathbf{5} \cdot \frac{4x}{5} = \mathbf{5} \cdot 84 \qquad \text{Clear fractions.}$$

$$10x + 4x = 420$$
$$14x = 420$$
$$x = 30 \text{ centimeters} \qquad \text{Length}$$
$$\frac{2x}{5} = 12 \text{ centimeters} \qquad \text{Width}$$

Checking is left to you.

Problem 2 Find the dimensions of a rectangle with perimeter 176 centimeters if its width is three-eighths its length. Write an equation and solve.

RATIO AND PROPORTION PROBLEMS

The ratio of two quantities is the first quantity divided by the second quantity. Symbolically:

The Ratio of *a* to *b*

The ratio of a to b, $b \neq 0$, is $\frac{a}{b}$

Example 3 If a parking meter has 45 nickels, 30 dimes, and 15 quarters, then the ratio of nickels to quarters is

$$\frac{45}{15} = \frac{3}{1}$$

(which is also written 3:1 or 3/1 and is read "3 to 1").

Problem 3 In Example 3 what is the ratio of quarters to nickels? Of dimes to nickels?

In addition to providing a way of comparing known quantities, ratios also provide a way of finding unknown quantities.

Example 4 Suppose you are told that the ratio of quarters to dimes in a parking meter is 3/5 and that there are 40 dimes in the meter. How many quarters are in the meter?

Solution Let

q = Number of quarters

Then the ratio of quarters to dimes is $q/40$. We also know that this ratio is 3/5. Thus,

$$\frac{q}{40} = \frac{3}{5} \quad \text{To isolate } q \text{, multiply both sides by 40.}$$

$$q = 40 \cdot \frac{3}{5}$$

$$= 24 \text{ quarters}$$

Problem 4 If the ratio of dimes to quarters in a meter is 3/2 and there are 24 quarters in the meter, how many dimes are there?

A statement of equality between two ratios, as in Example 4, is called a **proportion**; that is,

A Proportion

$$\frac{a}{b} = \frac{c}{d} \qquad b, d \neq 0$$

Example 5 If a car can travel 192 kilometers on 32 liters of gas, how far will it go on 60 liters?

Solution Let

x = Distance traveled on 60 liters

Then

$$\frac{x}{60} = \frac{192}{32} \qquad \frac{\text{km}}{\text{L}} = \frac{\text{km}}{\text{L}} \quad \text{(kilometers per liter)}$$

$$x = 60 \cdot \frac{192}{32}$$ We isolate x by multiplying both sides by 60—we do not need to use the LCM of 60 and 32.

$$= 360 \text{ kilometers}$$

Problem 5 If there are 24 milliliters of sulfuric acid in 64 milliliters of solution, how many milliliters are in 48 milliliters of the same solution? Set up a proportion and solve.

We will show how the concept of proportion can be used to convert metric units to English units, and vice versa. (A summary of metric units is located inside the back cover of the text.)

Example 6 If there is 0.45 kilogram in 1 pound, how many pounds are in 90 kilograms?

Solution Let x = Number of pounds in 90 kilograms. Set up a proportion (preferably with x in the numerator on the left side, since you have a choice). That is, set up a proportion of the form

$$\frac{\text{Pounds}}{\text{Kilograms}} = \frac{\text{Pounds}}{\text{Kilograms}}$$ Each ratio represents pounds per kilogram.

Thus,

$$\frac{x}{90} = \frac{1}{0.45}$$

$$x = 90 \cdot \frac{1}{0.45}$$

$$= 200 \text{ pounds}$$

Problem 6 If there are 2.2 pounds in 1 kilogram, how many kilograms are in 100 pounds? Set up a proportion and solve to two decimal places.

Example 7 If there are 3.76 liters in 1 gallon, how many gallons are in 50 liters? Set up a proportion (with the variable in the numerator on the left side) and solve to two decimal places.

Solution Let x = Number of gallons in 50 liters. We set up a proportion of the form

$$\frac{\text{Gallons}}{\text{Liters}} = \frac{\text{Gallons}}{\text{Liters}}$$ Each side gives gallons per liter.

Thus,

$$\frac{x}{50} = \frac{1}{3.76}$$

$$x = 50 \cdot \frac{1}{3.76}$$

$$= 13.30 \text{ gallons}$$

Problem 7 If there are 1.09 yards in 1 meter, how many meters are in 80 yards? Set up a proportion and solve to two decimal places.

If you are having trouble with word problems (many people do), return to the worked-out examples. Cover the solutions, proceed with your own solution until you get stuck, and then uncover only enough of the solution to get you started again. After completing an example in this way, immediately work the matched problem following the example and then work similar problems in the exercise set. A technique introduced in Appendix A may also be helpful.

ANSWERS TO MATCHED PROBLEMS

1. $P = \frac{P}{3} + 7 + \frac{P}{5}$; $P = 15$ centimeters

2. $2x + 2 \cdot \frac{3x}{8} = 176$; 64 by 24 centimeters

3. 1/3, 2/3 **4.** 36 dimes **5.** $\frac{x}{48} = \frac{24}{64}$; $x = 18$ milliliters

6. $\frac{x}{100} = \frac{1}{2.2}$; $x = 45.45$ kilograms

7. $\frac{x}{80} = \frac{1}{1.09}$; $x = 73.39$ meters

EXERCISE 4-1 A

Write as a ratio.

1. 33 dimes to 22 nickels

2. 17 quarters to 51 dimes

3. 25 centimeters to 10 centimeters

4. 30 meters to 18 meters

5. 300 kilometers to 24 liters

6. 320 miles to 12 gallons

Solve each proportion.

7. $\frac{m}{16} = \frac{5}{4}$ **8.** $\frac{n}{12} = \frac{2}{3}$ **9.** $\frac{x}{13} = \frac{21}{39}$ **10.** $\frac{x}{12} = \frac{27}{18}$

Set up appropriate equations and solve.

11. A 12-foot steel rod is cut into two pieces so that one piece is 3 feet less than twice the length of the other piece. How long is each piece?

12. A 32-centimeter string is cut into two pieces so that one piece is 4 centimeters more than 3 times the length of the other piece. How long is each piece?

13. Find the dimensions of a rectangle with perimeter 36 feet if the width is 6 feet shorter than the length.

14. Find the dimensions of a rectangle with perimeter 54 meters if the length is 7 meters longer than the width.

Set up appropriate proportions and solve. Compute decimal answers to two decimal places.

15. If in a pay telephone the ratio of quarters to dimes is 5/8 and there are 96 dimes, how many quarters are there?

16. If in a parking meter the ratio of pennies to nickels is 13/6 and there are 78 nickels, how many pennies are there?

17. If the ratio of the length of a rectangle to its width is 5/3 and its width is 24 meters, how long is it?

18. If the ratio of the width of a rectangle to its length is 4/7 and its length is 56 centimeters, how wide is it?

19. If a car can travel 108 kilometers on 12 liters of gas, how far will it go on 18 liters?

20. If a boat can travel 72 miles on 18 gallons of diesel fuel, how far will it travel on 15 gallons?

B *The following number, geometric, and ratio and proportion problems are mixed up. (Number problems were first introduced in Section 1-6.) Set up appropriate equations and solve. Compute decimal answers to two decimal places.*

21. Find the dimensions of a rectangle with perimeter 66 centimeters if its length is 3 centimeters more than twice its width.

22. Find the dimensions of a rectangle with perimeter 128 meters if its length is 6 meters less than 4 times the width.

23. Find a number such that 2 less than one-sixth the number is 1 more than one-fourth the number.

24. Find a number such that 5 less than half the number is 3 more than one-third the number.

25. If there are 9 milliliters of hydrochloric acid in 46 milliliters of solution, how many milliliters will be in 52 milliliters of solution?

26. If 0.75 cup of flour is needed in a recipe that will feed 6 people, how much flour will be needed in the recipe that will feed 9 people?

27. Find three consecutive odd numbers such that the sum of the first and second is 5 more than the third.

28. Find three consecutive odd numbers such that the sum of the second and third is 1 more than 3 times the first.

29. If there are 1.06 quarts in 1 liter, how many liters are in 1 gallon (4 quarts)?

30. If there are 2.2 pounds in 1 kilogram, how many kilograms are in 10 pounds?

31. If there is 0.62 mile in 1 kilometer, how many kilometers are in 1 mile?

32. If there is 0.91 meter in 1 yard, how many yards are in 1 meter?

33. Find the dimensions of a rectangle with perimeter 84 meters if its width is one-sixth its length.

34. Find the dimensions of a rectangle with perimeter 72 centimeters if its width is one-third its length.

35. If a commission of $240 is charged on the purchase of 200 shares of a stock, how much commission would be charged for 500 shares of the same stock?

36. If the price/earning ratio of a common stock is 8.4 and the stock earns $23.50 per share, what is the price of the stock per share? [*Note:* Ratios are often written as decimal fractions. In this case 8.4 (that is, 8.4/1) is the ratio of price per share to earnings per share. Thus, if x is the price per share, we obtain the proportion $x/23.5 = 8.4$.]

37. A 35 by 23 millimeter colored slide is used to make an enlargement whose longer side is 10 inches. How wide will the enlargement be if all of the slide is used?

38. A 3.25 by 4.25 inch negative is used to produce an enlargement whose shortest side is 12 inches. How long will the enlargement be if all of the negative is used?

39. Find a number such that 4 less than three-fifths the number is 8 more than one-third the number?

40. Find a number such that 5 more than two-thirds the number is 10 less than one-fourth the number.

41. If there is 0.26 gallon in 1 liter, how many liters are in 5 gallons?

42. If there is 0.94 liter in 1 quart, how many quarts are in 10 liters?

C **43.** If one side of a triangle is two-fifths the perimeter P, the second side is 70 centimeters, and the third side is one-fourth the perimeter, what is the perimeter?

44. If one side of a triangle is one-fourth the perimeter P, the second side is 3 meters, and the third side is one-third the perimeter, what is the perimeter?

45. Estimate the total number of trout in a lake if a sample of 300 is netted, marked, and released, and after a suitable period for mixing, a second sample of 250 produces 25 marked trout. (Assume that the ratio of the marked trout in the second sample to the total number of the sample is the same as the ratio of those marked in the first sample to the total lake population.)

46. Repeat the last problem with a first (marked) sample of 400 and a second sample of 264 with only 24 marked trout.

47. On a trip across the Grand Canyon in Arizona, a group traveled one-third the distance by mule, 6 kilometers by boat, and one-half the distance by foot. How long was the trip?

48. A high diving tower is located in a lake. If one-fifth the height of the tower is in sand, 6 meters in water, and one-half the total height in air, what is the total height of the tower?

49. If in the figure the diameter of the smaller pipe is 12 millimeters and the diameter of the larger pipe is 24 centimeters, how much force would be required to lift a 1,200-kilogram car? (Neglect the weight of the hydraulic lift equipment and use the proportion shown in the figure.)

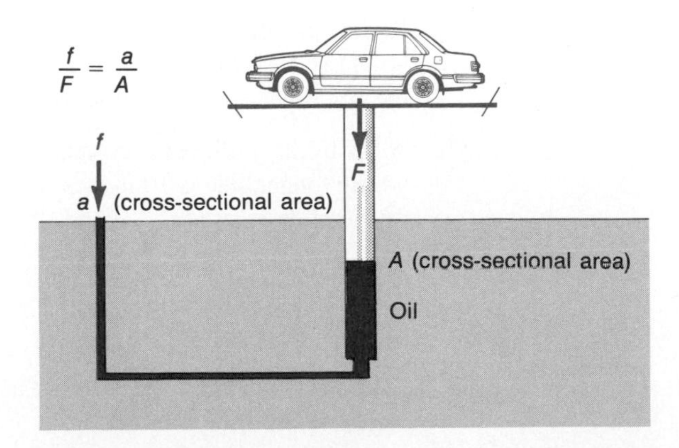

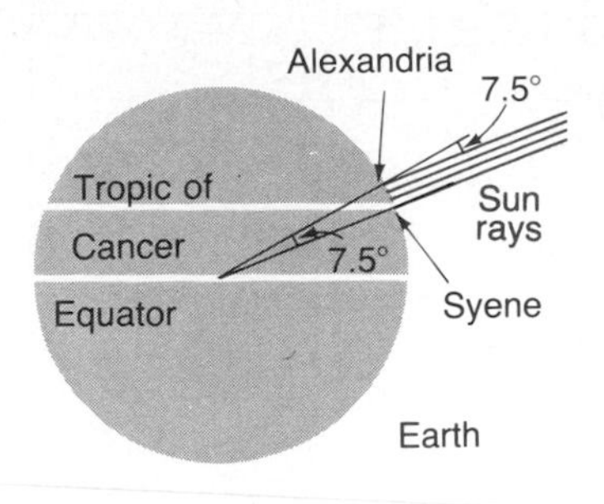

50. Do you have any idea how one might measure the circumference of the earth? In 240 B.C. Eratosthenes measured the size of the earth from its curvature. At Syene, Egypt (lying on the Tropic of Cancer), the sun was directly overhead at noon on June 21. At the same time in Alexandria, a town 500 miles directly north, the sun's rays fell at an angle of 7.5° to the vertical. Using this information and a little knowledge of geometry (see the figure), Eratosthenes was able to approximate the circumference of the earth using the following proportion: the circumference of the earth C is to 500 as 360 is to 7.5. Compute Eratosthenes' estimate.

4-2 APPLICATIONS: RATE–TIME AND MISCELLANEOUS PROBLEMS

- Rate–Time Problems
- Miscellaneous Problems

There are many types of rate–time problems in addition to the distance–rate–time problems with which you are probably familiar. In this section we will look at rate–time problems as a general class of problems that includes distance–rate–time problems as one of many special cases. We then consider some miscellaneous problems—problems where the structure is not suggested by a classification such as geometric problem, rate–time problem, etc.

RATE–TIME PROBLEMS

If a runner travels 20 kilometers in 2 hours, then the ratio

$$\frac{20 \text{ kilometers}}{2 \text{ hours}} \quad \text{or} \quad 10 \text{ kilometers per hour}$$

is called a **rate**. It is the number of kilometers produced (traveled) in each unit of time (each hour). Similarly, if an automatic bottling machine bottles 3,200 bottles of soft drinks in 20 minutes, then the ratio

$$\frac{3{,}200 \text{ bottles}}{20 \text{ minutes}} = 160 \text{ bottles per minute}$$

is also a rate. It is the number of bottles produced (filled) in each unit of time (each minute).

In general, if Q is the quantity of something produced (kilometers, words, parts, and so on) in T units of time (hours, years, minutes, seconds, and so on) and R is the rate (quantity produced in one unit of time), then Q, T, and R are related by the following formula:

Quantity–Rate†–Time Formula

$$Q = RT \qquad \text{Quantity} = (\text{Rate})(\text{Time}) \tag{1}$$

If Q is distance D, then

$$D = RT \qquad \text{Distance} = (\text{Rate})(\text{Time}) \tag{2}$$

Formulas (1) and (2) should be memorized. [Formula (2) is simply a special case of (1).] Two other forms of Formulas (1) and (2) (one for rate R and the other for time T) are easily derived from (1) and (2) by dividing both sides of each by either T or R:

$$Q = RT \qquad\qquad D = RT$$

$$RT = Q \qquad\qquad RT = D$$

$$T = \frac{Q}{R} \quad R = \frac{Q}{T} \qquad T = \frac{D}{R} \quad R = \frac{D}{T}$$

Example 8 A gas pump in a service station can deliver 36 liters in 4 minutes.

(A) What is its rate (liters per minute)?
(B) How much gas can be delivered in 5 minutes?
(C) How long will it take to deliver 54 liters?

Solution **(A)** $R = \dfrac{Q}{T} = \dfrac{36 \text{ liters}}{4 \text{ minutes}} = 9$ liters per minute

(B) $Q = RT = (9 \text{ liters per minute})(5 \text{ minutes}) = 45$ liters

(C) $T = \dfrac{Q}{R} = \dfrac{54 \text{ liters}}{9 \text{ liters per minute}} = 6$ minutes

Problem 8 A woman jogs 18 kilometers in 2 hours.

(A) What is her rate (kilometers per hour)?
(B) How far will she jog in 1.5 hours?
(C) How long will it take her to jog 12 kilometers?

Example 9 A jet plane leaves San Francisco and travels at 650 kilometers per hour toward Los Angeles. At the same time another plane leaves Los Angeles and travels at 800 kilometers per hour toward San Francisco. If the cities are 570 kilometers apart, how long will it take the jets to meet, and how far from San Francisco will they be at that time?

† These are also referred to as **average rates**.

Solution Let T = Number of hours until both planes meet. Draw a diagram and label known and unknown parts. Both planes will have traveled the same amount of time when they meet.

$$\begin{pmatrix}\text{Distance plane}\\ \text{from SF travels}\\ \text{to meeting point}\end{pmatrix} + \begin{pmatrix}\text{Distance plane}\\ \text{from LA travels}\\ \text{to meeting point}\end{pmatrix} = \begin{pmatrix}\text{Total distance}\\ \text{from}\\ \text{SF to LA}\end{pmatrix}$$

$$D_1 + D_2 = 570$$

$$650T + 800T = 570$$

$$1{,}450T = 570$$

$$T = \frac{570}{1{,}450} \approx 0.39 \text{ hour}$$

$$\text{Distance from SF} = (\text{Rate from SF})(\text{Time from SF})$$
$$\approx (650)(0.39) = 253.5 \text{ kilometers}$$

You may prefer to summarize the given information in a table:

	RATE	TIME	DISTANCE
FROM SF	650	T	$650T$
FROM LA	800	T	$800T$
			570

The sum of the distances is 570, so we obtain the equation

$$650T + 800T = 570$$

as above.

Problem 9 If an older printing press can print 45 fliers per minute and a newer press can print 80, how long will it take both presses together to print 4,500 fliers? How many will the older press have printed by then? [Here $Q = RT$ takes the following form: the number of fliers printed (Q) = the number of fliers per minutes (R) times the number of minutes (T).]

Example 10 Find the total amount of time to print the fliers in Problem 9 if the newer press is brought on the job 10 minutes later than the older press and both continue until the job is completed.

Solution Let

$$x = \text{Time to complete whole job}$$

Then

$$\begin{aligned} x &= \text{Time old press is on the job} \\ x - 10 &= \text{Time new press is on the job} \end{aligned}$$

$$\begin{pmatrix}\text{Quantity}\\ \text{printed}\\ \text{by old press}\end{pmatrix} + \begin{pmatrix}\text{Quantity}\\ \text{printed}\\ \text{by new press}\end{pmatrix} = \text{Total needed}$$

Remember: Quantity = (Rate)(Time)

$$\begin{aligned} 45x \quad + \quad 80(x - 10) &= \quad 4{,}500 \\ 45x + 80x - 800 &= 4{,}500 \\ 125x &= 5{,}300 \\ x &= 42.4 \text{ minutes} \end{aligned}$$

Problem 10 A car leaves a city traveling at 60 kilometers per hour. How long will it take a second car traveling at 80 kilometers per hour to catch up to the first car if it leaves 2 hours later? Set up an equation and solve. [*Hint:* Both cars will have traveled the same distance when the second car catches up to the first.]

Example 11 A speedboat takes 1.5 times longer to go 120 miles up a river than to return. If the boat cruises at 25 miles per hour in still water, what is the rate of the current?

Solution Let

$$\begin{aligned} x &= \text{Rate of current in miles per hour} \\ 25 - x &= \text{Rate of boat upstream} \\ 25 + x &= \text{Rate of boat downstream} \end{aligned}$$

$$\begin{aligned} \text{Time upstream} &= (1.5)(\text{Time downstream}) \\ \frac{\text{Distance upstream}}{\text{Rate upstream}} &= (1.5)\frac{\text{Distance downstream}}{\text{Rate downstream}} \\ \frac{120}{25 - x} &= (1.5)\frac{120}{25 + x} \\ \frac{120}{25 - x} &= \frac{180}{25 + x} \\ (25 + x)120 &= (25 - x)180 \\ 3{,}000 + 120x &= 4{,}500 - 180x \\ 300x &= 1{,}500 \\ x &= 5 \text{ miles per hour} \end{aligned}$$

Recall that $T = D/R$ from $D = RT$.

Rate of current

Check

$$\text{Time upstream} = \frac{\text{Distance upstream}}{\text{Rate upstream}} = \frac{120}{20} = 6 \text{ hours}$$

$$\text{Time downstream} = \frac{\text{Distance downstream}}{\text{Rate downstream}} = \frac{120}{30} = 4 \text{ hours}$$

Thus, time upstream is 1.5 times longer than time downstream.

Problem 11 A fishing boat takes twice as long to go 24 miles up a river than to return. If the boat cruises at 9 miles per hour in still water, what is the rate of the current?

Example 12 An advertising company has an automated mailing machine that can fold, stuff, and address a particular mailing in 6 hours. With the help of a newer machine the job can be completed in 2 hours. How long would it take the new machine to do the job alone?

Solution Let

x = Time for new machine to do the job alone

If a job can be completed in 6 hours (old machine), then the rate of completion is $\frac{1}{6}$ job per hour. That is, from the rate–time formula $Q = RT$, 1 job (Q) = the number of jobs per hour (R) times the number of hours (T), so $1 = R \cdot 6$ and $R = \frac{1}{6}$. Similarly, if a job can be completed in 2 hours (both machines together), then the rate of completion is $\frac{1}{2}$ job per hour. If a job can be completed in x hours (new machine alone), then the rate of completion is $1/x$ job per hour. Thus,

$$\begin{pmatrix}\text{Rate of old}\\ \text{machine}\end{pmatrix} + \begin{pmatrix}\text{Rate of new}\\ \text{machine}\end{pmatrix} = \begin{pmatrix}\text{Rate}\\ \text{together}\end{pmatrix}$$

$$\frac{1}{6} + \frac{1}{x} = \frac{1}{2} \qquad \text{Multiply by } 6x.$$

$$x + 6 = 3x$$
$$2x = 6$$
$$x = 3 \text{ hours} \qquad \text{New machine alone}$$

Check $\frac{1}{6} + \frac{1}{3} = \frac{1}{6} + \frac{2}{6} = \frac{3}{6} = \frac{1}{2}$

Problem 12 At a family cabin, water is pumped and stored in a large water tank. Two pumps are used for this purpose. One can fill the tank by itself in 6 hours, and the other can do the job in 9 hours. How long will it take both pumps operating together to fill the tank?

MISCELLANEOUS PROBLEMS

Thus far, when given an applied problem, you have usually also been told what type of problem you were dealing with—number, geometric, rate–time, and so forth. In a sense, having this information at the beginning provides a suggestion of how to set up and solve the problem. We now consider some problems without this extra information. Many more are provided in the exercise set for this section. If you have difficulty setting up the problems, the approach given in Appendix A may be helpful.

Example 13 Five people form a glider club and decide to share the cost of a glider equally. They find, however, that if they let three more join the club, the share for each of the original five will be reduced by $480. What is the total cost of the glider?

Solution Let x be the cost of the glider. With five shares, each share of the cost of the glider is $x/5$. With an additional three shares, each share is $x/8$. Each share for eight persons is $480 less than it is for five; that is,

$$\frac{x}{8} = \frac{x}{5} - 480$$

Solve by clearing fractions first.

$$\begin{aligned} 5x &= 8x - 40 \cdot 480 \\ 3x &= 19{,}200 \\ x &= 6{,}400 \end{aligned}$$

Check $6,400 divided into five shares is $1,280 each; divided into eight shares it is $800 each. The difference is $480 per share.

Problem 13 Three women bought a sailboat together. If they had taken in a fourth person, the cost for each would have been reduced by $400. What was the total cost of the boat?

Example 14 If a stock that you bought on Monday went up 10% on Tuesday and fell 10% on Wednesday, how much did you pay for the stock on Monday if you sold it on Wednesday for $99?

Solution Let x be the Monday price. Then the Tuesday price is $1.1x$. (Remember that a 10% increase means the new price is 100% + 10% = 110% of the old, that is, 1.1 times the old.) The Wednesday price is 0.9 times the Tuesday price, or $(0.9)(1.1x)$. Thus,

$$\begin{aligned} (0.9)(1.1)x &= 99 \\ 0.99x &= 99 \\ x &= 100 \end{aligned}$$

Check Monday price \$100, Tuesday price \$110, Wednesday price \$99.

Problem 14 A company's sales decreased 5% in 1987 and increased 5% in 1988. What were the sales in 1986 if sales for 1988 were \$9,975,000?

ANSWERS TO MATCHED PROBLEMS

8. **(A)** 9 kilometers per hour **(B)** 13.5 kilometers
(C) $1\frac{1}{3}$ hours
9. 36 minutes; 1,620 fliers
10. $60(x + 2) = 80x$; $x = 6$ hours
11. 3 miles per hour
12. 3.6 hours
13. \$4,800 **14.** \$10,000,000

EXERCISE 4-2

Set up appropriate equations and solve.

A

RATE–TIME PROBLEMS

1. Two cars leave Chicago at the same time and travel in opposite directions. If one travels at 62 kilometers per hour and the other at 88 kilometers per hour, how long will it take them to be 750 kilometers apart?

2. Two airplanes leave Miami at the same time and fly in opposite directions. If one flies at 840 kilometers per hour and the other at 510 kilometers per hour, how long will it take them to be 3,510 kilometers apart?

3. The distance between towns A and B is 750 kilometers. If a passenger train leaves town A and travels toward town B at 90 kilometers per hour at the same time a freight train leaves town B and travels toward A at 35 kilometers per hour, how long will it take the two trains to meet?

4. Repeat Problem 3 using 630 kilometers for the distance between the two towns, 100 kilometers per hour as the rate for the passenger train, and 40 kilometers per hour as the rate for the freight train.

5. An office worker can fold and stuff 14 envelopes per minute. If another office worker can fold and stuff 10 envelopes per minute, how long will it take them working together to fold and stuff 1,560 envelopes?

6. One file clerk can file 12 folders per minute and a second clerk 9. How long will it take them working together to file 672 folders?

7. A car leaves a town traveling at 50 kilometers per hour. How long will it take a second car traveling at 60 kilometers per hour to catch up to the first car if it leaves 1 hour later?

8. Repeat Problem 7 if the first car travels at 45 kilometers per hour and the second car leaves 2 hours later traveling at 75 kilometers per hour.

9. Find the total time to complete the job in Problem 5 if the second (slower) office worker is brought on the job 15 minutes after the first person has started.

10. Find the total time to complete the job in Problem 6 if the faster file clerk is brought on the job 14 minutes after the slower clerk has started.

B **11.** Pipe A can fill a tank in 8 hours and pipe B can fill the same tank in 6 hours. How long will it take both pipes together to fill the tank?

12. A typist can complete a mailing in 5 hours. If another typist requires 7 hours, how long will it take both working together to complete the mailing?

13. A painter can paint a house in 5 days. With the help of another painter, the house can be painted in 3 days. How long would it take the second painter to paint the house alone?

14. You are at a river resort and rent a motorboat for 5 hours at 7 A.M. You are told that the boat will travel at 8 kilometers per hour upstream and 12 kilometers per hour returning. You decide that you would like to go as far up the river as you can and still be back at noon. At what time should you turn back, and how far from the resort will you be at that time?

C **15.** Three seconds after a person fires a rifle at a target, she hears the sound of impact. If sound travels at 335 meters per second and the bullet at 670 meters per second, how far away is the target?

16. An explosion is set off on the surface of the water 11,000 feet from a ship. If the sound reaches the ship through the water 7.77 seconds before it arrives through the air and if sound travels through water 4.5 times faster than through air, how fast (to the nearest foot per second) does sound travel in air and in water?

The remaining problems in this set of exercises contain a variety of applications, including rate–time problems. The more difficult problems are marked with two stars (★★), the moderately difficult problems with one star (★), and the easier problems are not marked.

LIFE SCIENCES **17.** A good approximation for the normal weight w (in kilograms) of a person over 150 centimeters tall is given by the formula $w = 0.98h - 100$, where height h is in centimeters. What would be the normal height of a person with a normal weight of 76 kilograms?

18. Find the normal height of a person in Problem 17 with a normal weight of 55 kilograms.

19. A scuba diver knows that 1 atmosphere of pressure is the weight of a column of air 1 square inch extending straight up from the surface of the earth without end (14.7 pounds per square inch). Also, the water pressure below the surface increases 1 atmosphere for each 33 feet of depth. In terms of a formula,

$$P = 1 + \frac{D}{33}$$

where P is pressure in atmospheres and D is depth of the water in feet. At what depth will the pressure be 3.6 atmospheres?

20. A company selling water-resistant watches advertises that they are waterproof to 3 atmospheres of pressure. How deep could a diver go (see Problem 19) and safely use the watch?

★21. A wildlife management group approximated the number of chipmunks in a wildlife preserve by using the popular capture-mark-recapture technique. Using live traps, they captured and marked 600 chipmunks and then released them. After a period for mixing, they captured another 500 and found 60 marked ones among them. Assuming that the ratio of the total chipmunk population to the chipmunks marked in the first sample is the same as the ratio of all chipmunks in the second sample to those found marked, estimate the chipmunk population in the preserve.

★22. A naturalist for a fish-and-game department estimated the total number of rainbow trout in a lake by using the method described in Problem 21. The naturalist netted, marked, and released 200 rainbow trout. A week later, after thorough mixing, 200 more were netted and 8 marked trout were found among them. Estimate the total rainbow trout population in the lake.

DOMESTIC

23. A student needs at least 80% of all points on the tests in a class to get a B. There are three 100-point tests and a 250-point final. The student's test scores are 72, 85, and 78 for the 100-point tests. What is the least score the student can make on the final and still get a B?

24. Repeat Problem 23 but this time suppose the student scores 95, 87, and 66 on the three 100-point tests.

★25. The cruising speed of an airplane is 150 miles per hour (relative to ground). You wish to hire the plane for a 3-hour sightseeing trip. You instruct the pilot to fly north as far as possible and still return to the airport at the end of the allotted time.

(A) How far north should the pilot fly if there is a 30-mile-an-hour wind blowing from the north?

(B) How far north should the pilot fly if there is no wind blowing?

★26. Repeat Problem 25 for an airplane with a cruising speed of 350 kilometers per hour (in still air) and a wind blowing at 70 kilometers per hour from the north.

MUSIC

27. Starting with a string tuned to a given note, one can move up and down the scale simply by decreasing or increasing its length (while maintaining the same tension) according to simple whole-number ratios (see the figure). For example, $\frac{8}{9}$ of the C string gives the next higher note D, $\frac{2}{3}$ of the C string gives G, and $\frac{1}{2}$ of the C string gives C 1 octave higher. (The reciprocals of these fractions, $\frac{9}{8}$, $\frac{3}{2}$, and 2, respectively, are proportional to the frequencies of these notes.) Find the lengths of seven strings (each less than 30 inches) that will produce the following seven chords when paired with a 30-inch string:

(A)	Octave	1:2	**(B)**	Fifth	2:3
(C)	Fourth	3:4	**(D)**	Major third	4:5
(E)	Minor third	5:6	**(F)**	Major sixth	3:5
(G)	Minor sixth	5:8			

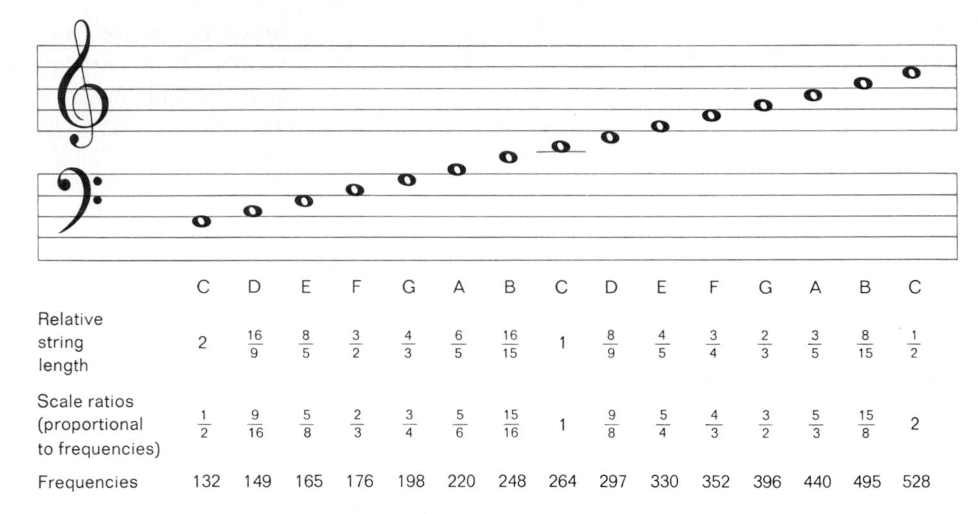

28. The three major chords in music are composed of notes whose frequencies are in the ratio 4:5:6. If the first note of a chord has a frequency of 264 hertz (middle C on the piano), find the frequencies of the other two notes. [*Hint:* Set up two proportions using 4:5 and 4:6, respectively.]

BUSINESS

29. If you paid \$160 for a camera after receiving a discount of 20%, what was the price of the camera before the discount?

30. A car rental company charges \$21 per day and 10 cents per mile. If a car was rented for 2 days, how far was it driven if the total bill came to \$53.20?

31. It costs a book publisher \$74,200 to prepare a book for publication (typesetting, art, editing, and so on); printing and binding costs are \$5.50 per book. If the book is sold to bookstores for \$19.50 per copy, how many copies must be sold for the publisher to break even?

32. A woman borrowed a sum of money from a bank at 18% simple interest. At the end of 10 months she repaid the bank \$1,380. How much was borrowed from the bank? [*Hint:* $A = P + Prt$, where A is the amount repaid, P is the amount borrowed, r is the interest rate expressed as a decimal, and t is time in years.]

PHYSICS AND ENGINEERING

33. If a small steel ball is thrown downward from a tower with an initial velocity of 15 meters per second, its velocity in meters per second after t seconds is given approximately by

$$v = 15 + 9.75t$$

How many seconds are required for the object to attain a velocity of 93 meters per second?

34. How long would it take the ball in Problem 33 to reach a velocity of 120 meters per second?

35. If the large cross-sectional area in a hydraulic lift (see the figure) is approximately 630 square centimeters and a person wants to lift 2,250 kilograms with a 25-kilogram force, how large should the small cross-sectional area be?

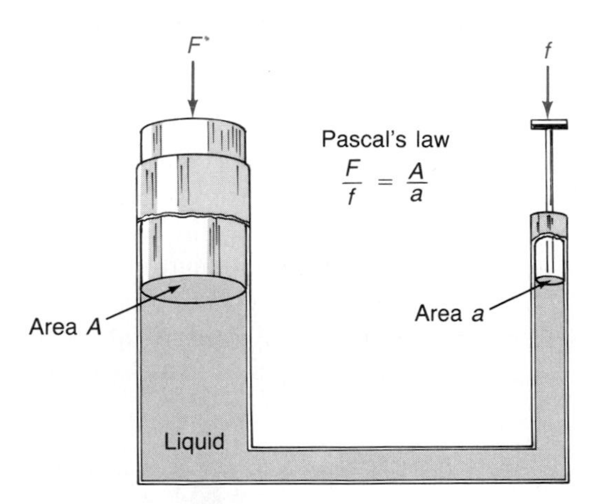

36. If the large cross-sectional area of the hydraulic lift shown in the figure accompanying Problem 35 is 560 square centimeters and the small cross-sectional area is 8 square centimeters, how much force f will be required to lift 2,100 kilograms?

37. A type of physics problem with wide applications is the *lever problem*. For a lever, relative to a fulcrum, to be in static equilibrium (balanced), the sum of the downward forces times their respective distances on one side of the fulcrum must equal the sum of the downward forces times their respective distances on the other side of the fulcrum (see the figure). If a person has a 200-centimeter steel wrecking bar and places a fulcrum 20 centimeters from one end, how much can be lifted with a force of 50 kilograms on the long end?

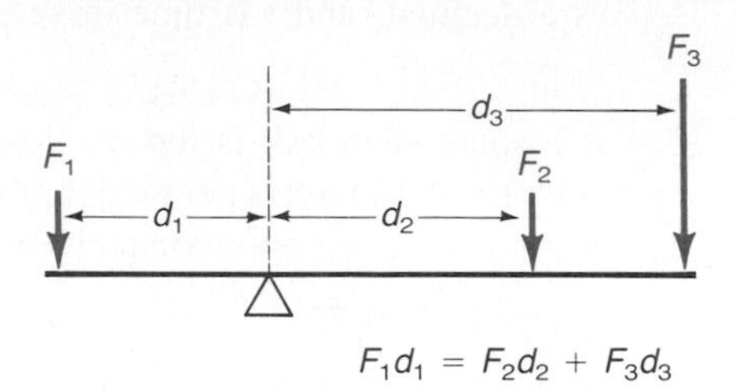

$$F_1d_1 = F_2d_2 + F_3d_3$$

38. Two people decided to move a 1,920-pound rock by use of a 9-foot steel bar (see the figure). If they place the fulcrum 1 foot from the rock and one of them applies a force of 150 pounds on the other end, how much force will the second person have to apply 2 feet from that end to lift the rock?

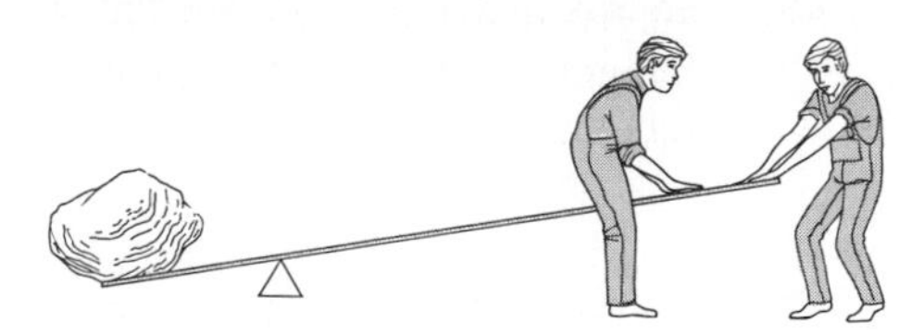

★39. In 1849, during a celebrated experiment, the French mathematician Fizeau made the first accurate approximation of the speed of light. By using a rotating disk with notches equally spaced on the circumference and a reflecting mirror 5 miles away (see the figure), he was able to measure the elapsed time for the light traveling to the mirror and back. Calculate his estimate for the speed of light (in miles per second) if his measurement for the elapsed time was $\frac{1}{20,000}$ second ($d = rt$).

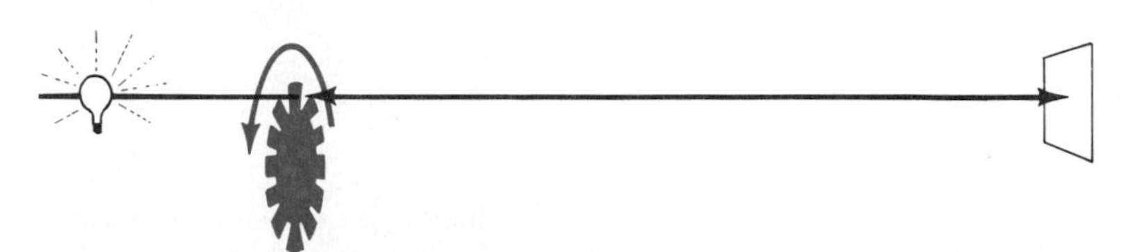

★★40. An earthquake emits a primary wave and a secondary wave. Near the surface of the earth, the primary wave travels at about 5 miles per second and the secondary wave at about 3 miles per second. From the time lag between the two waves arriving at a given seismic station, it is possible to estimate the distance to the quake. (The *epicenter* can be located by getting distance bearings at three or more stations.) Suppose a station measured a time difference of 12 seconds between the arrival of the two waves. How far would the earthquake be from the station?

PUZZLES

41. A pole is located in a pond. One-fifth of the length of the pole is in the sand, 4 meters is in water, and two-thirds of the length is in the air. How long is the pole?

★42. Diophantus, an early Greek algebraist (A.D. 280), was the subject for a famous ancient puzzle. See if you can find Diophantus' age at death from the following information: Diophantus was a boy for one-sixth of his life; after one-twelfth more he grew a beard; after one-seventh more he married, and after 5 years of marriage he was granted a son; the son lived one-half as long as his father; and Diophantus died 4 years after his son's death.

★★43. A classic problem is the courier problem. If a column of soldiers 3 miles long is marching at 5 miles per hour, how long will it take a courier on a motorcycle traveling at 25 miles per hour to deliver a message from the end of the column to the front and then return?

★★44. After 12:00 noon exactly, what time will the hands of a clock be together again?

4-3 FORMULAS AND LITERAL EQUATIONS

From the distance–rate–time formula $D = RT$, we have seen that we can express the rate or the time in terms of the remaining variables:

$$R = \frac{D}{T} \quad \text{or} \quad T = \frac{D}{R}$$

This process of changing a formula into an equivalent one by solving for a different variable is necessary in many applied situations. As long as we are solving for a variable that does not occur to a higher power than 1, the process is no different than solving a linear equation in one variable. This process was first considered in Section 1-6. The same strategy will apply:

Solving for a Particular Variable

1. Clear fractions.
2. Simplify both sides.
3. Get all terms with the variable in question to one side, all remaining terms to the other, and combine like terms.
4. Isolate the variable by dividing by its coefficient.

For example, to solve $D = RT$ for R, steps 1 to 3 are unnecessary. We simply divide by the coefficient T to obtain

$$\frac{D}{T} = R$$

and write this as

$$R = \frac{D}{T}$$

by using the symmetric property of equality from Section 1-2.

Example 15 Solve $F = \frac{9}{5}C + 32$, the Celsius-to-Farenheit-conversion formula, for C.

Solution

$F = \frac{9}{5}C + 32$ — Multiply by 5 to clear fractions

$5F = 9C + 160$ — Leave terms involving C on the right, shift all others to the left.

$5F - 160 = 9C$ — Divide by 9

$\frac{5}{9}F - \frac{160}{9} = C$ — To obtain the usual form of this equation, factor out $\frac{5}{9}$ on the left side.

$\frac{5}{9}(F - 32) = C$ or $C = \frac{5}{9}(F - 32)$

Problem 15 Solve $A = P + Prt$, the simple interest formula, for r.

Example 16 Solve $a_n = a_1 + (n - 1)d$, the formula for the nth term of an arithmetic sequence, for n. (Treat a_n as a fixed number.)

Solution

$a_n = a_1 + (n - 1)d$ — Simplify the right side.

$a_n = a_1 + nd - d$ — Move all terms not involving n to the left.

$a_n - a_1 + d = nd$ — Divide by the coefficient of n

$$\frac{a_n - a_1 + d}{d} = n \quad \text{or} \quad n = \frac{a_n - a_1 + d}{d}$$

Problem 16 Solve

$$S = \frac{n}{2}[2a + (n - 1)d]$$

the formula for the sum of an arithmetic sequence, for d.

Example 17 Solve the formula $A = P + Prt$ for P.

Solution

$A = P + Prt$

$P + Prt = A$

Since P is a common factor to both terms on the left, we factor P out and complete the problem:

$$P(1 + rt) = A$$

$$\frac{P(1 + rt)}{1 + rt} = \frac{A}{1 + rt}$$

$$P = \frac{A}{1 + rt}$$ Note that P appears only on the left side.

If we write $P = A - Prt$, we have not solved for P. To solve for P is to isolate P on the left side with a coefficient of 1. In general, if the variable for which we are solving appears on both sides of the equation, we have not solved for it!

Problem 17 Solve $A = xy + xz$ for x.

ANSWERS TO MATCHED PROBLEMS

15. $r = \dfrac{A - P}{Pt}$ **16.** $d = \dfrac{2S - 2an}{n(n - 1)}$ **17.** $x = \dfrac{A}{y + z}$

EXERCISE 4-3

A

1. Solve $A = \frac{1}{2}bh$ for h. *Area of a triangle*

2. Solve $V = \frac{1}{3}\pi r^2 h$ for h. *Volume of a cone*

3. Solve $A = \frac{1}{2}(a + b)h$ for a. *Area of a trapezoid*

4. Solve $A = \frac{1}{2}(a + b)h$ for h. *Area of a trapezoid*

5. Solve $y = 3x + 7$ for x. *Equation of a line*

6. Solve $y = -7x + 3$ for x. *Equation of a line*

7. Solve $l = \dfrac{\pi}{180}r\theta$ for θ. *Arc length on a circle*

8. Solve $A = \dfrac{\pi}{360}r^2\theta$ for θ. *Area of a circular sector*

9. Solve $P = S(1 - dt)$ for t. *Simple discount*

10. Solve $V = V_0(1 + Bt)$ for t. *Gas expansion*

11. Solve $V = \frac{4}{3}\pi ab^2$ for a. *Volume of ellipsoid*

12. Solve $S_n = \dfrac{n}{2}(a_1 + a_n)$ for n. *Arithmetic progression*

B

13. Solve $P = \dfrac{p}{p + q}$ for p. *Probability*

14. Solve $P = \dfrac{p}{p + q}$ for q. *Probability*

15. Solve $(\text{ERA}) = \dfrac{9R}{I}$ for I. *Baseball—earned run average*

16. Solve $(\text{IQ}) = \dfrac{100(\text{MA})}{(\text{CA})}$ for (CA). *Psychology—intelligence quotient*

17. Solve $F = m\dfrac{v - v_0}{t}$ for v. *Momentum*

18. Solve $z = \dfrac{x - \mu}{\sigma}$ for x. *Statistics*

19. Solve $y = Q(P - V) - F$ for V. *Profit analysis*

20. Solve $F_2 = \alpha Y_1 + (1 - \alpha)F_1$ for F_1. *Forecasting*

C 21. Solve $\dfrac{1}{f} = \dfrac{1}{a} + \dfrac{1}{b}$ for f. *Optics—focal length*

22. Solve $\dfrac{1}{R} = \dfrac{1}{R_1} + \dfrac{1}{R_2}$ for R. *Electric circuits*

23. Solve $C = \frac{1}{2}QC_h + \dfrac{D}{Q}C_0$ for D. *Inventory*

24. Solve $H = \dfrac{kA(t_2 - t_1)}{L_1}$ for A. *Heat flow*

25. Solve $\dfrac{P_1V_1}{T_1} = \dfrac{P_2V_2}{T_2}$ for T_2. *Gas law*

26. Solve $\dfrac{P_1V_1}{T_1} = \dfrac{P_2V_2}{T_2}$ for V_1. *Gas law*

27. Solve $y = \dfrac{2x - 3}{3x - 5}$ for x. *Rational equation*

28. Solve $y = \dfrac{3x + 2}{2x - 4}$ for x. *Rational equation*

4-4 SOLVING INEQUALITIES

- Interval Notation
- Solving Linear Inequalities
- Applications

In Section 1-2 we introduced simple inequality statements of the form

$$x > 2 \qquad -4 < x \leq 3 \qquad x \leq -1$$

which have obvious solutions. In this section we will consider inequality statements that do not have obvious solutions. Try to guess the real number solutions for

$$3(x - 2) + 1 < 3x - (x + 7)$$

By the end of this section you will be able to solve this type of inequality almost as easily as you solved first-degree equations. First, however, we digress for a moment to introduce interval notation.

INTERVAL NOTATION

In Section 1-2 we used parentheses ( ) and brackets [ ] in the graphic representation of certain inequality statements. For example,

$-3 < x \leq 4$ and (line graph from -3, open, to 4, closed; labels -3, 0, 4, x)

are two ways of indicating that x is between -3 and 4, including 4 but excluding -3. Another convenient way of representing this fact is in terms of the interval notation

$$(-3, 4]$$

Table 1 shows the use of **interval notation** in its most common forms.

TABLE 1

INTERVAL NOTATION	INEQUALITY NOTATION	LINE GRAPH
$[a, b]$	$a \leq x \leq b$	[a, b]
$[a, b)$	$a \leq x < b$	[a, b)
$(a, b]$	$a < x \leq b$	(a, b]
(a, b)	$a < x < b$	(a, b)
$[b, \infty)$†	$x \geq b$	[b →
(b, ∞)	$x > b$	(b →
$(-\infty, a]$	$x \leq a$	← a]
$(-\infty, a)$	$x < a$	← a)

† The symbol ∞ (read "infinity") is not a number. When we write $[b, \infty)$, we are simply referring to the interval starting at b and continuing indefinitely to the right. We would never write $[b, \infty]$.

Example 18 Write each of the following in inequality notation and graph on a real number line:

(A) $[-2, 3)$ **(B)** $(-4, 2)$ **(C)** $[-2, \infty)$ **(D)** $(-\infty, 3)$

Solution **(A)** $-2 \le x < 3$

(B) $-4 < x < 2$

(C) $x \ge -2$

(D) $x < 3$

Problem 18 Write each of the following in interval notation and graph on a real number line:

(A) $-3 < x \le 3$ **(B)** $-1 \le x \le 2$ **(C)** $x > 1$ **(D)** $x \le 2$

SOLVING LINEAR INEQUALITIES

We now turn to the problem of solving linear inequalities in one variable, such as

$$2(2x + 3) < 6(x - 2) + 10 \qquad \text{and} \qquad -3 < 2x + 3 \le 9$$

The **solution set** for an inequality is the set of elements from its replacement set that make the inequality a true statement. Any element of the solution set is called a **solution** of the inequality. To **solve an inequality** is to find its solution set. Two inequalities are **equivalent** if they have the same solution set. Just as with equations, we try to perform operations on inequalities that produce simpler equivalent inequalities. We continue the process until an inequality is reached whose solution is obvious. The six properties of inequalities listed in the box on the next page produce equivalent inequalities when applied. These properties are similar to the addition, subtraction, multiplication, and division properties that we use to solve equations.

Similar properties hold if each inequality sign is reversed or if $<$ is replaced with $\le$ and $>$ is replaced with $\ge$. Thus, we find that we can perform essentially the same operations on inequalities that we perform on equations. When working with inequalities, we have to be particularly careful of the use of the multiplication and division properties.

The sense of the inequality reverses if we multiply or divide both sides of an inequality statement by a negative number.

Inequality Properties

For a, b, and c any real numbers:

1. If $a < b$, then $a + c < b + c$. — ADDITION PROPERTY

   $-2 < 4 \qquad -2 + 3 < 4 + 3$

2. If $a < b$, then $a - c < b - c$. — SUBTRACTION PROPERTY

   $-2 < 4 \qquad -2 - 3 < 4 - 3$

3. If $a < b$ and c is positive, then $ca < cb$.

   $-2 < 4 \qquad 3(-2) < 3(4)$

4. If $a < b$ and c is negative, then $ca > cb$.

   $-2 < 4 \qquad (-3)(-2) > (-3)(4)$

   (3 and 4:) MULTIPLICATION PROPERTY (Note the difference between **3** and **4**.)

5. If $a < b$ and c is positive, than $\dfrac{a}{c} < \dfrac{b}{c}$.

   $-2 < 4 \qquad \dfrac{-2}{2} < \dfrac{4}{2}$

6. If $a < b$ and c is negative, then $\dfrac{a}{c} > \dfrac{b}{c}$.

   $-2 < 4 \qquad \dfrac{-2}{-2} > \dfrac{4}{-2}$

   (5 and 6:) DIVISION PROPERTY (Note the difference between **5** and **6**.)

Let us sketch a proof of the multiplication property: if $a < b$, then by definition of $<$, there exists a positive number p such that $a + p = b$. If we multiply both sides of $a + p = b$ by a positive number c, we obtain $ca + cp = cb$, where cp is positive. Thus, by definition of $<$, we see that $ca < cb$. Now if we multiply both sides of $a + p = b$ by a negative number c, we obtain $ca + cp = cb$ or $ca = cb - cp = cb + (-cp)$, where $-cp$ is positive. Hence, by definition of $<$, we see that $cb < ca$ or $ca > cb$.

Now let us see how the inequality properties are used to solve linear inequalities. Several examples will illustrate the process.

Example 19 Solve and graph: $2(2x + 3) - 10 < 6(x - 2)$

Solution

$$2(2x + 3) - 10 < 6(x - 2) \quad \text{Simplify left and right sides.}$$
$$4x + 6 - 10 < 6x - 12$$
$$4x - 4 < 6x - 12$$

$$\boxed{4x - 4 + \mathbf{4} < 6x - 12 + \mathbf{4}} \quad \text{Addition property}$$

$$4x < 6x - 8$$

$$\boxed{4x - \mathbf{6x} < 6x - 8 - \mathbf{6x}}$$ Subtraction property

$$-2x < -8$$

$$\boxed{\frac{-2x}{\mathbf{-2}} > \frac{-8}{\mathbf{-2}}}$$ Division property—note that the sense reverses. (Why?)

$$x > 4 \quad \text{or} \quad (4, \infty)$$

Problem 19 Solve and graph: $3(x - 1) \geq 5(x + 2) - 5$

Example 20 Solve and graph: $\dfrac{2x - 3}{4} + 6 \geq 2 + \dfrac{4x}{3}$

Solution

$$\frac{2x - 3}{4} + 6 \geq 2 + \frac{4x}{3}$$ Multiply both sides by 12, the LCM of 4 and 3.

$$\boxed{\mathbf{12} \cdot \frac{2x - 3}{4} + \mathbf{12} \cdot 6 \geq \mathbf{12} \cdot 2 + \mathbf{12} \cdot \frac{4x}{3}}$$ Sense does not reverse. (Why?)

$$3(2x - 3) + 72 \geq 24 + 4 \cdot 4x$$
$$6x - 9 + 72 \geq 24 + 16x$$
$$6x + 63 \geq 24 + 16x$$
$$-10x \geq -39$$
$$x \leq 3.9 \quad \text{or} \quad (-\infty, 3.9]$$ Sense reverses. (Why?)

Problem 20 Solve and graph: $\dfrac{4x - 3}{3} + 8 < 6 + \dfrac{3x}{2}$

Example 21 Solve and graph: $-3 \leq 4 - 7x < 18$

Solution We proceed as in the preceding examples, except that we try to isolate x in the middle with a coefficient of 1. That is, we try to solve the two inequalities $-3 \leq 4 - 7x$ and $4 - 7x < 18$ at the same time:

$$-3 \leq 4 - 7x < 18$$ Subtract 4 from each member.

$$\boxed{-3 - \mathbf{4} \leq 4 - 7x - \mathbf{4} < 18 - \mathbf{4}}$$

$$-7 \le -7x < 14$$ Divide each member by −7.

$$\frac{-7}{-7} \ge \frac{-7x}{-7} > \frac{14}{-7}$$ Sense reverses. (Why?)

$$1 \ge x > -2 \quad \text{or} \quad -2 < x \le 1 \quad \text{or} \quad (-2, 1]$$

Problem 21 Solve and graph: $-3 < 7 - 2x \le 7$

APPLICATIONS

Example 22 In a chemistry experiment a solution of hydrochloric acid is to be kept between 30° and 35° Celsius—that is, $30 \le C \le 35$. What is the range in temperature in degrees Fahrenheit? [$C = \frac{5}{9}(F - 32)$]

Solution

$$30 \le C \le 35$$ Replace C with $\frac{5}{9}(F - 32)$.

$$30 \le \frac{5}{9}(F - 32) \le 35$$ Multiply each member by $\frac{9}{5}$. (Why?)

$$\frac{9}{5} \cdot 30 \le \frac{9}{5} \cdot \frac{5}{9}(F - 32) \le \frac{9}{5} \cdot 35$$

$$54 \le F - 32 \le 63$$ Add 32 to each member.

$$54 + 32 \le F - 32 + 32 \le 63 + 32$$

$$86 \le F \le 95 \quad \text{or} \quad [86, 95]$$

Problem 22 A film developer is to be kept between 68° and 77° Fahrenheit—that is, $68 \le F \le 77$. What is the range in temperature in degrees Celsius? ($F = \frac{9}{5}C + 32$)

ANSWERS TO MATCHED PROBLEMS

18. **(A)** $(-3, 3]$

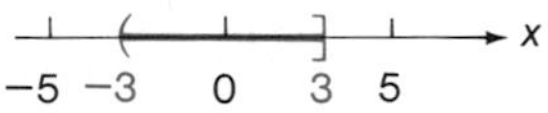

(B) $[-1, 2]$

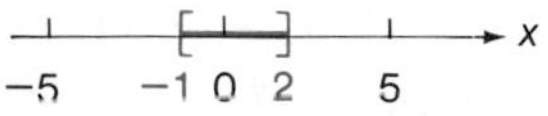

(C) $(1, \infty)$

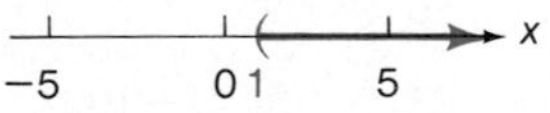

(D) $(-\infty, 2]$

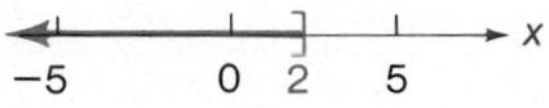

19. $x \le -4$ or $(-\infty, -4]$

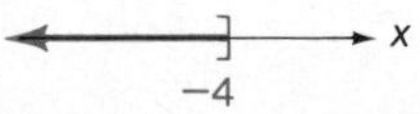

20. $x > 6$ or $(6, \infty)$

6 x

21. $5 > x \geq 0$ or $0 \leq x < 5$ or $[0, 5)$

0 5 x

22. $20 \leq C \leq 25$ or $[20, 25]$

EXERCISE 4-4

The replacement set for all variables is the set of real numbers.

A *Write in inequality notation and graph on a real number line.*

1. $[-8, 7]$ **2.** $(-4, 8)$ **3.** $[-6, 6)$

4. $(-3, 3]$ **5.** $[-6, \infty)$ **6.** $(-\infty, 7)$

Write in interval notation and graph on a real number line.

7. $-2 < x \leq 6$ **8.** $-5 \leq x \leq 5$ **9.** $-7 < x < 8$

10. $-4 \leq x < 5$ **11.** $x \leq -2$ **12.** $x > 3$

Write in interval and inequality notation.

13. −10 −5 0 5 10 x

14. −10 −5 0 5 10 x

15. −10 −5 0 5 10 x

16. −10 −5 0 5 10 x

Solve and graph.

17. $7x - 8 < 4x + 7$ **18.** $4x + 8 \geq x - 1$

19. $3 - x \geq 5(3 - x)$ **20.** $2(x - 3) + 5 < 5 - x$

21. $\dfrac{N}{-2} > 4$ **22.** $\dfrac{M}{-3} \leq -2$

23. $-5t < -10$ **24.** $-7n \geq 21$

25. $3 - m < 4(m - 3)$ **26.** $2(1 - u) \geq 5u$

27. $-2 - \dfrac{B}{4} \leq \dfrac{1 + B}{3}$ **28.** $\dfrac{y - 3}{4} - 1 > \dfrac{y}{2}$

29. $-4 < 5t + 6 \leq 21$ **30.** $2 \leq 3m - 7 < 14$

B *What numbers satisfy the given conditions? Solve using inequality methods.*

31. 3 less than twice the number is greater than or equal to -6.

32. 5 more than twice the number is less than or equal to 7.

33. 15 reduced by 3 times the number is less than 6.

34. 5 less than 3 times the number is less than or equal to 4 times the number.

Solve and graph.

35. $\dfrac{q}{7} - 3 > \dfrac{q-4}{3} + 1$

36. $\dfrac{p}{3} - \dfrac{p-2}{2} \le \dfrac{p}{4} - 4$

37. $\dfrac{2x}{5} - \dfrac{1}{2}(x-3) \le \dfrac{2x}{3} - \dfrac{3}{10}(x+2)$

38. $\dfrac{2}{3}(x+7) - \dfrac{x}{4} > \dfrac{1}{2}(3-x) + \dfrac{x}{6}$

39. $-4 \le \frac{9}{5}x + 32 \le 68$

40. $-1 \le \frac{2}{3}A + 5 \le 11$

41. $-12 < \frac{3}{4}(2-x) \le 24$

42. $24 \le \frac{2}{3}(x-5) < 36$

43. $16 < 7 - 3x \le 31$

44. $-1 \le 9 - 2x < 5$

45. $-6 < -\frac{2}{5}(1-x) \le 4$

46. $15 \le 7 - \frac{2}{5}x \le 21$

C **47.** If both a and b are negative numbers and b/a is greater than 1, then is $a - b$ positive or negative?

48. If both a and b are positive numbers and b/a is greater than 1, then is $a - b$ positive or negative?

49. Indicate true (T) or false (F):
(A) If $p > q$ and $m > 0$, then $mp < mq$.
(B) If $p < q$ and $m < 0$, then $mp > mq$.
(C) If $p > 0$ and $q < 0$, then $p + q > q$.

50. Assume that $m > n > 0$; then

$$\begin{aligned} mn &> n^2 \\ mn - m^2 &> n^2 - m^2 \\ m(n-m) &> (n+m)(n-m) \\ m &> n + m \\ 0 &> n \end{aligned}$$

But it was assumed $n > 0$. Can you find the error?

APPLICATIONS *Set up inequalities and solve.*

51. ***Earth science*** As dry air moves upward it expands, and in so doing cools at a rate of about 5.5°F for each 1,000-foot rise up to about 40,000 feet. If the ground temperature is 70°F, then the temperature T at height h is given approximately by $T = 70 - 0.0055h$. For what range in altitude will the temperature be between 26° and −40°F?

★52. ***Energy*** If the power demands in a 110-volt electric circuit in a home vary between 220 and 2,750 watts, what is the range of current flowing through the circuit? ($W = EI$, where W = Power in watts, E = Pressure in volts, and I = Current in amperes.)

53. ***Business and economics*** For a business to make a profit it is clear that revenue R must be greater than cost C; in short, a profit will result only if $R > C$. If a company manufactures records and its cost equation for a week is $C = 300 + 1.5x$ and its revenue equation is $R = 2x$, where x is the number of records sold in a week, how many records must be sold for the company to realize a profit?

54. ***Psychology*** IQ is given by the formula

$$\text{IQ} = \frac{\text{MA}}{\text{CA}} 100$$

where MA is mental age and CA is chronological age. If

$$80 \leq \text{IQ} \leq 140$$

for a group of 12-year-old children, find the range of their mental ages.

★★55. ***Puzzle*** A railroad worker is walking through a train tunnel (see the figure) when he notices an unscheduled train approaching him. If he is three-quarters of the way through the tunnel and the train is one tunnel length ahead of him, which way should he run to maximize his chances of escaping?

4-5 ABSOLUTE VALUE IN EQUATIONS AND INEQUALITIES

- Absolute Value and Distance
- Absolute Value in Equations and Inequalities

Equations and inequalities may also involve absolute values of algebraic expressions. Such problems can be solved by geometric and algebraic approaches. We consider both in this section after reviewing the absolute-value concept.

ABSOLUTE VALUE AND DISTANCE

We start with a review of the geometric and algebraic definitions of absolute value (Section 1-4). If a is the coordinate of a point on a real number line, then the (nondirected) distance from the origin to a, a nonnegative quantity, is represented by $|a|$ and is referred to as the **absolute value** of a (Figure 1). Thus, if $|x| = 5$, then x can be either -5 or 5.

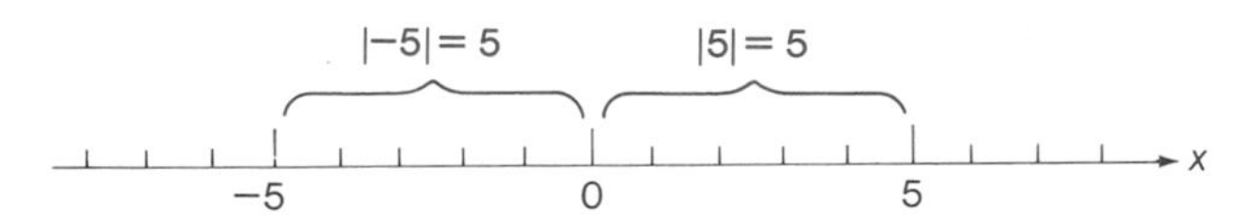

FIGURE 1 Absolute value

Algebraically, recall that we defined absolute value as follows:

Absolute Value

$$|x| = \begin{cases} x & \text{if } x \text{ is positive} \\ 0 & \text{if } x \text{ is } 0 \\ -x & \text{if } x \text{ is negative} \end{cases}$$

[*Note:* $-x$ is positive if x is negative.]

Both the geometric and algebraic definitions of absolute value are useful, as will be seen in the material that follows. Remember:

The absolute value of a number is never negative.

Example 23 Write without the absolute-value sign:

(A) $|7|$ **(B)** $|\pi - 3|$ **(C)** $|-7|$ **(D)** $|3 - \pi|$

Solution

(A) $|7| = 7$

(B) $|\pi - 3| = \pi - 3$ Since $\pi - 3$ is nonnegative ($\pi > 3$)

(C) $|-7| = -(-7) = 7$

(D) $|3 - \pi| = -(3 - \pi) = \pi - 3$ Since $3 - \pi$ is negative

Problem 23 Write without the absolute-value sign:

(A) $|8|$ **(B)** $|\sqrt{5} - 2|$ **(C)** $|-\sqrt{2}|$ **(D)** $|2 - \sqrt{5}|$

Following the same reasoning used in Example 23(B) and (D), it can be shown (see Problem 61 in Exercise 4-5) that:

For all real numbers a and b:

$$|b - a| = |a - b| \qquad |7 - 4| = |3| = 3 = |-3| = |4 - 7|$$

We use this result in defining the distance between two points on a real number line.

Distance between Points A and B

Let A and B be two points on a real number line with coordinates a and b, respectively. The **distance between** A **and** B (also called the **length of the line segment** joining A and B) is given by

$$d(A, B) = |b - a|$$

A B
a b

Example 24 Find the distance between points A and B with coordinates a and b, respectively, as given:

(A) $a = 4,\ b = 9$ **(B)** $a = 9,\ b = 4$
(C) $a = 0,\ b = 6$ **(D)** $a = -3,\ b = 5$

Solution **(A)** $d(A, B) = |9 - 4| = |5| = 5$

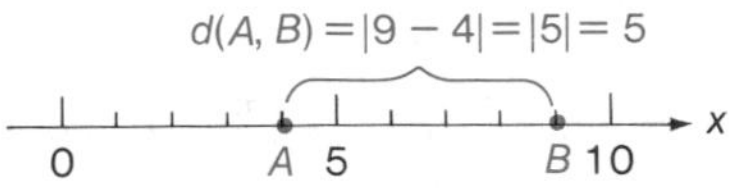

0 A 5 B 10 x

(B) $d(A, B) = |4 - 9| = |-5| = 5$

0 B 5 A 10 x

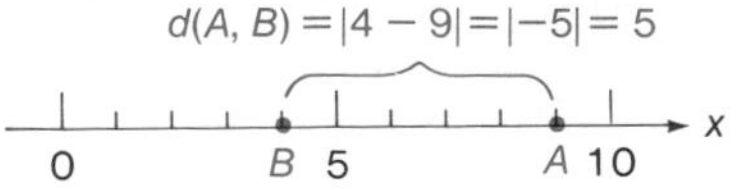

(C) $d(A, B) = |6 - 0| = |6| = 6$

0 A 5 B 10 x

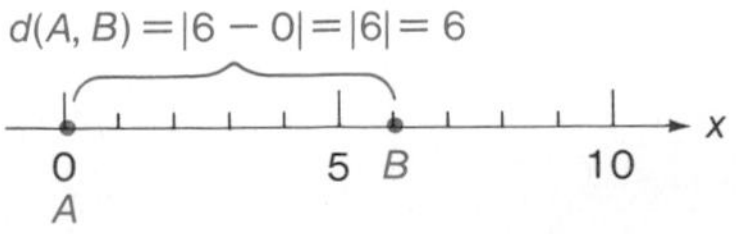

$d(A, B) = |5 - (-3)| = |8| = 8$

(D) −5 A 0 5 B x

It is clear, since $|b - a| = |a - b|$, that

$$d(A, B) = d(B, A)$$

Hence, in computing the distance between two points on a real number line, it does not matter how the two points are labeled—point A can be to the left or to the right of point B. Note also that if A is at the origin O, then

$$d(O, B) = |b - 0| = |b|$$

Problem 24 Find the indicated distances given:

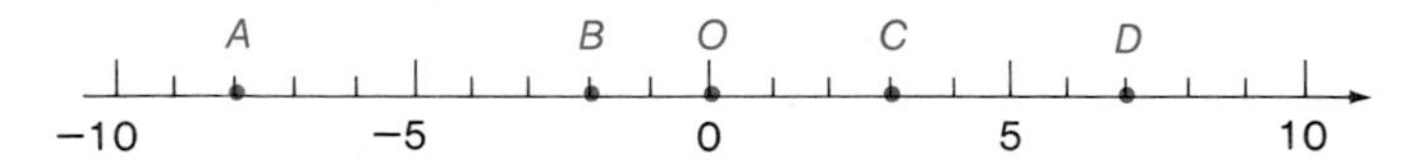

(A) $d(C, D)$ **(B)** $d(D, C)$ **(C)** $d(A, B)$
(D) $d(A, C)$ **(E)** $d(O, A)$ **(F)** $d(D, A)$

ABSOLUTE VALUE IN EQUATIONS AND INEQUALITIES

Absolute value is frequently encountered in equations and inequalities. Some of these forms have immediate geometric interpretation.

Example 25 Solve geometrically and graph:

(A) $|x - 3| = 5$ **(B)** $|x - 3| < 5$
(C) $0 < |x - 3| < 5$ **(D)** $|x - 3| > 5$

Solution **(A)** Geometrically, $|x - 3|$ represents the distance between x and 3; thus, in $|x - 3| = 5$, x is a number whose distance from 3 is 5. That is, x is 5 units to the left of 3 or 5 units to the right of 3:

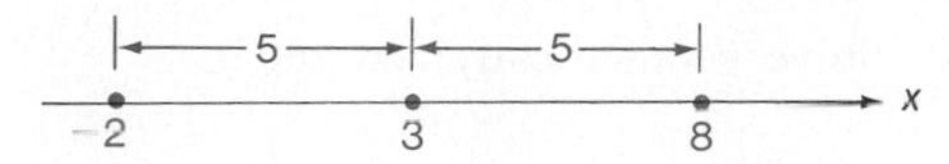

$x = 3 - 5$ or $x = 3 + 5$ More compactly: $x = 3 \pm 5 = -2$ or 8
$x = -2$ $\qquad x = 8$

(B) Geometrically, in $|x - 3| < 5$, x is a number whose distance from 3 is less than 5; that is, x is within 5 units of 3:

$-2 \quad 3 \quad 8 \quad x$

$-2 < x < 8$ or $(-2, 8)$

(C) The form $0 < |x - 3| < 5$ is encountered in calculus and advanced mathematics. Geometrically, x is a number whose distance from 3 is less than 5, but x cannot equal 3. Thus,

$-2 \quad 3 \quad 8 \quad x$

$-2 < x < 8 \qquad x \neq 3$

(D) Geometrically, in $|x - 3| > 5$, x is a number whose distance from 3 is greater than 5; that is,

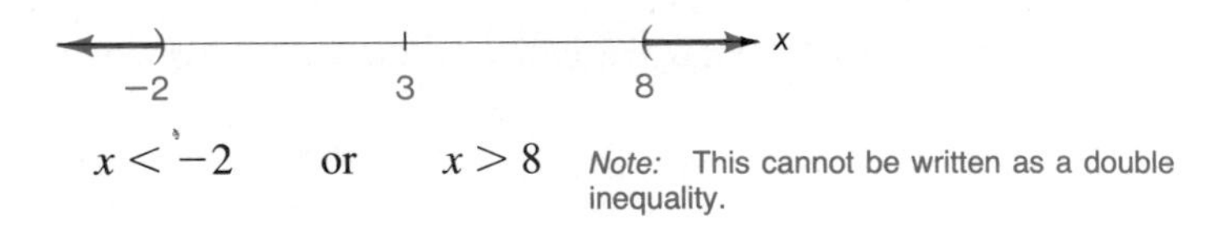

$x < -2$ or $x > 8$ *Note:* This cannot be written as a double inequality.

We summarize the preceding results in Table 2.

TABLE 2

FORM ($d > 0$)	GEOMETRIC INTERPRETATION	GRAPH
$\lvert x - c\rvert = d$	Distance between x and c is equal to d.	d, d; $c - d$, c, $c + d$, x
$\lvert x - c\rvert < d$	Distance between x and c is less than d.	$c - d$, c, $c + d$, x
$0 < \lvert x - c\rvert < d$	Distance between x and c is less than d, but $x \neq c$.	$c - d$, c, $c + d$, x
$\lvert x - c\rvert > d$	Distance between x and c is greater than d.	$c - d$, c, $c + d$, x

Problem 25 Solve geometrically and graph:

(A) $|x + 2| = 6$ **(B)** $|x + 2| < 6$
(C) $0 < |x + 2| < 6$ **(D)** $|x + 2| > 6$

[*Hint:* $|x + 2| = |x - (-2)|$]

Reasoning geometrically as before (noting that $|x| = |x - 0|$), we can establish this result:

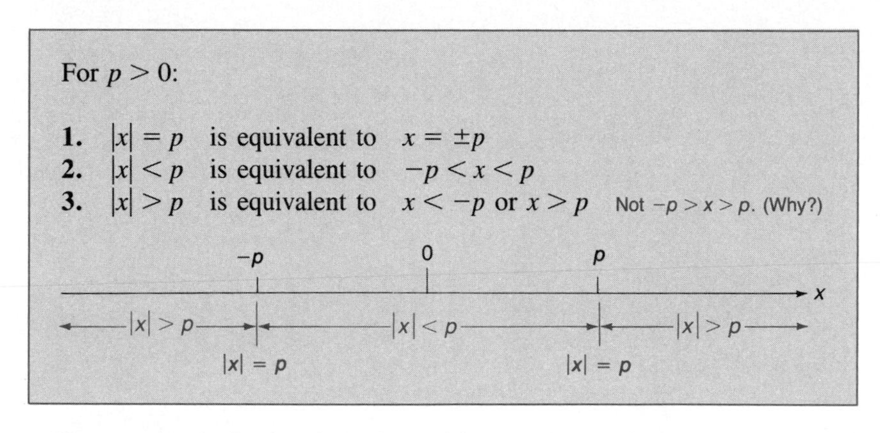
For $p > 0$:

1. $|x| = p$ is equivalent to $x = \pm p$
2. $|x| < p$ is equivalent to $-p < x < p$
3. $|x| > p$ is equivalent to $x < -p$ or $x > p$ Not $-p > x > p$. (Why?)

If we replace x in the above box with $ax + b$, we obtain a more general result:

For $p > 0$:

1. $|ax + b| = p$ is equivalent to $ax + b = \pm p$
2. $|ax + b| < p$ is equivalent to $-p < ax + b < p$
3. $|ax + b| > p$ is equivalent to $ax + b < -p$ or $ax + b > p$

In fact, if we replace x with any algebraic expression, we will obtain other variations of these results.

Example 26 Solve:

(A) $|3x + 5| = 4$ **(B)** $|x| < 5$

(C) $|2x - 1| < 3$ **(D)** $|7 - 3x| \le 2$

Solution

(A)
$$\begin{aligned} |3x + 5| &= 4 \\ 3x + 5 &= \pm 4 \\ 3x &= -5 \pm 4 \\ x &= \frac{-5 \pm 4}{3} \\ &= -3, -\tfrac{1}{3} \end{aligned}$$

(B)
$$|x| < 5$$
$$-5 < x < 5$$

(C)
$$|2x - 1| < 3$$
$$-3 < 2x - 1 < 3$$
$$-2 < 2x < 4$$
$$-1 < x < 2$$

(D)
$$|7 - 3x| \le 2$$
$$-2 \le 7 - 3x \le 2$$
$$-9 \le -3x \le -5$$
$$3 \ge x \ge \tfrac{5}{3}$$
$$\tfrac{5}{3} \le x \le 3$$

Problem 26 Solve:

(A) $|2x - 1| = 8$ **(B)** $|x| \le 7$
(C) $|3x + 3| \le 9$ **(D)** $|5 - 2x| < 9$

Example 27 Solve:

(A) $|x| > 3$ **(B)** $|2x - 1| \ge 3$ **(C)** $|7 - 3x| > 2$

Solution **(A)** $|x| > 3$

$$x < -3 \quad \text{or} \quad x > 3$$

(B) $|2x - 1| \ge 3$

$$\begin{aligned} 2x - 1 \le -3 \quad &\text{or} \quad 2x - 1 \ge 3 \\ 2x \le -2 \quad &\text{or} \quad 2x \ge 4 \\ x \le -1 \quad &\text{or} \quad x \ge 2 \end{aligned}$$

(C) $|7 - 3x| > 2$

$$\begin{aligned} 7 - 3x < -2 \quad &\text{or} \quad 7 - 3x > 2 \\ -3x < -9 \quad &\text{or} \quad -3x > -5 \\ x > 3 \quad &\text{or} \quad x < \tfrac{5}{3} \end{aligned}$$

Problem 27 Solve:

(A) $|x| \ge 5$ **(B)** $|4x - 3| > 5$ **(C)** $|6 - 5x| > 16$

ANSWERS TO MATCHED PROBLEMS

23. **(A)** 8 **(B)** $\sqrt{5} - 2$ **(C)** $\sqrt{2}$ **(D)** $\sqrt{5} - 2$

24. **(A)** 4 **(B)** 4 **(C)** 6 **(D)** 11 **(E)** 8 **(F)** 15

25. **(A)** $x = -8, 4$

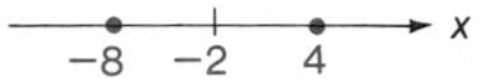

(B) $-8 < x < 4$ or $(-8, 4)$

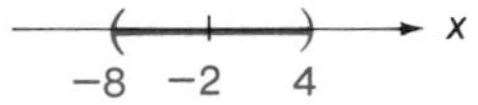

(C) $-8 < x < 4$, $x \ne -2$

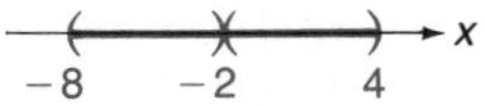

(D) $x < -8$ or $x > 4$

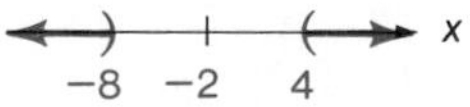

26. **(A)** $-\frac{7}{2}, \frac{9}{2}$ **(B)** $-7 \le x \le 7$ **(C)** $-4 \le x \le 2$
(D) $-2 < x < 7$

27. **(A)** $x \le -5$ or $x \ge 5$ **(B)** $x < -\frac{1}{2}$ or $x > 2$
(C) $x < -2$ or $x > \frac{22}{5}$

EXERCISE 4-5

A *Simplify, and write without absolute-value signs. Leave radicals in radical form (see Problem 23 following Example 23).*

1. $|\sqrt{5}|$ **2.** $|-\frac{3}{4}|$ **3.** $|(-6) - (-2)|$

4. $|(-2) - (-6)|$ **5.** $|5 - \sqrt{5}|$ **6.** $|\sqrt{7} - 2|$

7. $|\sqrt{5} - 5|$ **8.** $|2 - \sqrt{7}|$

Find the distance between points A and B with coordinates a and b, respectively, as given.

9. $a = -7, \quad b = 5$ **10.** $a = 3, \quad b = 12$

11. $a = 5, \quad b = -7$ **12.** $a = 12, \quad b = 3$

13. $a = -16, \quad b = -25$ **14.** $a = -9, \quad b = -17$

Find the indicated distances, given

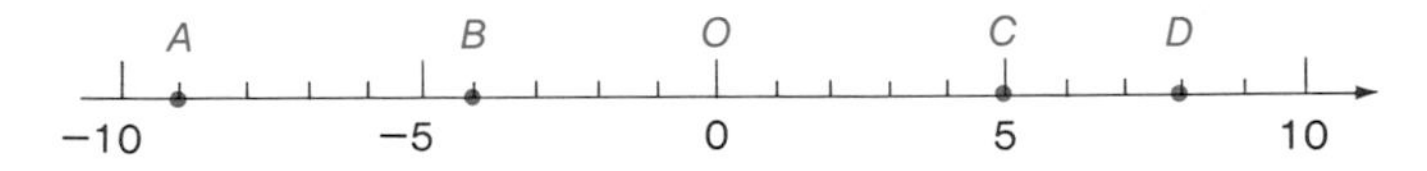

15. $d(B, O)$ **16.** $d(A, B)$ **17.** $d(O, B)$

18. $d(B, A)$ **19.** $d(B, C)$ **20.** $d(D, C)$

Solve and graph.

21. $|x| = 7$ **22.** $|x| = 5$ **23.** $|x| \leq 7$

24. $|t| \leq 5$ **25.** $|x| \geq 7$ **26.** $|x| \geq 5$

27. $|y - 5| = 3$ **28.** $|t - 3| = 4$ **29.** $|y - 5| < 3$

30. $|t - 3| < 4$ **31.** $|y - 5| > 3$ **32.** $|t - 3| > 4$

33. $|u + 8| = 3$ **34.** $|x + 1| = 5$ **35.** $|u + 8| \leq 3$

36. $|x + 1| \leq 5$ **37.** $|u + 8| \geq 3$ **38.** $|x + 1| \geq 5$

B *Solve.*

39. $|3x + 4| = 8$ **40.** $|2x - 3| = 5$ **41.** $|5x - 3| \leq 12$

42. $|2x - 3| \leq 5$ **43.** $|2y - 8| > 2$ **44.** $|3u + 4| > 3$

45. $|5t - 7| = 11$ **46.** $|6m + 9| = 13$ **47.** $|9 - 7u| < 14$

48. $|7 - 9M| < 15$ **49.** $|1 - \frac{2}{3}x| \geq 5$ **50.** $|\frac{3}{4}x + 3| \geq 9$

51. $|\frac{9}{5}C + 32| < 31$ **52.** $|\frac{5}{9}(F - 32)| < 40$

C *For what values of x does each of the following hold?*

53. $|x - 5| = x - 5$

54. $|x + 7| = x + 7$

55. $|x + 8| = -(x + 8)$

56. $|x - 11| = -(x - 11)$

57. $|4x + 3| = 4x + 3$

58. $|5x - 9| = (5x - 9)$

59. $|5x - 2| = -(5x - 2)$

60. $|3x + 7| = -(3x + 7)$

61. Show that $|b - a| = |a - b|$ for all real numbers a and b.

4-6 CHAPTER REVIEW

The strategy for solving word problems can be applied to geometric problems and ratio and proportion problems. The **ratio** of a to b is a/b; a **proportion** is a statement that two ratios are equal. *(4-1)*

The method may also be applied to rate–time and other problems. The formula Quantity = Rate × Time involves average rates over time. *(4-2)*

The **solution set** of a **linear inequality**—an inequality in one variable with no exponents—is the set of all values (**solutions**) that make the inequality true. Inequalities are **equivalent** if they have the same solution set. These **inequality properties** produce equivalent inequalities:

If $a < b$, then $a + c < b + c$ and $a - c < b - c$.

If $a < b$ and $c > 0$, then $a \cdot c < b \cdot c$ and $a/c < b/c$.

If $a < b$ and $c < 0$, then $a \cdot c > b \cdot c$ and $a/c > b/c$.

Solutions to inequalities can be written in **interval notation**, summarized in Table 1. *(4-4)*

The distance between two points A and B on the real number line is given by $d(A, B) = |A - B| = |B - A|$. Equations and inequalities involving absolute value can thus sometimes be solved geometrically. They are solved algebraically by translating the statement to one without absolute values using:

1. $|x| = p$ means $x = \pm p$.
2. $|x| < p$ means $-p < x < p$.
3. $|x| > p$ means $x < -p$ or $x > p$. *(4-5)*

REVIEW EXERCISE 4-6

Work through all the problems in this chapter review and check answers in the back of the book. (Answers to all problems are there, and following each answer is a number in italics indicating the section in which that type of problem is discussed.) Where weaknesses show up, review appropriate sections in the text.

A *Solve.*

1. $4x - 9 = x - 15$

2. $2x + 3(x - 1) = 5 - (x - 4)$

3. $4x - 9 < x - 15$

4. $-3 < 2x - 5 < 7$

5. $|x| = 6$

6. $|x| < 6$

7. $|x| > 6$

8. $|y + 9| = 5$

9. $|y + 9| < 5$

10. $|y + 9| > 5$

11. $\frac{x}{4} - 1 \geq \frac{x}{3}$

12. Solve $A = bh/2$ for b. *Area of a triangle*

13. If the coordinates of A and B on a real number line are -8 and -2, respectively, find:
(A) $d(A, B)$ **(B)** $d(B, A)$

B *Solve each and graph each inequality or absolute-value statement.*

14. $-14 \leq 3x - 2 < 7$

15. $-3 \leq 5 - 2x < 3$

16. $3(2 - x) - 2 \leq 2x - 1$

17. $0.4x - 0.3(x - 3) = 5$

18. $|4x - 7| = 5$

19. $|4x - 7| \leq 5$

20. $|4x - 7| > 5$

21. $\frac{x + 3}{8} \leq 5 - \frac{2 - x}{3}$

22. $-6 < \frac{3}{5}(x - 4) \leq -3$

23. Solve $S = \frac{n(a + L)}{2}$ for L. *Arithmetic progression*

24. Solve $P = M - Mdt$ for M. *Mathematics of finance*

25. Indicate true (T) or false (F):
(A) If $x < y$ and $a > 0$, then $ax < ay$.
(B) If $x < y$ and $a < 0$, then $ax > ay$.

C *Solve each and graph each inequality or absolute-value statement.*

26. $|3 - 2x| \leq 5$

27. $\frac{3x}{5} - \frac{1}{2}(x - 3) \leq \frac{1}{3}(x + 2)$

28. $-4 \leq \frac{2}{3}(6 - 2x) \leq 8$

29. $|2x - 3| < -2$

30. Solve $y = \dfrac{4x + 3}{2x - 5}$ for x in terms of y.

31. Solve $\dfrac{1}{f} = \dfrac{1}{f_1} + \dfrac{1}{f_2}$ for f_1. *Optics*

32. $|2x - 3| = 2x - 3$ for what values of x?

33. $|2x - 3| = -(2x - 3)$ for what values of x?

APPLICATIONS *Set up equations or inequalities and solve.*

34. If the width of a rectangle with perimeter 76 centimeters is 2 centimeters less than three-fifths the length, what are the dimensions of the rectangle?

★35. If one car leaves a town traveling at 56 kilometers per hour, how long will it take a second car traveling at 76 kilometers per hour to catch up, if the second car leaves 1.5 hours later?

36. If you paid $210 for a stereo that was on sale for 30% off list price, what was the price before the sale?

★37. Suppose one printing press can print 45 brochures per minute and a newer press can print 55. How long will it take to print 3,000 brochures if the newer press is brought on the job 10 minutes after the older press has started and both continue until finished?

38. If there are 2.54 centimeters in 1 inch, how many inches are in 127 centimeters?

39. If 50 milliliters of a solution contains 18 milliliters of alcohol, how many milliliters of alcohol are in 70 milliliters of the same solution?

40. A student received grades of 65 and 80 on two tests. What must the student receive on a third test to have an average not lower than 75?

41. If the ratio of all the squirrels in a forest to the ones that were captured, marked, and released is 55/6, and there are 360 marked squirrels, how many squirrels are in the forest?

42. A chemical solution is to be kept between 10 and 15°C, inclusive; that is, $10 \le C \le 15$. What is the temperature range in Fahrenheit degrees? [$C = \frac{5}{9}(F - 32)$.]

5

EXPONENTS, RADICALS, AND COMPLEX NUMBERS

The concept of exponents was first introduced in Chapter 2. There we considered only positive-integer exponents:

$$a^n = \underbrace{a \cdot a \cdot \cdots \cdot a}_{n \text{ factors}}$$

These were the only exponents needed to deal with polynomials. In this chapter, we will extend the concept to include any integer (positive, zero, or negative) as an exponent, and then any rational number. The latter requires a consideration of roots and radicals, and these in turn will lead us to investigate a new number system, the complex numbers.

5-1 POSITIVE-INTEGER EXPONENTS

- Positive-Integer Exponents and Exponent Properties
- Use of the Exponent Properties

In Section 2-1, we introduced the concept of a positive-integer (natural number) exponent and one property that we used repeatedly when dealing with polynomials:

Property 1

$$a^m a^n = a^{m+n} \qquad a^5a^2 = a^{5+2} = a^7$$

In this section, we will discuss four more exponent properties. The five properties will dictate how, in subsequent sections, we extend the exponent concept beyond positive integers.

POSITIVE-INTEGER EXPONENTS AND EXPONENT PROPERTIES

Property 1 of positive-integer exponents was based upon the following observation:

$$a^3a^4 = \underbrace{(a \cdot a \cdot a)}_{\substack{3\\ \text{factors}}}\underbrace{(a \cdot a \cdot a \cdot a)}_{\substack{4\\ \text{factors}}} = \underbrace{(a \cdot a \cdot a \cdot a \cdot a \cdot a \cdot a)}_{\substack{7\\ \text{factors}}} = a^{3+4} = a^7$$

Some other properties can be obtained similarly.

Consider:

$$(a^3)^4 = a^3 \cdot a^3 \cdot a^3 \cdot a^3 = \overbrace{(a \cdot a \cdot a)(a \cdot a \cdot a)(a \cdot a \cdot a)(a \cdot a \cdot a)}^{\text{4 groups of 3 factors each}}$$

$$= \overbrace{(a \cdot a \cdot a \cdot a \cdot a \cdot a \cdot a \cdot a \cdot a \cdot a \cdot a \cdot a)}^{4 \cdot 3 \text{ factors}} = a^{4 \cdot 3} = a^{12}$$

which suggests:

Property 2

$$(a^n)^m = a^{mn}$$

$$(a^2)^5 = a^{5 \cdot 2} = a^{10}$$

Consider:

$$(ab)^4 = \overbrace{(ab)(ab)(ab)(ab)}^{\text{4 factors of } (ab)} = \overbrace{(a \cdot a \cdot a \cdot a)}^{\text{4 factors of } a}\overbrace{(b \cdot b \cdot b \cdot b)}^{\text{4 factors of } b} = a^4b^4$$

which suggests:

Property 3

$$(ab)^m = a^m b^m$$

$$(ab)^7 = a^7b^7$$

Consider:

$$\left(\frac{a}{b}\right)^5 = \overbrace{\left(\frac{a}{b} \cdot \frac{a}{b} \cdot \frac{a}{b} \cdot \frac{a}{b} \cdot \frac{a}{b}\right)}^{\text{5 factors of } a/b} = \frac{\overbrace{a \cdot a \cdot a \cdot a \cdot a}^{\text{5 factors of } a}}{\underbrace{b \cdot b \cdot b \cdot b \cdot b}_{\text{5 factors of } b}} = \frac{a^5}{b^5}$$

which suggests:

Property 4

$$\left(\frac{a}{b}\right)^m = \frac{a^m}{b^m}$$

$$\left(\frac{a}{b}\right)^3 = \frac{a^3}{b^3}$$

Consider:

(A) $$\frac{a^7}{a^3} = \frac{a \cdot a \cdot a \cdot a \cdot a \cdot a \cdot a}{a \cdot a \cdot a}$$

$$= \frac{\cancel{(a \cdot a \cdot a)}(a \cdot a \cdot a \cdot a)}{\cancel{(a \cdot a \cdot a)}} = a^{7-3} = a^4$$

(B) $$\frac{a^3}{a^3} = \frac{a \cdot a \cdot a}{a \cdot a \cdot a} = 1$$

(C) $$\frac{a^4}{a^7} = \frac{a \cdot a \cdot a \cdot a}{a \cdot a \cdot a \cdot a \cdot a \cdot a \cdot a}$$

$$= \frac{\cancel{(a \cdot a \cdot a \cdot a)}}{\cancel{(a \cdot a \cdot a \cdot a)}(a \cdot a \cdot a)} = \frac{1}{a^{7-4}} = \frac{1}{a^3}$$

which suggests:

Property 5

$$\frac{a^m}{a^n} = \begin{cases} a^{m-n} & \text{if } m \text{ is larger than } n \\ 1 & \text{if } m = n \\ \dfrac{1}{a^{n-m}} & \text{if } n \text{ is larger than } m \end{cases}$$

$$\frac{a^8}{a^3} = a^{8-3} = a^5 \qquad \frac{a^8}{a^8} = 1 \qquad \frac{a^3}{a^8} = \frac{1}{a^{8-3}} = \frac{1}{a^5}$$

Remember that the properties of exponents apply to products and quotients, not to sums and differences. Many mistakes are made in algebra by people applying a property of exponents to the wrong algebraic form. For example:

$$(ab)^2 = a^2b^2 \qquad \text{but} \qquad (a + b)^2 \neq a^2 + b^2$$

The exponent properties are summarized here for m and n positive integers:

Exponent Properties

For m and n positive integers:

1. $a^m a^n = a^{m+n}$
2. $(a^n)^m = a^{mn}$
3. $(ab)^m = a^m b^m$
4. $\left(\dfrac{a}{b}\right)^m = \dfrac{a^m}{b^m}$
5. $\dfrac{a^m}{a^n} = \begin{cases} a^{m-n} & \text{if } m \text{ is larger than } n \\ 1 & \text{if } m = n \\ \dfrac{1}{a^{n-m}} & \text{if } n \text{ is larger than } m \end{cases}$

USE OF THE EXPONENT PROPERTIES

Example 1 Simplify, using natural number exponents only; that is, rewrite the expression so that each variable occurs only once and has only a single natural number exponent applied to it.

(A) $x^{12}x^{13}$ **(B)** $(t^7)^5$ **(C)** $(xy)^5$

(D) $\left(\dfrac{u}{v}\right)^3$ **(E)** $\dfrac{x^{12}}{x^4}$ **(F)** $\dfrac{t^4}{t^9}$

Solution **(A)** $x^{12}x^{13} \boxed{= x^{12+13}} = x^{25}$ **(B)** $(t^7)^5 \boxed{= t^{7\cdot 5}} = t^{35}$

(C) $(xy)^5 = x^5y^5$ **(D)** $\left(\dfrac{u}{v}\right)^3 = \dfrac{u^3}{v^3}$

(E) $\dfrac{x^{12}}{x^4} \boxed{= x^{12-4}} = x^8$ **(F)** $\dfrac{t^4}{t^9} \boxed{= \dfrac{1}{t^{9-4}}} = \dfrac{1}{t^5}$

Problem 1 Simplify, using natural number exponents only:

(A) x^8x^6 **(B)** $(u^4)^5$ **(C)** $(xy)^9$

(D) $\left(\dfrac{x}{y}\right)^4$ **(E)** $\dfrac{x^{10}}{x^3}$ **(F)** $\dfrac{x^3}{x^{10}}$

Example 2 Simplify, using natural number exponents only:

(A) $(x^2y^3)^4$ **(B)** $\left(\dfrac{u^3}{v^4}\right)^3$ **(C)** $\dfrac{2x^9y^{11}}{4x^{12}y^7}$

Solution **(A)** $(x^2y^3)^4 \boxed{= (x^2)^4(y^3)^4} = x^8y^{12}$ **(B)** $\left(\dfrac{u^3}{v^4}\right)^3 \boxed{= \dfrac{(u^3)^3}{(v^4)^3}} = \dfrac{u^9}{v^{12}}$

(C) $\dfrac{2x^9y^{11}}{4x^{12}y^7} \boxed{= \dfrac{2}{4}\cdot\dfrac{x^9}{x^{12}}\cdot\dfrac{y^{11}}{y^7} = \dfrac{1}{2}\cdot\dfrac{1}{x^3}\cdot\dfrac{y^4}{1}} = \dfrac{y^4}{2x^3}$

Problem 2 Simplify, using natural number exponents only:

(A) $(u^3v^4)^2$ **(B)** $\left(\dfrac{x^4}{y^3}\right)^2$ **(C)** $\dfrac{9x^7y^2}{3x^5y^3}$ **(D)** $\dfrac{(2x^2y)^3}{(4xy^3)^2}$

Knowing the rules of the game of chess doesn't make you skilled at playing chess; similarly, memorizing the properties of exponents doesn't necessarily make you skilled at using these properties. To acquire skill in their use, you must use these properties in a fairly large variety of problems. The following exercises should help you acquire this skill.

ANSWERS TO MATCHED PROBLEMS

1. **(A)** x^{14} **(B)** u^{20} **(C)** x^9y^9 **(D)** $\dfrac{x^4}{y^4}$ **(E)** x^7 **(F)** $\dfrac{1}{x^7}$

2. **(A)** u^6v^8 **(B)** $\dfrac{x^8}{y^6}$ **(C)** $\dfrac{3x^2}{y}$ **(D)** $\dfrac{x^4}{2y^3}$

EXERCISE 5-1

A *Replace the question marks with appropriate symbols.*

1. $y^2y^8 = y^?$ **2.** $x^8x^5 = x^?$ **3.** $y^9 = y^3y^?$

4. $x^{11} = x^?x^6$ **5.** $(u^4)^3 = u^?$ **6.** $(v^2)^3 = ?$

7. $x^{10} = (x^?)^5$ **8.** $y^{12} = (y^6)^?$ **9.** $(uv)^7 = ?$

10. $(xy)^5 = x^5y^?$ **11.** $p^4q^4 = (pq)^?$ **12.** $m^3n^3 = (mn)^?$

13. $\left(\dfrac{a}{b}\right)^8 = ?$ **14.** $\left(\dfrac{x}{y}\right)^4 = \dfrac{x^?}{y^4}$ **15.** $\dfrac{m^3}{n^3} = \left(\dfrac{m}{n}\right)^?$

16. $\dfrac{x^7}{y^7} = \left(\dfrac{x}{y}\right)^?$ **17.** $\dfrac{n^{14}}{n^8} = n^?$ **18.** $\dfrac{x^7}{x^3} = x^?$

19. $m^6 = \dfrac{m^8}{m^?}$ **20.** $x^3 = \dfrac{x^?}{x^4}$ **21.** $\dfrac{x^4}{x^{11}} = \dfrac{1}{x^?}$

22. $\dfrac{a^5}{a^9} = \dfrac{1}{a^?}$ **23.** $\dfrac{1}{x^8} = \dfrac{x^4}{x^?}$ **24.** $\dfrac{1}{u^2} = \dfrac{u^?}{u^9}$

Simplify, using appropriate properties of exponents.

25. $(4x^2)(2x^{10})$ **26.** $(2x^4)(4x^7)$ **27.** $\dfrac{9x^6}{3x^4}$

28. $\dfrac{4x^8}{2x^6}$ **29.** $\dfrac{6m^5}{8m^7}$ **30.** $\dfrac{4u^3}{2u^7}$

31. $(xy)^{10}$ **32.** $(cd)^{12}$ **33.** $\left(\dfrac{m}{n}\right)^5$

34. $\left(\dfrac{x}{y}\right)^6$

B **35.** $(4y^3)(5y)(y^7)$ **36.** $(4x^2)(3x^5)(x^4)$

37. $(6 \times 10^8)(8 \times 10^9)$ **38.** $(3 \times 10^3)(4 \times 10^{13})$

39. $(10^7)^2$ **40.** $(10^4)^5$ **41.** $(x^3)^2$ **42.** $(y^4)^5$

43. $(m^2n^5)^3$ **44.** $(x^2y^3)^4$ **45.** $\left(\frac{c^2}{d^5}\right)^3$ **46.** $\left(\frac{a^3}{b^2}\right)^4$

47. $\frac{9u^8v^6}{3u^4v^8}$ **48.** $\frac{2x^3y^8}{6x^7y^2}$ **49.** $(2s^2t^4)^4$ **50.** $(3a^3b^2)^3$

51. $6(xy^3)^5$ **52.** $2(x^2y)^4$ **53.** $\left(\frac{mn^3}{p^2q}\right)^4$ **54.** $\left(\frac{x^2y}{2w^2}\right)^3$

55. $\frac{(4u^3v)^3}{(2uv^2)^6}$ **56.** $\frac{(2xy^3)^2}{(4x^2y)^3}$ **57.** $\frac{(9x^3)^2}{(-3x)^2}$ **58.** $\frac{(-2x^2)^3}{(2^2x)^4}$

59. $\frac{-x^2}{(-x)^2}$ **60.** $\frac{-2^2}{(-2)^2}$ **61.** $\frac{(-x^2)^2}{(-x^3)^3}$ **62.** $\frac{-2^4}{(-2a^2)^4}$

63. $\left(-\frac{x}{y}\right)^3\left(\frac{y^2}{w}\right)^2\left(\frac{w}{x^2}\right)^3$ **64.** $\left(-\frac{a^2b}{c}\right)^2\left(\frac{c}{b^2}\right)^3\left(\frac{1}{a^3}\right)^2$

65. $\frac{3(x+y)^3(x-y)^4}{(x-y)^2 6(x+y)^5}$ **66.** $\frac{10(u-v+w)^8}{5(u-v+w)^{11}}$

C *Simplify, assuming n is restricted so that each exponent represents a positive integer.*

67. $x^{5-n}x^{n+3}$ **68.** $y^{2n+2}y^{n-3}$ **69.** $\frac{x^{2n}}{x^n}$

70. $\frac{x^{n+2}}{x^n}$ **71.** $(x^{n+1})^2$ **72.** $(x^{n+1})^n$

73. $\frac{u^{n+3}v^n}{u^{n+1}v^{n+4}}$ **74.** $\frac{(x^ny^{n+1})^2}{x^{2n+1}y^{2n}}$

5-2 INTEGER EXPONENTS

- Zero Exponents
- Negative-Integer Exponents

In this section we will extend the exponent concept to include zero and negative integers as exponents. Typical scientific expressions such as the following will then make sense:

The diameter of a red corpuscle is approximately 8×10^{-5} centimeter.
The amount of water found in the air as vapor is about 9×10^{-6} times the amount of water found in the sea.
The focal length of a thin lens is given by $f^{-1} = a^{-1} + b^{-1}$.

In extending the concept of exponent beyond the natural numbers, we will require that any new exponent symbol be defined in such a way that all five laws of exponents for natural numbers continue to hold. We will need only one set of laws for all types of exponents rather than a new set for each new exponent.

ZERO EXPONENTS

We will start by defining the 0 exponent. If all the exponent laws must hold even if some of the exponents are 0, then a^0 ($a \neq 0$) should be defined so that when the first law of exponents is applied,

$$a^0 \cdot a^2 = a^{0+2} = a^2$$

This suggests that a^0 should be defined as 1 for all nonzero real numbers a, since 1 is the only real number that gives a^2 when multiplied by a^2. If we let $a = 0$ and follow the same reasoning, we find that we require

$$0^0 \cdot 0^2 = 0^{0+2} = 0^2 = 0$$

However, since $0^0 \cdot 0^2 = 0^0 \cdot 0 = 0$ is true whatever value is assigned to 0^0, 0^0 could be any real number and is not uniquely determined. For this reason and others, we choose not to define 0^0.

Definition of 0 Exponent

For all real numbers $a \neq 0$:

$a^0 = 1$

0^0 is not defined.

Example 3 Simplify:

(A) 5^0 **(B)** 325^0 **(C)** $(\frac{1}{3})^0$ **(D)** t^0, $t \neq 0$
(E) $(x^2y^3)^0$, $x, y \neq 0$

Solution All are equal to 1.

Problem 3 Simplify:

(A) 12^0 **(B)** 999^0 **(C)** $(\frac{2}{7})^0$
(D) x^0, $x \neq 0$ **(E)** $(m^3n^3)^0$, $m, n \neq 0$

NEGATIVE-INTEGER EXPONENTS

To get an idea of how a negative-integer exponent should be defined, we can proceed as above. If the first law of exponents is to hold, then a^{-2} $(a \neq 0)$ must be defined so that

$$a^{-2} \cdot a^2 = a^{-2+2} = a^0 = 1$$

Thus a^{-2} must be the reciprocal of a^2; that is,

$$a^{-2} = \frac{1}{a^2}$$

This kind of reasoning leads us to the following general definition:

Definition of Negative-Integer Exponents

If n is a positive integer and a is a nonzero real number, then

$$a^{-n} = \frac{1}{a^n}$$

It follows, using equality properties,† that

$$a^n = \frac{1}{a^{-n}}$$

Example 4 Write using positive exponents or no exponents:

(A) a^{-7} **(B)** $\dfrac{1}{x^{-8}}$ **(C)** 10^{-3} **(D)** $\dfrac{x^{-3}}{y^{-5}}$

Solution **(A)** $a^{-7} = \dfrac{1}{a^7}$ **(B)** $\dfrac{1}{x^{-8}} = x^8$

(C) $10^{-3} = \dfrac{1}{10^3}$ or $\dfrac{1}{1{,}000}$ or 0.001

(D) $\dfrac{x^{-3}}{y^{-5}} \left[= \dfrac{x^{-3}}{1} \cdot \dfrac{1}{y^{-5}} = \dfrac{1}{x^3} \cdot \dfrac{y^5}{1}\right] = \dfrac{y^5}{x^3}$

† Multiply both sides of $a^{-n} = 1/a^n$ by a^n/a^{-n} to obtain

$$\frac{a^n}{a^{-n}} \cdot a^{-n} = \frac{a^n}{a^{-n}} \cdot \frac{1}{a^n}$$

$$a^n = \frac{1}{a^{-n}}$$

Problem 4 Write using positive exponents or no exponents:

(A) x^{-5} **(B)** $\dfrac{1}{y^{-4}}$ **(C)** 10^{-2} **(D)** $\dfrac{m^{-2}}{n^{-3}}$

With the definition of negative exponent and 0 exponent behind us, we can now replace the fifth law of exponents with a simpler form that does not have any restrictions on the relative size of the exponents. Thus,

$$\frac{a^m}{a^n} = a^{m-n} = \frac{1}{a^{n-m}}$$

Example 5 Simplify, leaving answers with negative exponents:

(A) $\dfrac{2^5}{2^8}$ **(B)** $\dfrac{10^{-3}}{10^6}$

Simplify, leaving answers with positive exponents:

(C) $\dfrac{2^5}{2^8}$ **(B)** $\dfrac{10^{-3}}{10^6}$

Solution **(A)** $\dfrac{2^5}{2^8} = 2^{5-8} = 2^{-3}$ **(B)** $\dfrac{10^{-3}}{10^6} = 10^{-3-6} = 10^{-9}$

(C) $\dfrac{2^5}{2^8} = \dfrac{1}{2^{8-5}} = \dfrac{1}{2^3}$ **(D)** $\dfrac{10^{-3}}{10^6} = \dfrac{1}{10^{6-(-3)}} = \dfrac{1}{10^9}$

Problem 5 Simplify, leaving answers with negative exponents:

(A) $\dfrac{3^4}{3^9}$ **(B)** $\dfrac{x^{-2}}{x^3}$

Simplify, leaving answers with positive exponents:

(C) $\dfrac{3^4}{3^9}$ **(D)** $\dfrac{x^{-2}}{x^3}$

Table 1 provides a summary of all of our work on exponents to this point.

TABLE 1 INTEGER EXPONENTS AND THEIR LAWS (SUMMARY)

DEFINITION OF a^p p an integer, a a real number	LAWS OF EXPONENTS n and m integers, a and b real numbers
1. If p is a positive integer, then $a^p = a \cdot a \cdot \cdots \cdot a \quad p$ factors of a *Example:* $3^5 = 3 \cdot 3 \cdot 3 \cdot 3 \cdot 3$	**1.** $a^m a^n = a^{m+n}$ **2.** $(a^n)^m = a^{mn}$ **3.** $(ab)^m = a^m b^m$
2. If $p = 0$, then $a^p = 1 \quad a \neq 0$ *Example:* $3^0 = 1$	**4.** $\left(\frac{a}{b}\right)^m = \frac{a^m}{b^m}$ **5.** $\frac{a^m}{a^n} = a^{m-n} = \frac{1}{a^{n-m}}$
3. If p is a negative integer, then $a^p = \frac{1}{a^{-p}} \quad a \neq 0$ *Example:* $3^{-4} \boxed{= \frac{1}{3^{-(-4)}}} = \frac{1}{3^4}$	

Example 6 Simplify and express answers using positive exponents only:†

(A) a^5a^{-2} (B) $(a^{-3}b^2)^{-2}$ (C) $\left(\frac{a^{-5}}{a^{-2}}\right)^{-1}$

(D) $\frac{4x^{-3}y^{-5}}{6x^{-4}y^3}$ (E) $\frac{10^{-4} \cdot 10^2}{10^{-3} \cdot 10^5}$ (F) $\left(\frac{m^{-3}m^3}{n^{-2}}\right)^{-2}$

Solutions

(A) $a^5a^{-2} \boxed{= a^{5-2}} = a^3$

(B) $(a^{-3}b^2)^{-2} \boxed{= (a^{-3})^{-2}(b^2)^{-2}} = a^6b^{-4} = \frac{a^6}{b^4}$

(C) $\left(\frac{a^{-5}}{a^{-2}}\right)^{-1} \boxed{= \frac{(a^{-5})^{-1}}{(a^{-2})^{-1}}} = \frac{a^5}{a^2} = a^3$

(D) $\frac{4x^{-3}y^{-5}}{6x^{-4}y^3} \boxed{= \frac{2x^{-3-(-4)}}{3y^{3-(-5)}} = \frac{2x^{-3+4}}{3y^{3+5}}} = \frac{2x}{3y^8}$

† It is important to realize that there are situations where it is desirable to allow negative exponents in an answer (see Section 5-3, for example). In this section we ask you to write answers using positive exponents only so that problems in the exercise set will have unique answer forms.

or, changing to positive exponents first,

$$\frac{4x^{-3}y^{-5}}{6x^{-4}y^3} = \frac{2x^4}{3x^3y^3y^5} = \frac{2x}{3y^8}$$

(E) $$\frac{10^{-4} \cdot 10^2}{10^{-3} \cdot 10^5} = \frac{10^{-4+2}}{10^{-3+5}} = \frac{10^{-2}}{10^2} = \frac{1}{10^4} = \frac{1}{10{,}000} = 0.0001$$

(F) $$\left(\frac{m^{-3}m^3}{n^{-2}}\right)^{-2} = \left(\frac{m^{-3+3}}{n^{-2}}\right)^{-2} = \left(\frac{m^0}{n^{-2}}\right)^{-2}$$

$$= \left(\frac{1}{n^{-2}}\right)^{-2} = \frac{1^{-2}}{(n^{-2})^{-2}} = \frac{1}{n^4}$$

Problem 6 Simplify and express answers using positive exponents only:

(A) $x^{-2}x^6$ **(B)** $(x^3y^{-2})^{-2}$ **(C)** $\left(\frac{x^{-6}}{x^{-2}}\right)^{-1}$

(D) $\frac{8m^{-2}n^{-4}}{6m^{-5}n^2}$ **(E)** $\frac{10^{-3} \cdot 10^5}{10^{-2} \cdot 10^6}$

As stated earlier, laws of exponents involve products and quotients, not sums and differences. Consider:

$$\frac{a^{-2}y}{b} = \frac{y}{a^2b} \qquad \frac{a^{-2} + y}{b} \neq \frac{y}{a^2b}$$

The plus sign in the numerator of the second illustration makes a big difference. Actually, $\frac{a^{-2} + y}{b}$ represents a compact way of writing a complex fraction. To simplify, we replace a^{-2} with $1/a^2$, then proceed as in Section 3-5:

$$\frac{a^{-2} + y}{b} = \frac{\frac{1}{a^2} + y}{b} = \frac{a^2\left(\frac{1}{a^2} + y\right)}{a^2 \cdot b}$$

$$= \frac{1 + a^2y}{a^2b}$$

Also, consider the following:

$$(a^{-1}b^{-1})^2 = a^{-2}b^{-2} = \frac{1}{a^2b^2}$$

but

$$(a^{-1} + b^{-1})^2 \neq a^{-2} + b^{-2} \qquad \text{and} \qquad a^{-2} + b^{-2} \neq \frac{1}{a^2 + b^2}$$

The expression $(a^{-1} + b^{-1})^2$ is correctly simplified in the next example. You are asked to do the same for $a^{-2} + b^{-2}$ in the related problem.

Example 7 Simplify and express answers using positive exponents only:

(A) $\dfrac{3^{-2} + 2^{-1}}{11}$ **(B)** $(a^{-1} + b^{-1})^2$

Solution **(A)** $$\frac{3^{-2} + 2^{-1}}{11} = \frac{\frac{1}{3^2} + \frac{1}{2}}{11} = \frac{\frac{2}{18} + \frac{9}{18}}{11} \boxed{= \frac{11}{18} \div 11} = \frac{11}{18} \cdot \frac{1}{11} = \frac{1}{18}$$

(B) $$(a^{-1} + b^{-1})^2 = \left(\frac{1}{a} + \frac{1}{b}\right)^2 = \left(\frac{b + a}{ab}\right)^2 = \frac{b^2 + 2ab + a^2}{a^2b^2}$$

Alternatively,

$$(a^{-1} + b^{-1})^2 = a^{-2} + 2a^{-1}b^{-1} + b^{-2} = \frac{1}{a^2} + \frac{2}{ab} + \frac{1}{b^2} = \frac{b^2 + 2ab + a^2}{a^2b^2}$$

Problem 7 Simplify and express answers using positive exponents only:

(A) $\dfrac{2^{-2} + 3^{-1}}{5}$ **(B)** $a^{-2} + b^{-2}$

ANSWERS TO MATCHED PROBLEMS

3. All are equal to 1.

4. **(A)** $\dfrac{1}{x^5}$ **(B)** y^4 **(C)** $\dfrac{1}{10^2}$ or $\dfrac{1}{100}$ or 0.01 **(D)** $\dfrac{n^3}{m^2}$

5. **(A)** 3^{-5} **(B)** x^{-5} **(C)** $\frac{1}{3^5}$ **(D)** $\frac{1}{x^5}$

6. **(A)** x^4 **(B)** $\frac{y^4}{x^6}$ **(C)** x^4 **(D)** $\frac{4m^3}{3n^6}$

(E) $\frac{1}{10^2}$ or $\frac{1}{100}$ or 0.01 **7.** **(A)** $\frac{7}{60}$ **(B)** $\frac{a^2 + b^2}{a^2b^2}$

EXERCISE 5-2 *Simplify and write answers using positive exponents only.*

A

1. 23^0 **2.** 10^0 **3.** y^0 **4.** x^0

5. 3^{-3} **6.** 2^{-2} **7.** m^{-7} **8.** x^{-4}

9. $\frac{1}{4^{-3}}$ **10.** $\frac{1}{3^{-2}}$ **11.** $\frac{1}{y^{-5}}$ **12.** $\frac{1}{x^{-3}}$

13. $10^7 \cdot 10^{-5}$ **14.** $10^{-4} \cdot 10^6$ **15.** $y^{-3}y^4$ **16.** x^6x^{-2}

17. u^5u^{-5} **18.** $m^{-3}m^3$ **19.** $\frac{10^3}{10^{-7}}$ **20.** $\frac{10^8}{10^{-3}}$

21. $\frac{x^9}{x^{-2}}$ **22.** $\frac{a^8}{a^{-4}}$ **23.** $\frac{z^{-2}}{z^3}$ **24.** $\frac{b^{-3}}{b^5}$

25. $\frac{10^{-1}}{10^6}$ **26.** $\frac{10^{-4}}{10^2}$ **27.** $(10^{-4})^{-3}$ **28.** $(2^{-3})^{-2}$

29. $(y^{-2})^{-4}$ **30.** $(x^{-5})^{-2}$ **31.** $(u^{-5}v^{-3})^{-2}$ **32.** $(x^{-3}y^{-2})^{-1}$

33. $(x^2y^{-3})^2$ **34.** $(x^{-2}y^3)^2$ **35.** $(x^{-2}y^3)^{-1}$ **36.** $(x^2y^{-3})^{-1}$

B

37. $(m^2)^0$ **38.** $1{,}231^0$ **39.** $\frac{10^{-3}}{10^{-5}}$

40. $\frac{10^{-2}}{10^{-4}}$ **41.** $\frac{y^{-2}}{y^{-3}}$ **42.** $\frac{x^{-3}}{x^{-2}}$

43. $\frac{10^{-13} \cdot 10^{-4}}{10^{-21} \cdot 10^3}$ **44.** $\frac{10^{23} \cdot 10^{-11}}{10^{-3} \cdot 10^{-2}}$ **45.** $\frac{18 \times 10^{12}}{6 \times 10^{-4}}$

46. $\frac{8 \times 10^{-3}}{2 \times 10^{-5}}$ **47.** $\left(\frac{y}{y^{-2}}\right)^3$ **48.** $\left(\frac{x^2}{x^{-1}}\right)^2$

49. $\frac{1}{(3mn)^{-2}}$ **50.** $(2cd^2)^{-3}$ **51.** $(2mn^{-3})^3$

52. $(3x^3y^{-2})^2$ **53.** $(m^4n^{-5})^{-3}$ **54.** $(x^{-3}y^2)^{-2}$

55. $(2^2 3^{-3})^{-1}$ **56.** $(2^{-3}3^2)^{-2}$

57. $(10^{12} \cdot 10^{-12})^{-1}$

58. $(10^2 \cdot 3^0)^{-2}$

59. $\dfrac{8x^{-3}y^{-1}}{6x^2y^{-4}}$

60. $\dfrac{9m^{-4}n^3}{12m^{-1}n^{-1}}$

61. $\dfrac{2a^6b^{-2}}{16a^{-3}b^2}$

62. $\dfrac{4x^{-2}y^{-3}}{2x^{-3}y^{-1}}$

63. $\left(\dfrac{x^{-1}}{x^{-8}}\right)^{-1}$

64. $\left(\dfrac{n^{-3}}{n^{-2}}\right)^{-2}$

65. $\left(\dfrac{m^{-2}n^3}{m^4n^{-1}}\right)^2$

66. $\left(\dfrac{x^4y^{-1}}{x^{-2}y^3}\right)^2$

67. $\left(\dfrac{6nm^{-2}}{3m^{-1}n^2}\right)^{-3}$

68. $\left(\dfrac{2x^{-3}y^2}{4xy^{-1}}\right)^{-2}$

69. $\left[\left(\dfrac{x^{-2}y^3t}{x^{-3}y^{-2}t^2}\right)^2\right]^{-1}$

70. $\left[\left(\dfrac{u^3v^{-1}w^{-2}}{u^{-2}v^{-2}w}\right)^{-2}\right]^2$

71. $\left(\dfrac{2^2x^2y^0}{8x^{-1}}\right)^{-2}\left(\dfrac{x^{-3}}{x^{-5}}\right)^3$

72. $\left(\dfrac{3^3x^0y^{-2}}{2^3x^3y^{-5}}\right)^{-1}\left(\dfrac{3^3x^{-1}y}{2^2x^2y^{-2}}\right)^2$

C 73. $(a^2 - b^2)^{-1}$

74. $(x + 2)^{-2}$

75. $\dfrac{x^{-1} + y^{-1}}{x + y}$

76. $\dfrac{2^{-1} + 3^{-1}}{25}$

77. $\dfrac{c - d}{c^{-1} - d^{-1}}$

78. $\dfrac{12}{2^{-2} + 3^{-1}}$

79. $(x^{-1} + y^{-1})^{-1}$

80. $(2^{-2} + 3^{-2})^{-1}$

81. $(x^{-1} - y^{-1})^2$

82. $(10^{-2} + 10^{-3})^{-1}$

83. $\left(\dfrac{x^{-1}}{x^{-1} - y^{-1}}\right)^{-1}$

84. $\left[\dfrac{u^{-2} - v^{-2}}{(u^{-1} - v^{-1})^2}\right]^{-1}$

5-3 SCIENTIFIC NOTATION AND APPLICATIONS

Work in science and engineering often involves the use of very, very large numbers:

> The estimated free oxygen of the earth weighs approximately 1,500,000,000,000,000,000,000 grams.

Also involved is the use of very, very small numbers:

> The probable mass of a hydrogen atom is 0.000 000 000 000 000 000 000 001 7 gram.

Writing and working with numbers of this type in standard decimal notation is generally awkward. It is often convenient to represent numbers of this type in **scientific notation**—that is, as the product of a number in the interval [1, 10) and a power of 10. Any decimal fraction, however large or small, can be represented in scientific notation.

Here are some examples of decimal fractions and scientific notation:

$100 = 1 \times 10^2$	$0.001 = 1 \times 10^{-3}$
$1{,}000{,}000 = 1 \times 10^6$	$0.000\ 01 = 1 \times 10^{-5}$
$5 = 5 \times 10^0$	$0.7 = 7 \times 10^{-1}$
$35 = 3.5 \times 10$	$0.083 = 8.3 \times 10^{-2}$
$430 = 4.3 \times 10^2$	$0.004\ 3 = 4.3 \times 10^{-3}$
$5{,}870 = 5.87 \times 10^3$	$0.000\ 687 = 6.87 \times 10^{-4}$
$8{,}910{,}000 = 8.91 \times 10^6$	$0.000\ 000\ 36 = 3.6 \times 10^{-7}$

You can discover a simple mechanical rule that relates the number of decimal places the decimal is moved with the power of 10 that is used:

$$7{,}320{,}000 = 7.320\ 000. \times 10^6 = 7.32 \times 10^6$$

6 places left

Positive exponent

$$0.000\ 000\ 54 = 0.000\ 000\ 5.4 \times 10^{-7} = 5.4 \times 10^{-7}$$

7 places right

Negative exponent

Example 8 Write in scientific notation:

(A) 123 **(B)** 0.001 23 **(C)** 1.23

Solution **(A)** $123 = 1.23 \times 10^2 = 1.23 \times 10^2$

2 places left

Positive exponent

(B) $0.001\ 23 = 0001.23 \times 10^{-3} = 1.23 \times 10^{-3}$

3 places right

Negative exponent

(C) $1.23 = 1.23 \times 10^0 = 1.23 \times 10^0$

0 places moved

Zero exponent

Problem 8 Write in scientific notation:

(A) 450 **(B)** 27,000 **(C)** 0.05
(D) 0.000 006 3 **(E)** 0.000 1 **(F)** 10,000

Scientific notation may be used to evaluate complicated arithmetic problems.

Example 9 Convert to scientific notation and evaluate:

$$\frac{(0.000\ 000\ 000\ 000\ 026)(720)}{(48{,}000{,}000{,}000)(0.001\ 3)}$$

Solution

$$\frac{(0.000\ 000\ 000\ 000\ 026)(720)}{(48{,}000{,}000{,}000)(0.001\ 3)}$$

Write each number in scientific notation.

$$= \frac{(2.6 \times 10^{-14})(7.2 \times 10^{2})}{(4.8 \times 10^{10})(1.3 \times 10^{-3})}$$

Collect all powers of 10 together.

$$= \frac{(2.6)(7.2)}{(4.8)(1.3)} \cdot \frac{(10^{-14})(10^{2})}{(10^{10})(10^{-3})}$$

Calculate the power of 10 and numerical part separately.

$$= 3 \times 10^{-19}$$

If you try to work this problem directly using a hand calculator, you will find that some of the numbers will not fit unless they are first converted to scientific notation. If you have a calculator, try it. Some calculators can compute directly in scientific notation and read out in scientific notation.

Problem 9 Convert to scientific notation and evaluate:

$$\frac{(42{,}000)(0.000\ 000\ 000\ 09)}{(600{,}000{,}000{,}000)(0.000\ 21)}$$

Figure 1 on page 210 shows the relative size of a number of familiar objects on a power-of-10 scale. Note that 10^{10} is not just double 10^{5}.

We are able to look back into time by looking out into space. Since light travels at a fast but finite rate, we see heavenly bodies not as they exist now, but as they existed sometime in the past. If the distance between the sun and the earth is approximately 9.3×10^{7} miles and if light travels at the rate of approximately 1.86×10^{5} miles per second, we see the sun as it was how many minutes ago?

$$t = \frac{d}{r} \approx \frac{9.3 \times 10^{7}}{1.86 \times 10^{5}} = 5 \times 10^{2} = 500 \text{ seconds} \quad \text{or} \quad \frac{500}{60} \approx 8.3 \text{ minutes}$$

Hence, we always see the sun as it was approximately 8.3 minutes ago.

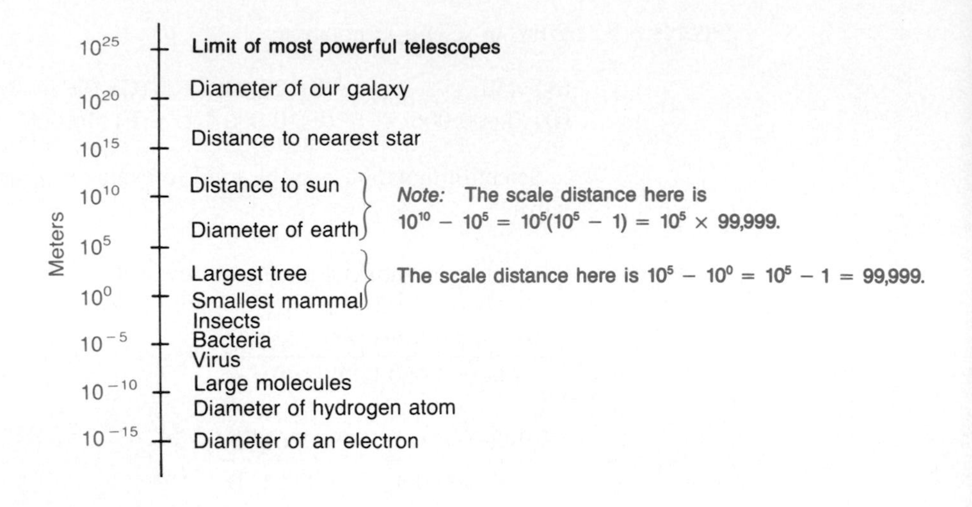

FIGURE 1

ANSWERS TO MATCHED PROBLEMS

8. **(A)** 4.5×10^2 **(B)** 2.7×10^4 **(C)** 5×10^{-2} **(D)** 6.3×10^{-6} **(E)** 1.0×10^{-4} **(F)** 1.0×10^4

9. 3×10^{-14}

EXERCISE 5-3

A *Write in scientific notation.*

1. 70 **2.** 50 **3.** 800 **4.** 600

5. 80,000 **6.** 600,000 **7.** 0.008 **8.** 0.06

9. 0.000 000 08 **10.** 0.000 06 **11.** 52

12. 35 **13.** 0.63 **14.** 0.72

15. 340 **16.** 270 **17.** 0.085

18. 0.032 **19.** 6,300 **20.** 5,200

21. 0.000 006 8 **22.** 0.000 72

Write as a decimal fraction.

23. 8×10^2 **24.** 5×10^2 **25.** 4×10^{-2}

26. 8×10^{-2} **27.** 3×10^5 **28.** 6×10^6

29. 9×10^{-4} **30.** 2×10^{-5} **31.** 5.6×10^4

32. 7.1×10^3 **33.** 9.7×10^{-3} **34.** 8.6×10^{-4}

35. 4.3×10^5 **36.** 8.8×10^6 **37.** 3.8×10^{-7}

38. 6.1×10^{-6}

B *Write in scientific notation.*

39. 5,460,000,000

40. 42,700,000

41. 0.000 000 072 9

42. 0.000 072 3

43. The energy of a laser beam can go as high as 10,000,000,000,000 watts.

44. The distance that light travels in 1 year is called a light-year. It is approximately 5,870,000,000,000 miles.

45. The nucleus of an atom has a diameter of a little more than 1/100,000 that of the whole atom.

46. The mass of one water molecule is 0.000 000 000 000 000 000 000 03 gram.

Write as a decimal fraction.

47. 8.35×10^{10}

48. 3.46×10^{9}

49. 6.14×10^{-12}

50. 6.23×10^{-7}

51. The diameter of the sun is approximately 8.65×10^{5} miles.

52. The distance from the earth to the sun is approximately 9.3×10^{7} miles.

53. The probable mass of a hydrogen atom is 1.7×10^{-24} gram.

54. The diameter of a red corpuscle is approximately 7.5×10^{-5} centimeter.

Simplify and express answer in scientific notation.

55. $(3 \times 10^{-6})(3 \times 10^{10})$

56. $(4 \times 10^{5})(2 \times 10^{-3})$

57. $(2 \times 10^{3})(3 \times 10^{-7})$

58. $(4 \times 10^{-8})(2 \times 10^{5})$

59. $\dfrac{6 \times 10^{12}}{2 \times 10^{7}}$

60. $\dfrac{9 \times 10^{8}}{3 \times 10^{5}}$

61. $\dfrac{15 \times 10^{-2}}{3 \times 10^{-6}}$

62. $\dfrac{12 \times 10^{3}}{4 \times 10^{-4}}$

Convert each numeral to scientific notation and simplify. Express your answer in scientific notation and as a decimal fraction.

63. $\dfrac{(90{,}000)(0.000\ 002)}{0.006}$

64. $\dfrac{(0.000\ 6)(4{,}000)}{0.000\ 12}$

65. $\dfrac{(60{,}000)(0.000\ 003)}{(0.000\ 4)(1{,}500{,}000)}$ **66.** $\dfrac{(0.000\ 039)(140)}{(130{,}000)(0.000\ 21)}$

C **67.** If the mass of the earth is 6×10^{27} grams and each gram is 1.1×10^{-6} ton, find the mass of the earth in tons.

68. In 1929 Vernadsky, a biologist, estimated that all the free oxygen of the earth weighs 1.5×10^{21} grams and is produced by life alone. If 1 gram is approximately 2.2×10^{-3} pound, what is the amount of free oxygen in pounds?

69. Designers of high-speed computers are currently thinking of single-addition times of 10^{-7} second (100 nanoseconds). How many additions would such a computer be able to perform in 1 second? In 1 minute?

70. If electricity travels in a computer circuit at the speed of light (1.86×10^5 miles per second), how far will it travel in the time it takes the computer in the preceding problem to complete a single addition? (Size of circuits is becoming a critical problem in computer design.) Give the answer in miles and in feet.

71. India has a population of 713,000,000 people and a land area of 1,269,000 square miles. What is the population density (people per square mile)?

72. The United States has a population (1980) of 227,000,000 and a land area of 3,539,000 square miles. What is the population density?

5-4 RATIONAL EXPONENTS

- Roots of Real Numbers
- Rational Exponents

We will now extend the exponent concept to include rational number exponents. Expressions such as $x^{1/3}$ and $x^{-2/5}$ will then be meaningful. We will insist that the five exponent properties continue to be true. If $x^{1/3}$ makes sense, $(x^{1/3})^3 = x^{(1/3)\cdot 3} = x^1 = x$, so $x^{1/3}$ is a number whose cube is x, that is, $x^{1/3}$ must name a cube root of x. Thus, we must first consider cube roots and other roots of real numbers before developing rational exponents further.

ROOTS OF REAL NUMBERS

What do we mean by a root of a number? Perhaps you recall that a square root of a number b is a number a such that $a^2 = b$ and a cube root of a number b is a number a such that $a^3 = b$.

What are the square roots of 4?

2 is a square root of 4, since $2^2 = 4$.

-2 is a square root of 4, since $(-2)^2 = 4$.

Thus, 4 has two real square roots, one the negative of the other.

What are the cube roots of 8?

2 is a cube root of 8, since $2^3 = 8$

and 2 is the only real number with this property. In general:

Definition of an *n*th Root

For n a natural number:

a is an nth root of b if $a^n = b$.

2 is a fourth root of 16, since $2^4 = 16$
-3 is a cube root of -27 since $(-3)^3 = -27$

How many real square roots of 9 exist? Of 7? Of -4? How many real fourth roots of 7 exist? Of -7? How many real cube roots of 27 are there? The following important result (which we state without proof) answers these questions completely.

NUMBER OF REAL *n*th ROOTS OF A REAL NUMBER *b*†

	n EVEN	*n* ODD
b positive	Two real *n*th roots -2 and 2 are both fourth roots of 16	One real *n*th root 2 is the only real cube root of 8
b negative	No real *n*th root -4 has no real square roots	One real *n*th root -2 is the only real cube root of -8

†In this section we limit our discussion to *real* roots of real numbers. After the real numbers are extended to the complex numbers (Section 5-8), then additional roots can be considered. For example, it turns out that 8 has three cube roots: in addition to the real number 2, there are two other cube roots in the complex number system. A thorough discussion of roots in the complex number system is reserved for advanced courses on the subject.

Thus:

5 has two real square roots, two real fourth roots, and so on.

7 has one real cube root, one real fifth root, and so on.

What symbols do we use to represent these roots? We turn to this question now.

RATIONAL EXPONENTS

If all exponent laws are to continue to hold even if some of the exponents are not integers, then

$$(5^{1/2})^2 = 5^{2/2} = 5 \qquad \text{and} \qquad (7^{1/3})^3 = 7^{3/3} = 7$$

Thus, $5^{1/2}$ is a number whose square is 5, so $5^{1/2}$ must name a square root of 5. Similarly, $7^{1/3}$ is a number whose cube is 7, so $7^{1/3}$ must name a cube root of 7.

More generally, if $b^{1/n}$ exists, it must be an nth root of b since

$$(b^{1/n})^n = b^{n/n} = b^1 = b$$

We need to be careful in our definition when n is an even number, because from the result above b will have either two real nth roots or none.

Definition of $b^{1/n}$

For n a positive integer,

$b^{1/n}$ is an nth root of b.

1. n odd: $b^{1/n}$ is the (only) nth root of b

   $32^{1/5} = 2 \qquad (-32)^{1/5} = -2 \qquad 0^{1/5} = 0$

2. n even:

   (A) b positive: $b^{1/n}$ is the positive nth root of b (sometimes called the **principal nth root of b**). The negative nth root of b is then $-b^{1/n}$.

   $16^{1/2} = 4 \qquad -16^{1/2} = -4$

   (B) b negative: $b^{1/n}$ does not represent a real number.

   $(-16)^{1/2}$ is not real.

The following table summarizes the definition of $b^{1/n}$:

	POWER $1/n$	
BASE b	n EVEN	n ODD
Positive	$b^{1/n}$ is the positive nth root of b.	$b^{1/n}$ is the unique nth root of b.
Negative	$b^{1/n}$ is not a real number.	$b^{1/n}$ is the unique nth root of b.

Example 10 Find each of the following:

(A) $4^{1/2}$ **(B)** $-4^{1/2}$ **(C)** $(-4)^{1/2}$
(D) $8^{1/3}$ **(E)** $(-8)^{1/3}$ **(F)** $0^{1/5}$

Solution **(A)** $4^{1/2} = 2$
(B) $-4^{1/2} = -2$ Note carefully the difference between parts (B) and (C).
(C) $(-4)^{1/2}$ is not a real number **(D)** $8^{1/3} = 2$
(E) $(-8)^{1/3} = -2$ **(F)** $0^{1/5} = 0$

Problem 10 Find each of the following:

(A) $9^{1/2}$ **(B)** $-9^{1/2}$ **(C)** $(-9)^{1/2}$
(D) $27^{1/3}$ **(E)** $(-27)^{1/3}$ **(F)** $0^{1/4}$

How should an expression such as $5^{2/3}$ be defined? If the properties of exponents are to continue to hold for all rational exponents, then $5^{2/3} = (5^{1/3})^2$; that is, $5^{2/3}$ must represent the square of the cube root of 5. Thus, we are led to the following general definition:

Definition of $b^{m/n}$ and $b^{-m/n}$

For m and n natural numbers and b any real number (except b cannot be negative when n is even):

$$b^{m/n} = (b^{1/n})^m \quad \text{and} \quad b^{-m/n} = \frac{1}{b^{m/n}} \quad (b \neq 0)$$

$4^{3/2} = (4^{1/2})^3 = 2^3 = 8$ $\quad 4^{-3/2} = \frac{1}{4^{3/2}} = \frac{1}{8}$ $\quad (-4)^{3/2}$ is not real

$(-32)^{3/5} = [(-32)^{1/5}]^3 = (-2)^3 = -8$

We have now discussed $b^{m/n}$ for all rational numbers m/n and real numbers b. It can be shown, though we will not do so, that all five properties of exponents discussed in Section 5-1 continue to hold for rational exponents so long as we avoid even roots of negative numbers. With the latter restriction in effect, the following useful relationship is an immediate consequence of the exponent properties:

$$b^{m/n} = (b^{1/n})^m = (b^m)^{1/n}$$

Example 11 Simplify, and express answers using positive exponents only:

(A) $8^{2/3}$ (B) $(-8)^{5/3}$ (C) $(3x^{1/3})(2x^{1/2})$

(D) $(2x^{1/3}y^{-2/3})^3$ (E) $\left(\dfrac{4x^{1/3}}{x^{1/2}}\right)^{1/2}$ (F) $(2a^{1/2} + b^{1/2})(x^{1/2} + 3b^{1/2})$

Solution

(A) $8^{2/3} = (8^{1/3})^2 = 2^2 = 4$ or $8^{2/3} = (8^2)^{1/3} = 64^{1/3} = 4$

(B) $(-8)^{5/3} = [(-8)^{1/3}]^5 = (-2)^5 = -32$ Easier than computing $[(-8)^5]^{1/3}$

(C) $(3x^{1/3})(2x^{1/2}) = 6x^{1/3+1/2} = 6x^{5/6}$

(D) $(2x^{1/3}y^{-2/3})^3 = 8xy^{-2} = \dfrac{8x}{y^2}$

(E) $\left(\dfrac{4x^{1/3}}{x^{1/2}}\right)^{1/2} = \dfrac{4^{1/2}x^{1/6}}{x^{1/4}} = \dfrac{2}{x^{1/4-1/6}} = \dfrac{2}{x^{1/12}}$

(F) $(2a^{1/2} + b^{1/2})(a^{1/2} + 3b^{1/2}) = 2a + 7a^{1/2}b^{1/2} + 3b$

Problem 11 Simplify, and express the answers using positive exponents only:

(A) $9^{3/2}$ (B) $(-27)^{4/3}$ (C) $(5y^{3/4})(2y^{1/3})$

(D) $(2x^{-3/4}y^{1/4})^4$ (E) $\left(\dfrac{8x^{1/2}}{x^{2/3}}\right)^{1/3}$ (F) $(x^{1/2} - 2y^{1/2})(3x^{1/2} + y^{1/2})$

The properties of exponents can be used as long as we are dealing with symbols that name real numbers. Can you resolve the following contradiction?

$$-1 = (-1)^{2/2} = [(-1)^2]^{1/2} = 1^{1/2} = 1$$

The second member of the equality chain, $(-1)^{2/2}$, involves the even root of a negative number, which is not real. Thus we see that the properties of exponents do not necessarily hold when we are dealing with nonreal quantities unless further restrictions are imposed. One such restriction is to require all rational exponents to be reduced to lowest terms.

ANSWERS TO MATCHED PROBLEMS

10. **(A)** 3 **(B)** -3 **(C)** Not a real number **(D)** 3 **(E)** -3 **(F)** 0

11. **(A)** 27 **(B)** 81 **(C)** $10y^{13/12}$ **(D)** $16y/x^3$ **(E)** $2/x^{1/18}$ **(F)** $3x - 5x^{1/2}y^{1/2} - 2y$

EXERCISE 5-4

In Problems 1–72 all variables represent positive real numbers.

A *Most of the following are integers. Find them.*

1. $25^{1/2}$ **2.** $36^{1/2}$ **3.** $(-25)^{1/2}$ **4.** $(-36)^{1/2}$

5. $8^{1/3}$ **6.** $27^{1/3}$ **7.** $(-8)^{1/3}$ **8.** $(-27)^{1/3}$

9. $-8^{1/3}$ **10.** $-27^{1/3}$ **11.** $16^{3/2}$ **12.** $25^{3/2}$

13. $8^{2/3}$ **14.** $27^{2/3}$

Simplify, and express the answer using positive exponents only.

15. $x^{1/4}x^{3/4}$ **16.** $y^{1/5}y^{2/5}$ **17.** $\frac{x^{2/5}}{x^{3/5}}$ **18.** $\frac{a^{2/3}}{a^{1/3}}$

19. $(x^4)^{1/2}$ **20.** $(y^{1/2})^4$ **21.** $(a^3b^9)^{1/3}$ **22.** $(x^4y^2)^{1/2}$

23. $\left(\frac{x^9}{y^{12}}\right)^{1/3}$ **24.** $\left(\frac{m^{12}}{n^{16}}\right)^{1/4}$ **25.** $(x^{1/3}y^{1/2})^6$ **26.** $\left(\frac{u^{1/2}}{v^{1/3}}\right)^{12}$

B *Most of the following are rational numbers. Find them.*

27. $(\frac{4}{25})^{1/2}$ **28.** $(\frac{9}{4})^{1/2}$ **29.** $(\frac{4}{25})^{3/2}$ **30.** $(\frac{9}{4})^{3/2}$

31. $(\frac{1}{8})^{2/3}$ **32.** $(\frac{1}{27})^{2/3}$ **33.** $36^{-1/2}$ **34.** $25^{-1/2}$

35. $25^{-3/2}$ **36.** $16^{-3/2}$ **37.** $5^{3/2} \cdot 5^{1/2}$

38. $7^{2/3} \cdot 7^{4/3}$ **39.** $(3^6)^{-1/3}$ **40.** $(4^{-8})^{3/16}$

Simplify, and express the answer using positive exponents only.

41. $x^{1/4}x^{-3/4}$ **42.** $\frac{d^{2/3}}{d^{-1/3}}$ **43.** $n^{3/4}n^{-2/3}$

44. $m^{1/2}m^{-1/3}$ **45.** $(x^{-2/3})^{-6}$ **46.** $(y^{-8})^{1/16}$

47. $(4u^{-2}v^4)^{1/2}$ **48.** $(8x^3y^{-6})^{1/3}$ **49.** $(x^4y^6)^{-1/2}$

50. $(4x^{1/3}y^{3/2})^2$ **51.** $\left(\frac{x^{-2/3}}{y^{-1/2}}\right)^{-6}$ **52.** $\left(\frac{m^{-3}}{n^2}\right)^{-1/6}$

53. $\left(\frac{25x^5y^{-1}}{16x^{-3}y^{-5}}\right)^{1/2}$ **54.** $\left(\frac{8a^{-4}b^3}{27a^2b^{-3}}\right)^{1/3}$ **55.** $\left(\frac{8y^{1/3}y^{-1/4}}{y^{-1/12}}\right)^2$

56. $\left(\frac{9x^{1/3}x^{1/2}}{x^{-1/6}}\right)^{1/2}$

Multiply, and express the answer using positive exponents only.

57. $3m^{3/4}(4m^{1/4} - 2m^8)$ **58.** $2x^{1/3}(3x^{2/3} - x^6)$

59. $(2x^{1/2} + y^{1/2})(x^{1/2} + y^{1/2})$ **60.** $(x^{1/2} + y^{1/2})(x^{1/2} - y^{1/2})$

61. $(x^{1/2} + y^{1/2})^2$ **62.** $(x^{1/2} - y^{1/2})^2$

C *Simplify, and express the answer using positive exponents only.*

63. $(-16)^{-3/2}$

64. $-16^{-3/2}$

65. $(a^{-1/2} + 3b^{-1/2})(2a^{-1/2} - b^{-1/2})$

66. $(x^{-1/2} - y^{-1/2})^2$

67. $(a^{n/2}b^{n/3})^{1/n}$, $n > 0$

68. $(a^{3/n}b^{3/m})^{1/3}$, $n > 0$, $m > 0$

69. $\left(\dfrac{x^{m+2}}{x^m}\right)^{1/2}$, $m > 0$

70. $\left(\dfrac{a^m}{a^{m-2}}\right)^{1/2}$, $m > 0$

71. $(x^{m/4}x^{m/4})^{-2}$, $m > 0$

72. $(y^{m^2+1}y^{2m})^{1/(m+1)}$, $m > 0$

73. **(A)** Find a value of x such that $(x^2)^{1/2} \neq x$.
(B) Find a real number x and a natural number n such that $(x^n)^{1/n} \neq x$.

74. For which real numbers does $(x^2)^{1/2} = |x|$? (More will be said about this form in Section 5-6.)

5-5 RADICAL FORMS AND RATIONAL EXPONENTS

In the preceding section we introduced the symbol $b^{1/n}$ to represent an nth root of b and found that the symbol could be combined with other exponent forms by using the properties of exponents. Another symbol is also used to represent an nth root: the radical sign. Both symbols are widely used, and you should become familiar with them and their respective properties.

nth-Root Radical

For n a natural number greater than 1 and b any real number:

$$\sqrt[n]{b} = b^{1/n}$$

Thus, $\sqrt[n]{b}$ represents an nth root of b.

We can summarize the definition of $\sqrt[n]{b}$ in a table, just as we did for $b^{1/n}$ in Section 5-4:

	$\sqrt[n]{b}$	
BASE b	n EVEN	n ODD
Positive	$\sqrt[n]{b}$ is the positive nth root of b.	$\sqrt[n]{b}$ is the unique nth root of b.
Negative	$\sqrt[n]{b}$ is not a real number.	$\sqrt[n]{b}$ is the unique nth root of b.

Various parts of the form $\sqrt[n]{b}$ are given names: the symbol $\sqrt{\ }$ is called a **radical**, n is called the **index**, and b is called the **radicand**. If $n = 2$, we write

$$\sqrt{b} \quad \text{and not} \quad \sqrt[2]{b}$$

and refer to $\sqrt{b}$ as "the positive square root of b." Note the following radical forms and equivalent rational exponent forms:

$$\sqrt{3} = 3^{1/2} \qquad \sqrt[8]{5} = 5^{1/8}$$

$$\sqrt[3]{x^2} = (x^2)^{1/3} = x^{2/3} \qquad \left(\sqrt[3]{x}\right)^2 = (x^{1/3})^2 = x^{2/3}$$

There are occasions when it is more convenient to work with radicals than with rational exponents, and vice versa. It is often an advantage to be able to shift back and forth between the two forms. The following relationships, suggested by the preceding examples, are useful in this regard.

For b nonnegative when n is even:

$$b^{m/n} = (b^m)^{1/n} = \sqrt[n]{b^m} \quad \text{and} \quad b^{m/n} = (b^{1/n})^m = \left(\sqrt[n]{b}\right)^m$$

The following examples should make the process of changing from one form to the other clear. All variables represent *positive* real numbers.

Example 12 Convert from rational exponent form to radical form:

(A) $5^{1/2}$ **(B)** $x^{1/7}$ **(C)** $7m^{2/3}$
(D) $(3u^2v^3)^{3/5}$ **(E)** $y^{-2/3}$ **(F)** $(x^2 + y^2)^{1/2}$

Solution

(A) $5^{1/2} = \sqrt{5}$ — Positive square root of 5.

(B) $x^{1/7} = \sqrt[7]{x}$

(C) $7m^{2/3} = 7\sqrt[3]{m^2}$ or $7\left(\sqrt[3]{m}\right)^2$ — First form is usually more useful.

(D) $(3u^2v^3)^{3/5} = \sqrt[5]{(3u^2v^3)^3}$ or $\left(\sqrt[5]{3u^2v^3}\right)^3$ — First form is usually more useful.

(E) $y^{-2/3} = \dfrac{1}{y^{2/3}} = \dfrac{1}{\sqrt[3]{y^2}}$ or $\dfrac{1}{\left(\sqrt[3]{y}\right)^2}$ — Index of radical cannot be negative.

(F) $(x^2 + y^2)^{1/2} = \sqrt{x^2 + y^2}$ — $\neq x + y$ (Why?)

Problem 12 Convert to radical form:

(A) $7^{1/2}$ **(B)** $u^{1/5}$ **(C)** $3x^{3/5}$
(D) $(2x^3y^2)^{2/3}$ **(E)** $x^{-3/4}$ **(F)** $(x^3 + y^3)^{1/3}$

Example 13 Convert from radical form to rational exponent form:

(A) $\sqrt{13}$ (B) $\sqrt[5]{x}$ (C) $\sqrt[4]{w^3}$

(D) $\sqrt[5]{(3x^2y^2)^4}$ (E) $\dfrac{1}{\sqrt[3]{x^2}}$ (F) $\sqrt[4]{x^4 + y^4}$

Solution (A) $\sqrt{13} = 13^{1/2}$ Positive square root of 13

(B) $\sqrt[5]{x} = x^{1/5}$

(C) $\sqrt[4]{w^3} = w^{3/4}$

(D) $\sqrt[5]{(3x^2y^2)^4} = (3x^2y^2)^{4/5}$

(E) $\dfrac{1}{\sqrt[3]{x^2}} = \dfrac{1}{x^{2/3}} = x^{-2/3}$

(F) $\sqrt[4]{x^4 + y^4} = (x^4 + y^4)^{1/4}$ $\neq x + y$ (Why?)

Problem 13 Convert to rational exponent form:

(A) $\sqrt{17}$ (B) $\sqrt[7]{m}$ (C) $\sqrt[5]{x^2}$

(D) $\left(\sqrt[7]{5m^3n^4}\right)^3$ (E) $\dfrac{1}{\sqrt[6]{u^5}}$ (F) $\sqrt[5]{x^5 - y^5}$

ANSWERS TO MATCHED PROBLEMS

12. (A) $\sqrt{7}$ (B) $\sqrt[5]{u}$ (C) $3\sqrt[5]{x^3}$ or $3\left(\sqrt[5]{x}\right)^3$
(D) $\sqrt[3]{(2x^3y^2)^2}$ or $\left(\sqrt[3]{2x^3y^2}\right)^2$ (E) $\dfrac{1}{\sqrt[4]{x^3}}$ or $\dfrac{1}{\left(\sqrt[4]{x}\right)^3}$
(F) $\sqrt[3]{x^3 + y^3}$

13. (A) $17^{1/2}$ (B) $m^{1/7}$ (C) $x^{2/5}$ (D) $(5m^3n^4)^{3/7}$
(E) $u^{-5/6}$ (F) $(x^5 - y^5)^{1/5}$

EXERCISE 5-5

All variables are restricted to avoid even roots of negative numbers.

A *Change to radical form. (Do not simplify.)*

1. $11^{1/2}$
2. $7^{1/2}$
3. $5^{1/3}$
4. $6^{1/4}$
5. $u^{3/5}$
6. $x^{3/4}$
7. $4y^{3/7}$
8. $5m^{2/3}$
9. $(4y)^{3/7}$
10. $(5m)^{2/3}$
11. $(4ab^3)^{2/5}$
12. $(7x^2y)^{2/3}$
13. $(a + b)^{1/2}$
14. $(a^2 + b^2)^{1/2}$

Change to rational exponent form. (Do not simplify.)

15. $\sqrt{6}$
16. $\sqrt{3}$
17. $\sqrt[4]{m}$

18. $\sqrt[7]{m}$ **19.** $\sqrt[5]{y^3}$ **20.** $\sqrt[3]{a^2}$

21. $\sqrt[4]{(xy)^3}$ **22.** $\sqrt[5]{(7m^3n^3)^4}$ **23.** $\sqrt{x^2 - y^2}$

24. $\sqrt{1 + y^2}$

B *Change to radical form. (Do not simplify.)*

25. $-5y^{2/5}$ **26.** $-3x^{1/2}$ **27.** $(1 + m^2n^2)^{3/7}$

28. $(x^2y^2 - w^3)^{4/5}$ **29.** $w^{-2/3}$ **30.** $y^{-3/5}$

31. $(3m^2n^3)^{-3/5}$ **32.** $(2xy)^{-2/3}$ **33.** $a^{1/2} + b^{1/2}$

34. $x^{-1/2} + y^{-1/2}$ **35.** $(a^3 + b^3)^{2/3}$ **36.** $(x^{1/2} + y^{-1/2})^{1/3}$

Change to rational exponent form. (Do not simplify.)

37. $\sqrt[3]{(a + b)^2}$ **38.** $\sqrt[5]{(x - y)^2}$ **39.** $-3x\sqrt[4]{a^3b}$

40. $-5\sqrt[3]{2x^2y^2}$ **41.** $\sqrt[9]{-2x^3y^7}$ **42.** $\sqrt[5]{-4m^2n^3}$

43. $\dfrac{3}{\sqrt[3]{y}}$ **44.** $\dfrac{2x}{\sqrt{y}}$ **45.** $\dfrac{-2x}{\sqrt{x^2 + y^2}}$

46. $\dfrac{2}{\sqrt{x}} + \dfrac{3}{\sqrt{y}}$ **47.** $\sqrt[3]{m^2} - \sqrt{n}$ **48.** $\dfrac{-5u^2}{\sqrt{u} + \sqrt[5]{v^3}}$

C **49.** Show that $(x^2 + y^2)^{1/2} \neq x + y$.

50. Show that $\sqrt{a^2 + b^2} \neq a + b$.

51. **(A)** Find a value of x such that $\sqrt{x^2} \neq x$.
(B) Find a positive integer n and a real number x such that $\sqrt[n]{x^n} \neq x$.

52. Which of the following statements is true for all real x (n is a positive integer)?
(A) $\sqrt{x^2} = |x|$ **(B)** $\sqrt[2n]{x^{2n}} = |x|$

5-6 CHANGING AND SIMPLIFYING RADICAL EXPRESSIONS

- Properties of Radicals
- Simplest Radical Form
- Simplifying $\sqrt[n]{x^n}$ for All Real x

Changing and simplifying radical expressions is aided by the introduction of several properties of radicals that follow directly from the exponent properties considered earlier. In this section we develop these properties and then use them to rewrite radical expressions in alternative forms. One form, called *simplest radical form,* is useful for comparing radical expressions.

PROPERTIES OF RADICALS

Consider the following examples:

1. $\sqrt[5]{2^5} = (2^5)^{1/5} = 2^{5/5} = 2^{5/5} = 2^1 = 2$
2. $\sqrt{4 \cdot 9} = \sqrt{36} = 6$ and $\sqrt{4}\sqrt{9} = 2 \cdot 3 = 6$
3. $\sqrt{\dfrac{36}{4}} = \sqrt{9} = 3$ and $\dfrac{\sqrt{36}}{\sqrt{4}} = \dfrac{6}{2} = 3$
4. $\sqrt[6]{2^4} = (2^4)^{1/6} = 2^{4/6} = 2^{2/3} = (2^2)^{1/3} = \sqrt[3]{2^2}$

These examples suggest the following general properties of radicals:

Properties of Radicals

In the following n, m, and k are natural numbers ≥ 2; x and y are positive real numbers:

1. $\sqrt[n]{x^n} = x \qquad (\sqrt[n]{x})^n = x$ $\qquad \sqrt[3]{x^3} = x,\ (\sqrt[3]{x})^3 = x$
2. $\sqrt[n]{xy} = \sqrt[n]{x}\sqrt[n]{y}$ $\qquad \sqrt[4]{xy} = \sqrt[4]{x}\sqrt[4]{y}$
3. $\sqrt[n]{\dfrac{x}{y}} = \dfrac{\sqrt[n]{x}}{\sqrt[n]{y}}$ $\qquad \sqrt[5]{\dfrac{x}{y}} = \dfrac{\sqrt[5]{x}}{\sqrt[5]{y}}$
4. $\sqrt[kn]{x^{km}} = \sqrt[n]{x^m}$ $\qquad \sqrt[12]{x^8} = \sqrt[4\cdot 3]{x^{4\cdot 2}} = \sqrt[3]{x^2}$

It is important to remember that these four properties hold in general only if x and y are restricted to positive numbers. **In this section, unless otherwise stated, all variables are assumed to represent positive numbers.** Near the end of the section we will discuss what happens if we relax this restriction.

The properties of radicals are readily established using exponent properties:

1. $\sqrt[n]{x^n} = (x^n)^{1/n} = x^{n/n} = x$
2. $\sqrt[n]{xy} = (xy)^{1/n} = x^{1/n}y^{1/n} = \sqrt[n]{x}\sqrt[n]{y}$
3. $\sqrt[n]{\dfrac{x}{y}} = \left(\dfrac{x}{y}\right)^{1/n} = \dfrac{x^{1/n}}{y^{1/n}} = \dfrac{\sqrt[n]{x}}{\sqrt[n]{y}}$
4. $\sqrt[kn]{x^{km}} = (x^{km})^{1/kn} = x^{km/kn} = x^{m/n} = \sqrt[n]{x^m}$

The following example illustrates how these properties are used. Properties 2 and 3 are used from right to left as well as from left to right.

Example 14 Simplify by using the properties of radicals:

(A) $\sqrt[5]{(3x^2y)^5}$ **(B)** $\sqrt{10}\sqrt{5}$ **(C)** $\sqrt[3]{\dfrac{x}{27}}$ **(D)** $\sqrt[6]{x^4}$

Solution **(A)** $\sqrt[5]{(3x^2y)^5} = 3x^2y$ — Property 1

(B) $\sqrt{10}\sqrt{5} = \sqrt{50} = \sqrt{25 \cdot 2} = \sqrt{25}\sqrt{2} = 5\sqrt{2}$ — Property 2

(C) $\sqrt[3]{\dfrac{x}{27}} = \dfrac{\sqrt[3]{x}}{\sqrt[3]{27}} = \dfrac{\sqrt[3]{x}}{3}$ or $\frac{1}{3}\sqrt[3]{x}$ — Property 3

(D) $\sqrt[6]{x^4} = \sqrt[2\cdot 3]{x^{2\cdot 2}} = \sqrt[3]{x^2}$ — Property 4

Problem 14 Simplify as in Example 14:

(A) $\sqrt[7]{(u^2 + v^2)^7}$ **(B)** $\sqrt{6}\sqrt{2}$ **(C)** $\sqrt[3]{\dfrac{x^2}{8}}$ **(D)** $\dfrac{\sqrt[3]{54x^8}}{\sqrt[3]{2x^2}}$

(E) $\sqrt[8]{y^6}$

SIMPLEST RADICAL FORM

The properties of radicals provide us with the means for changing algebraic expressions containing radicals to a variety of equivalent forms. One form often useful is the simplest radical form. An algebraic expression that contains radicals is said to be in the **simplest radical form** if all four of the following conditions are satisfied:

Simplest Radical Form

1. A radicand (the expression within the radical sign) contains no polynomial factor to a power greater than or equal to the index of the radical. $\sqrt{x^3}$ violates this condition.
2. The power of the radicand and the index of the radical have no common factor other than 1. $\sqrt[6]{x^4}$ violates this condition.
3. No radical appears in a denominator. $3/\sqrt{5}$ violates this condition.
4. No fraction appears within a radical. $\sqrt{\frac{2}{3}}$ violates this condition.

It should be understood that forms other than the simplest radical form may be more useful on occasion. The choice depends on the situation. Simplest radical form, however, does provide a standard form for comparison purposes.

Example 15 Change to simplest radical form:

(A) $\sqrt{8x^3}$ **(B)** $\sqrt[3]{54x^5}$ **(C)** $\sqrt[9]{x^6}$ **(D)** $\dfrac{4x}{\sqrt{8x}}$ **(E)** $\sqrt[3]{\dfrac{y}{4x}}$

Solution

(A) $\sqrt{8x^3} = \sqrt{(4x^2)(2x)} = \sqrt{4x^2}\sqrt{2x} = 2x\sqrt{2x}$

Violates Condition 1. Factor $8x^3$ into a perfect-square part, $4x^2$, and what is left, $2x$; then use multiplication property 2.

(B) $\sqrt[3]{54x^5} = \sqrt[3]{(27x^3)(2x^2)} = \sqrt[3]{27x^3}\sqrt[3]{2x^2} = 3x\sqrt[3]{2x^2}$

Violates Condition 1. Factor $54x^5$ into a perfect-cube part, $27x^3$, and what is left, $2x^2$; then use multiplication property 2.

(C) $\sqrt[9]{x^6} = \sqrt[3]{x^2}$

Violates Condition 2. Index 9 and power of radicand 6 have the common factor 3.

(D) $\dfrac{4x}{\sqrt{8x}} = \dfrac{4x}{\sqrt{8x}} \cdot \dfrac{\sqrt{2x}}{\sqrt{2x}} = \dfrac{4x\sqrt{2x}}{\sqrt{16x^2}} = \dfrac{4x\sqrt{2x}}{4x} = \sqrt{2x}$

Violates Condition 3 (has a radical in the denominator). Multiply numerator and denominator by "smallest" or "simplest" expression that will make denominator the square root of a perfect square. Using $\sqrt{2x}$ rather than $\sqrt{8x}$ results in less work—try the latter to see why.

(E) $\sqrt[3]{\dfrac{y}{4x}} = \sqrt[3]{\dfrac{y}{4x} \cdot \dfrac{2x^2}{2x^2}} = \sqrt[3]{\dfrac{2x^2y}{8x^3}} = \dfrac{\sqrt[3]{2x^2y}}{\sqrt[3]{8x^3}} = \dfrac{\sqrt[3]{2x^2y}}{2x}$

Violates Condition 4. (There is a fraction within the radical.) Multiply numerator and denominator inside the radical by the "smallest" or "simplest" expression that will make the denominator a perfect cube. Using $2x^2$ rather than 4^2x^2 results in less work—try the latter to see why.

As mentioned above, the simplest radical form may not be the most useful form. In a product such as

$$\frac{4x}{\sqrt{8x}} \cdot \frac{\sqrt{8x}}{2}$$

it is easier to multiply directly rather than to change either factor to simplest form. On the other hand, if $4x/\sqrt{8x}$ is to be evaluated for several values of x, the work will be much easier if the expression is first changed to $\sqrt{2x}$ as in Example 15(D).

The process of removing radicals from a denominator as in Example 15(D) is called **rationalizing the denominator**. In general the process is to multiply both numerator and denominator by an expression so that the denominator is free of radicals. For example, if the denominator is $\sqrt{8x}$, multiply by $\sqrt{2x}$ as in Example 15(D):

$$\sqrt{8x} \cdot \sqrt{2x} = \sqrt{16x^2} = 4x$$

If the denominator is $\sqrt[3]{9x}$, multiply by $\sqrt[3]{3x^2}$:

$$\sqrt[3]{9x} \cdot \sqrt[3]{3x^2} = \sqrt[3]{27x^3} = 3x$$

Problem 15 Change to simplest radical form:

(A) $\sqrt{18y^5}$ **(B)** $\sqrt[3]{32m^8}$ **(C)** $\sqrt[12]{y^8}$ **(D)** $\dfrac{6u}{\sqrt[3]{4x}}$ **(E)** $\sqrt{\dfrac{3y}{8x}}$

Example 16 Change to simplest radical form:

(A) $\sqrt{12x^3y^5z^2}$ **(B)** $\sqrt[6]{16x^4y^2}$ **(C)** $\dfrac{3}{\sqrt{12}}$ **(D)** $\dfrac{6x^2}{\sqrt[3]{9x}}$ **(E)** $\sqrt[3]{\dfrac{2a^2}{3b^2}}$

Solution **(A)** $\sqrt{12x^3y^5z^2} = \sqrt{(2^2x^2y^4z^2)(3xy)} = \sqrt{2^2x^2y^4z^2}\sqrt{3xy} = 2xy^2z\sqrt{3xy}$

(B) $\sqrt[6]{16x^4y^2} = \sqrt[6]{(2^2x^2y)^2} = \sqrt[3]{4x^2y}$

(C) $\dfrac{3}{\sqrt{12}} = \dfrac{3}{\sqrt{12}} \cdot \dfrac{\sqrt{3}}{\sqrt{3}} = \dfrac{3\sqrt{3}}{\sqrt{36}} = \dfrac{3\sqrt{3}}{6} = \dfrac{\sqrt{3}}{2}$ or $\dfrac{1}{2}\sqrt{3}$

(D) $\dfrac{6x^2}{\sqrt[3]{9x}} = \dfrac{6x^2}{\sqrt[3]{9x}} \cdot \dfrac{\sqrt[3]{3x^2}}{\sqrt[3]{3x^2}} = \dfrac{6x^2\sqrt[3]{3x^2}}{\sqrt[3]{3^3x^3}} = \dfrac{6x^2\sqrt[3]{3x^2}}{3x} = 2x\sqrt[3]{3x^2}$

(E) $\sqrt[3]{\dfrac{2a^2}{3b^2}} = \sqrt[3]{\dfrac{(2a^2)(3^2b)}{(3b^2)(3^2b)}} = \sqrt[3]{\dfrac{18a^2b}{3^3b^3}} = \dfrac{\sqrt[3]{18a^2b}}{\sqrt[3]{3^3b^3}} = \dfrac{\sqrt[3]{18a^2b}}{3b}$

Problem 16 Change to simplest radical form:

(A) $\sqrt[3]{16}$ **(B)** $\sqrt[3]{16x^7y^4z^3}$ **(C)** $\sqrt[9]{8x^6y^3}$

(D) $\dfrac{6}{\sqrt{2x}}$ **(E)** $\dfrac{10x^3}{\sqrt[3]{4x^2}}$ **(F)** $\sqrt[3]{\dfrac{3y^2}{2x^4}}$

SIMPLIFYING $\sqrt[n]{x^n}$ FOR ALL REAL x

In the preceding discussion we restricted variables to nonnegative quantities. If we lift this restriction, then

$$\sqrt{x^2} = x$$

is true only for certain values of x and is not true for others. If x is positive or 0, then the equation is true; if x is negative, then the equation is false. For example, test the equation for $x = 2$ and for $x = -2$:

$x = 2$	$x = -2$
$\sqrt{x^2} = x$	$\sqrt{x^2} - x$
$\sqrt{2^2} \stackrel{?}{=} 2$	$\sqrt{(-2)^2} \stackrel{?}{=} (-2)$
$\sqrt{4} \stackrel{?}{=} 2$	$\sqrt{4} \stackrel{?}{=} -2$
$2 \stackrel{\checkmark}{=} 2$	$2 \neq -2$

Thus, we see that if x is negative, then we must write

$$\sqrt{x^2} = -x$$

Now both sides represent positive numbers. In summary, for x any real number

$$\sqrt{x^2} = \begin{cases} x & \text{if } x \text{ is positive} \\ 0 & \text{if } x \text{ is } 0 \\ -x & \text{if } x \text{ is negative} \end{cases}$$

Also, recall the definition of absolute value from Chapter 1:

$$|x| = \begin{cases} x & \text{if } x \text{ is positive} \\ 0 & \text{if } x \text{ is } 0 \\ -x & \text{if } x \text{ is negative} \end{cases}$$

We see that $\sqrt{x^2}$ and $|x|$ actually are the same, and we can write:

For x *any* real number:

$$\sqrt{x^2} = |x|$$

Thus, only if x is restricted to nonnegative real numbers can we drop the absolute-value sign.

Now let us consider $\sqrt[3]{x^3}$. Here we do not have the same kind of problem that we had above. It turns out that for *all* real numbers,

$$\sqrt[3]{x^3} = x$$

and we do not need the absolute-value sign on the right. As before, let us evaluate the equation for $x = 2$ and $x = -2$:

$x = 2$	$x = -2$
$\sqrt[3]{x^3} = x$	$\sqrt[3]{x^3} = x$
$\sqrt[3]{2^3} \stackrel{?}{=} 2$	$\sqrt[3]{(-2)^3} \stackrel{?}{=} (-2)$
$\sqrt[3]{8} \stackrel{?}{=} 2$	$\sqrt[3]{-8} \stackrel{?}{=} -2$
$2 \stackrel{\checkmark}{=} 2$	$-2 \stackrel{\checkmark}{=} -2$

If asked to simplify $\sqrt[3]{x^3} + \sqrt{x^2}$, many students would write

$$\sqrt[3]{x^3} + \sqrt{x^2} = x + x = 2x$$

and not think any more about it. But if we evaluate each side for $x = -2$, we find that

$$\sqrt[3]{(-2)^3} + \sqrt{(-2)^2} = \sqrt[3]{-8} + \sqrt{4} = -2 + 2 = 0 \quad \text{Left side}$$

and

$$2(-2) = -4 \quad \text{Right side}$$

Both sides are not equal! What is wrong? When x is not restricted to positive values or 0, we should write

$$\sqrt[3]{x^3} + \sqrt{x^2} = x + |x|$$

Then the right side will equal the left side for *all* real numbers. Consider the following example and related problem.

Example 17 Simplify $\sqrt[3]{x^3} + \sqrt{x^2}$:

(A) For x a positive number **(B)** For x a negative number

Solution **(A)** For x a positive number:

$$\sqrt[3]{x^3} + \sqrt{x^2} = x + |x| = x + x = 2x$$

(B) For x a negative number:

$$\sqrt[3]{x^3} + \sqrt{x^2} = x + |x| = x - x = 0$$

Problem 17 Simplify $2\sqrt[3]{x^3} - \sqrt{x^2}$:

(A) For x a positive number **(B)** For x a negative number

Following the same reasoning as above, we can obtain the more general result:

> In general, for x *any* real number and n a positive integer greater than 1:
>
> $$\sqrt[n]{x^n} = \begin{cases} |x| & \text{if } n \text{ is even} \quad \sqrt[4]{x^4} = |x| \\ x & \text{if } n \text{ is odd} \quad \sqrt[5]{x^5} = x \end{cases}$$

ANSWERS TO MATCHED PROBLEMS

14. **(A)** $u^2 + v^2$ **(B)** $2\sqrt{3}$ **(C)** $\frac{1}{2}\sqrt[3]{x^2}$ **(D)** $3x^2$
(E) $\sqrt[4]{y^3}$

15. **(A)** $3y^2\sqrt{2y}$ **(B)** $2m^2\sqrt[3]{4m^2}$ **(C)** $\sqrt[3]{y^2}$ **(D)** $\dfrac{3u\sqrt[3]{2x^2}}{x}$

(E) $\dfrac{\sqrt{6xy}}{4x}$

16. **(A)** $2\sqrt[3]{2}$ **(B)** $2x^2yz\sqrt[3]{2xy}$ **(C)** $\sqrt[3]{2x^2y}$ **(D)** $\dfrac{3\sqrt{2x}}{x}$

(E) $5x^2\sqrt[3]{2x}$ **(F)** $\dfrac{\sqrt[3]{12x^2y^2}}{2x^2}$

17. **(A)** x **(B)** $3x$

EXERCISE 5-6

In Problems 1–78, simplify, and write in simplest radical form. All variables represent positive real numbers.

A

1. $\sqrt{y^2}$
2. $\sqrt{x^2}$
3. $\sqrt{4u^2}$
4. $\sqrt{9m^2}$
5. $\sqrt{49x^4y^2}$
6. $\sqrt{25x^2y^4}$
7. $\sqrt{18}$
8. $\sqrt{8}$
9. $\sqrt{m^3}$
10. $\sqrt{x^3}$
11. $\sqrt{8x^3}$
12. $\sqrt{18y^3}$
13. $\sqrt{\frac{1}{9}}$
14. $\sqrt{\frac{1}{4}}$
15. $\dfrac{1}{\sqrt{y^2}}$
16. $\dfrac{1}{\sqrt{x^2}}$
17. $\dfrac{1}{\sqrt{5}}$
18. $\dfrac{1}{\sqrt{3}}$
19. $\sqrt{\frac{1}{5}}$
20. $\sqrt{\frac{1}{3}}$
21. $\dfrac{1}{\sqrt{y}}$
22. $\dfrac{1}{\sqrt{x}}$
23. $\sqrt{\dfrac{1}{y}}$
24. $\sqrt{\dfrac{1}{x}}$
25. $\sqrt{9x^3y^5}$
26. $\sqrt{4x^5y^3}$
27. $\sqrt{18x^8y^5}$
28. $\sqrt{8x^7y^6}$
29. $\dfrac{1}{\sqrt{2x}}$
30. $\dfrac{1}{\sqrt{3y}}$
31. $\dfrac{6x^2}{\sqrt{3x}}$
32. $\dfrac{4xy}{\sqrt{2y}}$
33. $\dfrac{3a}{\sqrt{2ab}}$
34. $\dfrac{2x^2y}{\sqrt{3xy}}$
35. $\sqrt{\dfrac{6x}{7y}}$
36. $\sqrt{\dfrac{3m}{2n}}$

B

37. $\sqrt{\dfrac{9m^5}{2n}}$
38. $\sqrt{\dfrac{4a^3}{3b}}$
39. $\sqrt[4]{16x^8y^4}$
40. $\sqrt[5]{32m^5n^{15}}$
41. $\sqrt[3]{2^4x^4y^7}$
42. $\sqrt[4]{2^4a^5b^8}$
43. $\sqrt[4]{x^2}$
44. $\sqrt[10]{x^6}$
45. $\sqrt{2}\sqrt{8}$
46. $\sqrt[3]{3}\sqrt[3]{9}$
47. $\sqrt{18m^3n^4}\sqrt{2m^3n^2}$
48. $\sqrt[3]{9x^2y}\sqrt[3]{3xy^2}$
49. $\dfrac{6}{\sqrt[3]{3}}$

50. $\dfrac{2}{\sqrt[3]{2}}$ **51.** $\dfrac{\sqrt{4a^3}}{\sqrt{3b}}$ **52.** $\dfrac{\sqrt{9m^5}}{\sqrt{2n}}$

53. $\sqrt{a^2 + b^2}$ **54.** $\sqrt[3]{x^3 + y^3}$ **55.** $\sqrt[3]{\dfrac{8x^3}{27y^6}}$

56. $\sqrt[4]{\dfrac{a^8b^4}{16c^{12}}}$ **57.** $-m\sqrt[5]{3^6m^7n^{11}}$ **58.** $-2x\sqrt[3]{8x^8y^{13}}$

59. $\sqrt[6]{x^4(x - y)^2}$ **60.** $\sqrt[8]{2^6(x + y)^6}$

61. $\sqrt[3]{2x^2y^3}\sqrt[3]{3x^5y}$ **62.** $\sqrt[4]{6u^3v^4}\sqrt[4]{4u^5v}$

63. $\dfrac{4x^3y^2}{\sqrt[3]{2xy^2}}$ **64.** $\dfrac{8u^3v^5}{\sqrt[3]{4u^2v^2}}$

65. $-2x\sqrt[3]{\dfrac{3y^2}{4x}}$ **66.** $6c\sqrt[3]{\dfrac{2ab}{9c^2}}$

C **67.** $\dfrac{x - y}{\sqrt[3]{x - y}}$ **68.** $\dfrac{1}{\sqrt[3]{(x - y)^2}}$ **69.** $\sqrt[4]{\dfrac{3y^3}{4x}}$

70. $\sqrt[5]{\dfrac{4n^2}{16m^3}}$ **71.** $-\sqrt{x^4 + 2x^2}$ **72.** $\sqrt[4]{m^4 + 4m^6}$

73. $\sqrt[4]{16x^4}\sqrt[3]{16x^{24}y^4}$ **74.** $\sqrt[3]{8\sqrt{16x^6y^4}}$

75. $\sqrt[3]{3m^2n^2}\sqrt[4]{3m^3n^2}$ **76.** $\sqrt{2x^5y^3}\sqrt[3]{16x^7y^7}$

77. $\sqrt[3]{x^{3n}(x + y)^{3n+6}}$ **78.** $\sqrt[n]{x^{2n}y^{n^2+n}}$

Simplify each of the following for **(A)** *x a positive number and* **(B)** *x a negative number.*

79. $2\sqrt[3]{x^3} + 4\sqrt{x^2}$ **80.** $\sqrt[3]{8x^3} - \sqrt{16x^2}$ **81.** $\sqrt[5]{x^5} + \sqrt[4]{x^4}$

82. $\sqrt[6]{x^6} + \sqrt[3]{x^3}$ **83.** $3\sqrt[4]{x^4} - 2\sqrt[5]{x^5}$ **84.** $5\sqrt[7]{x^7} - 3\sqrt[6]{x^6}$

5-7 BASIC OPERATIONS ON RADICALS

- Sums and Differences
- Products
- Quotients—Rationalizing Denominators

The algebraic expressions we have considered have become more complex in several stages by allowing additional operations. Polynomials were built up from addition, subtraction, and multiplication. Rational expressions allowed division. We can now add, subtract, multiply, and divide such expressions. In this section, we do the same for algebraic expressions that involve radicals.

SUMS AND DIFFERENCES

Algebraic expressions involving radicals can often be simplified by adding and subtracting terms that contain exactly the same radical expressions. We proceed in essentially the same way as we do when we combine like terms in polynomials. You will recall that the distributive property of real numbers plays a central role in this process. All variables represent positive real numbers.

Example 18 Combine as many terms as possible:

(A) $5\sqrt{3} + 4\sqrt{3}$ **(B)** $2\sqrt[3]{xy^2} - 7\sqrt[3]{xy^2}$
(C) $3\sqrt{xy} - 2\sqrt[3]{xy} + 4\sqrt{xy} - 7\sqrt[3]{xy}$

Solution **(A)** $5\sqrt{3} + 4\sqrt{3} \boxed{= (5 + 4)\sqrt{3}} = 9\sqrt{3}$

$5\sqrt{3}$ and $4\sqrt{3}$ are treated as like terms just as we would treat $5a$ and $4a$ in a sum $5a + 4a$.

(B) $2\sqrt[3]{xy^2} - 7\sqrt[3]{xy^2} \boxed{= (2 - 7)\sqrt[3]{xy^2}} = -5\sqrt[3]{xy^2}$

(C) $3\sqrt{xy} - 2\sqrt[3]{xy} + 4\sqrt{xy} - 7\sqrt[3]{xy} \boxed{= 3\sqrt{xy} + 4\sqrt{xy} - 2\sqrt[3]{xy} - 7\sqrt[3]{xy}}$

$= 7\sqrt{xy} - 9\sqrt[3]{xy}$

Problem 18 Combine as many terms as possible:

(A) $6\sqrt{2} + 2\sqrt{2}$
(B) $3\sqrt[5]{2x^2y^3} - 8\sqrt[5]{2x^2y^3}$
(C) $5\sqrt[3]{mn^2} - 3\sqrt{mn} - 2\sqrt[3]{mn^2} + 7\sqrt{mn}$

Thus, we see that if two terms contain exactly the same radical—having the same index and the same radicand—they can be combined into a single term. Occasionally, terms containing radicals can be combined after they have been expressed in simplest radical form.

Example 19 Express terms in simplest radical form and combine where possible:

(A) $4\sqrt{8} - 2\sqrt{18}$ **(B)** $2\sqrt{12} - \sqrt{\frac{1}{3}}$ **(C)** $\sqrt[3]{81} - \sqrt[3]{\frac{1}{9}}$

Solution **(A)** $4\sqrt{8} - 2\sqrt{18} = 4\sqrt{4 \cdot 2} - 2\sqrt{9 \cdot 2}$
$= 8\sqrt{2} - 6\sqrt{2}$
$= 2\sqrt{2}$

(B) $2\sqrt{12} - \sqrt{\frac{1}{3}} = 2 \cdot \sqrt{4} \cdot \sqrt{3} - \dfrac{1 \cdot \sqrt{3}}{\sqrt{3} \cdot \sqrt{3}}$

$= 4\sqrt{3} - \dfrac{\sqrt{3}}{3}$

$= (4 - \frac{1}{3})\sqrt{3}$

$= \dfrac{11}{3}\sqrt{3}$ or $\dfrac{11\sqrt{3}}{3}$

(C) $\sqrt[3]{81} - \sqrt[3]{\frac{1}{9}} = \sqrt[3]{3^3 \cdot 3} - \sqrt[3]{\dfrac{3}{3^3}} = 3\sqrt[3]{3} - \frac{1}{3}\sqrt[3]{3}$

$= (3 - \frac{1}{3})\sqrt[3]{3} = \frac{8}{3}\sqrt[3]{3}$

Problem 19 Express terms in simplest radical form and combine where possible:

(A) $\sqrt{12} - \sqrt{48}$ **(B)** $3\sqrt{8} - \sqrt{\frac{1}{2}}$ **(C)** $\sqrt[3]{\frac{1}{4}} - \sqrt[3]{16}$

PRODUCTS

We will now consider several types of special products and quotients that involve radicals. The distributive property of real numbers plays a central role in our approach to these problems. In the discussion that follows, all variables represent positive real numbers.

Example 20 Multiply and simplify:

(A) $\sqrt{2}(\sqrt{10} - 3)$ **(B)** $(\sqrt{2} - 3)(\sqrt{2} + 5)$

(C) $(\sqrt{x} - 3)(\sqrt{x} + 5)$ **(D)** $(\sqrt[3]{m} + \sqrt[3]{n^2})(\sqrt[3]{m^2} - \sqrt[3]{n})$

Solution **(A)** $\sqrt{2}(\sqrt{10} - 3) = \sqrt{2}\sqrt{10} - \sqrt{2} \cdot 3 = \sqrt{20} - 3\sqrt{2} = 2\sqrt{5} - 3\sqrt{2}$

(B) $(\sqrt{2} - 3)(\sqrt{2} + 5) = \sqrt{2}\sqrt{2} - 3\sqrt{2} + 5\sqrt{2} - 15$

$= 2 + 2\sqrt{2} - 15$

$= 2\sqrt{2} - 13$

(C) $(\sqrt{x} - 3)(\sqrt{x} + 5) = \sqrt{x}\sqrt{x} - 3\sqrt{x} + 5\sqrt{x} - 15$

$= x + 2\sqrt{x} - 15$

(D) $(\sqrt[3]{m} + \sqrt[3]{n^2})(\sqrt[3]{m^2} - \sqrt[3]{n}) = \sqrt[3]{m^3} + \sqrt[3]{m^2n^2} - \sqrt[3]{mn} - \sqrt[3]{n^3}$

$= m - \sqrt[3]{mn} + \sqrt[3]{m^2n^2} - n$

Problem 20 Multiply and simplify:

(A) $\sqrt{3}(\sqrt{6} - 4)$ **(B)** $(\sqrt{3} - 2)(\sqrt{3} + 4)$

(C) $(\sqrt{y} - 2)(\sqrt{y} + 4)$ **(D)** $(\sqrt[3]{x^2} - \sqrt[3]{y^2})(\sqrt[3]{x} + \sqrt[3]{y})$

Example 21 Show that $(2 - \sqrt{3})$ is a solution of the equation $x^2 - 4x + 1 = 0$.

Solution

$$x^2 - 4x + 1 = 0$$

$$(2 - \sqrt{3})^2 - 4(2 - \sqrt{3}) + 1 \stackrel{?}{=} 0$$

$$4 - 4\sqrt{3} + 3 - 8 + 4\sqrt{3} + 1 \stackrel{?}{=} 0$$

$$0 \stackrel{\checkmark}{=} 0$$

$(2 - \sqrt{3})^2$ can be squared as a binomial: $(2 - \sqrt{3})^2 = 2^2 - 2 \cdot 2\sqrt{3} + (\sqrt{3})^2$

Problem 21 Show that $(2 + \sqrt{3})$ is a solution of $x^2 - 4x + 1 = 0$.

QUOTIENTS—RATIONALIZING DENOMINATORS

Recall that to express $\sqrt{2}/\sqrt{3}$ in simplest radical form, we multiplied the numerator and denominator by $\sqrt{3}$ to clear the denominator of the radical:

$$\frac{\sqrt{2}}{\sqrt{3}} = \frac{\sqrt{2} \cdot \sqrt{3}}{\sqrt{3} \cdot \sqrt{3}} = \frac{\sqrt{6}}{3}$$

The denominator is thus converted to a rational number. Also recall that the process of converting irrational denominators to rational forms is called **rationalizing the denominator**.

How can we rationalize the binomial denominator in

$$\frac{1}{\sqrt{3} - \sqrt{2}}$$

Multiplying the numerator and denominator by $\sqrt{3}$ or $\sqrt{2}$ does not help. Try it!

Recalling the special product

$$(a - b)(a + b) = a^2 - b^2$$

suggests that we multiply the numerator and denominator by the denominator, only with the middle sign changed. Thus,

$$\frac{1}{\sqrt{3} - \sqrt{2}} = \frac{1(\sqrt{3} + \sqrt{2})}{(\sqrt{3} - \sqrt{2})(\sqrt{3} + \sqrt{2})} \quad \boxed{= \frac{\sqrt{3} + \sqrt{2}}{(\sqrt{3})^2 - (\sqrt{2})^2}}$$

$$= \frac{\sqrt{3} + \sqrt{2}}{3 - 2} = \sqrt{3} + \sqrt{2}$$

The expressions $\sqrt{3} + \sqrt{2}$ and $\sqrt{3} - \sqrt{2}$ are often called **conjugates** of each other. To rationalize the denominator where it is an expression like one of these, that is, the sum or difference of terms involving radicals, we multiply numerator and denominator by the conjugate of the denominator.

Example 22 Rationalize denominators and simplify:

(A) $\dfrac{\sqrt{2}}{\sqrt{6}-2}$ (B) $\dfrac{\sqrt{x}-\sqrt{y}}{\sqrt{x}+\sqrt{y}}$

Solution (A) $\dfrac{\sqrt{2}}{\sqrt{6}-2} = \dfrac{\sqrt{2}(\sqrt{6}+2)}{(\sqrt{6}-2)(\sqrt{6}+2)} = \dfrac{\sqrt{12}+2\sqrt{2}}{6-4}$

$$= \frac{2\sqrt{3}+2\sqrt{2}}{2} = \frac{2(\sqrt{3}+\sqrt{2})}{2} = \sqrt{3}+\sqrt{2}$$

(B) $\dfrac{\sqrt{x}-\sqrt{y}}{\sqrt{x}+\sqrt{y}} = \dfrac{(\sqrt{x}-\sqrt{y})(\sqrt{x}-\sqrt{y})}{(\sqrt{x}+\sqrt{y})(\sqrt{x}-\sqrt{y})} = \dfrac{x-2\sqrt{xy}+y}{x-y}$

Problem 22 Rationalize denominators and simplify: **(A)** $\dfrac{\sqrt{2}}{\sqrt{2}+3}$ **(B)** $\dfrac{\sqrt{x}+\sqrt{y}}{\sqrt{x}-\sqrt{y}}$

ANSWERS TO MATCHED PROBLEMS

18. **(A)** $8\sqrt{2}$ **(B)** $-5\sqrt[5]{2x^2y^3}$ **(C)** $3\sqrt[3]{mn^2}+4\sqrt{mn}$

19. **(A)** $-2\sqrt{3}$ **(B)** $\dfrac{11\sqrt{2}}{2}$ **(C)** $-\frac{3}{2}\sqrt[3]{2}$

20. **(A)** $3\sqrt{2}-4\sqrt{3}$ **(B)** $2\sqrt{3}-5$ **(C)** $y+2\sqrt{y}-8$

(D) $x+\sqrt[3]{x^2y}-\sqrt[3]{xy^2}-y$

21. $(2+\sqrt{3})^2-4(2+\sqrt{3})+1 = 4+4\sqrt{3}+3-8-4\sqrt{3}+1$
$= 0$

22. **(A)** $\dfrac{2-3\sqrt{2}}{-7}$ or $\dfrac{3\sqrt{2}-2}{7}$ **(B)** $\dfrac{x+2\sqrt{xy}+y}{x-y}$

EXERCISE 5-7

A *Express in simplest radical form and combine where possible.*

1. $7\sqrt{3}+2\sqrt{3}$

2. $5\sqrt{2}+3\sqrt{2}$

3. $2\sqrt{a}-7\sqrt{a}$

4. $\sqrt{y}-4\sqrt{y}$

5. $\sqrt{n}-4\sqrt{n}-2\sqrt{n}$

6. $2\sqrt{x}-\sqrt{x}+3\sqrt{x}$

7. $\sqrt{5}-2\sqrt{3}+3\sqrt{5}$

8. $3\sqrt{2}-2\sqrt{3}-\sqrt{2}$

9. $\sqrt{m}-\sqrt{n}-2\sqrt{n}$

10. $2\sqrt{x}-\sqrt{y}+3\sqrt{y}$

11. $\sqrt{18}+\sqrt{2}$

12. $\sqrt{8}-\sqrt{2}$

13. $\sqrt{8}-2\sqrt{32}$

14. $\sqrt{27}-3\sqrt{12}$

Multiply and simplify where possible.

15. $\sqrt{7}(\sqrt{7}-2)$ **16.** $\sqrt{5}(\sqrt{5}-2)$ **17.** $\sqrt{2}(3-\sqrt{2})$

18. $\sqrt{3}(2-\sqrt{3})$ **19.** $\sqrt{y}(\sqrt{y}-8)$ **20.** $\sqrt{x}(\sqrt{x}-3)$

21. $\sqrt{n}(4-\sqrt{n})$ **22.** $\sqrt{m}(3-\sqrt{m})$ **23.** $\sqrt{3}(\sqrt{3}+\sqrt{6})$

24. $\sqrt{5}(\sqrt{10}+\sqrt{5})$ **25.** $(2-\sqrt{3})(3+\sqrt{3})$

26. $(\sqrt{2}-1)(\sqrt{2}+3)$ **27.** $(\sqrt{5}+2)^2$

28. $(\sqrt{3}-3)^2$ **29.** $(\sqrt{m}-3)(\sqrt{m}-4)$

30. $(\sqrt{x}+2)(\sqrt{x}-3)$

Rationalize denominators and simplify.

31. $\dfrac{1}{\sqrt{5}+2}$ **32.** $\dfrac{1}{\sqrt{11}-3}$ **33.** $\dfrac{2}{\sqrt{5}+1}$

34. $\dfrac{4}{\sqrt{6}-2}$ **35.** $\dfrac{\sqrt{2}}{\sqrt{10}-2}$ **36.** $\dfrac{\sqrt{2}}{\sqrt{6}+2}$

37. $\dfrac{\sqrt{y}}{\sqrt{y}+3}$ **38.** $\dfrac{\sqrt{x}}{\sqrt{x}-2}$

B *Express in simplest radical form and combine where possible.*

39. $\sqrt{8mn}+2\sqrt{18mn}$ **40.** $\sqrt{4x}-\sqrt{9x}$

41. $\sqrt{8}-\sqrt{20}+4\sqrt{2}$ **42.** $\sqrt{24}-\sqrt{12}+3\sqrt{3}$

43. $\sqrt[5]{a}-4\sqrt[5]{a}+2\sqrt[5]{a}$ **44.** $3\sqrt[3]{u}-2\sqrt[3]{u}-2\sqrt[3]{u}$

45. $2\sqrt[3]{x}+3\sqrt[3]{x}-\sqrt{x}$ **46.** $5\sqrt[5]{y}-2\sqrt[5]{y}+3\sqrt[4]{y}$

47. $\sqrt{\frac{1}{8}}+\sqrt{8}$ **48.** $\sqrt{\frac{2}{3}}-\sqrt{\frac{3}{2}}$

49. $\sqrt{\dfrac{3uv}{2}}-\sqrt{24uv}$ **50.** $\sqrt{\dfrac{xy}{2}}+\sqrt{8xy}$

Multiply and simplify where possible.

51. $(4\sqrt{3}-1)(3\sqrt{3}-2)$ **52.** $(2\sqrt{7}-\sqrt{3})(2\sqrt{7}+\sqrt{3})$

53. $(\sqrt{x}-\sqrt{y})(\sqrt{x}+\sqrt{y})$ **54.** $(2\sqrt{x}+3)(2\sqrt{x}-3)$

55. $(5\sqrt{m}+2)(2\sqrt{m}-3)$ **56.** $(3\sqrt{u}-2)(2\sqrt{u}+4)$

57. $(\sqrt[3]{4}+\sqrt[3]{9})(\sqrt[3]{2}+\sqrt[3]{3})$ **58.** $\sqrt[3]{4}(\sqrt[3]{2}-\sqrt[3]{16})$

59. Show that $3 - \sqrt{2}$ is a solution to $x^2 - 6x + 7 = 0$.

60. Show that $3 + \sqrt{2}$ is a solution to $x^2 - 6x + 7 = 0$.

Rationalize denominators and simplify.

61. $\dfrac{\sqrt{3} + 2}{\sqrt{3} - 2}$ **62.** $\dfrac{\sqrt{2} - 1}{\sqrt{2} + 1}$ **63.** $\dfrac{\sqrt{2} + \sqrt{3}}{\sqrt{3} - \sqrt{2}}$

64. $\dfrac{3 - \sqrt{a}}{\sqrt{a} - 2}$ **65.** $\dfrac{2 + \sqrt{x}}{\sqrt{x} - 3}$ **66.** $\dfrac{\sqrt{5} - \sqrt{2}}{\sqrt{5} + \sqrt{2}}$

67. $\dfrac{3\sqrt{x}}{2\sqrt{x} - 3}$ **68.** $\dfrac{5\sqrt{a}}{3 - 2\sqrt{a}}$

C *Express in simplest radical form and combine where possible.*

69. $\dfrac{\sqrt{3}}{3} + 2\sqrt{\dfrac{1}{3}} + \sqrt{12}$ **70.** $\sqrt{\dfrac{1}{2}} + \dfrac{\sqrt{2}}{2} + \sqrt{8}$

71. $\sqrt[3]{\frac{1}{3}} + \sqrt[3]{3^5}$ **72.** $\sqrt[4]{32} - \sqrt[4]{\frac{1}{8}}$

Multiply and simplify where possible.

73. $(\sqrt[3]{x} - \sqrt[3]{y^2})(\sqrt[3]{x^2} + 2\sqrt[3]{y})$

74. $(\sqrt[5]{u^2} - \sqrt[5]{v^3})(\sqrt[5]{u^3} + \sqrt[5]{v^2})$

75. $(\sqrt[3]{x} + \sqrt[3]{y})(\sqrt[3]{x^2} - \sqrt[3]{x}\sqrt[3]{y} + \sqrt[3]{y^2})$

76. $(\sqrt[3]{x} - \sqrt[3]{y})(\sqrt[3]{x^2} + \sqrt[3]{x}\sqrt[3]{y} + \sqrt[3]{y^2})$

Rationalize denominators and simplify (see Problems 75 and 76 for Problems 79 and 80).

77. $\dfrac{2\sqrt{x} + 3\sqrt{y}}{4\sqrt{x} + 5\sqrt{y}}$ **78.** $\dfrac{3\sqrt{x} + 2\sqrt{y}}{2\sqrt{x} - 5\sqrt{y}}$

79. $\dfrac{1}{\sqrt[3]{x} + \sqrt[3]{y}}$ **80.** $\dfrac{1}{\sqrt[3]{x} - \sqrt[3]{y}}$

81. $\dfrac{1}{\sqrt{x} + \sqrt{y} - \sqrt{z}}$
[*Hint:* Start by multiplying numerator and denominator by $(\sqrt{x} + \sqrt{y}) + \sqrt{z}$.]

82. $\dfrac{1}{\sqrt{x} - \sqrt{y} + \sqrt{z}}$

5-8 COMPLEX NUMBERS

- Complex Numbers
- Complex Numbers and Radicals
- The Complex Number System

The Pythagoreans (500–275 B.C.) found that the simple equation

$$x^2 = 2 \tag{1}$$

had no rational number solutions. If Equation (1) were to have a solution, then a new kind of number had to be invented—the irrational numbers. The irrational numbers $\sqrt{2}$ and $-\sqrt{2}$ are both solutions to (1). Irrational numbers were not put on a firm mathematical foundation until the last century. The rational and irrational numbers together constitute the real number system.

Is there any need to extend the real number system further? Yes, since we find that another simple equation

$$x^2 = -1$$

has no real solutions. (What real number squared is negative?) Once again, we are forced to invent a new kind of number—a number that has the possibility of being negative when it is squared. This new system of numbers is called the **complex numbers**. The complex numbers evolved over a long period of time,† but, like the real numbers, it was not until the last century that they were placed on a firm mathematical basis.

COMPLEX NUMBERS

In order to extend the real numbers to a new system in which negative numbers have square roots, we first introduce one number whose square is -1. This number is called the **imaginary unit** and is usually denoted i:

$$i^2 = -1 \qquad i = \sqrt{-1}$$

† BRIEF HISTORY OF COMPLEX NUMBERS

APPROXIMATE DATE	PERSON	EVENT
50	Heron of Alexandria	First recorded encounter of a square root of a negative number
850	Mahavira of India	Said that a negative has no square root, since it is not a square
1545	Cardano of Italy	Found that solutions to cubic equations involved square roots of negative numbers
1637	Descartes of France	Introduced the terms "real" and "imaginary"
1748	Euler of Switzerland	Used i for $\sqrt{-1}$
1832	Gauss of Germany	Introduced the term "complex number"

Our new number system must also contain numbers like

$$2i \qquad 3 + i \qquad 4 + 7i \qquad 1 - i$$

and so forth. Our system will, in fact, consist of all numbers in the form

$$a + bi$$

where a and b are real numbers. Such a number is called a **complex number**.

Complex numbers can be added, subtracted, and multiplied using ordinary algebraic properties and keeping in mind that $i^2 = -1$. For example, to add $3 + 2i$ and $-4 + i$, we would add just as we would for polynomials, in essence collecting like terms:

$$\begin{aligned}(3 + 2i) + (-4 + i) &= (3 - 4) + (2i + i)\\ &= -1 + 3i\end{aligned}$$

Subtraction works similarly:

$$\begin{aligned}(3 + 2i) - (-4 + i) &= (3 + 2i) + 4 - i\\ &= 7 + i\end{aligned}$$

We multiply as in the FOIL method:

$$\begin{aligned}(3 + 2i)(-4 + i) &= \overset{\text{F}}{3(-4)} + \overset{\text{O}}{3i} - \overset{\text{I}}{8i} + \overset{\text{L}}{2i^2}\\ &= -12 - 5i + 2(-1) \qquad i^2 = -1\\ &= -14 - 5i\end{aligned}$$

We can state these definitions more precisely and also emphasize that for two complex numbers to be equal they must have the same real number term and the same real number coefficient of i:

Equality, Addition, and Multiplication of Complex Numbers

EQUALITY	$a + bi = c + di$ if and only if $a = c$ and $b = d$
ADDITION	$(a + bi) + (c + di) = (a + c) + (b + d)i$
MULTIPLICATION	$(a + bi)(c + di) = (ac - bd) + (ad + bc)i$

Example 23 Carry out the indicated operations and write each answer in the form $a + bi$:

(A) $(3 + 2i) + (2 - i)$ **(B)** $(3 + 2i) - (2 - i)$

(C) $(3 + 2i)(2 - i)$ **(D)** $\dfrac{3 + 2i}{2 - i}$

Solution **(A)** $(3 + 2i) + (2 - i) = 3 + 2i + 2 - i = 5 + i$ Remove parentheses and combine like terms.

(B) $(3 + 2i) - (2 - i) = 3 + 2i - 2 + i = 1 + 3i$ Remove parentheses and combine like terms.

(C) Multiply and replace i^2 with -1.

$$\begin{aligned}(3 + 2i)(2 - i) &= 6 + i - 2i^2\\ &= 6 + i - 2(-1)\\ &= 6 + i + 2\\ &= 8 + i\end{aligned}$$

(D) In order to eliminate i from the denominator, we multiply the numerator and denominator by $2 + i$. This is analogous to how we rationalized a denominator, and $2 + i$ is called the **conjugate** of $2 - i$:

$$\begin{aligned}\frac{3 + 2i}{2 - i} \cdot \frac{2 + i}{2 + i} = \frac{6 + 7i + 2i^2}{4 - i^2} &= \frac{6 + 7i + 2(-1)}{4 - (-1)}\\ &= \frac{4 + 7i}{5} = \frac{4}{5} + \frac{7}{5}i\end{aligned}$$

Recall that subtraction and division are defined, in general, as follows:

$A - B = C$ if and only if $A = B + C$

$A \div B = C$ if and only if $A = BC$, $B \neq 0$, and C is unique

The results obtained by the procedures illustrated in Examples 23(B) and (D) are consistent with these definitions, as can easily be checked. And with a little extra work, these procedures can be shown to hold in general (see the C-level problems in Exercise 5-8).

Problem 23 Carry out the indicated operations and write each answer in the form $a + bi$:

(A) $(3 + 2i) + (6 - 4i)$ **(B)** $(3 - 5i) - (1 - 3i)$

(C) $(2 - 4i)(3 + 2i)$ **(D)** $\dfrac{2 + 4i}{3 + 2i}$

Example 24 Carry out the indicated operations and write each answer in the form $a + bi$:

(A) $(2 - 3i)^2 - (4i)^2$ **(B)** $\dfrac{2 + i}{3i}$

Solution **(A)** $(2 - 3i)^2 - (4i)^2 = 4 - 12i + 9i^2 - 16i^2$

$$= 4 - 12i + 9(-1) - 16(-1)$$
$$= 4 - 12i - 9 + 16$$
$$= 11 - 12i$$

(B) $$\frac{2+i}{3i} = \frac{2+i}{3i} \cdot \frac{i}{i} = \frac{2i + i^2}{3i^2} = \frac{2i + (-1)}{3(-1)}$$

$$= \frac{2i - 1}{-3} = \frac{-1}{-3} + \frac{2}{-3}i = \frac{1}{3} - \frac{2}{3}i$$

Problem 24 Carry out the indicated operations and write each answer in the form $a + bi$:

(A) $(3i)^2 - (3 - 2i)^2$ **(B)** $\dfrac{3+i}{2i}$

COMPLEX NUMBERS AND RADICALS

Recall that we say y is a square root of x if $y^2 = x$. It can be shown that if x is a positive real number, then x has two real square roots, one the negative of the other; if x is negative, then x has two complex square roots, one also the negative of the other. In particular, if we let $x = -a$, $a > 0$, then one of the square roots of x is given by†

$$\sqrt{-a} = i\sqrt{a} \qquad a > 0 \qquad \sqrt{-9} = i\sqrt{9} = 3i$$

To check this, we square $i\sqrt{a}$ and obtain $-a$:

$$(i\sqrt{a})^2 = i^2(\sqrt{a})^2 = (-1)a = -a$$

The other square root is given by $-i\sqrt{a}$, since

$$(-i\sqrt{a})^2 = (-i)^2(\sqrt{a})^2 = i^2a = (-1)a = -a$$

Example 25 Write in the form $a + bi$:

(A) $\sqrt{-4}$ **(B)** $4 + \sqrt{-40}$ **(C)** $\dfrac{-3 - \sqrt{-7}}{2}$

Solution **(A)** $\sqrt{-4} = i\sqrt{4} = 2i$

† Note that if in $a + bi$, $b = \sqrt{k}$, then we often write $a + i\sqrt{k}$ instead of $a + \sqrt{k}i$ so that i will not accidentally end up under the radical sign.

(B) $4 + \sqrt{-40} = 4 + i\sqrt{40} = 4 + i\sqrt{4 \cdot 10} = 4 + 2i\sqrt{10}$

(C) $\dfrac{-3 - \sqrt{-7}}{2} = \dfrac{-3 - i\sqrt{7}}{2} = -\dfrac{3}{2} - \dfrac{\sqrt{7}}{2}i$

Problem 25 Write in the form $a + bi$:

(A) $\sqrt{-16}$ **(B)** $5 - \sqrt{-36}$ **(C)** $\dfrac{-5 - \sqrt{-2}}{2}$

Example 26 Convert square roots of negative numbers to complex form, perform the indicated operations, and express your answers in the form $a + bi$:

(A) $(3 + \sqrt{-4})(2 - \sqrt{-9})$ **(B)** $\dfrac{1}{3 - \sqrt{-4}}$

Solution **(A)**

$$\begin{aligned}(3 + \sqrt{-4})(2 - \sqrt{-9}) &= (3 + i\sqrt{4})(2 - i\sqrt{9}) \\ &= (3 + 2i)(2 - 3i) \\ &= 6 - 5i - 6i^2 \\ &= 6 - 5i - 6(-1) \\ &= 6 - 5i + 6 \\ &= 12 - 5i\end{aligned}$$

Note that $\sqrt{-4}\sqrt{-9} \neq \sqrt{(-4)(-9)}$ since $\sqrt{-4}\sqrt{-9} = (2i)(3i) = 6i^2 = -6$ while $\sqrt{(-4)(-9)} = \sqrt{36} = 6$

(B)

$$\begin{aligned}\frac{1}{3 - \sqrt{-4}} = \frac{1}{3 - i\sqrt{4}} &= \frac{1}{3 - 2i} \\ &= \frac{1}{3 - 2i} \cdot \frac{3 + 2i}{3 + 2i} = \frac{3 + 2i}{9 - 4i^2} \\ &= \frac{3 + 2i}{9 - 4(-1)} = \frac{3 + 2i}{9 + 4} \\ &= \frac{3 + 2i}{13} = \frac{3}{13} + \frac{2}{13}i\end{aligned}$$

Problem 26 Convert square roots of negative numbers to complex form, perform the indicated operations, and express your answers in the form $a + bi$:

(A) $(4 - \sqrt{-25})(3 + \sqrt{-49})$ **(B)** $\dfrac{1}{2 + \sqrt{-9}}$

THE COMPLEX NUMBER SYSTEM

Some additional terminology will help to clarify the relationship of the complex numbers to the other number systems studied thus far (see Figure 2 on page 241). If $b = 0$ in $a + bi$, the number is simply a, a real number. If

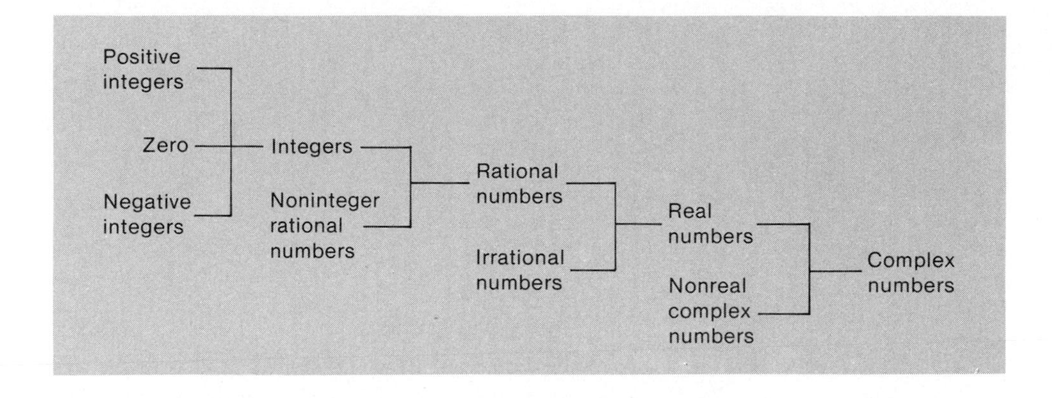

FIGURE 2

$b \neq 0$, $a + bi$ is called an **imaginary number**; if $a = 0$, it is called a **pure imaginary number**.

Complex numbers are used extensively by electrical, aeronautical, and space scientists, as well as chemists and physicists. The interpretation of complex numbers $a + bi$ relative to the real world is not readily seen until you have had more experience in some of these fields. We state only that a and b in the complex number $a + bi$ often represent real-world quantities. Our use of the complex numbers will be in connection with solutions to second-degree equations such as

$$x^2 - 4x + 5 = 0$$

which we will study in the next chapter.

ANSWERS TO MATCHED PROBLEMS

23. **(A)** $9 - 2i$ **(B)** $2 - 2i$ **(C)** $14 - 8i$ **(D)** $\frac{14}{13} + \frac{8}{13}i$

24. **(A)** $-14 + 12i$ **(B)** $\frac{1}{2} - \frac{3}{2}i$

25. **(A)** $4i$ **(B)** $5 - 6i$ **(C)** $-\frac{5}{2} - \frac{\sqrt{2}}{2}i$

26. **(A)** $47 + 13i$ **(B)** $\frac{2}{13} - \frac{3}{13}i$

EXERCISE 5-8

A *Perform the indicated operations and write each answer in the form $a + bi$.*

1. $(5 + 2i) + (3 + i)$
2. $(6 + i) + (2 + 3i)$
3. $(-8 + 5i) + (3 - 2i)$
4. $(2 - 3i) + (5 - 2i)$
5. $(8 + 5i) - (3 + 2i)$
6. $(9 + 7i) - (2 + 5i)$
7. $(4 + 7i) - (-2 - 6i)$
8. $(9 - 3i) - (12 - 5i)$
9. $(3 - 7i) + 5i$
10. $12 + (5 - 2i)$
11. $(5i)(3i)$
12. $(2i)(4i)$
13. $-2i(5 - 3i)$
14. $-3i(2 - 4i)$

15. $(2 - 3i)(3 + 3i)$

16. $(3 - 5i)(-2 - 3i)$

17. $(7 - 6i)(2 - 3i)$

18. $(2 - i)(3 + 2i)$

19. $(7 + 4i)(7 - 4i)$

20. $(5 - 3i)(5 + 3i)$

21. $\dfrac{1}{2 + i}$

22. $\dfrac{1}{3 - i}$

23. $\dfrac{3 + i}{2 - 3i}$

24. $\dfrac{2 - i}{3 + 2i}$

25. $\dfrac{13 + i}{2 - i}$

26. $\dfrac{15 - 3i}{2 - 3i}$

B *Convert square roots of negative numbers to complex form; perform the indicated operations; and express your answers in the form $a + bi$.*

27. $(5 - \sqrt{-9}) + (2 - \sqrt{-4})$

28. $(-8 + \sqrt{-25}) + (3 - \sqrt{-4})$

29. $(9 - \sqrt{-9}) - (12 - \sqrt{-25})$

30. $(4 + \sqrt{-49}) - (-2 - \sqrt{-36})$

31. $(-2 + \sqrt{-49})(3 - \sqrt{-4})$

32. $(5 + \sqrt{-9})(2 - \sqrt{-1})$

33. $\dfrac{5 - \sqrt{-4}}{3}$

34. $\dfrac{6 - \sqrt{-64}}{2}$

35. $\dfrac{1}{2 - \sqrt{-9}}$

36. $\dfrac{1}{3 - \sqrt{-16}}$

37. $\dfrac{2}{5i}$

38. $\dfrac{1}{3i}$

39. $\dfrac{1 + 3i}{2i}$

40. $\dfrac{2 - i}{3i}$

41. $(2 - i)^2 + 3(2 - i) - 5$

42. $(2 - 3i)^2 - 2(2 - 3i) + 9$

43. Evaluate $x^2 - 2x + 2$ for $x = 1 - i$.

44. Evaluate $x^2 - 2x + 2$ for $x = 1 + i$.

45. Evaluate $x^2 - 4x + 5$ for $x = 2 + i$.

46. Evaluate $x^2 - 4x + 5$ for $x = 2 - i$.

47. Simplify: i^2, i^3, i^4, i^5, i^6, i^7, and i^8

48. Simplify: i^{12}, i^{13}, i^{14}, i^{15}, and i^{16}

C *Perform the indicated operations and write each answer in the form $a + bi$.*

49. $(a + bi) + (c + di)$

50. $(a + bi) - (c + di)$

51. $(a + bi)(a - bi)$

52. $(u - vi)(u + vi)$

53. $(a + bi)(c + di)$

54. $\dfrac{a + bi}{c + di}$

55. $\left(-\dfrac{1}{2} - \dfrac{\sqrt{3}}{2}i\right)^3$

56. $\left(-\dfrac{1}{2} + \dfrac{\sqrt{3}}{2}i\right)^3$

Solve each equation.

57. $y^2 = -36$

58. $x^2 = -25$

59. $(x - 9)^2 = -9$

60. $(x - 3)^2 = -4$

61. Simplify: i^{-2}, i^{-3}, i^{-4}, i^{-5}, i^{-6}, i^{-7}, and i^{-8}

62. Simplify: i^{-12}, i^{-13}, i^{-14}, i^{-15}, and i^{-16}

63. For what values of x will $\sqrt{x - 10}$ be real?

64. When will

$$\frac{-b \pm \sqrt{b^2 - 4ac}}{2a}$$

represent a real number, assuming a, b, and c are all real numbers $(a \neq 0)$? When will it represent a nonreal complex number?

65. Evaluate

$$(a + bi)\left(\frac{a}{a^2 + b^2} - \frac{b}{a^2 + b^2}i\right) \qquad a \neq 0 \text{ and } b \neq 0$$

thus showing that each nonzero complex number $a + bi$ has an inverse relative to multiplication.

5-9 CHAPTER REVIEW

For a natural number n, the **base** a raised to the **exponent** n is $a^n = a \cdot a \cdot \cdots \cdot a$ (n factors). Natural number exponents satisfy these **five properties of exponents**:

1. $a^m a^n = a^{m+n}$
2. $(a^n)^m = a^{mn}$
3. $(ab)^m = a^m b^m$
4. $\left(\dfrac{a}{b}\right)^m = \dfrac{a^m}{b^m}$
5. $\dfrac{a^m}{a^n} = \begin{cases} a^{m-n} & \text{if } m > n \\ 1 & \text{if } m = n \\ \dfrac{1}{a^{n-m}} & \text{if } n > m \end{cases}$ *(5-1)*

The concept of exponent is extended to **0 and negative number exponents** by

$$a^0 = 1 \qquad a \neq 0,\ 0^0 \text{ not defined}$$

$$a^{-n} = \frac{1}{a^n} \qquad \text{for } n \text{ a positive integer}$$

The **five basic properties of exponents** continue to hold, and the fifth property can be rewritten as

$$\frac{a^m}{a^n} = a^{m-n} \qquad (5\text{-}2)$$

Any real number can be written in **scientific notation**—that is, in the form $a \cdot 10^n$, where a is in the interval $[1, 10)$ and n is an integer. *(5-3)*

For a natural number n, a is an **nth root** of b if $a^n = b$. For even n, b has two real nth roots when b is positive, none when b is negative; for odd n, b always has one real nth root. The concept of exponents is extended to **rational number exponents** by

$$b^{1/n} = \begin{cases} \text{The positive } n\text{th root of } b \text{ if } b \text{ is positive} \\ \text{The } n\text{th root of } b \text{ if } b \text{ is negative, } n \text{ odd} \\ \text{Not a real number if } b \text{ is negative, } n \text{ even} \end{cases}$$

and

$$b^{m/n} = (b^{1/n})^m = (b^m)^{1/n}$$

The five basic properties of exponents continue to hold for rational exponents as long as undefined roots are avoided. *(5-4)*

An nth root of b is also denoted by the **radical** form $\sqrt[n]{b} = b^{1/n}$; here n is called the **index** and b the **radicand**. In this notation $b^{m/n} = \left(\sqrt[n]{b}\right)^m = \sqrt[n]{b^m}$. *(5-5)*

For positive a and b, these properties of radicals hold:

1. $\sqrt[n]{b^n} = b$
2. $\sqrt[n]{ab} = \sqrt[n]{a} \cdot \sqrt[n]{b}$
3. $\sqrt[n]{\dfrac{a}{b}} = \dfrac{\sqrt[n]{a}}{\sqrt[n]{b}}$
4. $\sqrt[kn]{b^{km}} = \sqrt[n]{b^m}$

These properties can be used to convert an expression involving radicals to **simplest radical form** wherein the index of the radical is less than the power of any factor of the radicand, the index has no common factor with the power of the radicand, no radical occurs in a denominator, and no fraction appears in a radical. Property 1 can be extended to any real b:

$$\sqrt[n]{b^n} = \begin{cases} b & \text{if } n \text{ is odd} \\ |b| & \text{if } n \text{ is even} \end{cases} \quad \textit{(5-6)}$$

Expressions involving radicals can often be added, subtracted, or multiplied by making use of the distributive property for real numbers. Radicals can sometimes be removed from denominators by **rationalizing the denominator**—multiplying the fraction numerator and denominator by a factor that makes the denominator a rational number. *(5-7)*

A **complex number** is a number of the form $a + bi$, where i is called the **imaginary unit** and a and b are real numbers. If $b = 0$, the number is real. If $b \neq 0$, $a + bi$ is called an **imaginary number**; if $a = 0$, the number is **pure imaginary**. The **conjugate** of $a + bi$ is $a - bi$. Two complex numbers $a + bi$ and $c + di$ are **equal** when $a = c$ and $b = d$. Basic operations are defined by

$$(a + bi) + (c + di) = (a + c) + (b + d)i$$
$$(a + bi) - (c + di) = (a - c) + (b - d)i$$
$$(a + bi) \cdot (c + di) = (ac - bd) + (ad + bc)i$$

Division of complex numbers makes use of rationalizing the denominator by

$$\frac{1}{a + bi} = \frac{1}{a + bi} \cdot \frac{a - bi}{a - bi} = \frac{a}{a^2 + b^2} - \frac{b}{a^2 + b^2}i$$

Since $i^2 = -1$, i is also denoted $\sqrt{-1}$. Similarly, if b is positive then $\sqrt{-b} = i\sqrt{b}$. *(5-8)*

REVIEW EXERCISE 5-9

Work through all the problems in this chapter review and check answers in the back of the book. (Answers to all problems are there, and following each answer is a number in italics indicating the section in which that type of problem is discussed.) Where weaknesses show up, review appropriate sections in the text.

Unless otherwise stated, all variables represent positive real numbers.

A *Simplify, using natural number exponents only.*

1. $\dfrac{x^8}{x^3}$ **2.** $(xy)^3$ **3.** $\left(\dfrac{x}{y}\right)^3$

4. $\dfrac{x^3}{x^8}$ **5.** $(x^3)^8$ **6.** $\dfrac{x^3}{x^3}$

7. x^3x^8 **8.** $(-2x)^3$ **9.** $(-2x^3)(3x^8)$

Evaluate if possible, using only real numbers.

10. $\left(\frac{1}{3}\right)^0$ **11.** 3^{-2} **12.** $\frac{1}{2^{-3}}$

13. $4^{-1/2}$ **14.** $(-9)^{3/2}$ **15.** $(-8)^{2/3}$

16. Write in scientific notation: **(A)** 4,280,000,000 **(B)** 0.000 031 8

17. Write as a decimal fraction: **(A)** 7.29×10^5 **(B)** 6.03×10^{-4}

Simplify, and write answers using positive exponents only.

18. $(3x^3y^2)(2xy^5)$ **19.** $\frac{9u^8v^6}{3u^4v^8}$ **20.** $6(xy^3)^5$

21. $\left(\frac{c^2}{d^5}\right)^3$ **22.** $\left(\frac{2x^2}{3y^3}\right)^2$ **23.** $(x^{-3})^{-4}$

24. $\frac{y^{-3}}{y^{-5}}$ **25.** $(x^2y^{-3})^{-1}$ **26.** $(x^9)^{1/3}$

27. $(x^4)^{-1/2}$ **28.** $x^{1/3}x^{-2/3}$ **29.** $\frac{u^{5/3}}{u^{2/3}}$

30. Change to radical form: **(A)** $(3m)^{1/2}$ **(B)** $3m^{1/2}$

31. Change to rational exponent form: **(A)** $\sqrt{2x}$ **(B)** $\sqrt{a+b}$

Simplify, and write in simplest radical form.

32. $\sqrt{4x^2y^4}$ **33.** $\sqrt{\frac{25}{y^2}}$

34. $\sqrt{36x^4y^7}$ **35.** $\frac{1}{\sqrt{2y}}$

36. $\frac{6ab}{\sqrt{3a}}$ **37.** $\sqrt{2x^2y^5}\sqrt{18x^3y^2}$

38. $\sqrt{\frac{y}{2x}}$ **39.** $4\sqrt{x} - 7\sqrt{x}$

40. $\sqrt{7} + 2\sqrt{3} - 4\sqrt{3}$ **41.** $\sqrt{5}(\sqrt{5} + 2)$

42. $(\sqrt{3} - 1)(\sqrt{3} + 2)$ **43.** $\frac{\sqrt{5}}{3 - \sqrt{5}}$

Perform the indicated operations and write the answer in the form $a + bi$.

44. $(-3 + 2i) + (6 - 8i)$

45. $(3 - 3i)(2 + 3i)$

46. $\dfrac{13 - i}{5 - 3i}$

47. $\dfrac{2 - i}{2i}$

B *Simplify, using natural number exponents only.*

48. $\left(\dfrac{2x^3}{y^8}\right)^2$

49. $(-x^2y)^2(-xy^2)^3$

50. $\dfrac{-4(x^2y)^3}{(-2x)^2}$

51. $(3xy^3)^2(x^2y)^3$

52. $\left(\dfrac{-2x}{y^2}\right)^3$

53. $\left(\dfrac{3x^3y^2}{2x^2y^3}\right)^2$

54. Convert each number to scientific notation, simplify, and write your answer in scientific notation and as a decimal fraction:

$$\frac{0.000\ 052}{130(0.000\ 2)}$$

Simplify, and write answers using positive exponents only.

55. $\dfrac{3m^4n^{-7}}{6m^2n^{-2}}$

56. $(x^{-3}y^2)^{-2}$

57. $\dfrac{1}{(2x^2y^{-3})^{-2}}$

58. $\left(-\dfrac{a^2b}{c}\right)^2\left(\dfrac{c}{b^2}\right)^3\left(\dfrac{1}{a^3}\right)^2$

59. $\left(\dfrac{8u^{-1}}{2^2u^2v^0}\right)^{-2}\left(\dfrac{u^{-5}}{u^{-3}}\right)^3$

60. $\left(\dfrac{9m^3n^{-3}}{3m^{-2}n^2}\right)^{-2}$

61. $(x - y)^{-2}$

62. $(9a^4b^{-2})^{1/2}$

63. $\left(\dfrac{27x^2y^{-3}}{8x^{-4}y^3}\right)^{1/3}$

64. $\dfrac{m^{-1/4}}{m^{3/4}}$

65. $(2x^{1/2})(3x^{-1/3})$

66. $\dfrac{3x^{-1/4}}{6x^{-1/3}}$

67. $\dfrac{5^0}{3^2} + \dfrac{3^{-2}}{2^{-2}}$

68. $(x^{1/2} + y^{1/2})^2$

69. If a is a square root of b, then does $a^2 = b$ or does $b^2 = a$?

70. Change to radical form: **(A)** $(2mn)^{2/3}$ **(B)** $3x^{2/5}$

71. Change to rational exponent form: **(A)** $\sqrt[7]{x^5}$ **(B)** $-3\sqrt[3]{(xy)^2}$

Simplify, and write in simplest radical form.

72. $\sqrt[3]{(2x^2y)^3}$ **73.** $3x\sqrt[3]{x^5y^4}$ **74.** $\dfrac{\sqrt{8m^3n^4}}{\sqrt{12m^2}}$

75. $\sqrt[8]{y^6}$ **76.** $-2x\sqrt[5]{3^6x^7y^{11}}$ **77.** $\dfrac{2x^2}{\sqrt[3]{4x}}$

78. $\sqrt[5]{\dfrac{3y^2}{8x^2}}$ **79.** $(2\sqrt{x} - 5\sqrt{y})(\sqrt{x} + \sqrt{y})$

80. $\dfrac{\sqrt{x} - 2}{\sqrt{x} + 2}$ **81.** $\dfrac{3\sqrt{x}}{2\sqrt{x} - \sqrt{y}}$

82. $\sqrt{\frac{2}{3}} + \sqrt{\frac{3}{2}}$

Perform the indicated operations and write the answer in the form $a + bi$.

83. $(2 - 2\sqrt{-4}) - (3 - \sqrt{-9})$ **84.** $\dfrac{2 - \sqrt{-1}}{3 + \sqrt{-4}}$

85. $(3 + i)^2 - 2(3 + i) + 3$

Simplify, and write answers using positive exponents only.

86. $(x^{-1} + y^{-1})^{-1}$ **87.** $\left(\dfrac{a^{-2}}{b^{-1}} + \dfrac{b^{-2}}{a^{-1}}\right)^{-1}$

C *Simplify, and write in simplest radical form.*

88. $\sqrt[9]{8x^6y^{12}}$ **89.** $\sqrt[3]{3} - \dfrac{6}{\sqrt[3]{9}} + 3\sqrt[3]{\dfrac{1}{9}}$

90. Simplify $3\sqrt[3]{x^3} - 2\sqrt{x^2}$:
(A) For x a positive number **(B)** For x a negative number

6

SECOND-DEGREE EQUATIONS AND INEQUALITIES

The equation

$$\tfrac{1}{2}x - \tfrac{1}{3}(x + 3) = 2 - x$$

though complicated-looking, is actually a first-degree equation in one variable, since it can be transformed into the equivalent equation

$$7x - 18 = 0$$

which is a special case of

$$ax + b = 0 \qquad a \neq 0$$

We have solved many equations of this type and found that they always have a single solution. From a mathematical point of view we have essentially taken care of the problem of solving first-degree equations in one variable. We have done the same for linear inequalities in one variable.

In this chapter we will consider the next class of polynomial equations, called second-degree equations or quadratic equations. A **quadratic equation** in one variable is any equation that can be written in the form

Quadratic Equation (Standard Form)

$$ax^2 + bx + c = 0 \qquad a \neq 0 \qquad (1)$$

where x is a variable and a, b, and c are constants. We will refer to this form as the **standard form** for the quadratic equation. The equations

$$2x^2 - 3x + 5 = 0 \qquad \text{and} \qquad 15 = 180t - 16t^2$$

are both quadratic equations since they are either in the standard form or can be converted into this form.

Applications that give rise to quadratic equations are many and varied. A brief glance at Section 6-3 will give you some indication of the variety.

The chapter concludes with a consideration of quadratic and other nonlinear inequalities.

6-1 SOLVING QUADRATIC EQUATIONS BY SQUARE ROOTS AND BY COMPLETING THE SQUARE

- Solution by Factoring
- Solution by Square Roots
- Completing the Square
- Solution of Quadratic Equations by Completing the Square

The technique of solving equations by factoring was introduced in Section 2-7. The technique easily solves those standard quadratic equations that can be factored. In this section we develop a method applicable to all quadratic equations. The method depends, first, upon solving simple equations such as $x^2 = 16$ by taking square roots, and, second, upon a technique called completing the square that converts equations into a readily solved form. Although the method always works, it is cumbersome. We can, however, apply the method to produce a general formula for solving quadratic equations. This is done in the next section.

SOLUTION BY FACTORING

Recall the zero property for real numbers: the product of real numbers is zero if and only if one of the factors is zero. This allows us to solve equations of the form

$$\text{Factored polynomial} = 0$$

by setting each of the factors equal to 0. For example, to solve

$$(x - 2)(x + 1) = 0$$

we set the factors $x - 2 = 0$ and $x + 1 = 0$ to obtain

$$x = 2 \qquad \text{or} \qquad x = -1$$

Example 1 Solve by factoring if possible:

(A) $x^2 - 7x + 12 = 0$ **(B)** $x^2 + 6x - 2 = 0$

(C) $3 + \dfrac{5}{x} = \dfrac{2}{x^2}$

Solution **(A)** $x^2 - 7x + 12 = 0$

$(x - 3)(x - 4) = 0$ The product is 0 if and only if $x - 3 = 0$ or $x - 4 = 0$.

$$x - 3 = 0 \qquad \text{or} \qquad x - 4 = 0$$

$$x = 3 \qquad \text{or} \qquad x = 4$$

(B) The polynomial cannot be factored using integer coefficients. Another method must be used. This will be done later in this section.

(C) $3 + \frac{5}{x} = \frac{2}{x^2}$ — Multiply both sides by x^2, the LCM of the denominators ($x \neq 0$).

$3x^2 + 5x = 2$ — Write in standard form: $ax^2 + bx + c$.

$3x^2 + 5x - 2 = 0$ — Factor the left side, if possible.

$(3x - 1)(x + 2) = 0$

$3x - 1 = 0$ or $x + 2 = 0$

$3x = 1$ $\qquad x = -2$

$x = \frac{1}{3}$

Problem 1 Solve by factoring if possible:

(A) $x^2 + 2x - 15 = 0$ **(B)** $x^2 + 2x + 3$

(C) $x = \frac{3}{2x - 5}$

SOLUTION BY SQUARE ROOTS

The easiest type of quadratic equation to solve is the special form where the first-degree term is missing; that is, when Equation (1) is of the form

$$ax^2 + c = 0 \qquad a \neq 0$$

The method of solution makes direct use of the definition of square root. The process is illustrated in the following example.

Example 2 Solve by the square root method:

(A) $x^2 - 8 = 0$ **(B)** $2x^2 - 3 = 0$
(C) $3x^2 + 27 = 0$ **(D)** $(x + \frac{1}{2})^2 = \frac{5}{4}$

Solution **(A)** $x^2 - 8 = 0$

$x^2 = 8$ — What number squared is 8?

$x = \pm\sqrt{8}$ or $\pm 2\sqrt{2}$ — $\pm 2\sqrt{2}$ is a short way of writing $-2\sqrt{2}$ or $+2\sqrt{2}$.

(B) $2x^2 - 3 = 0$

$2x^2 = 3$ — Do not write $2x = \pm\sqrt{3}$ next since this would ignore the 2 when taking one square root.

$x^2 = \frac{3}{2}$ — What number squared is $\frac{3}{2}$?

$x = \pm\sqrt{\frac{3}{2}}$ or $\pm\frac{\sqrt{6}}{2}$

(C) $3x^2 + 27 = 0$

$$3x^2 = -27 \quad \text{Do not write } 3x = \pm\sqrt{-27} \text{ next. (Why?)}$$

$$x^2 = -9 \quad \text{What number squared is } -9?$$

$$x = \pm\sqrt{-9} = \pm 3i$$

(D) $(x + \frac{1}{2})^2 = \frac{5}{4}$ — Solve for $x + \frac{1}{2}$; then solve for x.

$$x + \tfrac{1}{2} = \pm\sqrt{\tfrac{5}{4}}$$

$$x = -\frac{1}{2} \pm \frac{\sqrt{5}}{2}$$

$$= \frac{-1 \pm \sqrt{5}}{2} \quad \text{Short for } \frac{-1 + \sqrt{5}}{2} \text{ or } \frac{-1 - \sqrt{5}}{2}$$

Problem 2 Solve by the square root method:

(A) $x^2 - 12 = 0$ **(B)** $3x^2 - 5 = 0$
(C) $2x^2 + 8 = 0$ **(D)** $(x + \frac{1}{3})^2 = \frac{2}{9}$

COMPLETING THE SQUARE

The factoring and square root methods are fast and easy to use when they apply. Unfortunately, many quadratic equations will not yield directly to either method. For example, the very simple-looking polynomial in the equation

$$x^2 + 6x - 2 = 0$$

cannot be solved by taking square roots and cannot be factored in the integers. The equation requires a new approach if it can be solved at all.

We will now discuss a method, called solution by completing the square, that will work for all quadratic equations. In the next section we will use this method to develop a general formula that will be used in the future whenever the square root or factoring method fails.

The method of completing the square is based on the process of transforming the standard quadratic equation

$$ax^2 + bx + c = 0$$

into the form

$$(x + A)^2 = B$$

where A and B are constants. This last equation can easily be solved by the square root method:

$$\begin{aligned}(x + A)^2 &= B\\ x + A &= \pm\sqrt{B}\\ x &= -A \pm \sqrt{B}\end{aligned}$$

Before considering how the transformation above is accomplished, let's pause for a moment and consider a related problem: what number must be added to $x^2 + 6x$ so that the result is the square of a linear expression? There is an easy mechanical rule for finding this number based on the squares of the following binomials:

$$(x + m)^2 = x^2 + 2mx + m^2 \qquad (x - m)^2 = x^2 - 2mx + m^2$$

In either case, we see that the third term on the right of each equation is the square of one-half of the coefficient of x in the second term on the right; that is, m^2 is the square of one-half of $2m$. This observation leads directly to the rule:

> To **complete the square** of a quadratic of the form
>
> $$x^2 + bx$$
>
> add the square of one-half of the coefficient of x, that is,
>
> $$\left(\frac{b}{2}\right)^2 \qquad \text{or} \qquad \frac{b^2}{4}$$
>
> Thus,
>
> $$x^2 + bx + \left(\frac{b}{2}\right)^2 = \left(x + \frac{b}{2}\right)^2$$

Example 3 Complete the square and factor:

(A) $x^2 + 6x$ **(B)** $x^2 - 3x$

Solution **(A)** To complete the square of $x^2 + 6x$, add $(\frac{6}{2})^2$, that is, 9. Thus,

$$x^2 + 6x + \mathbf{9} = (x + 3)^2$$

(B) To complete the square of $x^2 - 3x$, add $(-\frac{3}{2})^2$, that is, $\frac{9}{4}$. Thus,

$$x^2 - 3x + \tfrac{9}{4} = (x - \tfrac{3}{2})^2$$

Problem 3 Complete the square and factor:

(A) $x^2 + 10x$ **(B)** $x^2 - 5x$

Note: The rule stated above applies only to quadratic forms where the coefficient of the second-degree term is 1. When solving equations, we will divided through by the leading coefficient so that the rule may be applied (see Example 6 below).

We now use the method of completing the square to solve quadratic equations. In the next section we will use the method to develop a formula that will work for *all* quadratic equations.

SOLUTION OF QUADRATIC EQUATIONS BY COMPLETING THE SQUARE

Solving quadratic equations by the method of completing the square is best illustrated by examples.

Example 4 Solve $x^2 + 6x - 2 = 0$ by the method of completing the square.

Solution

$x^2 + 6x - 2 = 0$ Add 2 to both sides of the equation to remove -2 from the left side.

$x^2 + 6x = 2$ To complete the square of the left side, add the square of one-half of the coefficient of x, that is, $(\frac{6}{2})^2$, to each side of the equation.

$x^2 + 6x + \mathbf{9} = 2 + \mathbf{9}$ Factor the left side.

$(x + 3)^2 = 11$ Solve by square root method.

$x + 3 = \pm\sqrt{11}$

$x = -3 \pm \sqrt{11}$

Problem 4 Solve $x^2 - 8x + 10 = 0$ by completing the square.

Example 5 Solve $x^2 - 4x + 13 = 0$ by completing the square.

Solution

$x^2 - 4x + 13 = 0$

$x^2 - 4x = -13$ Add 4 to each side to complete the square on the left side.

$x^2 - 4x + \mathbf{4} = \mathbf{4} - 13$

$(x - 2)^2 = -9$

$x - 2 = \pm\sqrt{-9}$

$x - 2 = \pm 3i$

$x = 2 \pm 3i$

Problem 5 Solve $x^2 - 2x + 3 = 0$ by completing the square.

Example 6 Solve $2x^2 - 4x - 3 = 0$ by completing the square.

Solution

$$2x^2 - 4x - 3 = 0$$ Note that the coefficient of x^2 is not 1. Divide through by the leading coefficient and proceed as in the last example.

$$x^2 - 2x - \tfrac{3}{2} = 0$$
$$x^2 - 2x = \tfrac{3}{2}$$
$$x^2 - 2x + \mathbf{1} = \tfrac{3}{2} + \mathbf{1}$$
$$(x - 1)^2 = \tfrac{5}{2}$$
$$x - 1 = \pm\sqrt{\tfrac{5}{2}}$$
$$x = 1 \pm \frac{\sqrt{10}}{2}$$
$$= \frac{2 \pm \sqrt{10}}{2}$$

Problem 6 Solve $2x^2 + 8x + 3 = 0$ by completing the square.

The method of completing the square can be summarized as follows:

1. Write the equation in standard form:

$$ax^2 + bx + c = 0$$

2. Make the coefficient of x^2 equal to 1 by dividing both sides by the existing coefficient a.
3. Move the constant term to the right side.
4. Complete the square on the left side, adding the same amount to the right.
5. Solve by taking square roots.

ANSWERS TO MATCHED PROBLEMS

1. **(A)** $x = -5, 3$ **(B)** Cannot be solved by factoring.
   (C) $x = -\frac{1}{2}, 3$
2. **(A)** $x = \pm 2\sqrt{3}$ **(B)** $x = \pm\sqrt{\frac{5}{3}}$ or $\pm\dfrac{\sqrt{15}}{3}$
   (C) $x = \pm 2i$ **(D)** $x = (-1 \pm \sqrt{2})/3$
3. **(A)** $x^2 + 10x + 25 = (x + 5)^2$ **(B)** $x^2 - 5x + \frac{25}{4} = (x - \frac{5}{2})^2$
4. $x = 4 \pm \sqrt{6}$ **5.** $x = 1 \pm i\sqrt{2}$
6. $x = -2 \pm \sqrt{\frac{5}{2}}$ or $\dfrac{-4 \pm \sqrt{10}}{2}$

EXERCISE 6-1

A *Solve by factoring if possible.*

1. $x^2 + 5x - 6 = 0$
2. $x^2 - 4x + 5 = 0$
3. $x^2 - 7x + 10 = 0$
4. $x^2 + 4x - 21 = 0$
5. $2x^2 + 7x - 4 = 0$
6. $2x^2 - 7x - 15 = 0$

Solve by the square root method.

7. $x^2 - 16 = 0$ **8.** $x^2 - 25 = 0$ **9.** $x^2 + 16 = 0$

10. $x^2 + 25 = 0$ **11.** $y^2 - 45 = 0$ **12.** $m^2 - 12 = 0$

12. $4x^2 - 9 = 0$ **14.** $9y^2 - 16 = 0$ **15.** $16y^2 = 9$

16. $9x^2 = 4$

Complete the square and factor.

17. $x^2 + 4x$ **18.** $x^2 + 8x$ **19.** $x^2 - 6x$

20. $x^2 - 10x$ **21.** $x^2 + 12x$ **22.** $x^2 + 2x$

Solve by completing the square.

23. $x^2 + 4x + 2 = 0$ **24.** $x^2 + 8x + 3 = 0$

25. $x^2 - 6x - 3 = 0$ **26.** $x^2 - 10x - 3 = 0$

B *Solve by factoring if possible. (Write the equations in standard form first.)*

27. $\frac{t}{2} = \frac{2}{t}$ **28.** $y = \frac{9}{y}$

29. $\frac{m}{4}(m + 1) = 3$ **30.** $\frac{A^2}{2} = A + 4$

31. $2y = \frac{2}{y} + 3$ **32.** $L = \frac{15}{L - 2}$ **33.** $2 + \frac{2}{x^2} = \frac{5}{x}$

34. $1 - \frac{3}{x} = \frac{10}{x^2}$ **35.** $\frac{x}{6} = \frac{1}{x + 1}$ **36.** $x + 1 = \frac{2}{x}$

Solve by the square root method.

37. $y^2 = 2$ **38.** $x^2 = 3$ **39.** $16a^2 + 9 = 0$

40. $4x^2 + 25 = 0$ **41.** $9x^2 - 7 = 0$ **42.** $4t^2 - 3 = 0$

43. $(m - 3)^2 = 25$ **44.** $(n + 5)^2 = 9$ **45.** $(t + 1)^2 = -9$

46. $(d - 3)^2 = -4$ **47.** $(x - \frac{1}{3})^2 = \frac{4}{9}$ **48.** $(x - \frac{1}{2})^2 = \frac{9}{4}$

Complete the square and factor.

49. $x^2 + 3x$ **50.** $x^2 + x$

51. $u^2 - 5u$ **52.** $m^2 - 7m$

Solve by completing the square.

53. $x^2 + x - 1 = 0$ 　 **54.** $x^2 + 3x - 1 = 0$

55. $u^2 - 5u + 2 = 0$ 　 **56.** $n^2 - 3n - 1 = 0$

57. $m^2 - 4m + 8 = 0$ 　 **58.** $x^2 - 2x + 3 = 0$

59. $2y^2 - 4y + 1 = 0$ 　 **60.** $2x^2 - 6x + 3 = 0$

61. $2u^2 + 3u - 1 = 0$ 　 **62.** $3x^2 + x - 1 = 0$

C *Solve by factoring. (Write the equations in standard form first.)*

63. $x = \dfrac{1 + 3x}{x + 3}$ 　 **64.** $\dfrac{1}{x} - \dfrac{1}{x^2} = \dfrac{1}{x + 1}$

65. $\dfrac{x + 2}{x - 1} - \dfrac{6x}{x^2 - 1} = \dfrac{2x - 1}{x + 1}$ 　 **66.** $\dfrac{2(x - 1)}{x - 2} = \dfrac{1}{x} + \dfrac{1}{x - 2}$

Solve by the square root method.

67. $(y + \frac{5}{2})^2 = \frac{5}{2}$ 　 **68.** $(x - \frac{3}{2})^2 = \frac{3}{2}$

69. $(x - 2)^2 = -1$ 　 **70.** $(x + \frac{1}{2})^2 = -\frac{3}{4}$

Solve by completing the square.

71. $2u^2 - 3u + 2 = 0$ 　 **72.** $3x^2 - 5x + 3 = 0$

73. $x^2 + x + 1 = 0$ 　 **74.** $2x^2 - 3x + 4 = 0$

75. $x^2 + 2\sqrt{2}x - 2 = 0$ 　 **76.** $x^2 - 2\sqrt{5}x + 5 = 0$

77. $x^2 - 4\sqrt{3}x + 13 = 0$ 　 **78.** $x^2 + 2\sqrt{2}x + 3 = 0$

79. $x^2 - 2ix - 4 = 0$ 　 **80.** $x^2 + 2ix + 2 = 0$

81. Solve for x: $x^2 + mx + n = 0$

82. Solve for x: $ax^2 + bx + c = 0,\ a \neq 0$

Solve for the indicated letters in terms of the other letters. Use positive square roots only.

83. $a^2 + b^2 = c^2$ Solve for a. 　 **84.** $s = \frac{1}{2}gt^2$ Solve for t.

85. In a given city on a given day, the demand equation for gasoline is $d = 900/p$ and the supply equation is $s = p - 80$, where d and s denote the number of gallons demanded and supplied (in thousands), respectively, at a price of p cents per gallon. Find the price at which supply is equal to demand.

86. To find the critical velocity at the top of the loop necessary to keep a steel ball on the track (see the figure), the centripetal force mv^2/r is equated to the force due to gravity mg. The mass m cancels out of the equation, and we are left with $v^2 = gr$. For a loop of radius 0.25 foot, find the critical velocity (in feet per second) at the top of the loop that is required to keep the ball on the track. Use $g = 32$ and compute your answer to two decimal places using a calculator.

6-2 THE QUADRATIC FORMULA

- Quadratic Formula
- The Discriminant
- Which Method?

The method of completing the square can be used to solve any quadratic equation, but the process is often tedious. If you had a very large number of quadratic equatons to solve by completing the square, before you finished you would probably ask yourself if the process could not be made more efficient. In this section we will take the general equation

$$ax^2 + bx + c = 0 \qquad a \neq 0$$

and solve it for x in terms of the coefficients a, b, and c by the method of completing the square—thus obtaining a formula that can be memorized and used to find the solution of any quadratic equation.

QUADRATIC FORMULA

We apply the method of completing the square to the general equation

$$ax^2 + bx + c = 0 \qquad a \neq 0$$

The following steps correspond to the summary of the method of completing the square given in Section 6-1.

1. The equation is given in standard form.
2. We make the leading coefficient 1 by dividing both sides of the equation by a to obtain

$$x^2 + \frac{b}{a}x + \frac{c}{a} = 0$$

3. Add $-c/a$ to both sides to clear c/a from the left side:

$$x^2 + \frac{b}{a}x = -\frac{c}{a}$$

4. Complete the square on the left side by adding the square of one-half the coefficient of x, that is, $(b/2a)^2$, to each side:

$$x^2 + \frac{b}{a}x + \frac{b^2}{4a^2} = \frac{b^2}{4a^2} - \frac{c}{a}$$

5. We now factor the left side and solve by the square root method:

$$\left(x + \frac{b}{2a}\right)^2 = \frac{b^2 - 4ac}{4a^2}$$

$$x + \frac{b}{2a} = \pm\sqrt{\frac{b^2 - 4ac}{4a^2}}$$

$$x = \frac{-b}{2a} \pm \sqrt{\frac{b^2 - 4ac}{4a^2}}$$

$$= \frac{-b}{2a} \pm \frac{\sqrt{b^2 - 4ac}}{\sqrt{4a^2}}$$

If $a > 0$, $\sqrt{4a^2} = 2a$ and

$$x = \frac{-b}{2a} \pm \frac{\sqrt{b^2 - 4ac}}{2a} = \frac{-b \pm \sqrt{b^2 - 4ac}}{2a}$$

If $a < 0$, $\sqrt{4a^2} = -2a$ and

$$x = \frac{-b}{2a} \pm \frac{\sqrt{b^2 - 4ac}}{-2a} = \frac{-b \pm \sqrt{b^2 - 4ac}}{2a}$$

In either case, we obtain the following formula:

Quadratic Formula
$x = \dfrac{-b \pm \sqrt{b^2 - 4ac}}{2a} \qquad a \neq 0$

The **quadratic formula** should be memorized and used to solve quadratic equations when simpler methods fail.

Example 7 Solve $2x^2 - 4x - 3 = 0$ by use of the quadratic formula.

Solution

$$2x^2 - 4x - 3 = 0$$

$$x = \frac{-b \pm \sqrt{b^2 - 4ac}}{2a}$$ Write down the quadratic formula and identify a, b, and c. Here, $a = 2$, $b = -4$, $c = -3$.

$$= \frac{-(-4) \pm \sqrt{(-4)^2 - 4(2)(-3)}}{2(2)}$$ Substitute into the formula and simplify. Be careful of sign errors.

$$= \frac{4 \pm \sqrt{40}}{4} = \frac{4 \pm 2\sqrt{10}}{4}$$

$$= \frac{2 \pm \sqrt{10}}{2}$$ The solutions are $\frac{2 + \sqrt{10}}{2}$ and $\frac{2 - \sqrt{10}}{2}$.

Problem 7 Solve $x^2 - 2x - 1 = 0$ using the quadratic formula.

Example 8 Solve $x^2 + 11 = 6x$ using the quadratic formula.

Solution

$$x^2 + 11 = 6x$$ Write in standard form.

$$x^2 - 6x + 11 = 0$$

$$x = \frac{-b \pm \sqrt{b^2 - 4ac}}{2a}$$ $a = 1$, $b = -6$, $c = 11$

$$= \frac{-(-6) \pm \sqrt{(-6)^2 - 4(1)(11)}}{2(1)}$$ Be careful of sign errors here.

$$= \frac{6 \pm \sqrt{-8}}{2}$$

$$= \frac{6 \pm 2i\sqrt{2}}{2} = 3 \pm i\sqrt{2}$$ The solutions are $3 + i\sqrt{2}$ and $3 - i\sqrt{2}$.

Problem 8 Solve $2x^2 + 3 = 4x$ using the quadratic formula.

THE DISCRIMINANT

The expression $b^2 - 4ac$ that occurs under the radical in the quadratic formula is called the **discriminant**. It provides useful information about the number and nature of the roots.

Discriminant Test

$ax^2 + bx + c = 0 \qquad a, b, c,$ real numbers; $a \neq 0$

$b^2 - 4ac$	ROOTS
Positive	Two real roots
0	One real root
Negative	Two nonreal roots (the roots will be complex conjugates)

Example 9 Apply the discriminant test to determine the number and nature of the roots:

(A) $3x^2 - 4x + 1 = 0$ **(B)** $9x^2 - 6x + 1 = 0$
(C) $x^2 + 5x + 7 = 0$

Solution **(A)** The discriminant $b^2 - 4ac = 4^2 - 4 \cdot 3 \cdot 1 = 4$ is positive. The equation has two real roots. (Check that they are 1 and $\frac{1}{3}$.)
(B) The discriminant $b^2 - 4ac = 6^2 - 4 \cdot 9 \cdot 1 = 0$, so there is one real root. (Check that it is $\frac{1}{3}$.)
(C) The discriminant $b^2 - 4ac = 25 - 4 \cdot 1 \cdot 7 = -3$ is negative, so there are two nonreal complex conjugate roots. (Check that these roots are $-\frac{5}{2} \pm i\sqrt{3}/2$.)

Problem 9 Apply the discriminant test to determine the number and nature of the roots:

(A) $4x^2 - 20x + 25 = 0$ **(B)** $2x^2 + x - 1 = 0$
(C) $x^2 - 6x + 10 = 0$

WHICH METHOD?

In normal practice the quadratic formula is used whenever the square root method or the factoring method does not produce results easily. These latter methods are generally faster when they apply and should be used when possible.†

Note that any equation of the form

$$ax^2 + c = 0$$

† The process of completing the square, in addition to producing the quadratic formula, is used in many other places in mathematics. See Section 7-5, for example.

can always be solved by the square root method. And any equation of the form

$$ax^2 + bx = 0$$

can always be solved by factoring since $ax^2 + bx = x(ax + b)$.

To summarize:

To Solve $ax^2 + bx + c = 0$

1. If $b = 0$, solve by square roots.
   To solve $2x^2 - 5 = 0$,
   $x^2 = \frac{5}{2}$, $x = \pm\sqrt{\frac{5}{2}}$
2. If $c = 0$, solve by factoring.
   To solve $2x^2 - 5x = 0$,
   $x(2x - 5) = 0$
   $x = 0$ or $x = \frac{5}{2}$
3. Otherwise, try:
   (A) Factoring
   (B) Or completing the square
   (C) Or the quadratic formula (which always works)

It is important to realize that the quadratic formula can always be used and will produce the same results as any other method.

Example 10 Solve $\dfrac{30}{8 + x} + 2 = \dfrac{30}{8 - x}$ by the most efficient method.

Solution

$$\frac{30}{8 + x} + 2 = \frac{30}{8 - x}$$

Multiply both sides by the LCM of the denominators $(8 + x)(8 - x)$. Note that $x \neq -8, 8$.

$$30(8 - x) + 2(8 + x)(8 - x) = 30(8 + x)$$

$$240 - 30x + 128 - 2x^2 = 240 + 30x$$

$$-2x^2 - 60x + 128 = 0$$

Divide both sides by -2.

$$x^2 + 30x - 64 = 0$$

Factor the left side, if possible.

$$(x + 32)(x - 2) = 0$$

$$x + 32 = 0 \quad \text{or} \quad x - 2 = 0$$

$$x = -32 \qquad\qquad x = 2$$

We could also have solved $x^2 + 30x - 64 = 0$ by using the quadratic formula:

$$x^2 + 30x - 64 = 0$$

$$x = \frac{-b \pm \sqrt{b^2 - 4ac}}{2a} \qquad a = 1,\ b = 30,\ c = -64$$

$$= \frac{-30 \pm \sqrt{30^2 - 4(1)(-64)}}{2(1)}$$

$$= \frac{-30 \pm \sqrt{1{,}156}}{2}$$

$$= \frac{-30 \pm 34}{2}$$

Thus, $x = -32$ or $x = 2$.

It is clear that the factoring method was much easier in Example 10. Nevertheless, we got the same result, as expected.

Problem 10 Solve $\dfrac{6}{x-2} + 2 = \dfrac{4}{x}$ by the most efficient method.

ANSWERS TO MATCHED PROBLEMS

7. $x = 1 \pm \sqrt{2}$ **8.** $x = 1 \pm \dfrac{i\sqrt{2}}{2}$ or $\dfrac{2 \pm i\sqrt{2}}{2}$

9. **(A)** One real root **(B)** Two real roots
(C) Two nonreal complex conjugate roots

10. $x = \dfrac{1}{2} \pm \dfrac{\sqrt{15}}{2}i$

EXERCISE 6-2

A *Specify the constants a, b, and c for each quadratic equation when written in the standard form* $ax^2 + bx + c = 0$.

1. $2x^2 - 5x + 3 = 0$ **2.** $3x^2 - 2x + 1 = 0$

3. $m = 1 - 3m^2$ **4.** $2u^2 = 1 - 3u$

5. $3y^2 - 5 = 0$ **6.** $2x^2 - 5x = 0$

Solve by use of the quadratic formula.

7. $x^2 + 8x + 3 = 0$ **8.** $x^2 + 4x + 2 = 0$

9. $y^2 - 10y - 3 = 0$ **10.** $y^2 - 6y - 3 = 0$

B **11.** $u^2 = 1 - 3u$ **12.** $t^2 = 1 - t$

13. $y^2 + 3 = 2y$ **14.** $x^2 + 8 = 4x$

15. $2m^2 + 3 = 6m$ **16.** $2x^2 + 1 = 4x$

17. $p = 1 - 3p^2$ **18.** $3q + 2q^2 = 1$

Apply the discriminant test to determine the number and nature of the roots.

19. $4x^2 + 5x - 6 = 0$ **20.** $3x^2 + 2x + 1 = 0$

21. $9x^2 - 24x + 16 = 0$ **22.** $25x^2 + 10x + 1 = 0$

23. $x^2 - 8x + 17 = 0$ **24.** $x^2 + 11x + 30 = 0$

Solve each of the following equations by any method, excluding completing the square.

25. $(x - 5)^2 = 7$ **26.** $(y + 4)^2 = 11$ **27.** $x^2 + 2x = 2$

28. $x^2 - 1 = 3x$ **29.** $2u^2 + 3u = 0$ **30.** $2n^2 = 4n$

31. $x^2 - 2x + 9 = 2x - 4$ **32.** $x^2 + 15 = 2 - 6x$

33. $y^2 = 10y + 3$ **34.** $3(2x + 1) = x^2$

35. $2d^2 + 1 = 4d$ **36.** $2y(3 - y) = 3$

37. $\dfrac{2}{u} = \dfrac{3}{u^2} + 1$ **38.** $1 + \dfrac{8}{x^2} = \dfrac{4}{x}$

39. $\dfrac{1.2}{y - 1} + \dfrac{1.2}{y} = 1$ **40.** $\dfrac{24}{10 + m} + 1 = \dfrac{24}{10 - m}$

Solve for the indicated letter in terms of the other letters.

41. $d = \frac{1}{2}gt^2$ for t (positive) **42.** $a^2 + b^2 = c^2$ for a (positive)

43. $A = P(1 + r)^2$ for r (positive) **44.** $P = EI - RI^2$ for I

C *Solve by use of the quadratic formula.*

45. $x^2 - \sqrt{7}x + 2 = 0$ **46.** $3x^2 - 2\sqrt{15}x + 5 = 0$

47. $\sqrt{3}x^2 + 4x + \sqrt{3} = 0$ **48.** $\sqrt{2}x^2 + 2\sqrt{3}x + \sqrt{2} = 0$

49. $2x^2 + 3ix + 2 = 0$ **50.** $x^2 - ix + 6 = 0$

51. $x^2 + ix - 1 = 0$ **52.** $3x^2 - 5ix + 2 = 0$

Solve to two decimal places using a hand calculator.

53. $2.07x^2 - 3.79x + 1.34 = 0$ **54.** $0.61x^2 - 4.28x + 2.93 = 0$

55. $4.83x^2 + 2.04x - 3.18 = 0$ **56.** $5.13x^2 + 7.27x - 4.32 = 0$

Solve for x in terms of the remaining variables.

57. $y^2 + xy - x^2 = 0$

58. $x^2 + y^2 = x + y$

59. $\dfrac{x + y}{x - y} = \dfrac{x}{y}$

60. $x^2 + 3xy + y^2 - 2x + y + 1 = 0$

Use the discriminant to determine which equations have real solutions.

61. $0.013\,4x^2 + 0.041\,4x + 0.030\,4 = 0$

62. $0.543x^2 - 0.182x + 0.003\,12 = 0$

63. $0.013\,4x^2 + 0.021\,4x + 0.030\,4 = 0$

64. $0.543x^2 - 0.182x + 0.031\,2 = 0$

65. For what values of c does $2x^2 - 3x + c = 0$ have exactly one solution?

66. For what values of a does $ax^2 + 6x + 5 = 0$ have two real solutions?

67. Show that if r_1 and r_2 are the two roots of $ax^2 + bx + c = 0$, then $r_1 r_2 = c/a$.

68. For r_1 and r_2 in Problem 67, show that $r_1 + r_2 = -b/a$.

6-3 APPLICATIONS

We will now consider a number of applications from several fields. Since quadratic equations often have two solutions, it is important to check both solutions in the original problem to see if one or the other must be rejected. Also, a review of the strategy of solving word problems in Sections 1-6 and 4-1 should prove helpful.

Example 11 The sum of a number and its reciprocal is $\frac{5}{2}$. Find the number.

Solution Let x = The number. Its reciprocal is $1/x$. Then

$$x + \frac{1}{x} = \frac{5}{2} \qquad \text{Clear fractions.}$$

$$2x^2 + 2 = 5x \qquad \text{Write in standard form.}$$

$$2x^2 - 5x + 2 = 0 \qquad \text{Solve by factoring.}$$

$$(2x - 1)(x - 2) = 0$$

$$x = \tfrac{1}{2} \quad \text{or} \quad 2 \qquad \text{Both answers are solutions to the problem as can easily be checked.}$$

Problem 11 If the reciprocal of a number is subtracted from the original number, the difference is $\frac{8}{3}$. Find the number.

Example 12 A tank can be filled in 4 hours by two pipes when both are used. How many hours are required for each pipe to fill the tank alone if the smaller pipe requires 3 hours more than the larger one?

Solution Let

$$4 = \text{Time for both pipes to fill the tank together}$$
$$x = \text{Time for the larger pipe to fill the tank alone}$$
$$x + 3 = \text{Time for the smaller pipe to fill the tank alone}$$

Use the rate–time formula to find the rates at which the tank is filled:

$$\begin{pmatrix}\text{Amount of tank}\\ \text{filled per hour}\end{pmatrix} \times \begin{pmatrix}\text{Number of}\\ \text{hours}\end{pmatrix} = \begin{pmatrix}\text{Amount of}\\ \text{tank filled}\end{pmatrix}$$

Thus the rate at which the tank is filled when both pipes are used can be found:

$$\text{Rate} \cdot 4 \text{ hours} = 1 \text{ tank filled}$$
$$\text{Rate} = \tfrac{1}{4} \text{ tank per hour}$$

Similarly,

$$\frac{1}{x} = \text{Rate for larger pipe} \qquad \frac{1}{x} \text{ tank per hour}$$

$$\frac{1}{x+3} = \text{Rate for smaller pipe} \qquad \frac{1}{x+3} \text{ tank per hour}$$

$$\text{Sum of individual rates} = \text{Rate together}$$

$$\frac{1}{x} + \frac{1}{x+3} = \frac{1}{4}$$

$$\mathbf{4x(x+3)} \cdot \frac{1}{x} + \mathbf{4x(x+3)} \cdot \frac{1}{x+3} = \mathbf{4x(x+3)} \cdot \frac{1}{4}$$ Clear fractions.

$$4(x+3) + 4x = x(x+3)$$
$$4x + 12 + 4x = x^2 + 3x$$
$$x^2 - 5x - 12 = 0$$ Use the quadratic formula.

$$x = \frac{5 \pm \sqrt{73}}{2}$$ Since the time must be positive, we discard the negative answer.

$$x = \frac{5 + \sqrt{73}}{2}$$

≈ 6.77 hours — Larger pipe

$x + 3 \approx 9.77$ hours — Smaller pipe

Problem 12 Two pipes can fill a tank in 3 hours when used together. Alone, one can fill the tank 2 hours faster than the other. How long will it take each pipe to fill the tank alone? Compute the answers to two decimal places using a calculator.

Example 13 For a car traveling at a speed of v miles per hour, the least number of feet d under the best possible conditions that is necessary to stop a car (including a reaction time) is given approximately by the formula $d = 0.044v^2 + 1.1v$. Estimate the speed of a car requiring 200 feet to stop after danger is realized. Compute the answer to two decimal places.

Solution

$$0.044v^2 + 1.1v = 200$$ Write in standard form.

$$0.044v^2 + 1.1v - 200 = 0$$ Use the quadratic formula.

$$v = \frac{-b \pm \sqrt{b^2 - 4ac}}{2a}$$ $a = 0.044$, $b = 1.1$, $c = -200$

$$= \frac{-1.1 \pm \sqrt{1.1^2 - 4(0.044)(-200)}}{2(0.044)}$$

$$= \frac{-1.1 \pm \sqrt{36.41}}{0.088}$$ Disregard the negative answer, since we are only interested in positive v.

$$= \frac{-1.1 + 6.03}{0.088} = 56.02 \text{ miles per hour}$$ Complete to two decimal places.

Note: Example 13 is typical of most significant real-world problems in that decimal quantities rather than convenient small numbers are involved.

Problem 13 Repeat Example 13 for a car requiring 300 feet to stop after danger is realized.

ANSWERS TO MATCHED PROBLEMS

11. $-\frac{1}{3}$, 3 **12.** 5.16 hours and 7.16 hours
13. 71.01 miles per hour

EXERCISE 6-3

These problems are not grouped from easy (A) to difficult or theoretical (C). They are grouped somewhat according to type. The most difficult problems are marked with two stars (★★), those of moderate difficulty are marked with one star (★), and the easier problems are not marked.

NUMBER PROBLEMS

1. Find two consecutive positive even integers whose product is 168.

2. Find two positive numbers having a sum of 21 and a product of 104.

3. Find all numbers with the property that when the number is added to itself the sum is the same as when the number is multiplied by itself.

4. The sum of a number and its reciprocal is $\frac{10}{3}$. Find the number.

GEOMETRY

The following theorem may be used where needed:

Pythagorean theorem: A triangle is a right triangle if and only if the square of the longest side is equal to the sum of the squares of the two shorter sides.

$c^2 = a^2 + b^2$

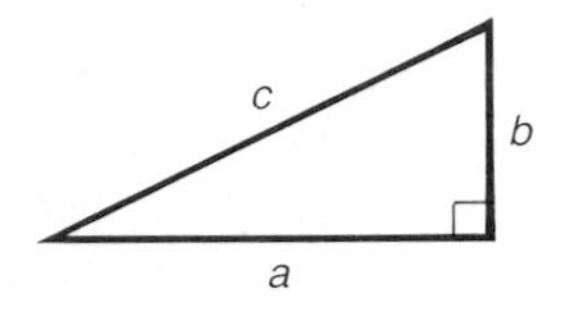

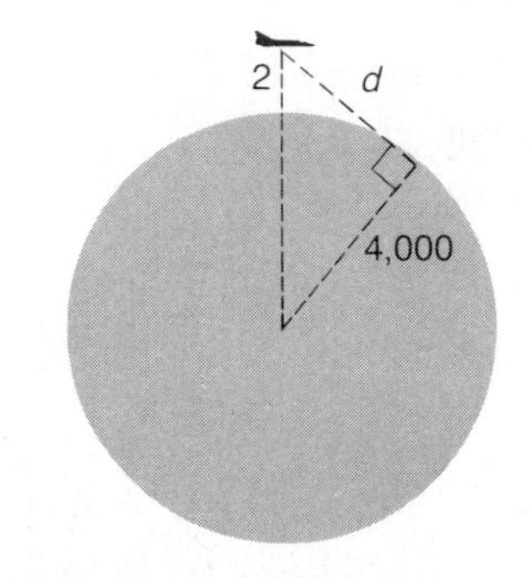

★5. Approximately how far would the horizon be from an airplane 2 miles high? Assume that the radius of the earth is 4,000 miles and use a calculator to estimate the answer to the nearest mile (see the figure).

6. Find the base and height of a triangle with area 2 square meters if its base is 3 meters longer than its height ($A = \frac{1}{2}bh$).

★7. If the length and width of a 4 by 2 centimeter rectangle are each increased by the same amount, the area of the new rectangle will be twice the old. What are the dimensions to two decimal places of the new rectangle?

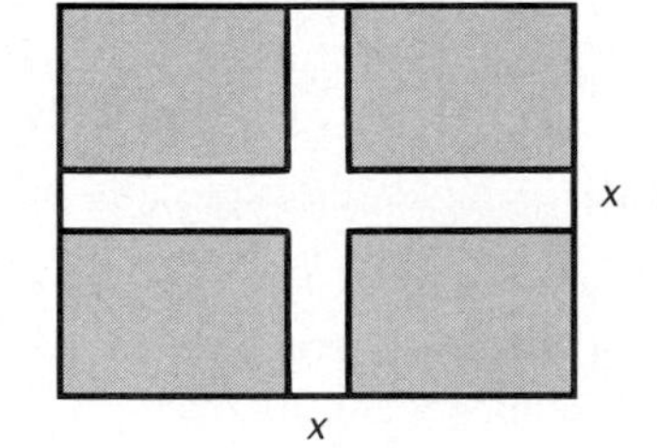

8. The width of a rectangle is 2 meters less than its length. Find its dimensions to two decimal places if its area is 12 square meters.

★9. A flag has a white cross of uniform width on a color background. Find the width of the cross so that it takes up exactly one-half the total area of a 4 by 3 foot flag.

PHYSICS AND ENGINEERING

10. The pressure p in pounds per square foot of wind blowing at v miles per hour is given by $p = 0.003v^2$. If a pressure gauge on a bridge registers a wind pressure of 14.7 pounds per square foot, what is the velocity of the wind?

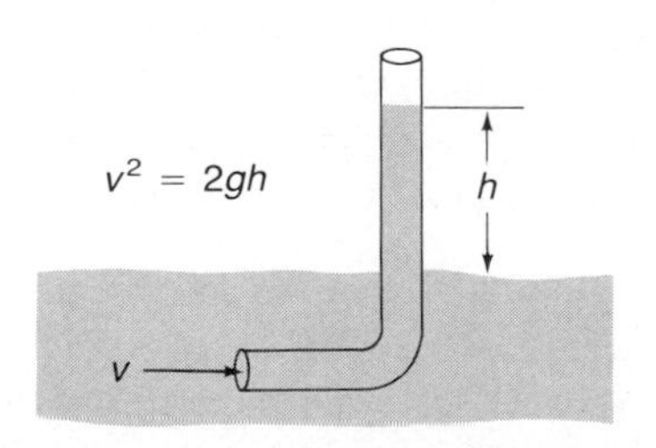

11. One method of measuring the velocity of water in a stream or river is to use an L-shaped tube, as indicated in the figure. Torricelli's law in physics tells us that the height (in feet) that the water is pushed up into the tube above the surface is related to the water's velocity (in feet per second) by the formula $v^2 = 2gh$, where g is approximately 32 feet per second per second. [*Note:* The device can also be used as a simple speedometer for a boat.] How fast is a stream flowing if $h = 0.5$ foot? Find the answer to two decimal points.

12. At 20 miles per hour a car collides with a stationary object with the same force it would have if it had been dropped $13\frac{1}{2}$ feet—that is, if it had been pushed off the roof of an average one-story house. In general, a car moving at r miles per hour hits a stationary object with a force of impact that is equivalent to that force with which it would hit the ground when falling from a certain height h given by the formula $h = 0.0336r^2$. Approximately how fast would a car have to be moving if it crashed as hard as if it had been pushed off the top of a 12-story building 121 feet hight?

★13. For a car traveling at a speed of v miles per hour, the least number of feet d under the best possible conditions that is necessary to stop a car (including a reaction time) is given by the empirical formula $d = 0.044v^2 + 1.1v$. Estimate the speed of a car requiring 165 feet to stop after danger is realized. (See Example 13.)

★14. If an arrow is shot vertically in the air (from the ground) with an initial velocity of 176 feet per second, its distance y above the ground t seconds after it is released (neglecting air resistance) is given by $y = 176t - 16t^2$.

(A) Find the time when y is 0, and interpret physically.

(B) Find the times when the arrow is 16 feet off the ground. Compute answers to two decimal places.

★15. A barrel 2 feet in diameter and 4 feet in height has a 1-inch-diameter drainpipe in the bottom. It can be shown that the height h of the surface of the water above the bottom of the barrel at time t minutes after the drain has been opened is given by the formula $h = (\sqrt{h_0} - \frac{5}{12}t)^2$, where h_0 is the water level above the drain at time $t = 0$. If the barrel is full and the drain opened, how long will it take to empty one-half of the contents?

RATE–TIME PROBLEMS **★★16.** One pipe can fill a tank in 5 hours less than another; together they fill the tank in 5 hours. How long would it take each alone to fill the tank? Compute the answer to two decimal places.

★★17. A new printing press can do a job in 1 hour less than an older press. Together they can do the same job in 1.2 hours. How long would it take each alone to do the job?

18. Two boats travel at right angles to each other after leaving the same dock at the same time; 1 hour later they are 13 kilometers apart. If one travels 7 kilometers per hour faster than the other, what is the rate of each? [*Hint:* Use the Pythagorean theorem stated on page 268.]

★19. A speedboat takes 1 hour longer to go 24 kilometers up a river than to return. If the boat cruises at 10 kilometers per hour in still water, what is the rate of the current?

ECONOMICS AND BUSINESS

20. If P dollars is invested at r percent compounded annually, at the end of 2 years it will grow to $A = P(1 + r)^2$. At what interest rate will \$100 grow to \$144 in 2 years? [*Note:* If $A = 144$ and $P = 100$, find r.]

★21. In a certain city the demand equation for popular records is $d = 3{,}000/p$, where d would be the quantity of records demanded on a given day if the selling price were p dollars per record. (Notice that as the price goes up, the number of records the people are willing to buy goes down, and vice versa.) On the other hand, the supply equation is $s = 200p - 700$, where s is the quantity of records a supplier is willing to supply at p dollars per record. (Notice that as the price goes up, the number of records a supplier is willing to sell goes up, and vice versa.) At what price will supply equal demand; that is, at what price will $d = s$? In economic theory the price at which supply equals demand is called the **equilibrium point**, the point at which the price ceases to change.

6-4 RADICAL EQUATIONS AND OTHER EQUATIONS REDUCIBLE TO QUADRATIC FORM

- Radical Equations
- Other Equations Reducible to Quadratic Form

Consider the equation

$$x - 1 = \sqrt{x + 11}$$

Such an equation is called a **radical equation** since it contains a variable in the radicand. In this section we will solve such equations and other equations that can be rewritten in quadratic form.

RADICAL EQUATIONS

To solve the equation

$$x - 1 = \sqrt{x + 11}$$

we must do something to the equation to eliminate the radical. What? Let us square both members to see what happens—certainly if $a = b$, then $a^2 = b^2$. Thus,

$$\begin{aligned} (x - 1)^2 &= (\sqrt{x + 11})^2 \\ x^2 - 2x + 1 &= x + 11 \\ x^2 - 3x - 10 &= 0 \\ (x + 2)(x - 5) &= 0 \\ x &= -2, 5 \end{aligned}$$

Check $x = -2$: $-2 - 1 \stackrel{?}{=} \sqrt{-2 + 11}$

$-3 \stackrel{?}{=} \sqrt{9}$ Recall that $\sqrt{9}$ names the positive square root of 9.

$-3 \neq 3$

Hence, $x = -2$ is not a solution.

$x = 5$: $5 - 1 \stackrel{?}{=} \sqrt{5 + 11}$

$4 \stackrel{?}{=} \sqrt{16}$

$4 \stackrel{\checkmark}{=} 4$

Hence, $x = 5$ is a solution.

Therefore, 5 is a solution and -2 is not. The process of squaring introduced an ''extraneous'' solution. In general, one can prove the following important result:

> If both members of an equation are raised to a natural number power, then the solution set of the original equation is a subset of the solution set of the new equation.

Thus, any new equation obtained by raising both members of an equation to the same natural number power may have solutions (called **extraneous solutions**) that are not solutions of the original equation. On the other hand, any solution of the original equation must be among those of the new equation. We need only check all of the solutions at the end of the process to eliminate the so-called extraneous ones.

Example 14 Solve: $x + \sqrt{x - 4} = 4$

Solution

$x + \sqrt{x - 4} = 4$ Isolate the radical on one side.

$\sqrt{x - 4} = 4 - x$ Square both sides.

$x - 4 = 16 - 8x + x^2$ Write in standard form.

$x^2 - 9x + 20 = 0$

$(x - 5)(x - 4) = 0$

$x = 4, 5$

Check Substitute in the *original* equation.

$x = 4$: $4 + \sqrt{4 - 4} = 4$, so 4 is a solution

$x = 5$: $5 + \sqrt{5 - 4} = 5 + 1 \neq 4$, so 5 is extraneous

Problem 14 Solve: $x = 5 + \sqrt{x - 3}$

Radical equations involving two radicals are usually easiest to solve if the radicals are on opposite sides of the equation.

Example 15 Solve: $\sqrt{2x + 3} - \sqrt{x - 2} = 2$

Solution

$$\sqrt{2x + 3} - \sqrt{x - 2} = 2$$

Easier to solve with a radical on each side.

$$\sqrt{2x + 3} = \sqrt{x - 2} + 2$$

Square both sides.

$$2x + 3 = x - 2 + 4\sqrt{x - 2} + 4$$

Isolate the radical on one side.

$$x + 1 = 4\sqrt{x - 2}$$

Square both sides again.

$$x^2 + 2x + 1 = 16(x - 2)$$
$$x^2 - 14x + 33 = 0$$
$$(x - 11)(x - 3) = 0$$
$$x = 3, 11$$

Check $x = 3$: $\sqrt{2 \cdot 3 + 3} - \sqrt{3 - 2} = \sqrt{9} - \sqrt{1} = 3 - 1 = 2$

so 3 is a solution.

$x = 11$: $\sqrt{2 \cdot 11 + 3} - \sqrt{11 - 2} = \sqrt{25} - \sqrt{9} = 5 - 3 = 2$

so 11 is a solution.

Problem 15 Solve: $\sqrt{2x + 7} + \sqrt{x + 3} = 1$

OTHER EQUATIONS REDUCIBLE TO QUADRATIC FORM

Many equations that are not immediately recognizable as quadratic can be transformed into a quadratic form and then solved. Let's look at some examples.

Example 16 Solve: $x^4 - x^2 - 12 = 0$

Solution If you recognize that the equation is quadratic in x^2, you can solve for x^2 first, then solve for x. You might find it convenient to make the substitution $u = x^2$ and then solve the equation

$$u^2 - u - 12 = 0$$
$$(u - 4)(u + 3) = 0$$
$$u = 4, -3$$

Replacing u with x^2, we obtain

$$x^2 = 4 \qquad x^2 = -3$$
$$x = \pm 2 \qquad x = \pm i\sqrt{3}$$

You can check that each of these four values is a solution of the original equation. Remember, however, that the only operation that might introduce extraneous roots is raising to powers. Since this solution did not involve raising both members of the equation to a higher power, extraneous roots will not occur.

Problem 16 Solve: $x^6 + 6x^3 - 16 = 0$ for real solutions only

In general, if an equation that is not quadratic can be transformed into the form

$$au^2 + bu + c = 0$$

where u is an expression in some other variable, then the equation is said to be in **quadratic form**. Once recognized as a quadratic form, an equation can often be solved using one of the quadratic methods.

Example 17 Solve: $x^{2/3} - x^{1/3} - 6 = 0$

Solution Let $u = x^{1/3}$, then $u^2 = x^{2/3}$. After substitution, the original equation becomes

$$u^2 - u - 6 = 0$$
$$(u - 3)(u + 2) = 0$$
$$u = 3, -2$$

Replacing u with $x^{1/3}$, we obtain

$$x^{1/3} = 3 \qquad x^{1/3} = -2$$
$$(x^{1/3})^3 = 3^3 \qquad (x^{1/3})^3 = (-2)^3$$
$$x = 27 \qquad x = -8$$

Raise both sides to the third power.

Note: $x \neq \sqrt[3]{3}$ and $x \neq \sqrt[3]{-2}$ (common errors).

Problem 17 Solve: $x^{2/3} - x^{1/3} - 12 = 0$

ANSWERS TO MATCHED PROBLEMS

14. 7 **15.** -3 **16.** $-2, \sqrt[3]{2}$ **17.** 64, -27

EXERCISE 6-4

Solve.

A

1. $x - 2 = \sqrt{x}$

2. $\sqrt{x} = x - 6$

3. $m - 13 = \sqrt{m + 7}$

4. $\sqrt{5n + 9} = n - 1$

5. $x^4 - 10x^2 + 9 = 0$

6. $x^4 - 13x^2 + 36 = 0$

7. $x^4 - 7x^2 - 18 = 0$

8. $y^4 - 2y^2 - 8 = 0$

9. $\sqrt{x^2 - 3x} = 2$

10. $\sqrt{x^2 + 8x} = 3$

B **11.** $m - 7\sqrt{m} + 12 = 0$

12. $t - 11\sqrt{t} + 18 = 0$

13. $1 + \sqrt{x + 5} = x$

14. $x - \sqrt{x + 10} = 2$

15. $\sqrt{3x + 1} = \sqrt{x} - 1$

16. $\sqrt{3x + 4} = 2 + \sqrt{x}$

17. $\sqrt{3t + 4} + \sqrt{t} = -3$

18. $\sqrt{3w - 2} - \sqrt{w} = 2$

19. $\sqrt{u - 2} = 2 + \sqrt{2u + 3}$

20. $\sqrt{3y - 2} = 3 - \sqrt{3y + 1}$

21. $\sqrt{2x - 1} - \sqrt{x - 4} = 2$

22. $\sqrt{y - 2} - \sqrt{5y + 1} = -3$

23. $x^6 - 7x^3 - 8 = 0$ (Find real solutions only.)

24. $x^6 + 3x^3 - 10 = 0$ (Find real solutions only.)

25. $y^8 - 17y^4 + 16 = 0$

26. $3m^4 - 4m^2 - 7 = 0$

27. $x^{2/3} - 3x^{1/3} - 10 = 0$

28. $2x^{2/3} + 3x^{1/3} - 2 = 0$

29. $y^{1/2} - 3y^{1/4} + 2 = 0$

30. $y^{1/2} - 5y^{1/4} + 6 = 0$

31. $6x^{-2} - 5x^{-1} - 6 = 0$

32. $3n^{-2} - 11n^{-1} - 20 = 0$

C **33.** $4x^{-4} - 17x^{-2} + 4 = 0$

34. $9y^{-4} - 10y^{-2} + 1 = 0$

35. $(m^2 - m)^2 - 4(m^2 - m) = 12$

36. $(x^2 + 2x)^2 - (x^2 + 2x) = 6$

37. $(x - 3)^4 + 3(x - 3)^2 = 4$

38. $(m - 5)^4 + 36 = 13(m - 5)^2$

39. $\sqrt{3x + 6} - \sqrt{x + 4} = \sqrt{2}$

40. $\sqrt{7x - 2} - \sqrt{x + 1} = \sqrt{3}$

41. $\dfrac{1}{\sqrt{x - 2}} + \dfrac{2}{3} = 1$

42. $\dfrac{1}{\sqrt{x + 5}} = \dfrac{\sqrt{x}}{6}$

43. $\dfrac{x}{3} + \dfrac{2}{x} = \dfrac{6x + 1}{3x}$

44. $x + \dfrac{1}{x} = \dfrac{x + 3}{2}$

45. $\dfrac{1}{x - 1} + \dfrac{1}{x - 2} = \dfrac{5}{6}$

46. $\sqrt{x} = 3 - \dfrac{2}{\sqrt{x}}$

6-5 NONLINEAR INEQUALITIES

- Quadratic Inequalities
- Set Union and Intersection
- Other Inequalities

We know how to solve linear inequalities in one variable, that is, inequalities such as

$$x + 2 < 8$$

where the variable only occurs to the first power. If we allow the variable to be squared, we obtain inequalities such as

$$x^2 + 2x < 8$$

Such an inequality is called a **quadratic inequality**. This section introduces a process for solving quadratic inequalities. The solutions will often be combinations of intervals on the number line for which set unions and intersections provide a convenient way of describing them. The same process can be applied to other inequalities that are not quadratic.

QUADRATIC INEQUALITIES

How do we solve an inequality such as $x^2 + 2x < 8$? If we move all terms to the left and factor, then we will be able to observe something that will lead to a solution. Thus,

$$x^2 + 2x - 8 < 0$$
$$(x + 4)(x - 2) < 0$$

We are looking for values of x that will make the left side less than 0—that is, negative. What will the signs of $(x + 4)$ and $(x - 2)$ have to be so that their product is negative? They must have opposite signs!

Let us see where each of the factors is positive, negative, and 0. The point at which either factor is 0 is called a **critical point**. We will see why shortly.

Sign analysis for $(x + 4)$:

Critical point	$(x + 4)$ is positive when	$(x + 4)$ is negative when
$x + 4 = 0$	$x + 4 > 0$	$x + 4 < 0$
$x = -4$	$x > -4$	$x < -4$

Geometrically:

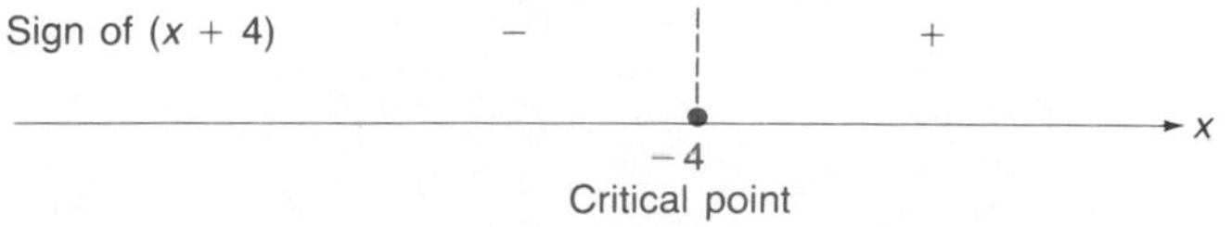

Sign of $(x + 4)$ − | +

x axis: −4 (Critical point)

Thus, $(x + 4)$ is negative for values of x to the left of the critical point and is positive for values of x to the right of the critical point.

Sign analysis for $(x - 2)$:

Critical point	$(x-2)$ is positive when	$(x-2)$ is negative when
$x - 2 = 0$	$x - 2 > 0$	$x - 2 < 0$
$x = 2$	$x > 2$	$x < 2$

Geometrically:

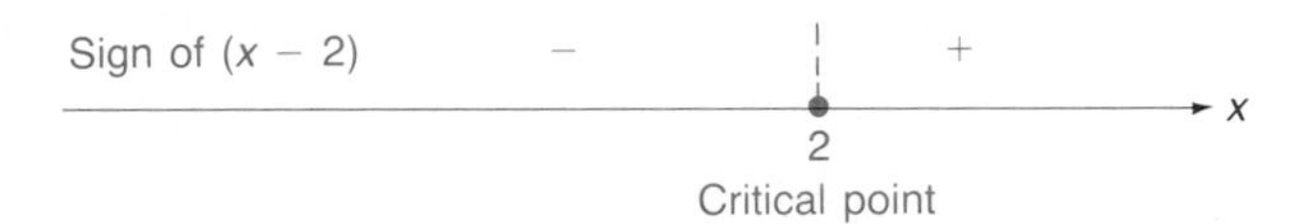

Thus, $(x - 2)$ is negative for values of x to the left of the critical point and is positive for values of x to the right of the critical point.

Combining these results in a single geometric representation leads to a simple solution of the original problem:

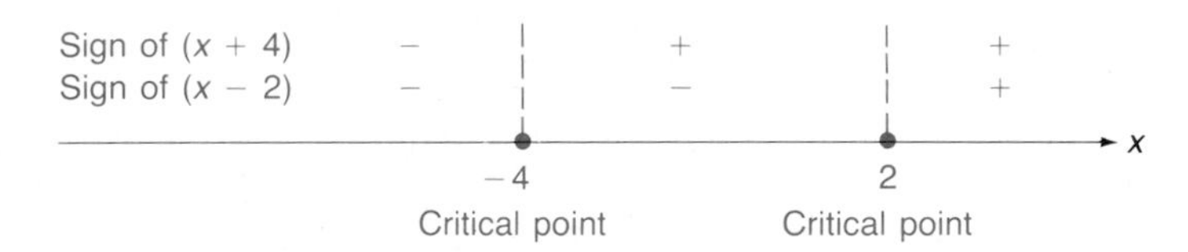

Now it is very easy to see that the factors have opposite signs for x between -4 and 2. Thus, the solution of $x^2 + 2x < 8$ is

$$-4 < x < 2$$

This discussion leads to the general result which is behind the sign-analysis method of solving quadratic inequalities:

> The value of x at which the linear expression $(ax + b)$ is 0 is called a **critical point**. To the left of the critical point, on the real number line, $(ax + b)$ has one sign and to the right of the critical point the opposite sign.

Example 18 Solve and graph: $x^2 \geq x + 6$

Solution

$$x^2 \geq x + 6$$
$$x^2 - x - 6 \geq 0$$
$$(x - 3)(x + 2) \geq 0$$

Critical points: -2 and 3

Locate these points on a real number line and indicate the sign of each factor to the left and to the right of its critical point:

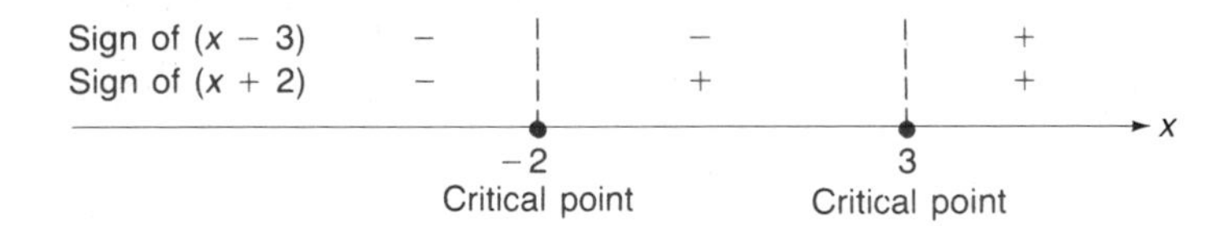

The inequality statement is satisfied when both factors have the same sign or when one or the other factor is 0. The factors have the same sign when x is to the left of -2 or to the right of 3; one or the other factor is 0 at the critical points. Thus,

Solution: $x \leq -2$ or $x \geq 3$

Graph:

−2 3 x

Problem 18 Solve and graph:

(A) $x^2 < x + 12$ **(B)** $x^2 \geq x + 12$

It is important to recognize that in the sign-analysis method we need to obtain an expression that we are to compare to zero. The method will work for an expression such as $(x - 3)(x + 2) \geq 0$ but will not work for an expression such as $(x - 3)(x + 2) \geq 5$.

The solution $x \leq -2$ or $x \geq 3$ to Example 18 can also be described as the interval $(-\infty, -2]$ combined with the interval $[3, \infty)$. There is a convenient notation used to describe such a combined set.

SET UNION AND INTERSECTION

The **union** of sets A and B, denoted by $A \cup B$, is the set of all elements formed by combining all the elements of A and all the elements of B into one set. The **intersection** of sets A and B, denoted by $A \cap B$, is the set of elements in A that are also in B. Symbolically:

UNION: $A \cup B = \{x \mid x \in A \text{ or } x \in B\}$

$\{2, 3\} \cup \{3, 4\} = \{2, 3, 4\}$

INTERSECTION: $A \cap B = \{x \mid x \in A \text{ and } x \in B\}$

$\{2, 3\} \cap \{3, 4\} = \{3\}$

The word ''or'' is used in the way it is generally used in mathematics; that is, x may be an element of set A or set B or both. If $A \cap B = \varnothing$ (that is, if the set of all elements common to both A and B is empty), then sets A and B are said to be **disjoint**.

Venn diagrams are useful aids in visualizing set relationships. Union and intersection of sets are illustrated in Figure 1.

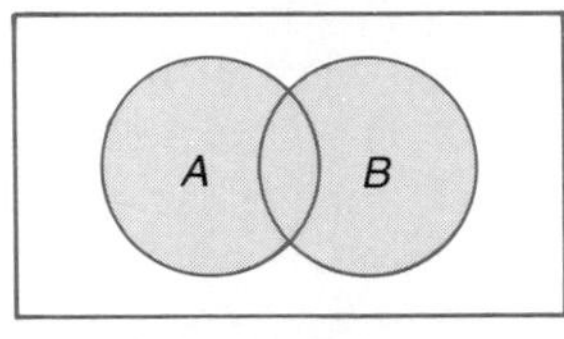

(a) Union of two sets

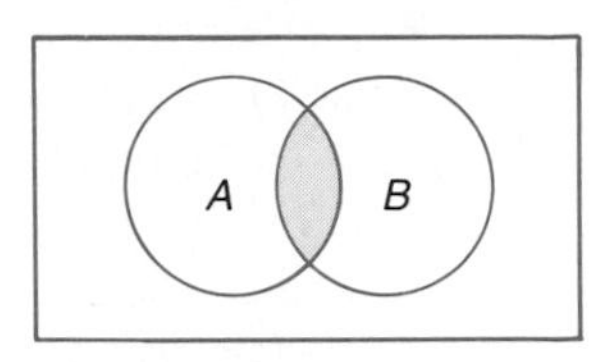

(b) Intersection of two sets

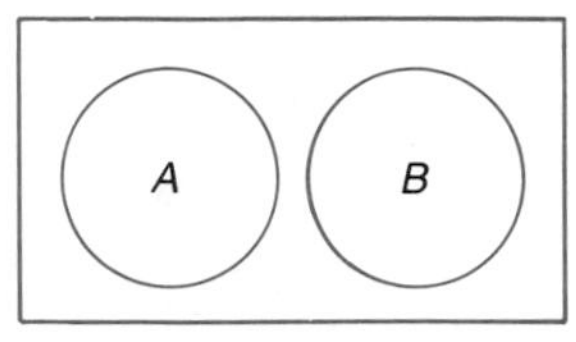

(c) Two disjoint sets

FIGURE 1
Venn diagrams

Example 19 If $A = \{1, 2, 3, 4\}$, $B = \{1, 3, 5, 7\}$, and $C = \{2, 4, 6\}$, then

$A \cup B = \{1, 2, 3, 4, 5, 7\}$ Elements in A combined with elements in B

$A \cap B = \{1, 3\}$ Elements in A that are also in B

$B \cap C = \varnothing$ Sets B and C are disjoint.

Problem 19 If $A = \{3, 6, 9\}$, $B = \{3, 4, 5, 6, 7\}$, and $C = \{4, 5, 7, 8\}$, find:

(A) $A \cup B$ **(B)** $A \cap B$ **(C)** $A \cap C$

Example 20 If $P = [-4, 2)$, $Q = (-1, 6]$, and $R = [3, \infty)$, find:

(A) $P \cup Q$ and $P \cap Q$ **(B)** $P \cup R$ and $P \cap R$

Solution **(A)** Graphically:

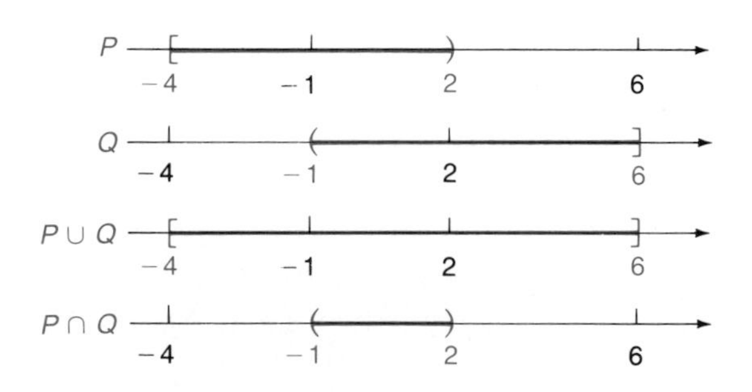

Symbolically:

$P \cup Q = [-4, 6]$ Combine elements in P with those in Q.

$P \cap Q = (-1, 2)$ Elements in P that are also in Q

(B) Graphically:

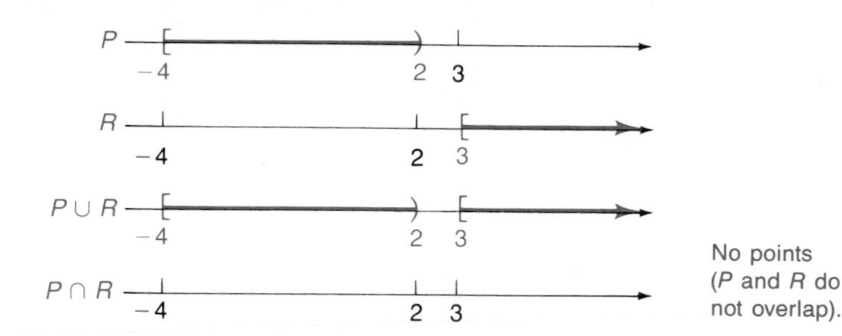

Symbolically:

$P \cup R = [-4, 2) \cup [3, \infty)$

$P \cap R = \varnothing$ Null or empty set (P and R are disjoint).

Problem 20 If $A = [0, 3]$, $B = (-2, 2)$, and $C = (-1, \infty)$, find:

(A) $A \cup B$ **(B)** $A \cap B$ **(C)** $B \cup C$ **(D)** $B \cap C$

OTHER INEQUALITIES

The procedures discussed above can also be used on some inequalities that are not quadratic.

Example 21 Solve and graph: $\dfrac{x^2 - x + 1}{2 - x} \geq 1$

Solution

$$\frac{x^2 - x + 1}{2 - x} \geq 1$$

We do not know the sign of $2 - x$. Thus we do not multiply both sides by it since we do not know whether to reverse the inequality. Instead we subtract 1 from each side to obtain an expression compared to 0.

$$\frac{x^2 - x + 1}{2 - x} - 1 \geq 0$$

Combine terms on the left side into a single fraction.

$$\frac{x^2 - x + 1}{2 - x} - \frac{2 - x}{2 - x} \geq 0$$

$$\frac{x^2 - 1}{2 - x} \geq 0$$

Factor the numerator.

$$\frac{(x - 1)(x + 1)}{2 - x} \geq 0$$

Proceed as in Example 18 to determine which values make the expression positive or zero.

Critical points: $-1, 1, 2$

Locate these points on a real number line and indicate the sign of each first-degree form to the left and to the right of its critical point:

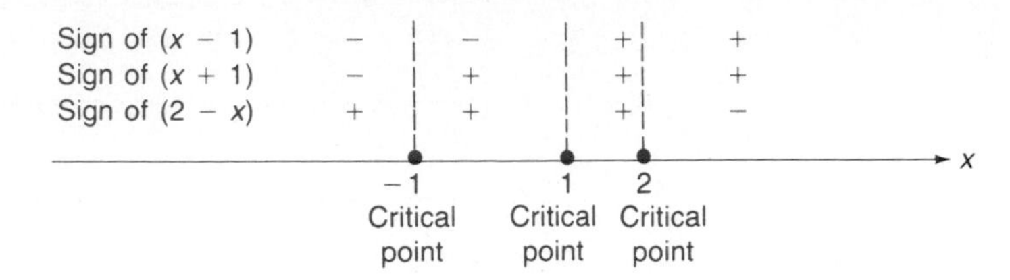

The rational expression is greater than or equal to 0 when $(x - 1)$, $(x + 1)$, and $(2 - x)$ are all positive, when two are negative and one is positive, or when the numerator is 0. Two factors are negative to the left of -1 and all are positive between 1 and 2. The equality part of the inequality holds when x is 1 or -1, but not when $x = 2$. Thus,

Solution: $x \le -1$ or $1 \le x < 2$ That is, $(-\infty, -1] \cup [1, 2)$.

Graph:

Problem 21 Solve and graph: $\dfrac{3}{2 - x} \le \dfrac{1}{x + 4}$

ANSWERS TO MATCHED PROBLEMS

18. **(A)** $-3 < x < 4$

(B) $x \le -3$ or $x \ge 4$

19. **(A)** $\{3, 4, 5, 6, 7, 9\}$ **(B)** $\{3, 6\}$ **(C)** $\varnothing$ (null set)

20. **(A)** $(-2, 3]$ **(B)** $[0, 2)$ **(C)** $(-2, \infty)$ **(D)** $(-1, 2)$

21. $(-4, -\frac{5}{2}] \cap (2, \infty)$

EXERCISE 6-5

A *Solve and graph.*

1. $(x - 3)(x + 4) < 0$

2. $(x + 2)(x - 4) < 0$

3. $(x - 3)(x + 4) \ge 0$

4. $(x + 2)(x - 4) > 0$

5. $x^2 + x < 12$

6. $x^2 < 10 - 3x$

7. $x^2 + 21 > 10x$

8. $x^2 + 7x + 10 > 0$

Write each set in Problems 9–14, using the listing method; that is, list the elements between braces. If the set is empty, write $\varnothing$.

9. $\{1, 3, 5\} \cup \{2, 3, 4\}$

10. $\{3, 4, 6, 7\} \cup \{3, 4, 5\}$

11. $\{1, 3, 5\} \cap \{2, 3, 4\}$

12. $\{3, 4, 6, 7\} \cap \{3, 4, 5\}$

13. $\{1, 5, 9\} \cap \{3, 4, 6, 8\}$

14. $\{6, 8, 9\} \cap \{4, 5, 7\}$

Write each set in Problems 15–20 as a single interval, if possible.

15. $(-3, 0] \cap [-2, 1)$

16. $(-3, 4) \cup [0, 8]$

17. $(-1, 1) \cup (0, \infty)$

18. $(-\infty, -1) \cap [-3, 2]$

19. $(-\infty, 3) \cup (2, 5]$

20. $[0, 2) \cup [2, 5)$

Solve and graph.

B

21. $x(x + 6) \geq 0$

22. $x(x - 8) \leq 0$

23. $x^2 \geq 9$

24. $x^2 > 4$

25. $\frac{x - 5}{x + 2} \leq 0$

26. $\frac{x + 2}{x - 3} < 0$

27. $\frac{x - 5}{x + 2} > 0$

28. $\frac{x + 2}{x - 3} \geq 0$

29. $\frac{x - 4}{x(x + 2)} \leq 0$

30. $\frac{x(x + 5)}{x - 3} \geq 0$

31. $\frac{1}{x} < 4$

32. $\frac{5}{x} > 3$

33. $x^2 + 4 \geq 4x$

34. $6x \leq x^2 + 9$

35. $x^2 + 9 < 6x$

36. $x^2 + 4 < 4x$

37. $x^2 \geq 3$

38. $x^2 < 2$

39. $\frac{2}{x - 3} \leq -2$

40. $\frac{2x}{x + 3} \geq 1$

C

41. $\frac{2}{x - 3} \leq \frac{2}{x + 2}$

42. $\frac{2}{x + 1} \geq \frac{1}{x - 2}$

43. $\frac{(x - 1)(x + 3)}{x} > 0$

44. $\frac{x(x - 3)}{x + 2} < 0$

45. $\frac{1}{x + 1} + \frac{1}{x - 2} \geq 0$

46. $\frac{x^2 - 6x + 8}{x + 2} > 0$

47. $\frac{(x + 1)^2}{x^2 + 2x - 3} \leq 0$

48. $\frac{(x - 1)(x + 3)}{x(x + 2)} > 0$

49. For what values of x will $\sqrt{x^2 - 3x + 2}$ by a real number?

50. For what values of x will $\sqrt{\frac{x - 3}{x + 5}}$ be a real number?

51. If an object is shot straight up from the ground with an initial velocity of 160 feet per second, its distance d in feet above the ground at the end of t seconds (neglecting air resistance) is given by $d = 160t - 16t^2$. Find the duration of time for which $d \geq 256$.

52. Repeat Problem 51 for $d \geq 0$.

6-6 CHAPTER REVIEW

The **standard form** of a **quadratic equation** is $ax^2 + bx + c = 0$ with $a \neq 0$. When $b = 0$, a quadratic equation can be solved by taking square roots. When $ax^2 + bx + c$ can be factored into two first-degree factors, the equation can be solved by setting each factor equal to 0.

Any quadratic equation can be solved by **completing the square**; to complete the square in $x^2 + bx$, add the square of one-half the coefficient of x to obtain a perfect square:

$$x^2 + bx + \left(\frac{b}{2}\right)^2 = \left(x + \frac{b}{2}\right)^2 \quad \textit{(6-1)}$$

This method leads to a general solution, the **quadratic formula**:

$$x = \frac{-b \pm \sqrt{b^2 - 4ac}}{2a}$$

The expression $b^2 - 4ac$ is called the **discriminant** and provides information about the roots: if positive, two real solutions; if negative, two nonreal, complex conjugate solutions; if 0, one real solution. *(6-2)*

Many applications lead to quadratic equations. *(6-3)*. Other equations may sometimes be reduced to quadratic form and solved. Radicals can be removed from equations by raising both sides to a power, but this procedure may introduce **extraneous solutions**. *(6-4)*

A **quadratic inequality** comparing $ax^2 + bx + c$ to 0 may be solved by factoring the quadratic and checking the signs of the factors. The points where the factors are 0 are called **critical points**. Sign analysis can also be applied to other inequalities comparing a product or quotient of linear factors to 0. Solutions of such inequalities can be written compactly in interval notation using set unions and intersections. The **union** $A \cup B$ of two sets A and B is the set of all elements in A or B (or both); the **intersection** $A \cap B$ is the set of all elements common to both A and B. *(6-5)*

REVIEW EXERCISE 6-6

Work through all the problems in this chapter review and check answers in the back of the book. (Answers to all problems are there, and following each answer is a number in italics indicating the section in which that type of problem is discussed.) Where weaknesses show up, review appropriate sections in the text.

A *Find all solutions by factoring or square root methods.*

1. $x^2 - 3x = 0$

2. $x^2 = 25$

3. $x^2 - 5x + 6 = 0$

4. $x^2 - 2x - 15 = 0$

5. $x^2 - 7 = 0$

6. Write $4x = 2 - 3x^2$ in standard form, $ax^2 + bx + c = 0$, and identify a, b, and c.

7. Write down the quadratic formula associated with the standard form $ax^2 + bx + c = 0$.

8. Use the quadratic formula to solve $x^2 + 3x + 1 = 0$.

9. Find two positive numbers whose product is 27 if one is 6 more than the other.

10. Find the discriminant of $7x^2 - 11x + 5 = 0$ and use it to determine the nature and number of roots.

In Problems 11–14, let $A = \{1, 2, 3, 4, 5\}$, $B = \{1, 3, 5, 7, 9\}$, $C = [0, 5)$, *and* $D = (-1, 3]$. *Find the set indicated.*

11. $A \cap B$

12. $A \cup B$

13. $C \cap D$

14. $C \cup D$

Solve and graph.

15. $x^2 + x < 20$

16. $x^2 + x \geq 20$

B *Find all solutions by factoring or square root methods.*

17. $10x^2 = 20x$

18. $3x^2 = 36$

19. $3x^2 + 27 = 0$

20. $(x - 2)^2 = 16$

21. $3t^2 - 8t - 3 = 0$

22. $2x = \dfrac{3}{x} - 5$

Solve using the quadratic formula.

23. $3x^2 = 2(x + 1)$

24. $2x(x - 1) = 3$

Solve using any method.

25. $2x^2 - 2x = 40$

26. $\dfrac{8m^2 + 15}{2m} = 13$

27. $m^2 + m - 1 = 0$

28. $u + \frac{3}{u} = 2$

29. $\sqrt{5x - 6} - x = 0$

30. $8\sqrt{x} = x + 15$

31. $m^4 + 5m^2 - 36 = 0$

32. $2x^{2/3} - 5x^{1/3} - 12 = 0$

Solve and graph.

33. $x^2 \geq 4x + 21$

34. $\frac{1}{x} < 2$

35. $10x > x^2 + 25$

36. $x^2 + 16 \geq 8x$

37. The perimeter of a rectangle is 22 inches. If its area is 30 square inches, find the length of each side.

C **38.** Solve $x^2 - 6x - 3 = 0$ by completing the square.

Solve using any method.

39. $\left(t - \frac{3}{2}\right)^2 = -\frac{3}{2}$

40. $3x - 1 = \frac{2(x + 1)}{x + 2}$

41. $y^8 - 17y^4 + 16 = 0$

42. $\sqrt{y - 2} - \sqrt{5y + 1} = -3$

43. $\frac{3}{x - 4} \leq \frac{2}{x - 3}$ Graph the solution.

44. If the hypotenuse of a right triangle is 15 centimeters and its area is 54 square centimeters, what are the lengths of the two sides? [*Hint:* If x represents one side, use the Pythagorean theorem to express the other side in terms of x; then use the formula for the area of a triangle, $A = \frac{1}{2}bh$.]

45. Cost equations for manufacturing companies are often quadratic in nature. (At very high or very low outputs the costs are more per unit because of inefficiency of plant operation at these extremes.) If the cost equation for producing paint is $C = x^2 - 10x + 31$, where C is the cost of producing x gallons per week (both in thousands), find:

(A) The output for a \$15,000 weekly cost

(B) The output for a \$6,000 weekly cost

7

GRAPHING INVOLVING TWO VARIABLES

In Chapters 4 and 6 we graphed equations and inequalities in one variable on a real number line. Recall:

STATEMENT

$-3 < x \leq 5$

GRAPH

Every real number can be associated with a unique point on a line, and, conversely, every point on a line can be associated with a unique real number. In this chapter we develop a system that will enable us to graph equations and inequalities in two variables such as

$3x - 2y = 5 \qquad y = x^2$

$x^2 + y^2 = 1 \qquad y \leq 4x - 1$

7-1 GRAPHING LINEAR EQUATIONS

- Cartesian Coordinate System
- Graphing a First-Degree Equation in Two Variables
- Vertical and Horizontal Lines
- Use of Different Scales for Each Coordinate Axis

In graphing relationships involving one variable, we used the identification of numbers with points on a number line, a one-dimensional figure. To graph relationships involving two variables, we must extend this identification of numbers with points on a plane, a two-dimensional figure. To do this, we will introduce a coordinate system whereby each point in a plane is uniquely named by a pair of real numbers.

CARTESIAN COORDINATE SYSTEM

To form a cartesian coordinate system we select two real number lines, one vertical and one horizontal, and let them cross through their origins (0's) as indicated in Figure 1a.† Up and to the right are the traditional choices for the positive directions. These two number lines are called the **vertical axis** and the **horizontal axis** or (together) the **coordinate axes**. The coordinate axes divide the plane into four parts called **quadrants**. The quadrants are numbered counterclockwise from I to IV. All points in the plane lie in one of the four quadrants except for points on the coordinate axes.

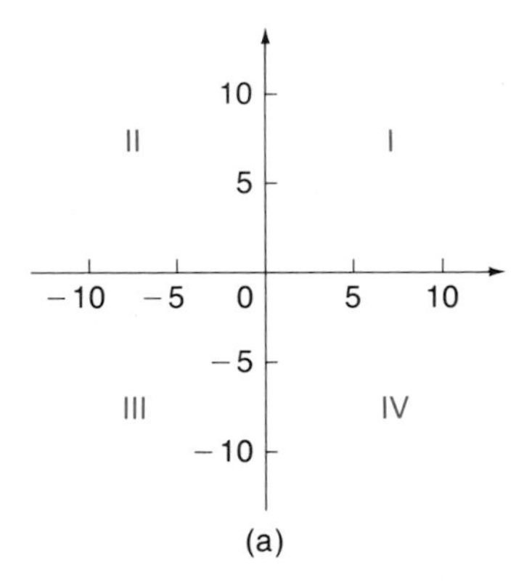

(a)

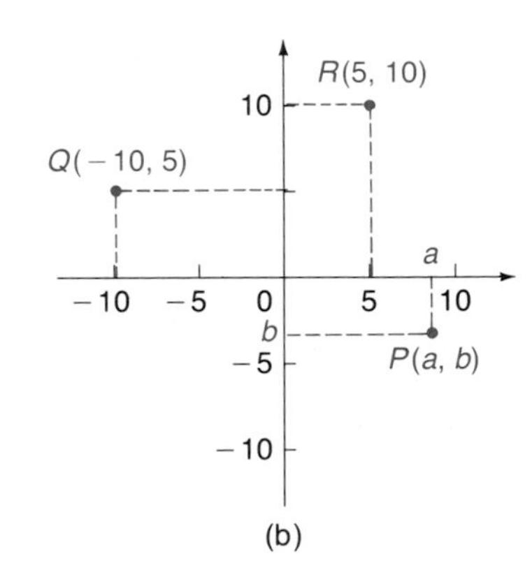

(b)

FIGURE 1

Pick a point P in the plane at random (see Figure 1b). Pass horizontal and vertical lines through the point. The vertical line will intersect the horizontal axis at a point with coordinate a, and the horizontal line will intersect the vertical axis at a point with coordinate b. These two numbers form the **coordinates**

$$(a, b)$$

† Here we use the same scale on each real number line. Later in this section we will consider a particular type of problem where it is useful to use a different scale on each real number line.

of the point P in the plane. In particular, the coordinates of the point Q are $(-10, 5)$ and those of the point R are $(5, 10)$.

The first coordinate a of the coordinates of point P is also called the **abscissa** of P; the second coordinate b of the coordinates of point P is also called the **ordinate** of P. The abscissa for Q in Figure 1b is -10 and the ordinate for Q is 5. The point with coordinates $(0, 0)$ is called the **origin**.

We know that coordinates (a, b) exist for each point in the plane since every point on each axis has a real number associated with it. Hence, by the procedure described, each point located in the plane can be labeled with a unique pair of real numbers. Conversely, by reversing the process, each pair of real numbers can be associated with a unique point in the plane.

The system that we have just defined is called a **cartesian coordinate system** (sometimes referred to as a **rectangular coordinate system**).

Example 1 Find the coordinates of each of the points A, B, C, and D.

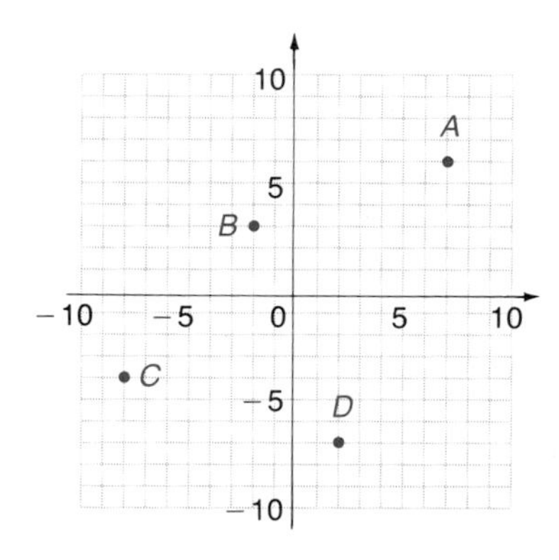

Solution $A(7, 6)$ $\quad B(-2, 3)$ $\quad C(-8, -4)$ $\quad D(2, -7)$

Problem 1 Find the coordinates, using the figure in Example 1, for each of the following points:

(A) 2 units to the right and 1 unit up from A
(B) 2 units to the left and 2 units down from C
(C) 1 unit up and 1 unit to the left of D
(D) 2 units to the right of B

Example 2 Plot (associate each ordered pair of numbers with a point in the cartesian coordinate system):

$(2, 7)$ $\quad (7, 2)$ $\quad (-8, 4)$ $\quad (4, -8)$ $\quad (-8, -4)$ $\quad (-4, -8)$
$(3, 0)$ $\quad (0, 3)$

Solution

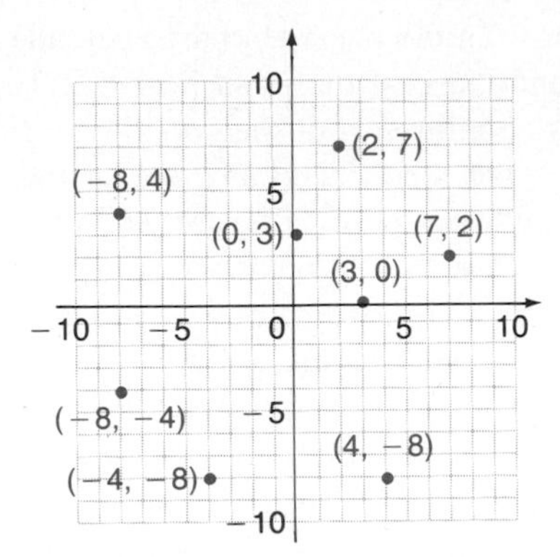

To plot (−8, 4), for example, start at the origin and count 8 units to the left; then go straight up 4 units.

It is very important to note that the ordered pair (2, 7) and the set {2, 7} are not the same thing; {2, 7} = {7, 2}, but (2, 7) ≠ (7, 2).

Problem 2 Plot: (3, 4), (−3, 2), (−2, −2), (4, −2), (0, 1), (−4, 0)

The development of the cartesian coordinate system represented a very important advance in mathematics. It was through the use of this system that René Descartes (1596–1650), a French philosopher-mathematician, was able to transform geometric problems requiring long tedious reasoning into algebraic problems that could be solved almost mechanically. This joining of algebra and geometry has now become known as **analytic geometry**.

Two fundamental problems of analytic geometry are the following:

1. Given an equation, find its graph.
2. Given a geometric figure, such as a straight line, circle, or ellipse, find an equation that has this figure as its graph.

In this section we will be mainly interested in the first problem. In other parts of this chapter we will consider the second problem. Before we take up the first problem, however, let us refresh your memory on what is meant by "the graph of an equation." In general:

The Graph of an Equation

The graph of an equation in two variables in a rectangular coordinate system must meet the following two conditions:

1. If an ordered pair of numbers is a solution to the equation, the corresponding point must be on the graph of the equation.
2. If a point is on the graph of an equation, its coordinates must satisfy the equation.

GRAPHING A FIRST-DEGREE EQUATION IN TWO VARIABLES

Suppose we are interested in graphing

$$y = 2x - 4$$

We start by finding some of its solutions. A **solution** of an equation in two variables is an ordered pair of real numbers, one for x and one for y, that satisfies the equation. If we agree that the first element in the ordered pair will replace x and the second y, then

$$(0, -4)$$

is a solution of $y = 2x - 4$, as can easily be checked. How do we find other solutions? The answer is easy: we simply assign to x in $y = 2x - 4$ any convenient value and solve for y. For example, if $x = 3$, then

$$y = 2(3) - 4 = 2$$

Hence,

$$(3, 2)$$

is another solution of $y = 2x - 4$. By proceeding in this manner we can get solutions to this equation without end. Thus, the solution set is infinite. Table 1 lists some solutions, and we have graphed these solutions in a cartesian coordinate system in Figure 2 on page 290, identifying the horizontal axis with x and the vertical axis with y.

TABLE 1

CHOOSE x	COMPUTE $2x - 4 = y$	WRITE ORDERED PAIR (x, y)
-4	$2(-4) - 4 = -12$	$(-4, -12)$
-2	$2(-2) - 4 = -8$	$(-2, -8)$
0	$2(0) - 4 = -4$	$(0, -4)$
2	$2(2) - 4 = 0$	$(2, 0)$
4	$2(4) - 4 = 4$	$(4, 4)$
6	$2(6) - 4 = 8$	$(6, 8)$
8	$2(8) - 4 = 12$	$(8, 12)$

It appears in Figure 2 that the graph of the equation is a straight line. If we knew this for a fact, then graphing $y = 2x - 4$ would be easy. We would simply find two solutions of the equation, plot them, then graph as much of

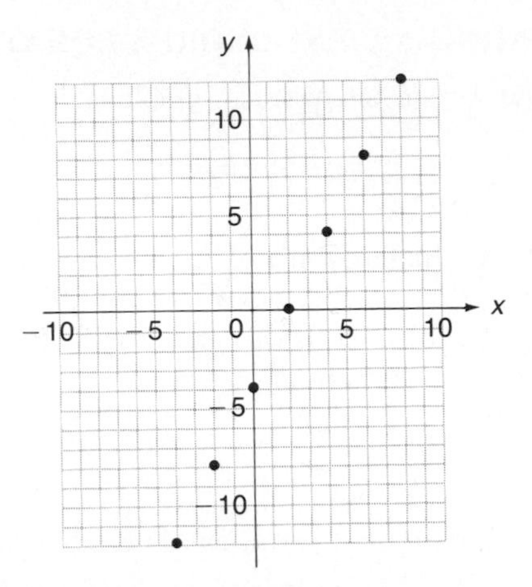

FIGURE 2

$y = 2x - 4$ as we like by drawing a line through the two points using a straightedge. It turns out that it is true that the graph of $y = 2x - 4$ is a straight line. In fact, we have the following result, which we state without proof:

> The graph of any equation of the form
>
> $y = mx + b$ or $Ax + By = C$
>
> where m, b, A, B, and C are constants (A and B not both 0) and x and y are variables is a straight line.

Thus, the graphs of

$y = \frac{2}{3}x - 5$ and $2x - 3y = 12$

are straight lines, since the first is of the form $y = mx + b$ and the second is of the form $Ax + By = C$.

Graphing Equations of the Form $y = mx + b$ or $Ax + By = C$

Step 1: Find two solutions of the equation. (A third solution is sometimes useful as a checkpoint.)

Step 2: Plot the solutions in a coordinate system.

Step 3: Using a straightedge, draw a line through the points plotted in step 2.

Note: The third solution provides a checkpoint, since if the line does not pass through all three points, a mistake has been made in finding the solutions.

Example 3 **(A)** Graph $y = 2x - 4$. **(B)** Graph $x + 3y = 6$.

Solution **(A)** Make up a table of at least two solutions (ordered pairs of numbers that satisfy the equation), plot them, then draw a line through these points with a straightedge.

x	$2x - 4 = y$	(x, y)
0	$2(0) - 4 = -4$	$(0, -4)$
2	$2(2) - 4 = 0$	$(2, 0)$
4	$2(4) - 4 = 4$	$(4, 4)$

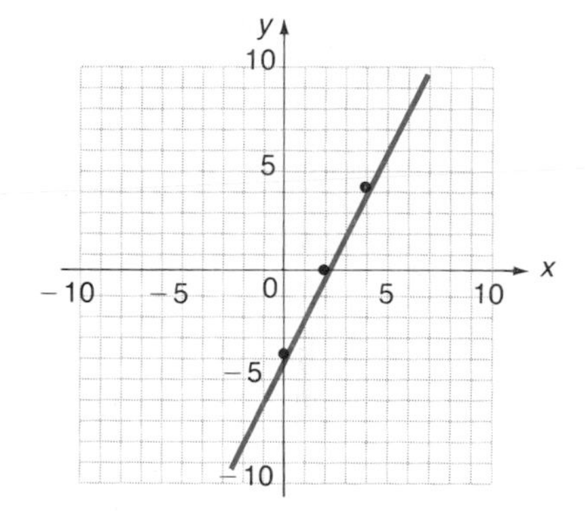

(B) To graph $x + 3y = 6$, assign to either x or y any convenient value and solve for the other variable. If we let $x = 0$, a convenient value, then

$$0 + 3y = 6$$
$$3y = 6$$
$$y = 2$$

Thus, $(0, 2)$ is a solution.

If we let $y = 0$, another convenient choice, then

$$x + 3(0) = 6$$
$$x + 0 = 6$$
$$x = 6$$

Thus, $(6, 0)$ is a solution.

To find a checkpoint, choose another value for x or y, say $x = -6$. Then

$$-6 + 3y = 6$$
$$3y = 12$$
$$y = 4$$

Thus, $(-6, 4)$ is also a solution.

We summarize these results in a table and then draw the graph. The first two solutions indicate where the graph crosses the coordinate axes and are called the **y and x intercepts**, respectively. That is, the y intercept is the y value where the line crosses the y axis; the x intercept is the x value where the line crosses the x axis. The intercepts are often the easiest points to find. To find the y intercept we let $x = 0$ and solve for y; to find the x intercept let $y = 0$ and solve for x. This is called the **intercept method** of graphing a straight line.

(x, y)	
(0, 2)	y intercept
(6, 0)	x intercept
(−6, 4)	Checkpoint

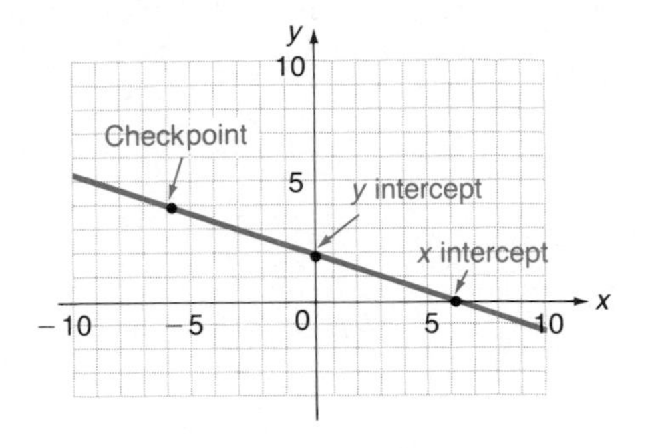

If a straight line does not pass through all three points, then we know we have made a mistake and must go back and check our work.

Problem 3 Graph: **(A)** $y = 2x - 6$ **(B)** $3x + y = 6$

VERTICAL AND HORIZONTAL LINES

Vertical and horizontal lines in rectangular coordinate systems have particularly simple equations.

Example 4 Graph the equations $y = 4$ and $x = 3$ in a rectangular coordinate system.

Solution To graph $y = 4$ or $x = 3$ in a rectangular coordinate system, each equation must be provided with the missing variable (usually done mentally) as follows:

$y = 4$	is equivalent to	$0x + y = 4$
$x = 3$	is equivalent to	$x + 0y = 3$

In the first case, we see that no matter what value is assigned to x, $0x = 0$; thus, as long as $y = 4$, x can assume any value:

x	y
5	4
0	4
−3	4

Thus, the graph of $y = 4$ is a horizontal line crossing the y axis at 4. Similarly, in the second case y can assume any value as long as $x = 3$:

x	y
3	−2
3	0
3	6

Thus, the graph of $x = 3$ is a vertical line crossing the x axis at 3. Thus:

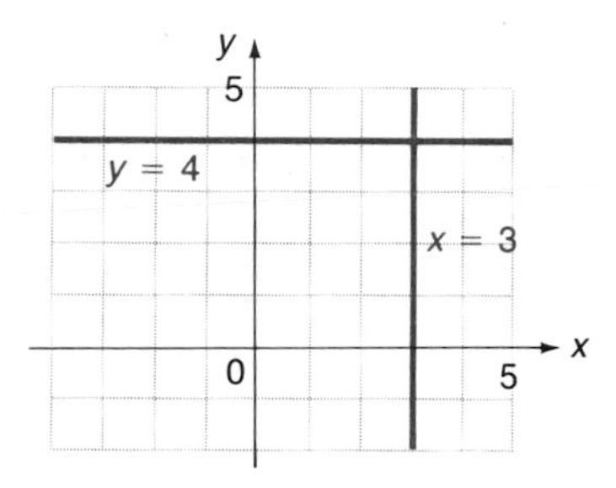

Problem 4 Graph $y = -3$ and $x = -4$ in a rectangular coordinate system.

It should now be clear why equations of the form $Ax + By = C$ and $y = mx + b$ are called **linear equations**: their graphs are straight lines.

USE OF DIFFERENT SCALES FOR EACH COORDINATE AXIS

Equations arising from applications often have restrictions put on the variables. They also may involve significant differences in the magnitude of the variables: to graph such equations we may want to use different scales for the two axes.

Example 5 Graph $y = 50 + 5x$ for $0 \leq x \leq 10$.

Solution We first note that x is restricted to values from 0 to 10. Let us find y for three values in this interval. We choose each end value and the middle value:

x	0	5	10
y	50	75	100

We see that as x varies from 0 to 10, y varies from 50 to 100. To keep the graph on the paper, we choose a different scale on the vertical axis (representing y) than that used on the horizontal axis (representing x). Thus, we obtain Figure 3, on page 294. The graph is a line segment joining the two points (0, 50) and (10, 100). The dashed lines are intended to guide one's eyes to the endpoint and are not part of the graph.

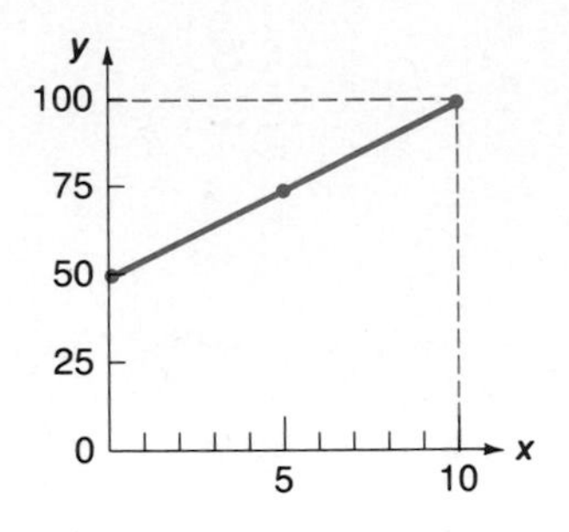

FIGURE 3

Problem 5 Graph $y = 100 - 6x$ for $2 \le x \le 15$.

ANSWERS TO MATCHED PROBLEMS

1. **(A)** (9, 7) **(B)** (−10, −6) **(C)** (1, −6) **(D)** (0, 3)

2.

(−3, 2)
(3, 4)
(−4, 0)
(0, 1)
(−2, −2)
(4, −2)

3. **(A)**

(x, y)
(0, −6)
(3, 0)
(2, −2)

(B)

(x, y)
(0, 6)
(2, 0)
(1, 3)

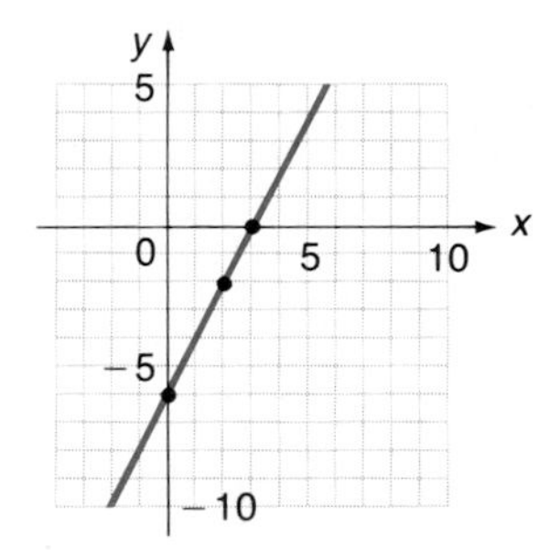

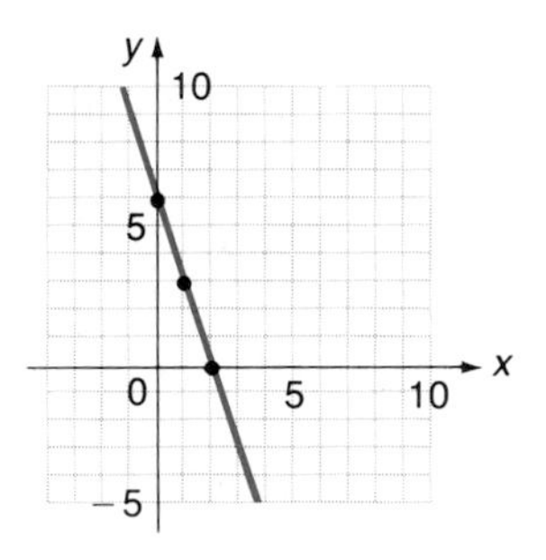

4.

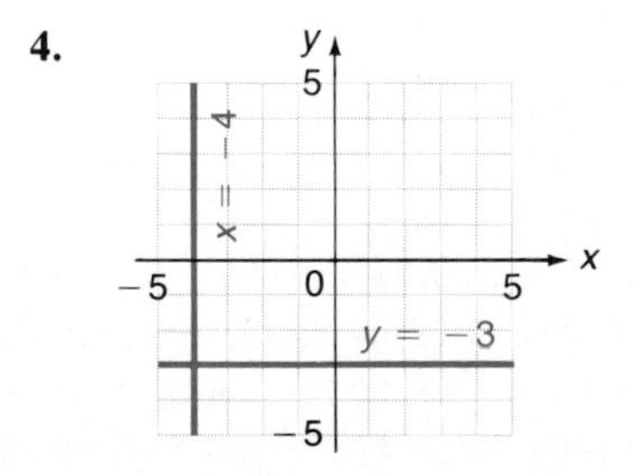

5.

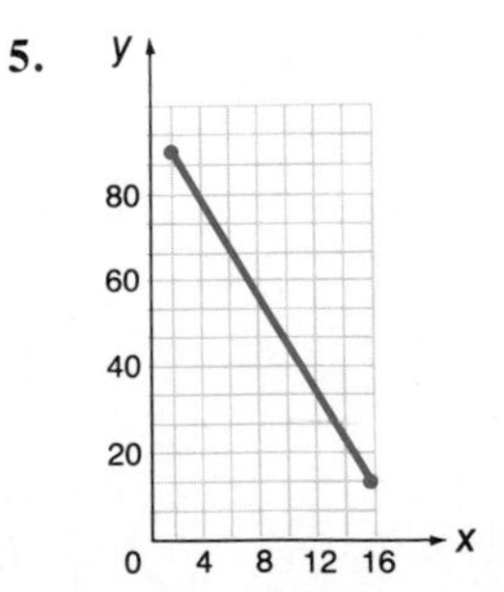

EXERCISE 7-1

A *Write down the coordinates of each labeled point.*

1.

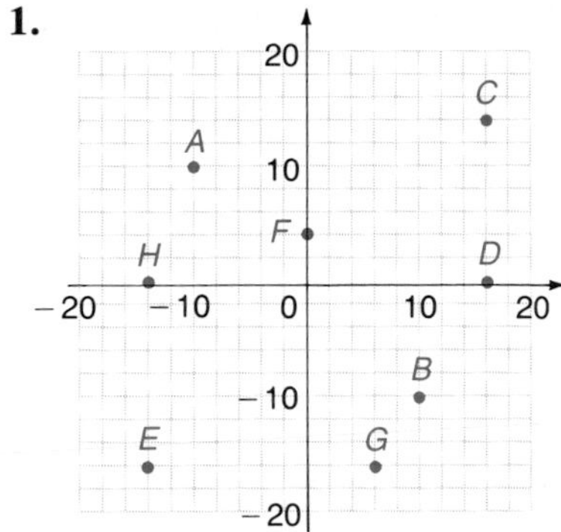

2.

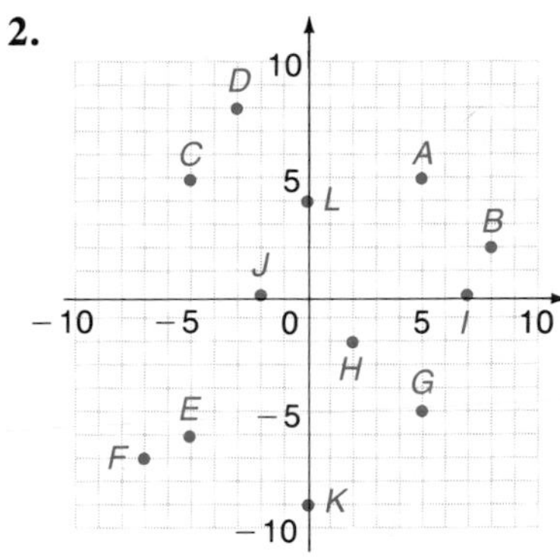

Plot each set of ordered pairs of numbers on the same coordinate system.

3. $(4, 4), (-4, 1), (-3, -3), (5, -1), (0, 2), (-2, 0)$

4. $(3, 1), (-2, 3), (-5, -1), (2, -1), (4, 0), (0, -5)$

5. $(2, 7), (7, 2), (-6, 3), (-4, -7), (2, 3), (0, -8), (9, 0)$

6. $(-9, 8), (8, -9), (0, 5), (4, -8), (-3, 0), (7, 7), (-6, -6)$

Write down the coordinates of each labeled point to the nearest quarter of a unit.

7.

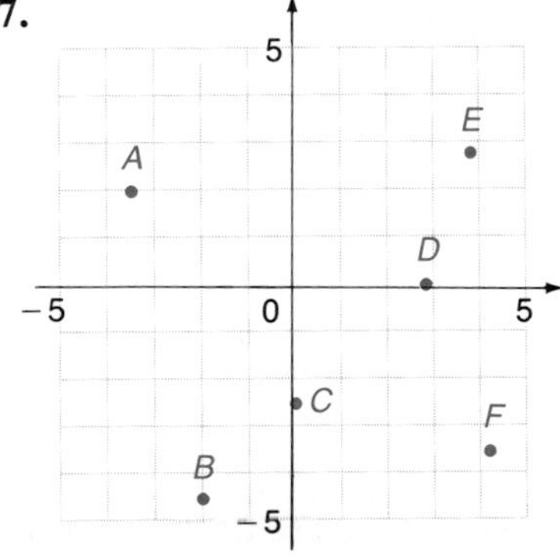

8.

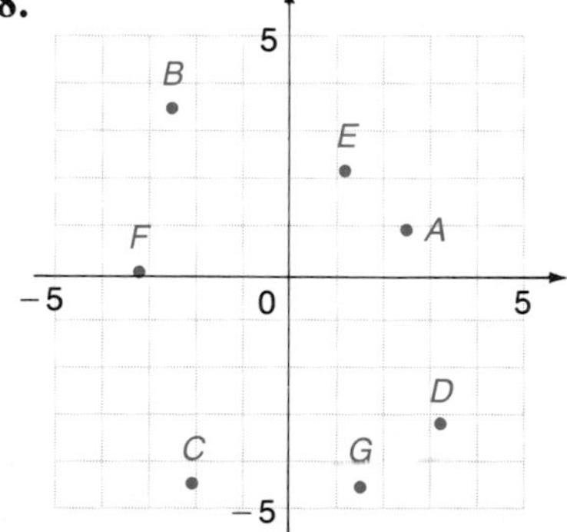

9. Plot the following ordered pairs of numbers on the same coordinate system: $A(3\frac{1}{2}, 2\frac{1}{2})$, $B(-4\frac{1}{2}, 3)$, $C(0, -3\frac{3}{4})$, $D(-2\frac{3}{4}, -3\frac{3}{4})$, $E(4\frac{1}{4}, -3\frac{3}{4})$

10. Plot the following ordered pairs of numbers on the same coordinate system: $A(1\frac{1}{2}, 3\frac{1}{2})$, $B(-3\frac{1}{4}, 0)$, $C(3, -2\frac{1}{2})$, $D(-4\frac{1}{2}, 1\frac{3}{4})$, $E(-2\frac{1}{2}, -4\frac{1}{4})$

Graph in a rectangular coordinate system.

11. $y = 2x$ **12.** $y = x$ **13.** $y = 2x - 2$

14. $y = x - 1$ **15.** $y = \dfrac{x}{3}$ **16.** $y = \dfrac{x}{2}$

17. $y = \dfrac{x}{3} + 2$ **18.** $y = \dfrac{x}{2} + 1$ **19.** $x + y = -4$

20. $x + y = 6$ **21.** $x - y = 3$ **22.** $x - y = 5$

23. $3x + 4y = 12$ **24.** $2x + 3y = 12$ **25.** $8x - 3y = 24$

26. $3x - 5y = 15$ **27.** $y = 3$ **28.** $x = 2$

29. $x = -4$ **30.** $y = 3$

B **31.** $y = \frac{1}{2}x$ **32.** $y = \frac{1}{4}x$ **33.** $y = \frac{1}{2}x - 1$

34. $y = \frac{1}{4}x + 1$ **35.** $y = -x + 2$ **33.** $y = -2x + 6$

37. $y = \frac{1}{3}x - 1$ **38.** $y = -\frac{1}{2}x + 2$ **39.** $3x + 2y = 10$

40. $2x + y = 7$ **41.** $5x - 6y = 15$ **42.** $7x - 4y = 21$

43. $y = 0$ **44.** $x = 0$

Write in the form $y = mx + b$ and graph.

45. $x + 6 = 3x + 2 - y$ **46.** $y - x - 2 = x + 1$

Write in the form $Ax + By = C$, $A > 0$, and graph.

47. $y + 8 = 2 - x - y$ **48.** $6x - 3 + y = 2y + 4x + 5$

Graph each of the following using a different scale on the vertical axis to keep the size of the graph within reason.

49. $I = 6t$, $0 \le t \le 10$ **50.** $d = 60t$, $0 \le t \le 10$

51. $v = 10 + 32t$, $0 \le t \le 5$ **52.** $A = 100 + 10t$, $0 \le t \le 10$

53. Graph $x + y = 3$ and $2x - y = 0$ on the same coordinate system. Determine by inspection the coordinates of the point where the two graphs cross. Show that the coordinates of the point of intersection satisfy both equations.

54. Repeat Problem 53 with the equations $2x - 3y = -6$ and $x + 2y = 11$.

C **55.** Graph $y = mx - 2$ for $m = 2$, $m = \frac{1}{2}$, $m = 0$, $m = -\frac{1}{2}$, and $m = -2$, all on the same coordinate system.

56. Graph $y = -\frac{1}{2}x + b$ for $b = -6$, $b = 0$, and $b = 6$, all on the same coordinate system.

57. Graph $y = |x|$. [*Hint*: Graph $y = x$ for $x \geq 0$, and graph $y = -x$ for $x < 0$.]

58. Graph $y = |2x|$ and $y = |\frac{1}{2}x|$ on the same coordinate system (see Problem 57).

APPLICATIONS

59. ***Psychology*** In 1948 Professor Brown, a psychologist, trained a group of rats (in an experiment on motivation) to run down a narrow passage in a cage to receive food in a box. A harness was put on each rat and the harness was then connected to an overhead wire that was attached to a scale. In this way the rat could be placed at different distances (in centimeters) from the food and Professor Brown could then measure the pull (in grams) of the rat toward the food. It was found that a relation between motivation (pull) and position was given approximately by the equation $p = -\frac{1}{5}d + 70$, $30 \leq d \leq 175$. Graph this equation for the indicated values of d.

60. ***Electronics*** In a simple electric circuit, such as found in a flashlight, if the resistance is 30 ohms, the current in the circuit I (in amperes) and the electromotive force E (in volts) are related by the equation $E = 30I$. Graph this equation for $0 \leq I \leq 1$.

61. ***Biology*** In biology there is an approximate rule, called the bioclimatic rule, for temperate climates. This rule states that in spring and early summer, periodic phenomena such as blossoming for a given species, appearance of certain insects, and ripening of fruit usually come about 4 days later for each 500 feet of altitude. Stated as a formula,

$$d = 4\left(\frac{h}{500}\right)$$

where d = Change in days and h = Change in altitude in feet. Graph the equation for $0 \leq h \leq 4{,}000$.

7-2 SLOPE AND EQUATIONS OF A LINE

- Slope of a Line
- Slope–Intercept Form
- Point–Slope Form
- Vertical and Horizontal Lines
- Parallel and Perpendicular Lines

In the preceding section we considered this problem: given a linear equation of the form

$$Ax + By = C \qquad \text{or} \qquad y = mx + b$$

find its graph. Now we will consider the reverse problem: given certain information about a straight line in a rectangular coordinate system, find its equation. We start by introducing a measure of the steepness of a line called slope.

SLOPE OF A LINE

If we take two points (x_1, y_1) and (x_2, y_2) on a line, then the ratio of the change in y to the change in x as we move from point P_1 to P_2 is called the **slope** of the line.

Slope Formula

If a line passes through $P_1(x_1, y_1)$ and $P_2(x_2, y_2)$, then its slope is given by the formula

$$m = \frac{y_2 - y_1}{x_2 - x_1} \qquad x_1 \neq x_2$$

$$= \frac{\text{Vertical change (rise)}}{\text{Horizontal change (run)}}$$

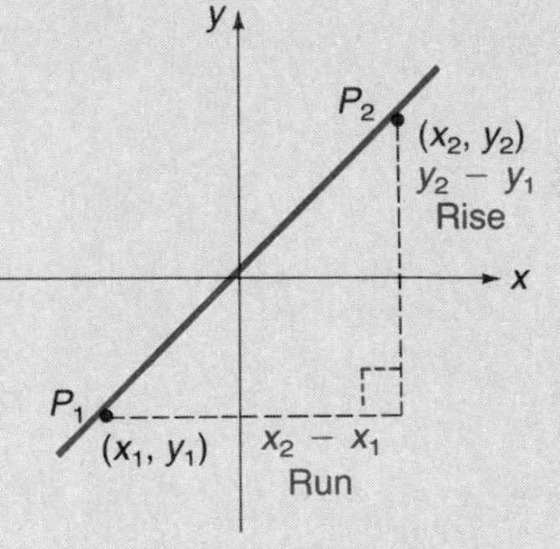

The slope is a measure of the direction and steepness of a line.

Example 6 Find the slope of the line passing through $(-3, -2)$ and $(3, 4)$.

Solution Let $(x_1, y_1) = (-3, -2)$ and $(x_2, y_2) = (3, 4)$; then

$$m = \frac{y_2 - y_1}{x_2 - x_1} = \frac{4 - (-2)}{3 - (-3)} = \frac{6}{6} = 1$$

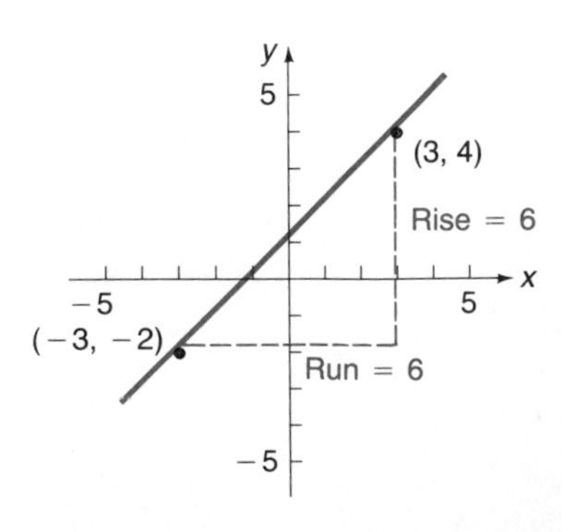

Note: It does not matter which point we call P_1 or P_2 as long as we stick to the choice once it is made. If we reverse the choice above, we obtain the same value for the slope, since the sign of the numerator and the denominator both change:

$$m = \frac{(-2) - 4}{(-3) - 3} = \frac{-6}{-6} = 1$$

Problem 6 Graph the line passing through $(-2, 7)$ and $(3, -3)$; then compute its slope.

For a horizontal line, y does not change as x changes; hence, its slope is 0. On the other hand, for a vertical line, x does not change as y changes; hence, $x_1 = x_2$ and

$$m = \frac{y_2 - y_1}{x_2 - x_1} = \frac{y_2 - y_1}{0} \quad \text{Vertical-line slope is not defined.}$$

In general, the slope of a line may be positive, negative, 0, or not defined. Each of these cases is interpreted geometrically as shown in Table 2.

TABLE 2 GOING FROM LEFT TO RIGHT

LINE	SLOPE	EXAMPLE
Rising	Positive	
Falling	Negative	
Horizontal	0	
Vertical	Not defined	

SLOPE–INTERCEPT FORM

Any equation of the form $Ax + By = C$, $B \neq 0$, can always be written in the form

$$y = mx + b$$

where m and b are constants. For example, starting with

$2x + 3y = 6$ Form $Ax + By = C$.

we solve for y to obtain

$$3y = -2x + 6 \qquad By = -Ax + C$$

$$\tfrac{1}{3}(3y) = \tfrac{1}{3}(-2x + 6) \qquad y = -\frac{A}{B}x + \frac{C}{B}$$

$$y = -\tfrac{2}{3}x + 2 \qquad \text{Form } y = mx + b.$$

The constants m and b in $y = mx + b$ have special geometric meaning. If we let $x = 0$, then

$$\begin{aligned} y &= m \cdot 0 + b \\ &= 0 + b \\ &= b \end{aligned}$$

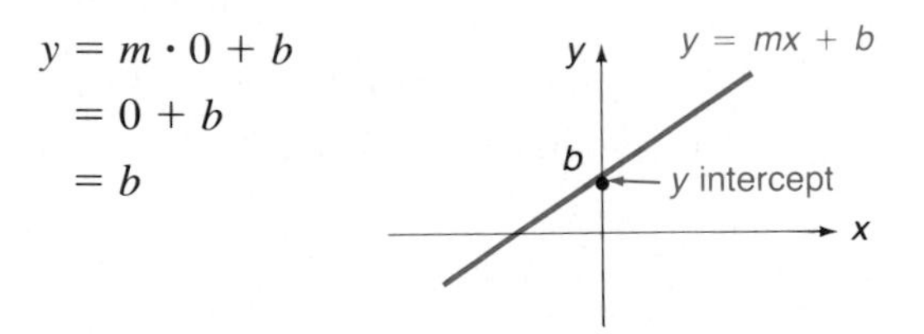

Thus, b is the y coordinate of the y intercept, the point where the graph crosses the y axis. For brevity, the value b is also called the **y intercept**. In the example given above, $y = -\frac{2}{3}x + 2$, the y intercept is 2.

Now let us determine the geometric significance of m in $y = mx + b$. We choose two points (x_1, y_1) and (x_2, y_2) on the graph of $y = mx + b$ (Figure 4). Since the two points are on the graph, they are solutions to the equations $y = mx + b$. Thus,

$$y_1 = mx_1 + b \qquad \text{and} \qquad y_2 = mx_2 + b$$

FIGURE 4

Solving both equations for b, we obtain

$$b = y_1 - mx_1 \qquad \text{and} \qquad b = y_2 - mx_2$$

Since $y_1 - mx_1$ and $y_2 - mx_2$ are both equal to b, they are equal to each other. Thus,

$$\begin{aligned} y_1 - mx_1 &= y_2 - mx_2 && \text{Now solve for } m. \\ mx_2 - mx_1 &= y_2 - y_1 \\ (x_2 - x_1)m &= y_2 - y_1 \\ m &= \frac{y_2 - y_1}{x_2 - x_1} && x_1 \neq x_2 \quad \text{Recall that } \frac{y_2 - y_1}{x_2 - x_1} \text{ is the slope.} \end{aligned}$$

Thus, ***m* is the slope** of the graph of $y = mx + b$.

In summary, if an equation of a line is written in the form $y = mx + b$, then b is the y intercept and m is the slope. Conversely, if we know the slope m and y intercept b of a line, we can write its equation in the form $y = mx + b$.

Example 7 **(A)** Find the slope and y intercept of the line $y = \frac{1}{3}x + 2$.
(B) Find the equation of a line with slope -2 and y intercept 3.

Solution **(A)** $y = \frac{1}{3}x + \mathbf{2}$ (Slope: $\frac{1}{3}$; y intercept: 2)

(B) Since $m = -2$ and $b = 3$, then $y = mx + b = -2x + 3$ is the equation.

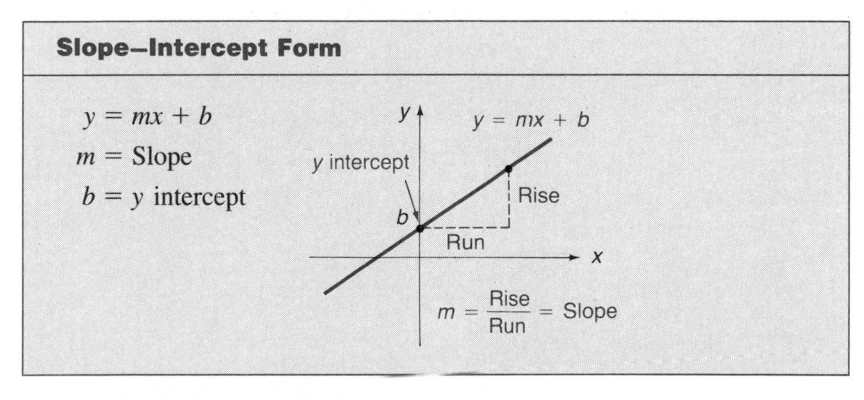

Problem 7 **(A)** Find the slope and y intercept of the line $y = \dfrac{x}{2} - 7$.

(B) Find the equation of a line with slope $-\frac{1}{3}$ and y intercept 6.

The slope–intercept form allows us to graph a linear equation very efficiently.

Example 8 Graph $y = -\frac{1}{2}x + 3$.

Solution The y intercept is 3, so the point $(0, 3)$ is on the graph. The slope of the line is $-\frac{1}{2}$, so if the x coordinate is increased (run) by 2 units, the y coordinate changes (rises) by -1. The resulting point is easily graphed, and the two points yield the graph of the line.

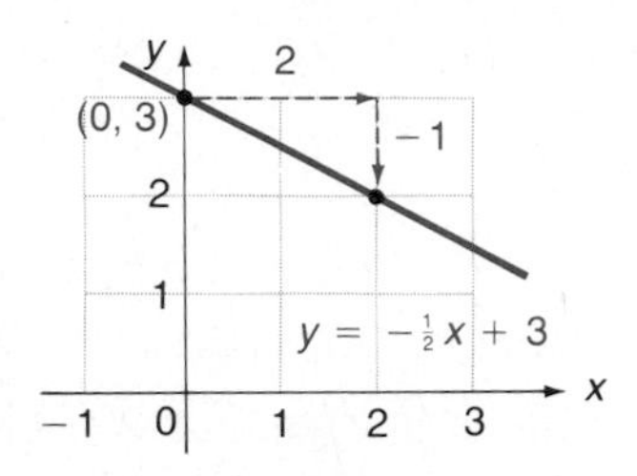

Problem 8 Graph $y = \frac{2}{3}x - 2$.

POINT–SLOPE FORM

In Example 7(B) we found the equation of a line given its slope and y intercept. Often it is necessary to find the equation of a line given its slope and the coordinates of a point through which it passes, or to find the equation of a line given the coordinates of two points through which it passes.

Let a line have slope m and pass through the fixed point (x_1, y_1). If the variable point (x, y) is to be a point on the line, the slope of the line passing through (x, y) and (x_1, y_1) must be m (see Figure 5).

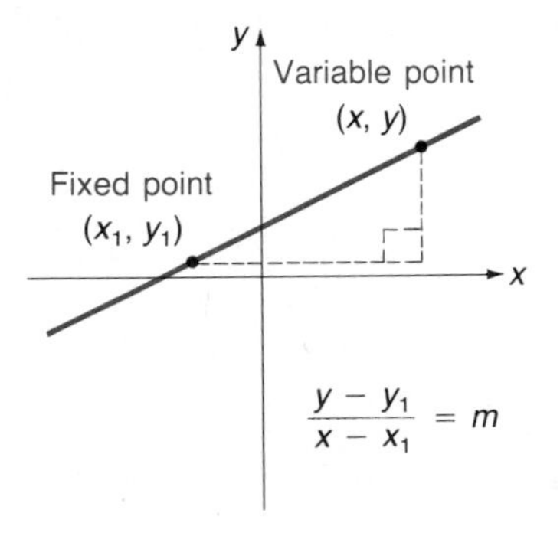

FIGURE 5

Thus, the equation

$$\frac{y - y_1}{x - x_1} = m$$

restricts the variable point (x, y), so that only those points in the plane lying on the line will have coordinates that satisfy the equation, and vice versa. This equation is usually written in the following form:

Point–Slope Form

The equation of a line passing through (x_1, y_1) with slope m is given by

$$y - y_1 = m(x - x_1)$$

It is referred to as the **point–slope form of the equation of a line**. Using this equation in conjunction with the slope formula, we can also find the equation of a line knowing only the coordinates of two points through which it passes.

Example 9 **(A)** Find an equation of a line with slope $-\frac{1}{3}$ that passes through $(6, -3)$. Write the resulting equation in the form $y = mx + b$.

(B) Find an equation of a line that passes through the two points $(-2, -6)$ and $(2, 2)$.

Solution **(A)**

$$\begin{aligned} y - y_1 &= m(x - x_1) \\ y - (-3) &= -\tfrac{1}{3}(x - 6) \\ y + 3 &= -\tfrac{1}{3}(x - 6) \\ y + 3 &= -\frac{x}{3} + 2 \\ y &= -\tfrac{1}{3}x - 1 \end{aligned}$$

(B) First find the slope of the line using the slope formula

$$m = \frac{y_2 - y_1}{x_2 - x_1} = \frac{2 - (-6)}{2 - (-2)} = 2$$

Now proceed as in part (A), using the coordinates of either point for (x_1, y_1).

Use $(x_1, y_1) = (-2, -6)$

$$\begin{aligned} y - y_1 &= m(x - x_1) \\ y - (-6) &= 2[x - (-2)] \\ y + 6 &= 2(x + 2) \\ y + 6 &= 2x + 4 \\ y &= 2x - 2 \end{aligned}$$

or

Use $(x_1, y_1) = (2, 2)$

$$\begin{aligned} y - y_1 &= m(x - x_1) \\ y - 2 &= 2(x - 2) \\ y - 2 &= 2x - 4 \\ y &= 2x - 2 \end{aligned}$$

Problem 9 **(A)** Find the equation of a line with slope $\frac{2}{3}$ that passes through $(-3, 4)$.
(B) Find the equation of a line that passes through the two points $(6, -1)$ and $(-2, 3)$. Transform the equation into the form $y = mx + b$.

The slope–intercept form can also be used to find the equation of the line given the slope and one point. In Example 9(A), for instance, given the slope $-\frac{1}{3}$, we know the equation has the form

$$y = -\tfrac{1}{3}x + b$$

Since $(6, -3)$ must satisfy the equation, we can solve for b:

$$-3 = -\tfrac{1}{3}(6) + b$$
$$-3 = -2 + b$$
$$-1 = b$$

Thus, the equation of the line is $y = -\frac{1}{3}x - 1$.

VERTICAL AND HORIZONTAL LINES

If a line is vertical, its slope is not defined. Since points on a vertical line have constant x coordinates and arbitrary y coordinates, the equation of a vertical line is of the form

$$x = c$$

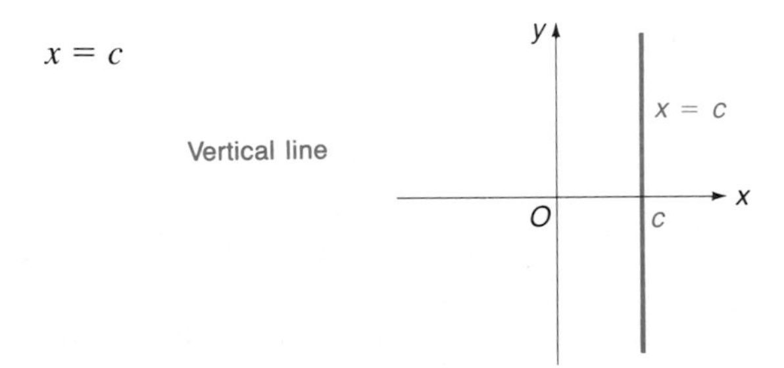

Vertical line

where c is the abscissa of each point on the line. Similarly, if a line is horizontal (slope 0), then every point on the line has constant y coordinates and arbitrary x coordinates. Thus, the equation of a horizontal line is of the form

$$y = c$$

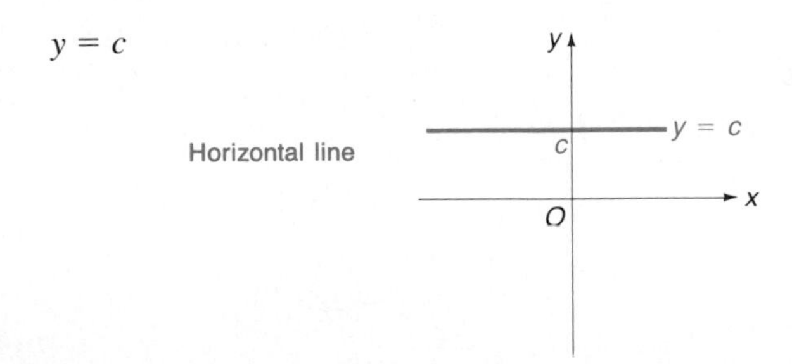

Horizontal line

where c is the ordinate of each point on the line. Also, since a horizontal line has slope 0 ($m = 0$), then, using the slope–intercept form, we obtain

$$\begin{aligned} y &= mx + b \\ &= 0x + c \\ &= c \end{aligned}$$

Example 10 What are the equations of vertical and horizontal lines through $(-2, -4)$?

Solution

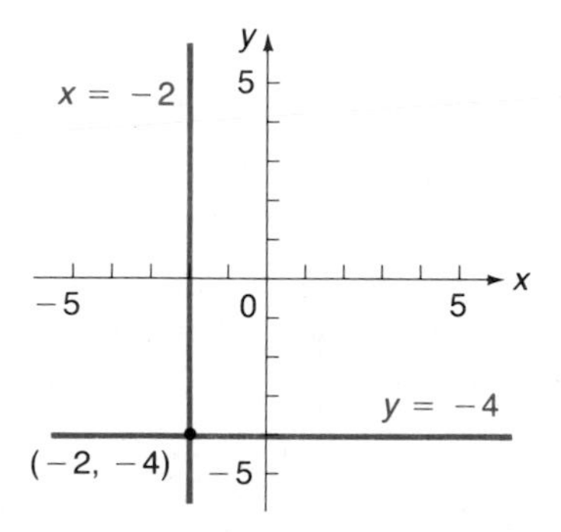

The equation of a vertical line through $(-2, -4)$ is $x = -2$, and the equation of a horizontal line through the same point is $y = -4$.

Problem 10 What are the equations of vertical and horizontal lines through $(3, -8)$?

PARALLEL AND PERPENDICULAR LINES

It can be shown that if two nonvertical lines are parallel, then they have the same slope. And if two lines have the same slope, they are parallel. It can also be shown that if two nonvertical lines are perpendicular, then their slopes are the negative reciprocals of each other (that is, $m_2 = -1/m_1$, or, equivalently, $m_1 m_2 = -1$). And if the slopes of two lines are the negative reciprocals of each other, the lines are perpendicular. Symbolically:

Parallel and Perpendicular Lines

Given nonvertical lines L_1 and L_2 with slopes m_1 and m_2, respectively, then

$L_1 \parallel L_2$ if and only if $m_1 = m_2$

$L_1 \perp L_2$ if and only if $m_1 m_2 = -1$ or $m_2 = -\dfrac{1}{m_1}$

Note: $\parallel$ means "is parallel to" and $\perp$ means "is perpendicular to."

The lines determined by

$$y = \tfrac{2}{3}x - 5 \qquad \text{and} \qquad y = \tfrac{2}{3}x + 8$$

are parallel since both have the same slope, $\frac{2}{3}$. And the lines that are determined by

$$y = \frac{x}{3} + 5 \qquad \text{and} \qquad y = -3x - 7$$

are perpendicular, since the product of their slopes is -1; that is,

$$(\tfrac{1}{3})(-3) = -1$$

Example 11 Given the line $x - 2y = 4$, find the equation of a line that passes through $(2, -3)$ and is:

(A) Parallel to the given line **(B)** Perpendicular to the given line

Write final equations in the form $y = mx + b$.

Solution First find the slope of the given line by writing $x - 2y = 4$ in the form $y = mx + b$:

$$x - 2y = 4$$

$$-2y = 4 - x$$

$$\boxed{-\tfrac{1}{2}(-2y) = -\tfrac{1}{2}(4 - x)}$$

$$y = -2 + \tfrac{1}{2}x$$

Slope

$$= \tfrac{1}{2}x - 2$$

The slope of the given line is $\frac{1}{2}$.

(A) The slope of a line parallel to the given line is also $\frac{1}{2}$. We have only to find the equation of a line through $(2, -3)$ with slope $\frac{1}{2}$ to solve part (A):

$$y - y_1 = m(x - x_1) \qquad m = \tfrac{1}{2} \text{ and } (x_1, y_1) = (2, -3)$$

$$y - (-3) = \tfrac{1}{2}(x - 2)$$

$$y + 3 = \tfrac{1}{2}x - 1$$

$$y = \tfrac{1}{2}x - 4$$

(B) The slope of the line perpendicular to the given line is the negative reciprocal of $\frac{1}{2}$, that is, -2. We have only to find the equation of a line through $(2, -3)$ with slope -2 to solve part (B):

$$y - y_1 = m(x - x_1) \qquad m = -2 \text{ and } (x_1, y_1) = (2, -3)$$
$$y - (-3) = -2(x - 2)$$
$$y + 3 = -2x + 4$$
$$y = -2x + 1$$

Problem 11 Given the line $2x = 6 - 3y$, find the equation of a line that passes through $(-3, 9)$ and is:

(A) Parallel to the given line **(B)** Perpendicular to the given line

Write final equations in the form $y = mx + b$.

ANSWERS TO MATCHED PROBLEMS

6. $m = -2$

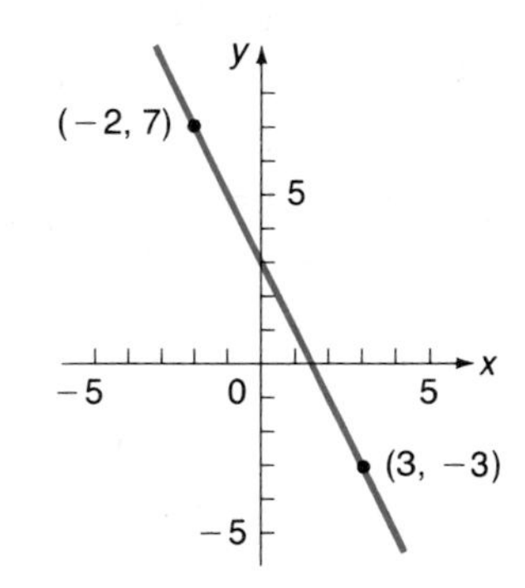

7. **(A)** $m = \frac{1}{2}$, $b = -7$ **(B)** $y = -\frac{1}{3}x + 6$

8.

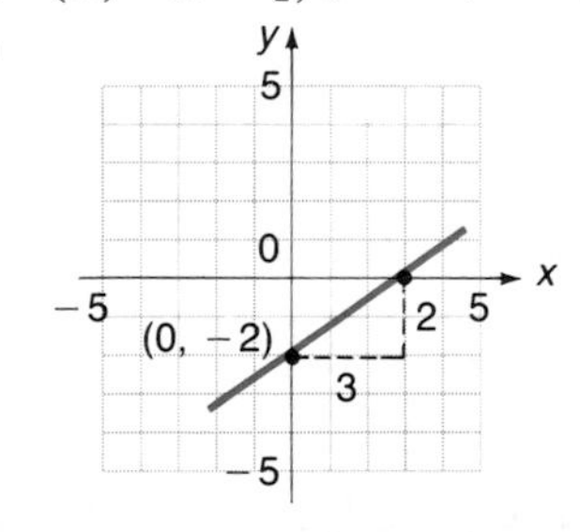

9. **(A)** $y - 4 = \frac{2}{3}(x + 3)$ or $y = \frac{2}{3}x + 6$ **(B)** $y = -\frac{1}{2}x + 2$

10. $x = 3$, $y = -8$

11. **(A)** $y = -\frac{2}{3}x + 7$ **(B)** $y = \frac{3}{2}x + \frac{27}{2}$

EXERCISE 7-2

A *Find the slope and y intercept, and graph each equation.*

1. $y = 2x - 3$

2. $y = x + 1$

3. $y = -x + 2$

4. $y = -2x + 1$

Write the equation of the line with slope and y intercept as indicated.

5. Slope = 5
y intercept = -2

6. Slope = 3
y intercept = -5

7. Slope = -2
y intercept = 4

8. Slope = -1
y intercept = 2

Write the equation of the line that passes through the given point with the indicated slope.

9. $m = 2$; $(5, 4)$

10. $m = 3$; $(2, 5)$

11. $m = -2$; $(2, 1)$

12. $m = -3$; $(1, 3)$

Find the slope of the line that passes through the given points.

13. $(3, 2)$ and $(5, 6)$

14. $(1, 3)$ and $(2, 4)$

15. $(2, 1)$ and $(10, 5)$

16. $(1, 3)$ and $(7, 5)$

Write the equation of the line through each indicated pair of points.

17. $(3, 2)$ and $(5, 6)$

18. $(1, 3)$ and $(2, 4)$

19. $(2, 1)$ and $(10, 5)$

20. $(1, 3)$ and $(7, 5)$

B *Find the slope and y intercept, and graph each equation.*

21. $y = -\dfrac{x}{3} + 2$

22. $y = -\dfrac{x}{4} - 1$

23. $x + 2y = 4$

24. $x - 3y = -6$

25. $2x + 3y = 6$

26. $3x + 4y = 12$

Write the equation of the line with slope and y intercept as given.

27. Slope = $-\frac{1}{2}$
y intercept = -2

28. Slope = $-\frac{1}{3}$
y intercept = -5

29. Slope = $\frac{2}{3}$
y intercept = $\frac{3}{2}$

30. Slope = $-\frac{3}{2}$
y intercept = $\frac{5}{2}$

Write the equation of the line that passes through the given point with the indicated slope. Transform the equation into the form $y = mx + b$.

31. $m = -2$; $(-3, 2)$

32. $m = -3$; $(4, -1)$

33. $m = \frac{1}{2}$; $(-4, 3)$

34. $m = \frac{2}{3}$; $(-6, -5)$

Find the slope of the line that passes through the given points.

35. $(3, 7)$ and $(-6, 4)$ **36.** $(-5, -2)$ and $(-5, -4)$

37. $(4, -2)$ and $(-4, 0)$ **38.** $(-3, 0)$ and $(3, -2)$

Write the equation of the line through each of the indicated pairs of points. Transform the equation into the form $y = mx + b$.

39. $(3, 7)$ and $(-6, 4)$ **40.** $(-5, -2)$ and $(5, -4)$

41. $(4, -2)$ and $(-4, 0)$ **42.** $(-3, 0)$ and $(3, -4)$

Write the equations of the vertical and horizontal lines through each point.

43. $(-3, 5)$ **44.** $(6, -2)$ **45.** $(-1, 22)$ **46.** $(5, 0)$

C *Given the indicated equation of a line and the indicated point, find the equation of the line through the given point that is* **(A)** *parallel to the given line and* **(B)** *perpendicular to the given line. Write the answers in the form $y = mx + b$.*

47. $y = \frac{3}{5}x + 1$; $(1, 4)$ **48.** $y = -\frac{2}{3}x + 3$; $(-1, -2)$

49. $y = -3x + \frac{1}{3}$; $(0, 2)$ **50.** $y = 5x - \frac{1}{4}$; $(-2, 0)$

51. $x + y = 3$; $(1, 1)$ **52.** $x - y = \frac{1}{7}$; $(2, 3)$

53. $y = 3$; $(-2, 5)$ **54.** $x = 4$; $(3, -2)$

APPLICATIONS

55. ***Business*** A sporting goods store sells a pair of cross-country ski boots costing \$20 for \$33 and a pair of cross-country skis costing \$60 for \$93.

(A) If the markup policy of the store for items costing over \$10 is assumed to be linear and is reflected in the pricing of these two items, write an equation that relates retail price R with cost C.

(B) Graph this equation for $10 \le C \le 300$.

(C) Use the equation to find the cost of a surfboard retailing for \$240.

56. ***Business*** The management of a company manufacturing ballpoint pens estimates costs for running the company to be \$200 per day at zero output and \$700 per day at an output of 1,000 pens.

(A) Assuming total cost per day C is linearly related to total output per day x, write an equation relating these two quantities.

(B) Graph the equation for $0 \le x \le 2{,}000$.

57. ***Physics*** Water freezes at 32° Fahrenheit and 0° Celsius and boils at 212° Fahrenheit and 100° Celsius. Find the linear relationship between the two scales.

58. ***Physics*** It is known from physics (Hooke's law) that the relationship between the stretch s of a spring and the weight w causing the stretch is linear (a principle upon which all spring scales are constructed). A 10-pound weight stretches a spring 1 inch, and with no weight the stretch of the spring is 0.

(A) Find a linear equation, $s = mw + b$, that represents this relationship. [*Hint:* Both points (10, 1) and (0, 0) are on its graph.]
(B) Find the stretch of the spring for 15-pound and 30-pound weights.
(C) What is the slope of the graph? (The slope indicates the increase in stretch for each pound increase in weight.)
(D) Graph the equation for $0 \le w \le 40$.

59. ***Business*** An electronic computer was purchased by a company for \$20,000 and is assumed to have a salvage value of \$2,000 after 10 years (for tax purposes). Its value is depreciated linearly from \$20,000 to \$2,000.

(A) Find the linear equation, $V = mt + b$, that relates value V in dollars to time t in years.
(B) Find the values of the computer after 4 and 8 years, respectively.
(C) Find the slope of the graph. (The slope indicates the decrease in value per year.)
(D) Graph the equation for $0 \le t \le 10$.

60. ***Biology*** A biologist needs to prepare a special diet for a group of experimental animals. Two food mixes, M and N, are available. If mix M contains 20% protein and mix N contains 10% protein, what combinations of each mix will provide exactly 20 grams of protein? Let x be the amount of M used and y the amount of N used. Then write a linear equation relating x, y, and 20. Graph this equation for $x \ge 0$ and $y \ge 0$.

7-3 GRAPHING LINEAR INEQUALITIES

We know how to graph first-degree equations such as

$$y = 2x - 3 \qquad \text{or} \qquad 2x - 3y = 5$$

but how do we graph first-degree inequalities such as

$$y \le 2x - 3 \qquad \text{or} \qquad 2x - 3y > 5$$

We will find that graphing inequalities is almost as easy as graphing equations. The following discussion leads to a simple solution to the problem.

A line in a cartesian coordinate system divides the plane into two **half-planes**. A vertical line divides the plane into left and right half-planes; a nonvertical line divides the plane into upper and lower half-planes (Figure 6, on page 311).

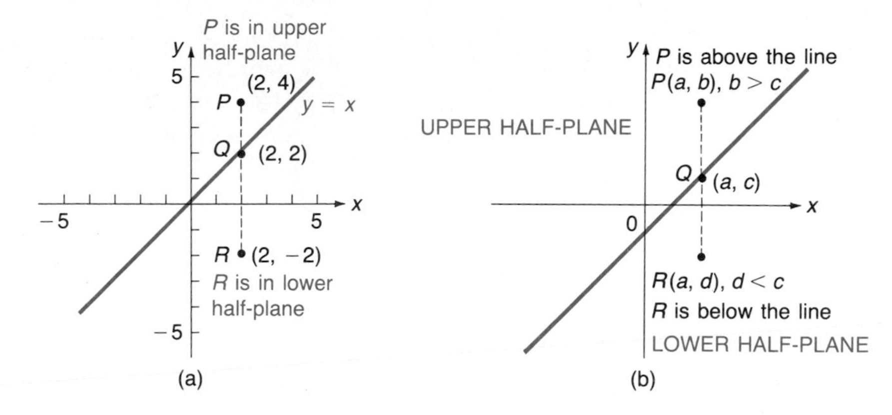

FIGURE 6

Now let us compare the graphs of

$$y < 2x - 3 \qquad y = 2x - 3 \qquad \text{and} \qquad y > 2x - 3$$

Consider, for example, the vertical line $x = 3$, and ask what the relationship of y is to $2 \cdot 3 - 3$ as we move $(3, y)$ up and down this vertical line (see Figure 7a). If we are at point Q, a point on the graph of $y = 2x - 3$, then $y = 2 \cdot 3 - 3$; if we move up the vertical line to P, the ordinate of $(3, y)$ increases and $y > 2 \cdot 3 - 3$; if we move down the line to R, the ordinate of $(3, y)$ decreases and $y < 2 \cdot 3 - 3$. Since the same results are obtained for each point x_0 on the x axis (see Figure 7b), we conclude that the graph of $y > 2x - 3$ is the upper half-plane determined by $y = 2x - 3$, and $y < 2x - 3$ is the lower half-plane.

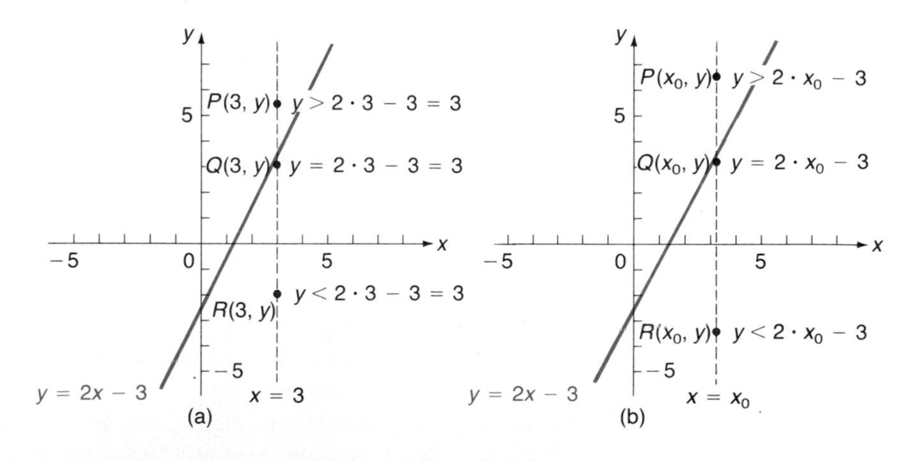

FIGURE 7

In graphing $y > 2x - 3$, we show the line $y = 2x - 3$ as a broken line, indicating that it is not part of the graph; in graphing $y \geq 2x - 3$, we show the line $y = 2x - 3$ as a solid line, indicating that it is part of the graph. Figure 8, on page 312, illustrates four typical cases.

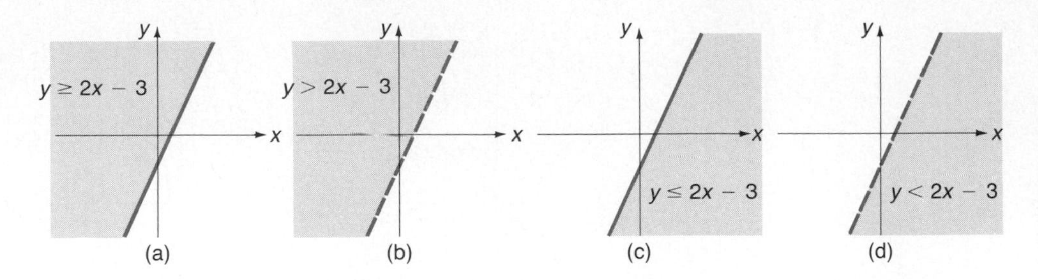

FIGURE 8

In this illustration the upper half-plane corresponded to the greater-than-or-equal ($\geq$) and greater-than ($>$) inequalities, the lower half-plane to less-than-or-equal ($\leq$) and less-than ($<$) inequalities. This is not a general rule, as will be seen in the next example. First, however, we state the following important result, which is based upon the preceding discussion:

The Graph of a Linear Inequality

The graph of a linear inequality

$$Ax + By < C \qquad \text{or} \qquad Ax + By > C$$

with $B \neq 0$, is either the upper half-plane or the lower half-plane (but not both) determined by the line $Ax + By = C$. If $B = 0$, the graph of

$$Ax < C \qquad \text{or} \qquad Ax > C$$

is either the left half-plane or the right half-plane (but not both) determined by the line $Ax = C$.

This result leads to a simple, fast procedure for graphing linear inequalities in two variables:

Steps in Graphing Linear Inequalities

1. Graph the corresponding equation $Ax + By = C$—as a broken line if equality is not included in original statement, as a solid line if equality is included in original statement.
2. Choose a test point in the plane not on the line—the origin is the best choice if it is not on the line—and substitute the coordinates into the inequality.
3. The graph of the original inequality includes:
   (A) The half-plane containing the test point if the inequality is satisfied by that point
   (B) The half-plane not containing the test point if the inequality is not satisfied by that point
4. Shade the half-plane to show the graph of the solution set.

Example 12 Graph $3x - 4y \leq 12$.

Solution *Step 1:* First graph the line $3x - 4y = 12$ as a solid line, since equality is included in $3x - 4y \leq 12$.

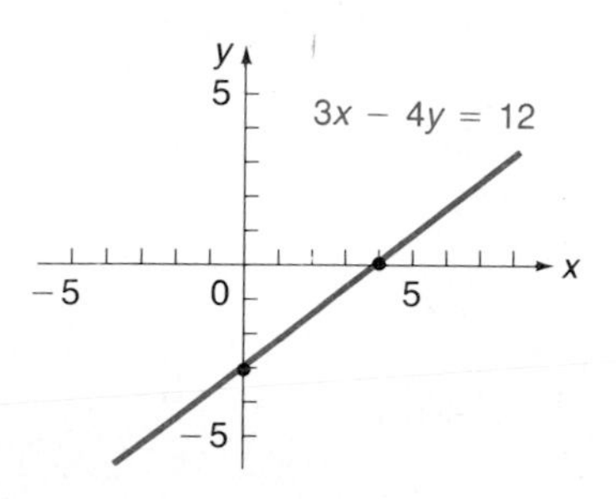

Step 2: Choose a convenient test point—any point not on the line will do. In this case the origin results in the simplest computation.

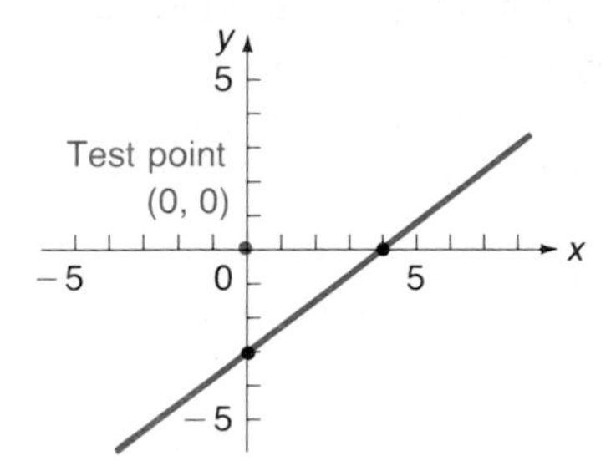

We see that the origin (0, 0) satisfies the original inequality:

$$3x - 4y \leq 12$$
$$3 \cdot 0 - 4 \cdot 0 \leq 12$$
$$0 \leq 12$$

Hence, all other points on the same side as the origin are also part of the graph. Thus, the graph is the upper half-plane.

Step 3: The final graph is the upper half-plane and the line:

Step 4: Shade the half-plane to show the graph of the solution set.

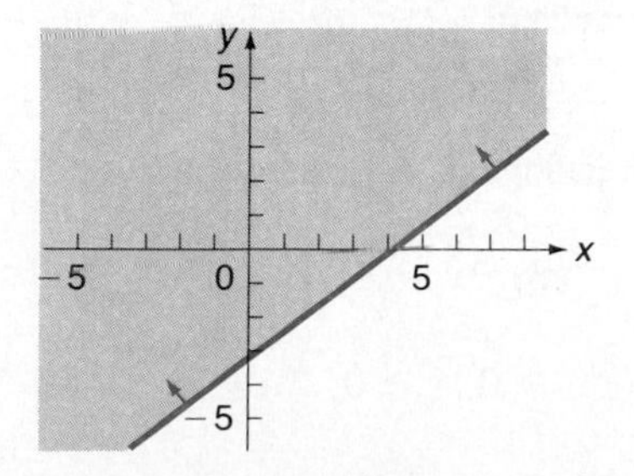

Problem 12 Graph $2x + 2y \le 5$.

Example 13 Graph in a rectangular coordinate system:

(A) $y > -3$ **(B)** $2x < 5$ **(C)** $-2 \le x \le 4$

Solution **(A)**

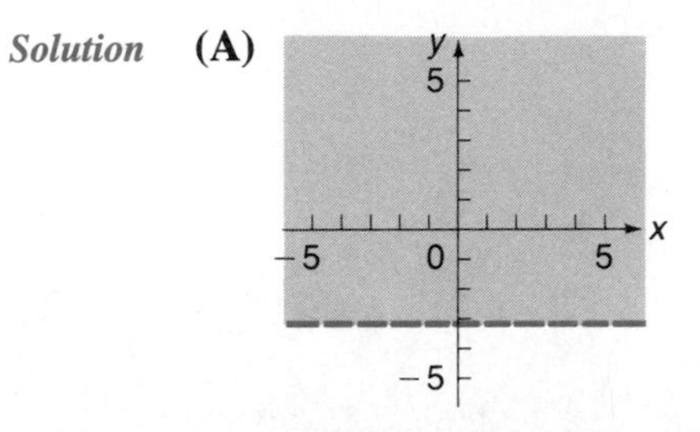

(B)

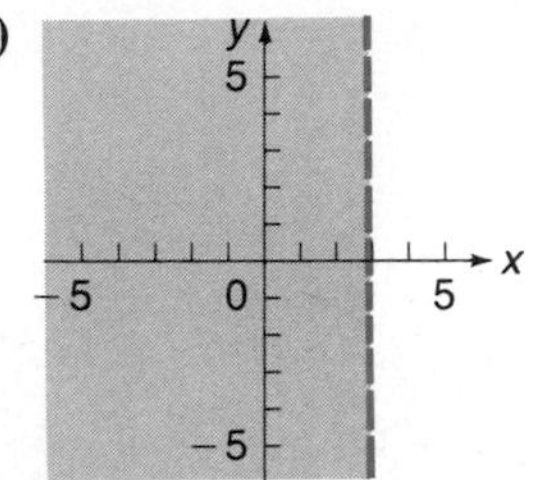

In both cases, we have graphed the boundary line and determined the correct half-plane using the origin as a test point.

(C) The solution to $-2 \le x$ is the half-plane to the right of $x = -2$. The solution to $x \le 4$ is the half-plane to the left of $x = 4$. Our solution is the region common to both half-planes:

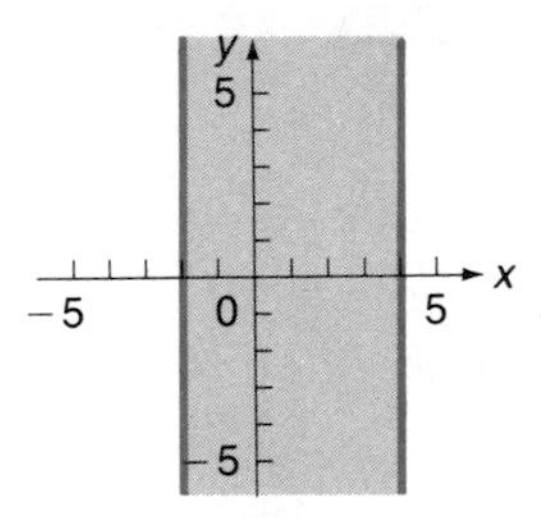

Remark If we are to graph an inequality involving only one variable, it is important to know the context in which the graph is to occur. In Example 13 the graphs are to occur in a rectangular coordinate system; hence, in each case the presence of a second variable is assumed, but with a 0 coefficient. For example, $y > -3$ is assumed to mean $0x + y > -3$. Thus, x can take on any value as long as y is greater than -3. The graph is the upper half-plane above the line $y = -3$. If we were to graph $y > -3$ on a single number line, however, then we would obtain

Problem 13 Graph in a rectangular coordinate system:

(A) $y < 2$ **(B)** $3x > -8$ **(C)** $1 \le y \le 5$

Example 14 Graph $\{(x, y) \mid x \ge 0,\ y \ge 0,\ 2x + 3y \le 18\}$.

Solution This is the set of all ordered pairs of real numbers (x, y) such that x and y are both nonnegative and satisfy $2x + 3y \leq 18$. We graph each inequality in the same coordinate system and take the intersection (see Section 6-5) of the graphs of these solution sets.

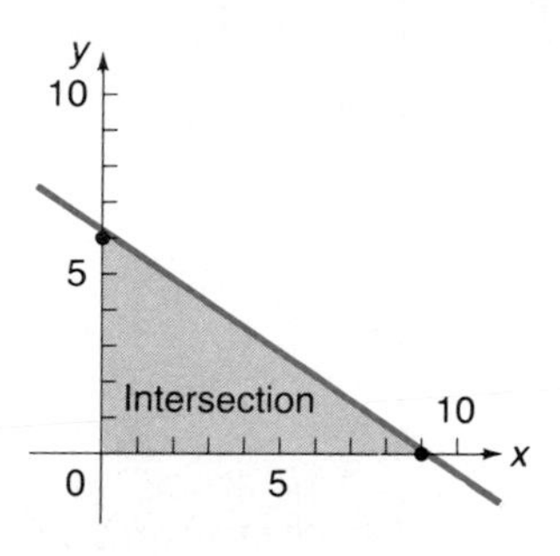

Problem 14 Graph $\{(x, y) \mid 1 \leq x \leq 4, \ -2 \leq y \leq 2\}$.

ANSWERS TO MATCHED PROBLEMS

12. $2x + 2y \leq 5$

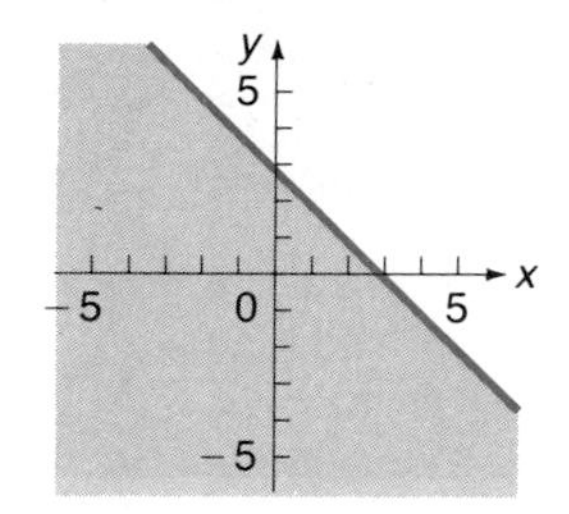

13. (A) $y < 2$

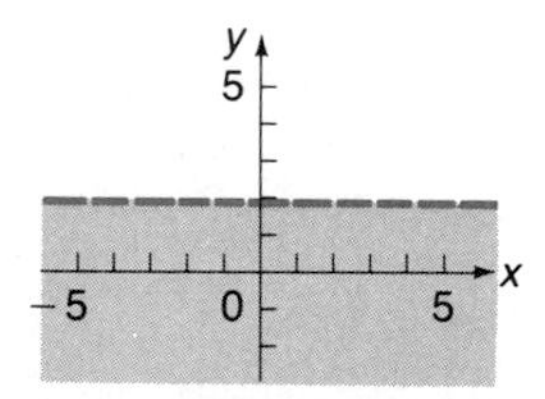

(B) $3x > -8$

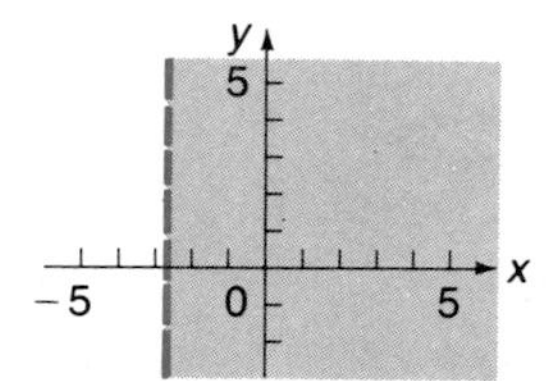

(C) $1 \leq y \leq 5$

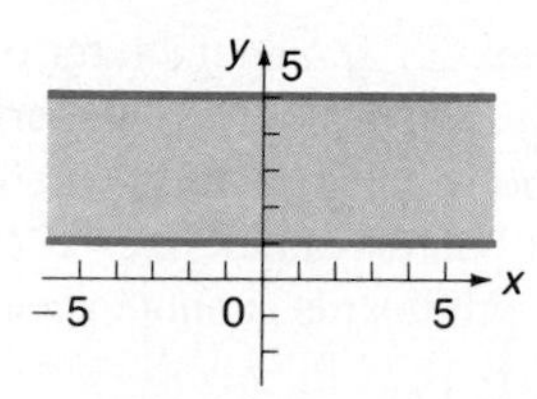

14. $\{(x, y) \mid 1 \le x \le 4, \ -2 \le y \le 2\}$

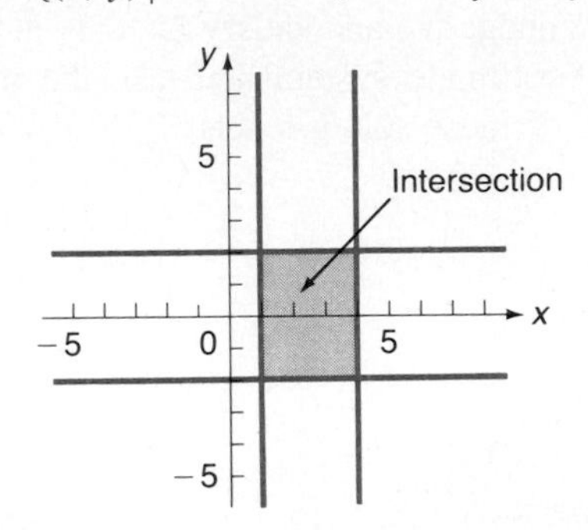

EXERCISE 7-3 *In Problems 1–22, graph each inequality in a rectangular coordinate system.*

A

1. $x + y \le 6$ **2.** $x + y \ge 4$ **3.** $x - y > 3$

4. $x - y < 5$ **5.** $y \ge x - 2$ **6.** $y \le x + 1$

B

7. $2x - 3y < 6$ **8.** $3x + 4y < 12$ **9.** $3y - 2x \ge 24$

10. $3x + 2y \ge 18$ **11.** $y \ge \frac{x}{3} - 2$ **12.** $y \le \frac{x}{2} - 4$

13. $y \le \frac{2}{3}x + 5$ **14.** $y > \frac{x}{3} + 2$ **15.** $x \ge -5$

16. $y \le 8$ **17.** $y < 0$ **18.** $x \ge 0$

19. $-1 < x \le 3$ **20.** $-3 \le y < 2$ **21.** $-2 \le y \le 2$

22. $-1 \le x \le 4$

C *Graph each set.*

23. $\{(x, y) \mid -3 \le x \le 3 \text{ and } -1 \le y \le 5\}$

24. $\{(x, y) \mid -1 \le x \le 5 \text{ and } -2 \le y \le 2\}$

25. $\{(x, y) \mid x \ge 0, \ y \ge 0, \text{ and } 3x + 4y \le 12\}$

26. $\{(x, y) \mid x \ge 0, \ y \ge 0, \text{ and } 3x + 2y \le 18\}$

APPLICATIONS

27. ***Business*** A manufacturer of sailboards makes a standard model and a competition model. The pertinent manufacturing data are summarized in the following table, where the total number of fabricating hours for both boards cannot exceed 120 and the total number of finishing hours for both boards cannot exceed 30. If x is the number of standard models and y is the number of competition models produced per week, write a system of inequalities that indicates the restrictions on x and y. Graph this system showing the region of permissible values for x and y.

	STANDARD MODEL (WORKHOURS PER BOARD)	COMPETITION MODEL (WORKHOURS PER BOARD)	MAXIMUM WORKHOURS AVAILABLE PER WEEK
FABRICATING	6	8	120
FINISHING	1	3	30

7-4 GRAPHING QUADRATIC POLYNOMIALS

- Graphing $y = x^2$ and Related Forms
- Graphing $y = ax^2 + bx + c$

We have seen that the graph of the linear equation $y = mx + b$, with y equal to a linear expression in x, is a straight line. In this section, we will consider the equation $y = ax^2 + bx + c$, with y equal to a quadratic expression in x. The resulting graph is called a *parabola*.

GRAPHING $y = x^2$ AND RELATED FORMS

The equation $y = x^2$ can be graphed by plotting a sufficient number of points and joining them with a smooth curve (Figure 9):

x	y
−4	16
−3	9
−2	4
−1	1
0	0
1	1
2	4
3	9
4	16

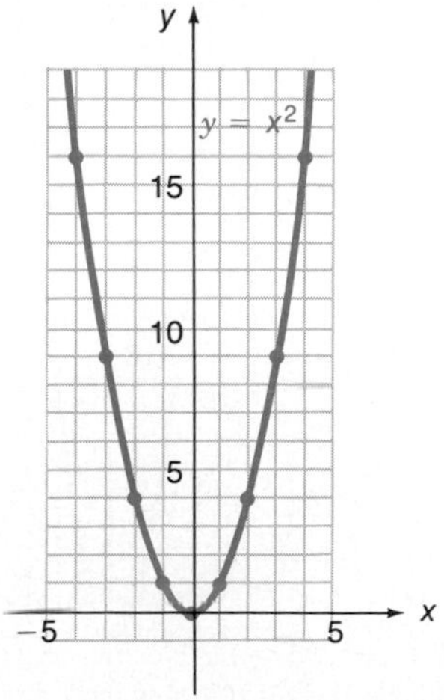

FIGURE 9

The related form $y = -x^2$ changes the sign of each y value in the preceding graph. The effect is to turn the graph upside down by reflecting it about the x axis (Figure 10):

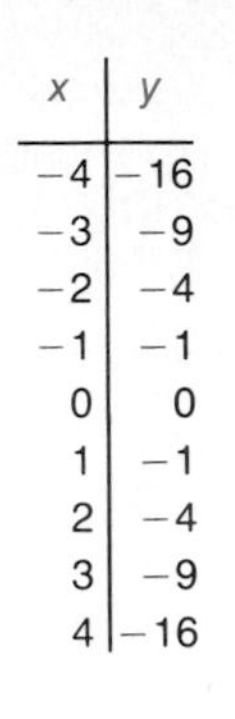

x	y
−4	−16
−3	−9
−2	−4
−1	−1
0	0
1	−1
2	−4
3	−9
4	−16

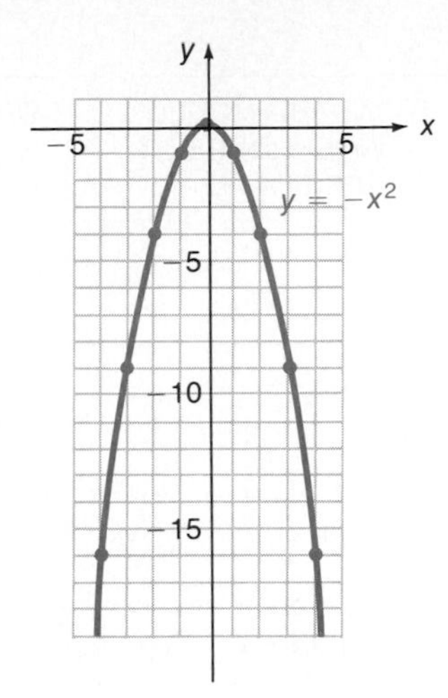

FIGURE 10

The related form $y = 2x^2$ doubles the y value of each point in the graph of $y = x^2$. The effect is to "stretch" the graph (Figure 11):

x	y
−4	32
−3	18
−2	8
−2	1
0	0
1	2
2	8
3	18
4	32

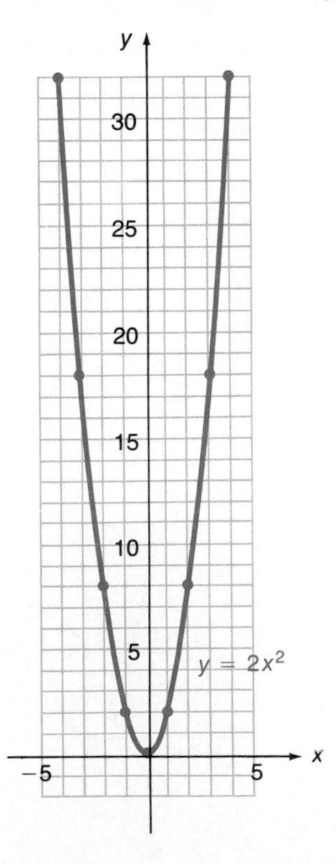

FIGURE 11

Multiplying by a number greater than 2 stretches it even more; multiplying by a number between 0 and 1 "flattens" the graph. Multiplying by a negative number a turns the graph over and either stretches or flattens it depending on $|a|$ (Figure 12):

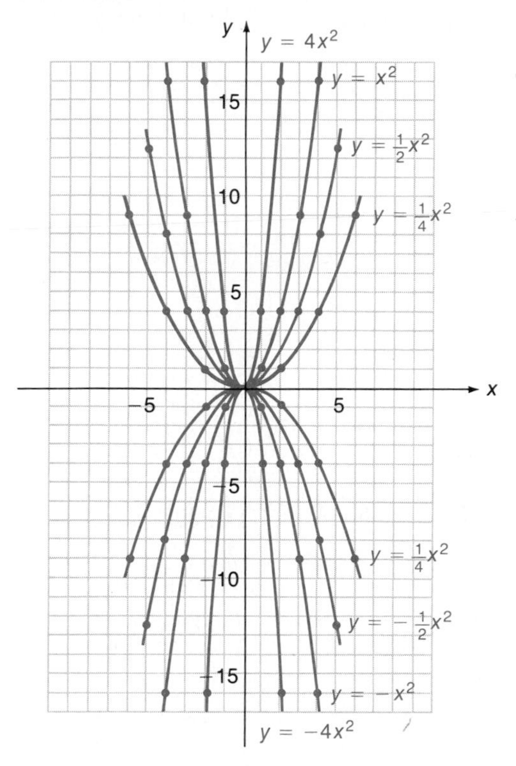

FIGURE 12

The related form $y = x^2 + 2$ has the same graph as $y = x^2$ but raised 2 units (Figure 13).

x	y
−4	18
−3	11
−2	6
−1	3
0	2
1	3
2	6
3	11
4	18

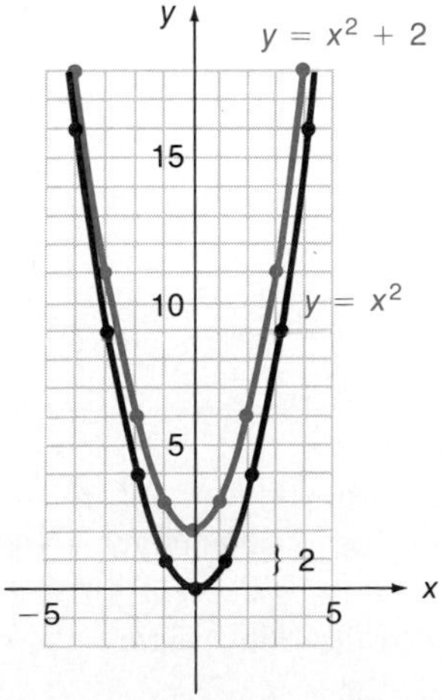

FIGURE 13

Example 15 Graph $y = -2x^2 + 3$.

Solution Begin with the graph of $y = x^2$. The graph of $y = 2x^2$ is this graph stretched, as we saw above, and shown in Figure 14a.

The graph of $y = -2x^2$ is this last graph turned over the x axis, as shown in Figure 14b.

Finally, the graph of $y = -2x^2 + 3$ is the preceding graph raised 3 units, as shown in Figure 14c.

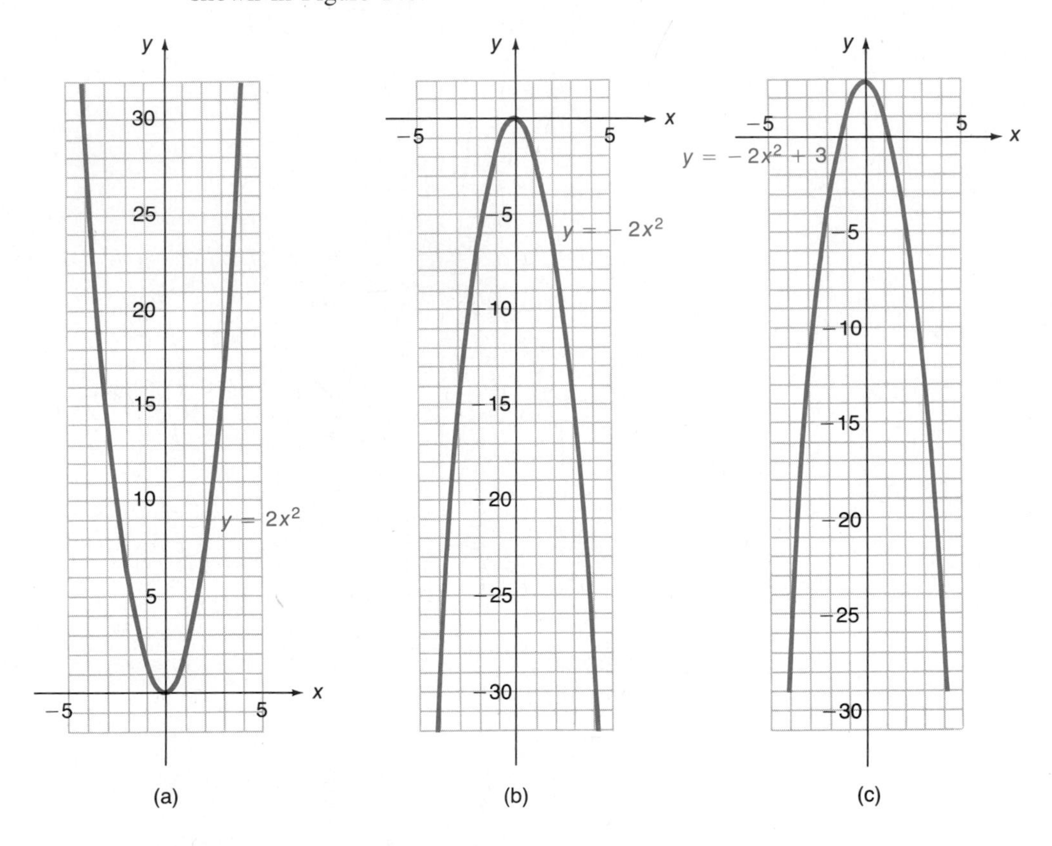

FIGURE 14

Problem 15 Graph $y = 3x^2 - 2$.

In the form $y = (x - 2)^2$, *the same* y values occur as in $y = x^2$, but they occur for x values 2 units larger. [Note in $y = (x - 2)^2$, $y = 0$ when $x = 2$, whereas in $y = x^2$, $y = 0$ when $x = 0$.] The effect is to shift the graph of $y = x^2$ two units to the right (Figure 15, on page 321):

x	y
−2	16
−1	9
0	4
1	1
2	0
3	1
4	4
5	9
6	16

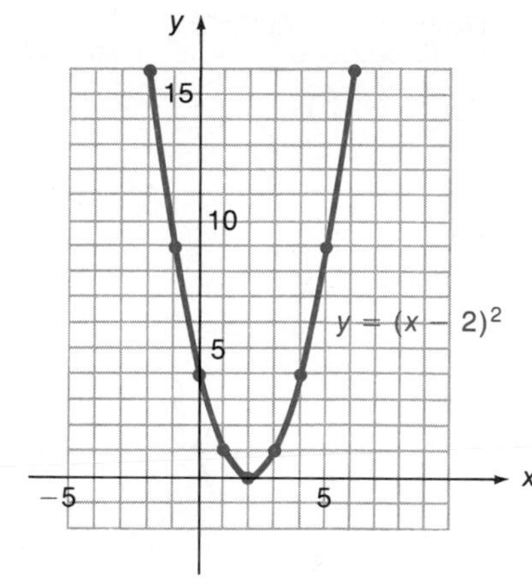

FIGURE 15

We have seen that knowing the graph of $y = x^2$ leads quickly to graphs of several related forms. In summary, we have the following:

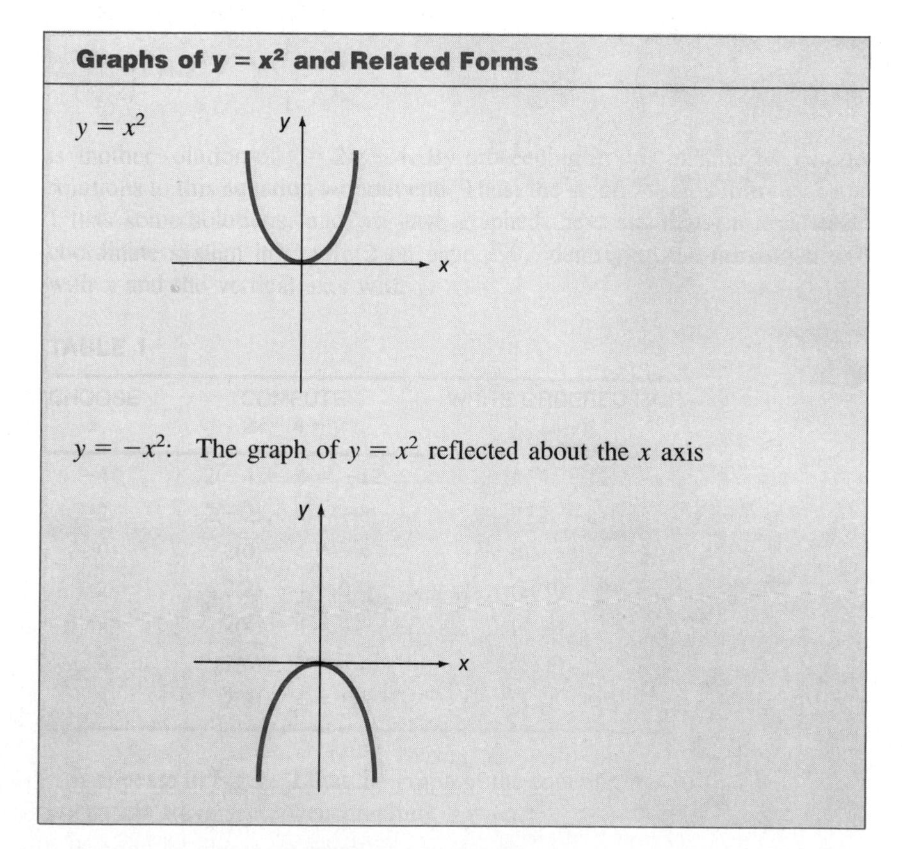

Graphs of $y = x^2$ and Related Forms *(Continued)*

$y = ax^2$, $a > 0$: The graph of $y = x^2$ stretched ($a > 1$) or flattened ($a < 1$)

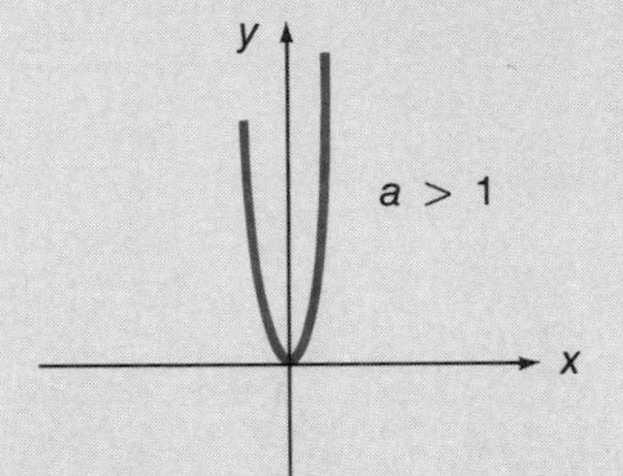

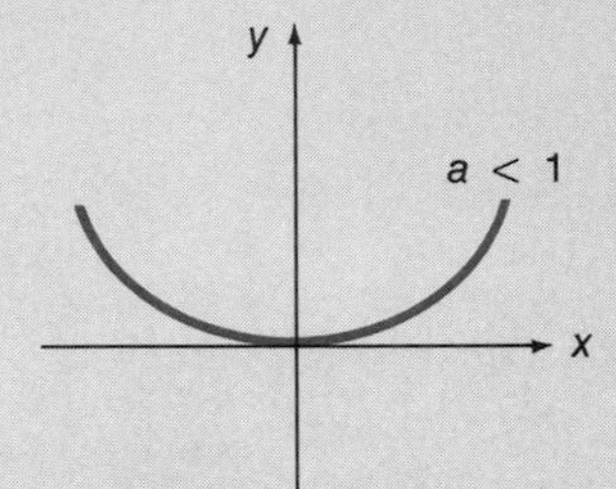

$y = x^2 + d$: The graph of $y = x^2$ raised d units if $d > 0$ or lowered $|d|$ units if $d < 0$

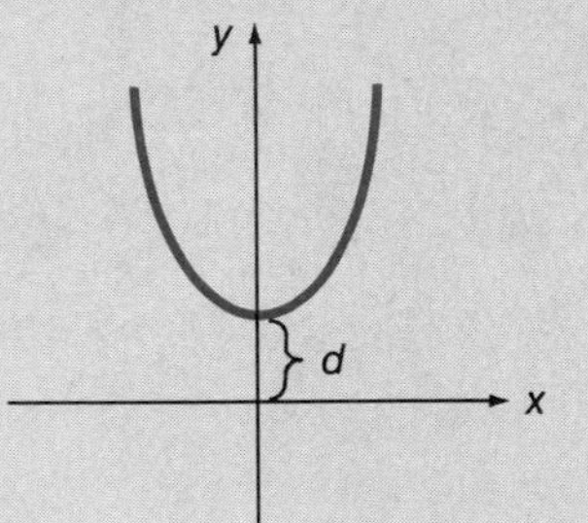

$y = (x - h)^2$: The graph of $y = x^2$ shifted h units to the right if $h > 0$ or $|h|$ units to the left if $h < 0$

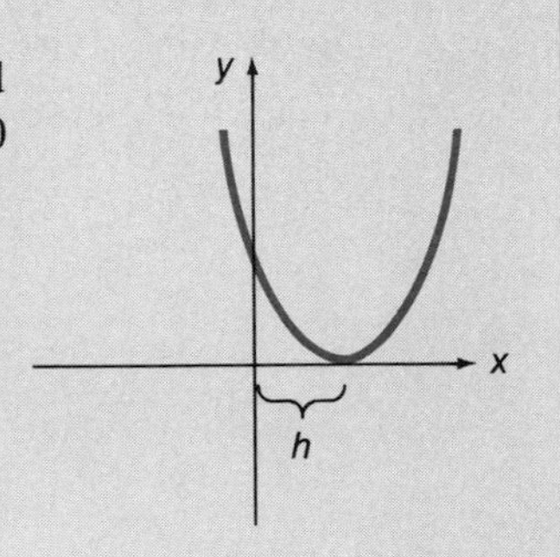

The related forms may occur in combination, such as $y = a(x - h)^2 + d$. We can see the effect of these several changes in sequence:

1. Start with the graph of $y = x^2$.
2. Stretch or flatten the graph using a; if a is negative also, turn the graph upside down.
3. Shift the graph right or left $|h|$ units. If h is negative, say -2 as in $(x + 2)^2 = [x - (-2)]^2$, this means move left $|h| = 2$ units.
4. Raise or lower the graph $|d|$ units. If d is negative, say -2, this means lower it $|d| = 2$ units.

Example 16 Graph $y = \frac{1}{2}(x - 2)^2 + 3$.

Solution The graph is that of $y = x^2$ except that it is flattened (using $a = \frac{1}{2}$), shifted right 2 units ($h = 2$), and raised 3 units ($d = 3$). Plotting very few points will now locate the graph. We choose points near 2 since $y = x^2$ is shifted right 2 units:

x	y
0	5
2	3
4	5

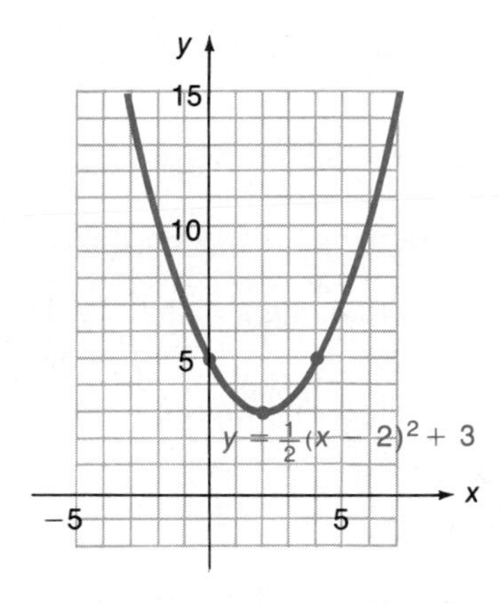

Problem 16 Graph $y = 2(x + 3)^2 - 1$.

GRAPHING $y = ax^2 + bx + c$

To graph

$$y = ax^2 + bx + c$$

we will rewrite the equation in the form

$$y = a(x - h)^2 + d$$

To do this we will use the technique of completing the square introduced in Section 6-1. We first rewrite.

$$ax^2 + bx = a\left(x^2 + \frac{b}{a}x\right) \qquad \text{Factor out } a.$$

$$= a\left(x^2 + \frac{b}{a}x + \frac{b^2}{4a^2} - \frac{b^2}{4a^2}\right) \qquad \text{Add the square of one-half the coefficient of } x\text{. To preserve the equality, we also subtract it.}$$

$$= a\left(x^2 + \frac{b}{a}x + \frac{b^2}{4a^2}\right) - \frac{b^2}{4a} \qquad \text{Regroup and simplify.}$$

$$= a\left(x + \frac{b}{2a}\right)^2 - \frac{b^2}{4a}$$

Therefore,

$$\begin{aligned} y &= ax^2 + bx + c \\ &= a\left(x + \frac{b}{2a}\right)^2 - \frac{b^2}{4a} + c \\ &= a\left(x + \frac{b}{2a}\right)^2 - \frac{b^2 - 4ac}{4a} \end{aligned}$$

Thus, the graph of $y = ax^2 + bx + c$ is the graph of $y = x^2$ changed thus:

1. Stretched or flattened using a; turned over the x axis if $a < 0$
2. Shifted right or left by $\left|-\frac{b}{2a}\right|$ units $\left[\text{since } x + \frac{b}{2a} = x - \left(-\frac{b}{2a}\right)\right]$
3. Raised or lowered $\left|-\frac{b^2 - 4ac}{4a}\right|$ units

Example 17 Graph $y = 3x^2 - 2x - 1$.

Solution The graph is that of $y = x^2$ except that it is stretched since $a = 3$; shifted right by $\frac{1}{3}$ unit, since

$$\left|-\frac{b}{2a}\right| = \frac{2}{6} = \frac{1}{3}$$

and lowered by $|-\frac{4}{3}| = \frac{4}{3}$ units, since

$$-\frac{b^2 - 4ac}{4a} = -\frac{4 + 12}{12} = \frac{16}{12} = -\frac{4}{3}$$

The graph, therefore, is shaped like ∪ and the lowest point is $(\frac{1}{3}, -\frac{4}{3})$. A couple of points plotted will locate the graph:

x	y
0	−1
$\frac{1}{3}$	$-\frac{4}{3}$
2	7

Problem 17 Graph $y = -2x^2 + 3x + 1$.

The numerator in the fraction

$$-\frac{b^2 - 4ac}{4a}$$

was previously called the discriminant (see Section 6-2) and used to determine the number of roots to the equation

$$ax^2 + bx + c = 0$$

Such roots are x values for which $y = 0$ in

$$y = ax^2 + bx + c$$

that is, x values where the graph of $y = ax^2 + bx + c$ crosses the x axis. The discriminant $b^2 - 4ac$, the fraction $-\frac{b^2 - 4ac}{4a}$, and the number of roots are closely connected.

If $b^2 - 4ac = 0$, the graph of $y = x^2$ is not raised, so it meets the x axis at a unique point, corresponding to the one root.

Suppose $b^2 - 4ac > 0$. If a is positive, then $-\frac{b^2 - 4ac}{4a}$ is negative, so the graph of $y = x^2$ is lowered and there are two roots:

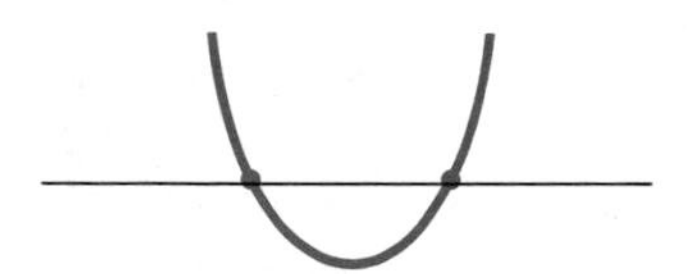

If a is negative, then $-\frac{b^2 - 4ac}{4a}$ is positive and the graph of $y = x^2$ is turned over and raised. Again there are two roots.

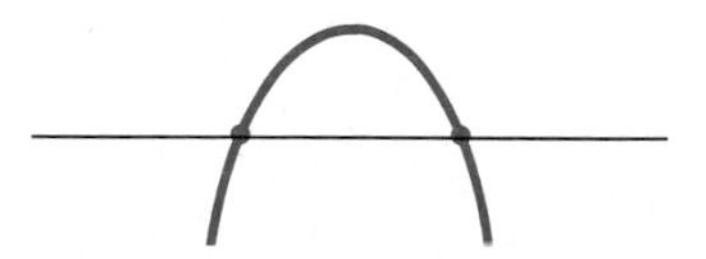

Suppose $b^2 - 4ac < 0$. If a is positive, then $-\frac{b^2 - 4ac}{4a}$ is positive and the graph of $y = x^2$ is raised so it does not cross the x axis. There are no roots:

If a is negative, $-\dfrac{b^2 - 4ac}{4a}$ is negative, so the graph of $y = x^2$ is turned over and lowered. Again there are no roots:

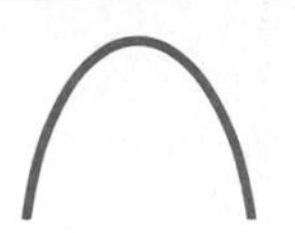

Example 18 Graph $y = x^2 - x - 6$. If the graph crosses the x axis, find the x values.

Solution The graph is that of $y = x^2$ except that it is shifted right by $\frac{1}{2}$ unit, since

$$-\frac{b}{2a} = \frac{1}{2}$$

and lowered by $\left|-\frac{25}{4}\right| = \frac{25}{4}$ units, since

$$-\frac{b^2 - 4ac}{4a} = -\frac{1 + 24}{4} = -\frac{25}{4}$$

Thus the graph is shaped like ◡ and the lowest point has coordinates $(\frac{1}{2}, -\frac{25}{4})$. We plot a few points to locate it:

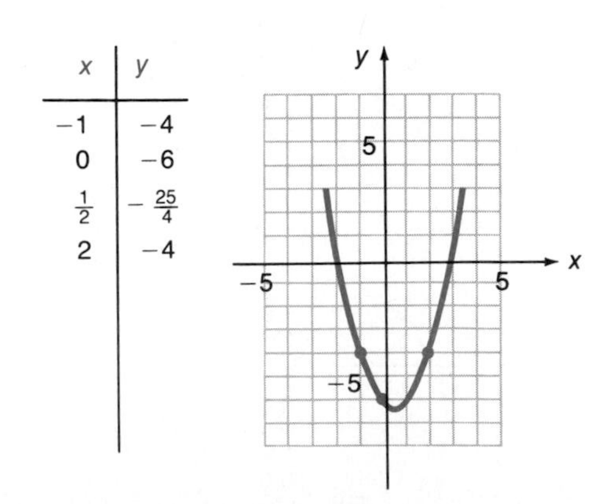

x	y
-1	-4
0	-6
$\frac{1}{2}$	$-\frac{25}{4}$
2	-4

The graph crosses the x axis twice. The values are the solutions to $y = x^2 - x + 6 = 0$. This quadratic equation can be solved by factoring or by the quadratic formula:

$$x = 3 \qquad \text{or} \qquad x = -2$$

Problem 18 Graph $y = -x^2 + 4x + 5$. If the graph crosses the x axis, find the x values.

ANSWERS TO MATCHED PROBLEMS

15.

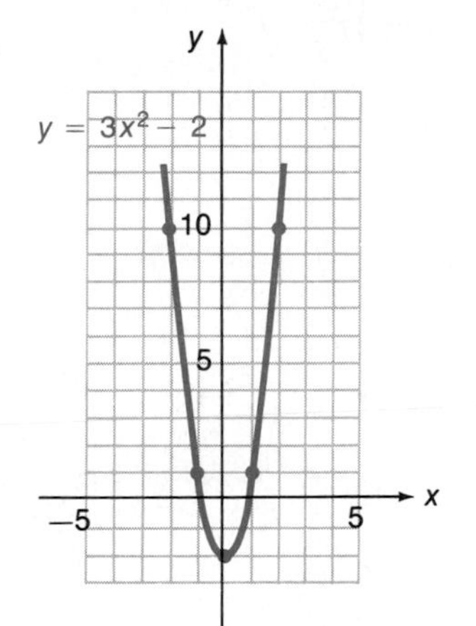

16.

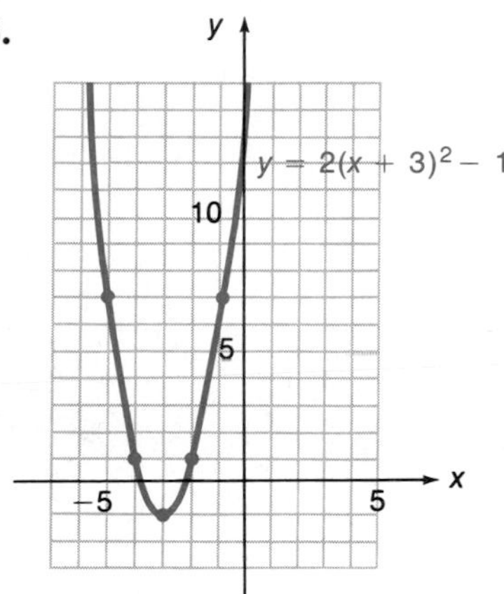

17.

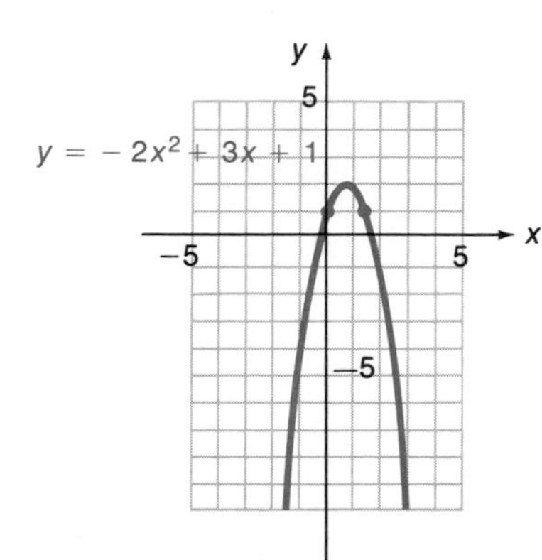

18.

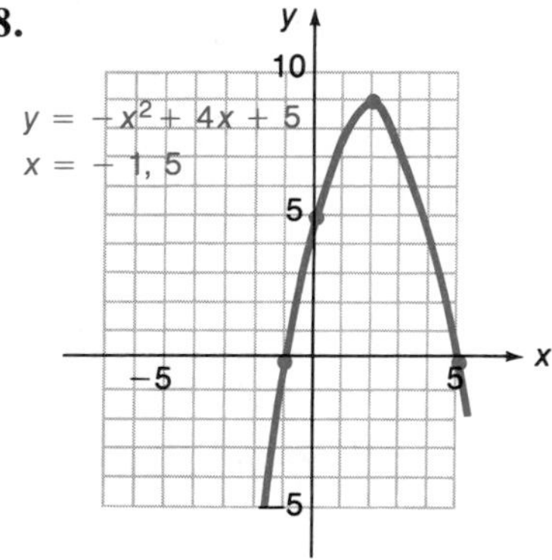

EXERCISE 7-4

A *In Problems 1–30, graph the equations.*

1. $y = 3x^2$ **2.** $y = \frac{1}{3}x^2$ **3.** $y = -\frac{1}{2}x^2$

4. $y = -2x^2$ **5.** $y = x^2 + 4$ **6.** $y = -x^2 + 2$

7. $y = -x^2 - 1$ **8.** $y = x^2 - 3$ **9.** $y = 2x^2 + 1$

10. $y = -3x^2 + 4$ **11.** $y = -\frac{1}{3}x^2 - 2$ **12.** $y = \frac{1}{2}x^2 - 1$

B **13.** $y = (x - 3)^2$ **14.** $y = (x - 1)^2$

15. $y = (x + 2)^2$ **16.** $y = (x + 3)^2$

17. $y = (x - 2)^2 + 1$ **18.** $y = (x + 1)^2 - 3$

19. $y = (x + 2)^2 - 4$ **20.** $y = (x - 3)^2 + 2$

21. $y = -(x + 4)^2 + 2$ **22.** $y = -(x + 3)^2 - 2$

23. $y = -(x - 1)^2 - 1$

24. $y = -(x - 4)^2 + 1$

25. $y = 2(x - 1)^2 + 1$

26. $y = 3(x + 1)^2 - 2$

27. $y = \frac{1}{2}(x + 1)^2 + 3$

28. $y = -\frac{1}{3}(x + 2)^2 - 2$

29. $y = -2(x + 2)^2 - 2$

30. $y = 3(x - 1)^2 + 2$

C *Graph. If the graph crosses the x axis, find the x value(s).*

31. $y = x^2 - 2x - 3$

32. $y = x^2 + 6x + 8$

33. $y = -x^2 + x + 2$

34. $y = -x^2 - 2x + 3$

35. $y = x^2 + x - 6$

36. $y = x^2 - 6x + 5$

37. $y = x^2 + 4x + 4$

38. $y = -x^2 + 6x - 9$

39. $y = x^2 - 3x + 3$

40. $y = x^2 + 4x + 5$

41. $y = -x^2 + 2x - 8$

42. $y = -x^2 - x + 12$

43. $y = -x^2 + 5x - 6$

44. $y = -x^2 + x - 1$

45. $y = 2x^2 - 7x + 5$

46. $y = 3x^2 + 11x - 4$

47. $y = -2x^2 - 3x + 2$

48. $y = -2x^2 - 5x + 12$

7-5 CONIC SECTIONS; CIRCLES AND PARABOLAS

- Distance-between-Two-Points Formula
- Circles
- Parabolas

In the first part of this chapter we discussed equations of a straight line, that is, equations such as

$$2x - 3y = 5 \qquad \text{and} \qquad y = -\tfrac{1}{3}x + 2$$

These are first-degree equations in two variables. If we increase the degree of the equations by 1, what kind of graphs will we get? That is, what kind of graphs will second-degree equations such as

$$y = x^2 \qquad x^2 + y^2 = 25 \qquad \frac{x^2}{4} - \frac{y^2}{16} = 1 \qquad x^2 + 4y^2 - 3x + 7y = 4$$

produce? The first of these, $y = x^2$, has as its graph a parabola, as we saw in Section 7-4. It can be shown that all the graphs will be one of the plane curves you would get by intersecting a plane and a general cone—thus the name **conic sections**. Some typical curves are shown in Figure 16, on page 329. The principal conic sections are circles, parabolas, ellipses, and hyperbolas.

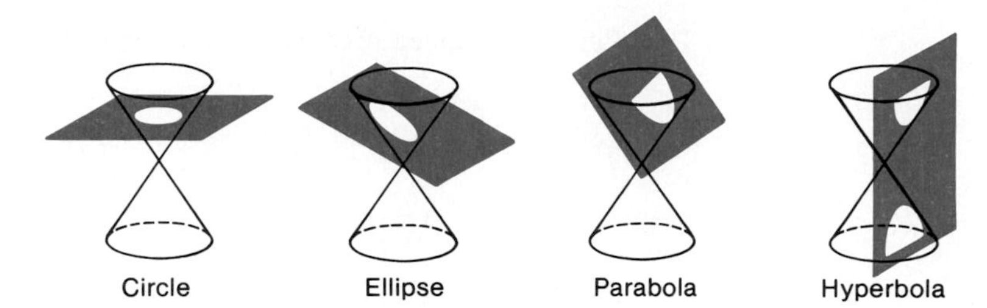

FIGURE 16 Conic sections

In this book we will consider a few interesting and useful second-degree equations in two variables. The subject is treated more thoroughly in a course in analytic geometry.

DISTANCE-BETWEEN-TWO-POINTS FORMULA

A basic tool in determining equations of conics is the distance-between-two-points formula. The derivation of the formula makes direct use of the Pythagorean theorem (see page 268). Let P_1 and P_2 have coordinates as indicated in Figure 17.

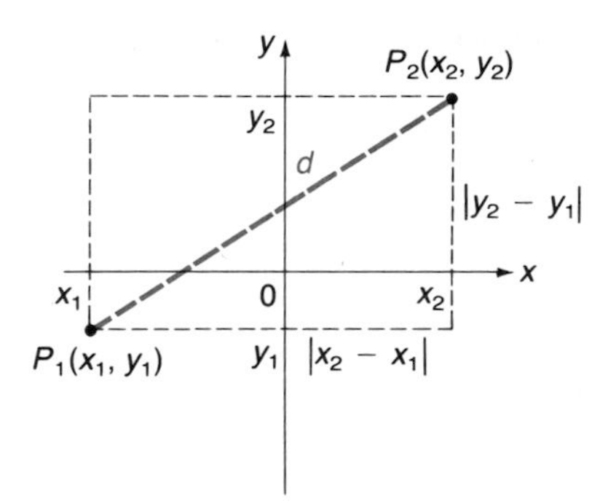

FIGURE 17

Using the Pythagorean theorem, we can write

$$d^2 = |x_2 - x_1|^2 + |y_2 - y_1|^2$$

or, since $|N|^2 = N^2$ and $d > 0$, the equation can be written in the following form:

Distance-between-Two-Points Formula

$$d = \sqrt{(x_2 - x_1)^2 + (y_2 - y_1)^2}$$

Note that it doesn't make any difference which point you call P_1 or P_2, since $(a - b)^2 = (b - a)^2$.

Example 19 Find the distance between (−3, 6) and (4, −2).

Solution Let $P_1 = (4, -2)$ and $P_2 = (-3, 6)$; then

$$\begin{aligned} d &= \sqrt{(x_2 - x_1)^2 + (y_2 - y_1)^2} \\ &= \sqrt{[(-3) - 4]^2 + [6 - (-2)]^2} \\ &= \sqrt{(-7)^2 + (8)^2} \\ &= \sqrt{113} \end{aligned}$$

Or let $P_1 = (-3, 6)$ and $P_2(4, -2)$; then

$$\begin{aligned} d &= \sqrt{[4 - (-3)]^2 + [(-2) - 6]^2} \\ &= \sqrt{(7)^2 + (-8)^2} \\ &= \sqrt{113} \end{aligned}$$

Problem 19 Find the distance between (4, −2) and (3, 1).

CIRCLES

We will start with the definition of a circle and then use the distance formula to find the standard equation of a circle in a plane.

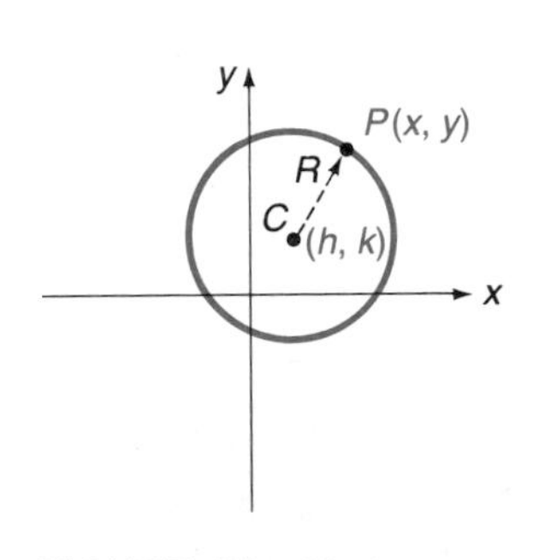

FIGURE 18 Circle

> **Definition of a Circle**
>
> A **circle** is the set of points equidistant from a fixed point. The fixed distance is called the **radius**, and the fixed point is called the **center**.

Let a circle have radius R and center at (h, k). Referring to Figure 18, we see that an arbitrary point $P(x, y)$ is on the circle if and only if

$$R = \sqrt{(x - h)^2 + (y - k)^2} \quad \text{Distance from } C \text{ to } P \text{ is constant}$$

or, equivalently,

$$(x - h)^2 + (y - k)^2 = R^2$$

Thus, we can state the following:

Equations of a Circle

1. Radius R and center (h, k):

$$(x - h)^2 + (y - k)^2 = R^2$$

2. Radius R and center at the origin:

$$x^2 + y^2 = R^2 \qquad \text{since } (h, k) = (0, 0)$$

Example 20 Graph: **(A)** $x^2 + y^2 = 25$ **(B)** $(x - 3)^2 + (y + 2)^2 = 9$

Solution **(A)** This is a circle with radius $\sqrt{25} = 5$ and center at the origin:

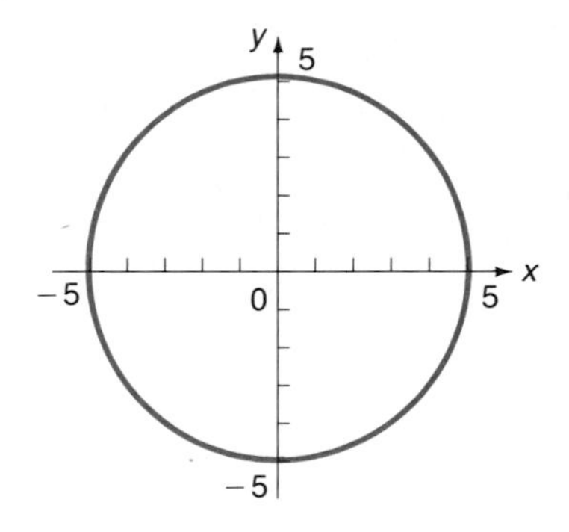

(B) We need the equation rewritten in the form $(x - h)^2 + (y - k)^2 = R^2$:

$$(x - 3)^2 + (y + 2)^2 = 9$$

is the same as

$$(x - 3)^2 + [y - (-2)]^2 = 3^2$$

which is the equation of a circle with radius 3 and center at $(h, k) = (3, -2)$. Thus:

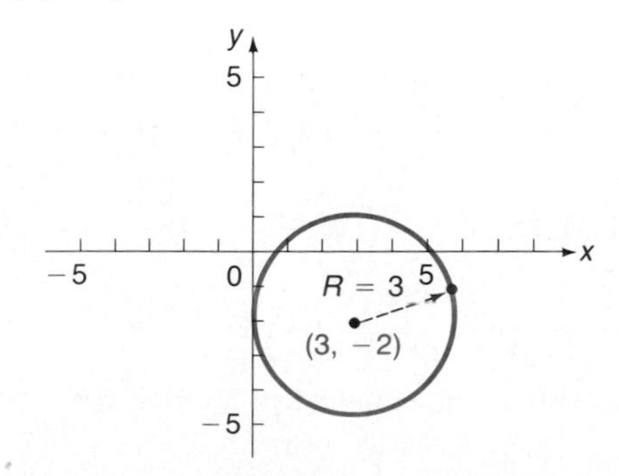

Problem 20 Graph: **(A)** $x^2 + y^2 = 4$ **(B)** $(x + 2)^2 + (y - 3)^2 = 16$

Example 21 What is the equation of a circle with radius 8 and center at the origin? Center at $(-5, 4)$?

Solution If the center is at the origin, then $(h, k) = (0, 0)$; thus, the equation is

$$x^2 + y^2 = 8^2 \quad \text{or} \quad x^2 + y^2 = 64$$

If the center is at $(-5, 4)$, then $h = -5$ and $k = 4$, and the equation is

$$[x - (-5)]^2 + (y - 4)^2 = 8^2$$

or

$$(x + 5)^2 + (y - 4)^2 = 64$$

Problem 21 What is the equation of a circle with radius $\sqrt{7}$ and center at the origin? Center at $(6, -4)$?

Example 22 Graph $x^2 + y^2 - 6x + 8y + 9 = 0$.

Solution We transform the equation into the form

$$(x - h)^2 + (y - k)^2 = R^2$$

by completing the square relative to x and relative to y (see Section 6-1).

$$\begin{aligned} x^2 - 6x + ? + y^2 + 8y + ? &= -9 + ? + ? \\ x^2 - 6x + \mathbf{9} + y^2 + 8y + \mathbf{16} &= -9 + \mathbf{9} + \mathbf{16} \\ (x - 3)^2 + (y + 4)^2 &= 16 \end{aligned}$$

or

$$(x - 3)^2 + [y - (-4)]^2 = 16 \quad \text{Thus, } (h, k) = (3, -4).$$

This is the equation of a circle with radius 4 and center at $(h, k) = (3, -4)$.

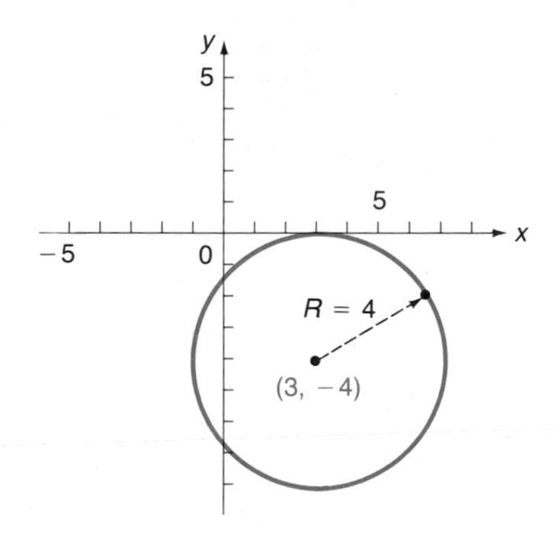

Problem 22 Graph $x^2 + y^2 + 10x - 4y + 20 = 0$.

PARABOLAS

In Section 7-4, the graph of the equation $y = ax^2 + bx + c$ was identified as a parabola. A parabola is defined geometrically as follows:

> **Defintion of a Parabola**
>
> A **parabola** is the set of all points equidistant from a fixed point and a fixed line. The fixed point is called the **focus**, and the fixed line is the **directrix**.

We will begin by finding an equation for the parabola with focus $(0, a)$, $a > 0$, and directrix $y = -a$ (see Figure 19).

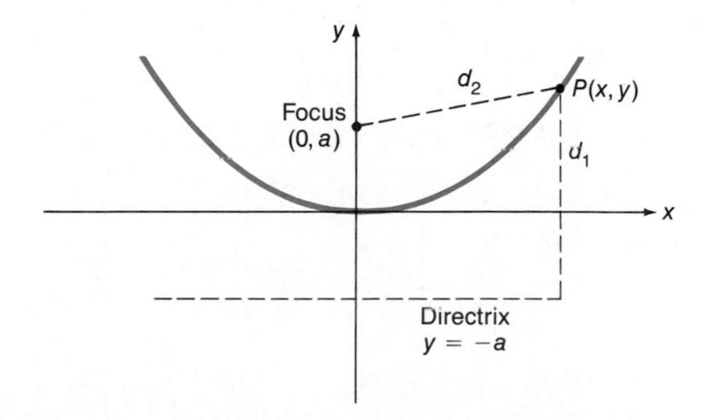

FIGURE 19

$$\begin{aligned}
d_1 &= d_2 \\
y + a &= \sqrt{(x - 0)^2 + (y - a)^2} \\
(y + a)^2 &= (x - 0)^2 + (y - a)^2 \\
y^2 + 2ay + a^2 &= x^2 + y^2 - 2ay + a^2 \\
4ay &= x^2 \\
y &= \frac{1}{4a} \cdot x^2 \\
&= kx^2 \qquad \text{where } k = \frac{1}{4a}
\end{aligned}$$

Thus $y = x^2$ and its related forms studied in Section 7-4 are, in fact, parabolas.

Proceeding in the same way, we can obtain similar equations for parabolas opening downward, to the right, and to the left. We summarize the four cases as follows:

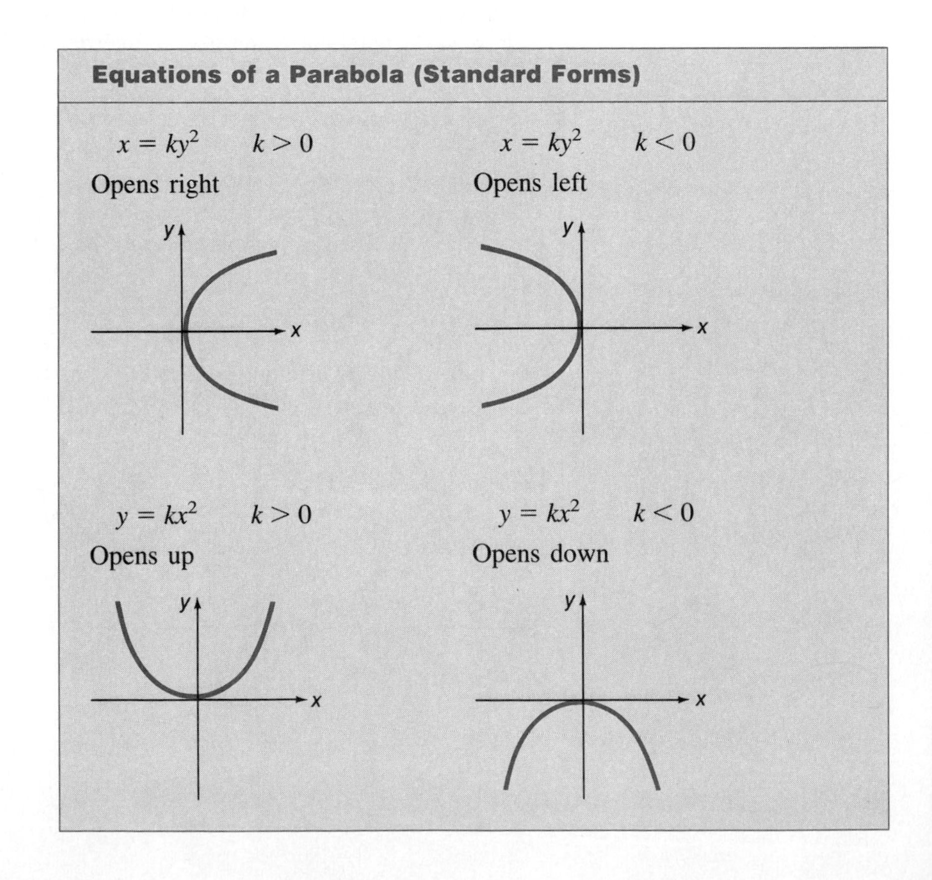

A line passing through the focus which is perpendicular to the directrix is called the **axis** of the parabola. The point on the axis midway between the focus and the directrix is called the **vertex**. In each of the four cases above, called **standard forms**, the vertex is the origin (0, 0). For $y = ax^2 + bx + c$, we found in Section 7-4 that the vertex is

$$\left(-\frac{b}{2a}, -\frac{b^2 - 4ac}{4a}\right)$$

Parabolas in standard form can be sketched rather quickly. We already know how to graph $y = kx^2$. The graph of $x = ky^2$ is a parabola opening left ($k < 0$) or right ($k > 0$) that can be sketched from a couple of points.

Example 23 Graph $x = -\frac{1}{8}y^2$.

Solution Since the x values must be negative, the parabola opens left. We look for values of y, in addition to $y = 0$, that make x easy to calculate and plot:

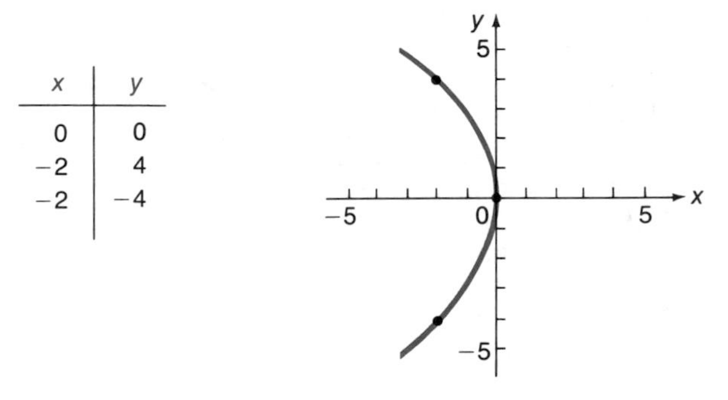

x	y
0	0
−2	4
−2	−4

Problem 23 Graph $x = \frac{3}{5}y^2$.

Parabolas are encountered frequently in the physical world. Suspension bridges, arch bridges, reflecting telescopes, radiotelescopes, radar equipment, solar furnaces, and searchlights all utilize parabolic forms in their design.

ANSWERS TO MATCHED PROBLEMS

19. $\sqrt{10}$

20. **(A)** **(B)**

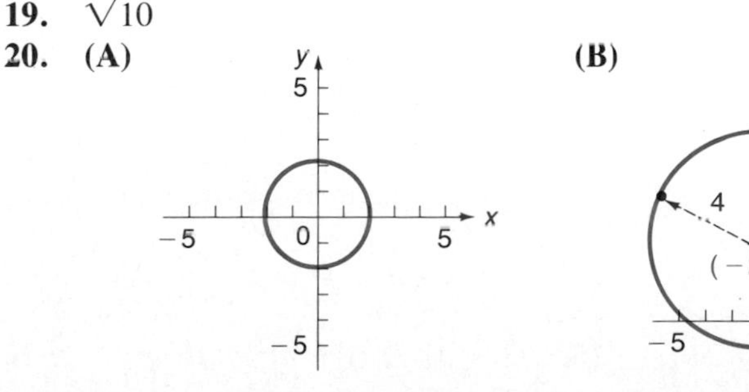

21. $x^2 + y^2 = 7$, $(x - 6)^2 + (y + 4)^2 = 7$

22. $(x + 5)^2 + (y - 2)^2 = 9$

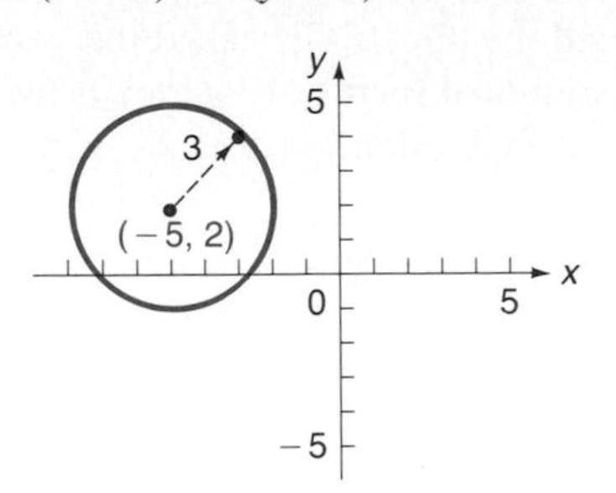

23.

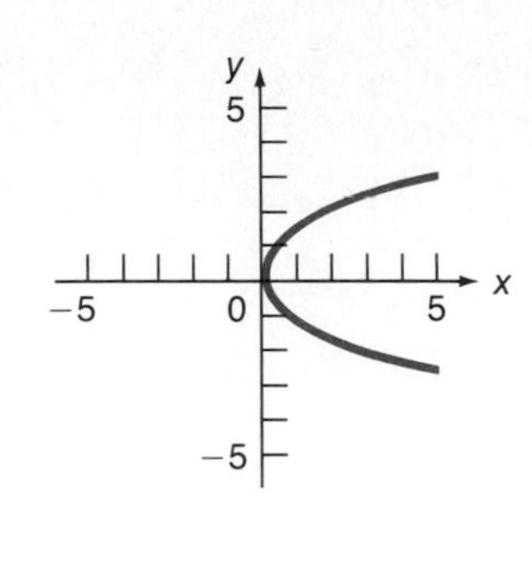

EXERCISE 7-5

A *Find the distance between each pair of points. Leave your answer in exact radical form.*

1. (2, 4), (4, 5) **2.** (7, 3), (8, 6)

3. (3, −7), (−2, 1) **4.** (−5, 3), (−1, −2)

5. (−8, −2), (−5,1) **6.** (2, −7), (−4, −3)

Group each equation in a cartesian coordinate system.

7. $x^2 + y^2 = 16$ **8.** $x^2 + y^2 = 36$

9. $x^2 + y^2 = 6$ **10.** $x^2 + y^2 = 10$

Write the equation of a circle with center at the origin and radius as given.

11. 7 **12.** 8 **13.** $\sqrt{5}$ **14.** $\sqrt{10}$

Graph each of the following parabolas:

15. $y^2 = 4x$ **16.** $y^2 = x$ **17.** $y^2 = -12x$

18. $y^2 = -16x$ **19.** $x^2 = y$ **20.** $x^2 = 12y$

21. $x^2 = -16y$ **22.** $x^2 = -4y$

B **23.** Is the triangle with vertices (−1, 2), (2, −1), (3, 3) an isosceles triangle?

24. Is the triangle in the preceding problem an equilateral triangle?

Graph each equation in a cartesian coordinate system.

25. $(x - 3)^2 + (y - 4)^2 = 16$ **26.** $(x - 4)^2 + (y - 2)^2 = 9$

27. $(x - 4)^2 + (y + 3)^2 = 9$

28. $(x + 4)^2 + (y - 2)^2 = 25$

29. $(x + 3)^2 + (y + 3)^2 = 16$

30. $(x + 2)^2 + (y + 4)^2 = 25$

Write the equation of a circle in the form

$$(x - h)^2 + (y - k)^2 = R^2$$

with radius and center as given.

31. 7, (3, 5)

32. 2, (4, 1)

33. 8, (−3, 3)

34. 6, (5, −2)

35. $\sqrt{3}$, (−4, −1)

36. $\sqrt{14}$, (−7, −5)

Graph the parabola after first rewriting the equation in standard form.

37. $4y^2 - 8x = 0$

38. $3x^2 + 9y = 0$

C *Use the method of completing the square to graph each of the following circles. Indicate the center and radius of each.*

39. $x^2 + y^2 - 4x - 6y + 4 = 0$

40. $x^2 + y^2 - 6x - 4y + 4 = 0$

41. $x^2 + y^2 - 6x + 6y + 2 = 0$

42. $x^2 + y^2 + 6x - 4y - 3 = 0$

43. $x^2 + y^2 + 6x + 4y + 4 = 0$

44. $x^2 + y^2 + 4x + 4y - 8 = 0$

45. Find the coordinates of the focus and the equation of the directrix for the parabola in Problem 15.

46. Find the coordinates of the focus and the equation of the directrix for the parabola in Problem 17.

47. Find x so that $(x, 8)$ is 13 units from (2, −4).

48. Find an equation of the set of points equidistant from (3, 3) and (6, 0).

49. Find the equation of the circle centered at the origin and passing through the point (4, −3).

50. Find the equation of the circle centered at (4, −3) and passing through the origin.

51. Use the definition of a parabola and the distance formula to find the equation of a parabola with directrix $y = 4$ and focus (2, 2).

52. Use the definition of a parabola and the distance formula to find the equation of a parabola with directrix $x = 2$ and focus at (6, 4).

APPLICATIONS

53. An ancient stone bridge in the form of a circular arc has a span of 80 feet (see the figure). If the height of the arch above its ends is 20 feet, find an equation of the circle containing the arch if its center is at the origin as indicated. [*Hint:* $(40, R - 20)$ must satisfy $x^2 + y^2 = R^2$.]

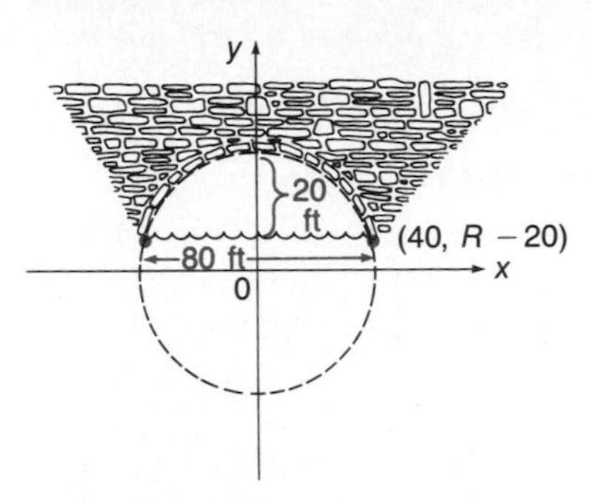

54. A cylindrical oil drum is cut to make a watering trough (see the figure). The end of the resulting trough is 20 centimeters high and 80 centimeters wide at the top. What was the radius R of the original drum?

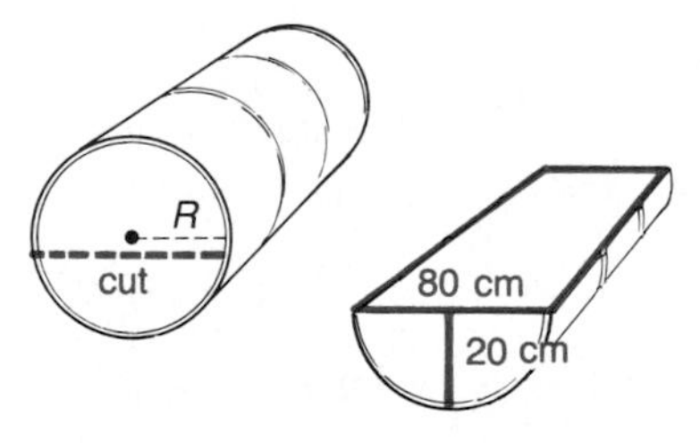

55. A parabolic concrete bridge is to span 100 meters. If the arch rises 25 meters above its ends, find the equation of the parabola, assuming it passes through the origin of a coordinate system and has its focus on the y axis. [*Hint:* $(50, -25)$ must satisfy $x^2 = -4ay$.]

56. A radar bowl, in the form of a rotated parabola, is 20 meters in diameter and 5 meters deep. Find the equation of the parabola, assuming it passes through the origin of a coordinate system and has its focus on the positive y axis. (See Problem 55.)

7-6 ELLIPSES AND HYPERBOLAS

- Ellipses
- Hyperbolas

This section provides a very brief glimpse at the other two conic sections. A brief exposure to these important curves at this time should help to increase your understanding of a more detailed development in a future course. In addition, you will have gained some concrete experience with graphs other than straight lines, circles, and parabolas.

ELLIPSES

You are no doubt aware of many uses or occurrences of elliptical forms; orbits of satellites, orbits of planets and comets, gears and cams, and domes in buildings are but a few examples. Formally, we define an ellipse as follows:

Definition of an Ellipse

An **ellipse** is the set of all points such that the sum of the distances of each to two fixed points is constant. The fixed points are called **foci**, and each separately is a **focus**.

An ellipse is easy to draw. Place two pins in a piece of cardboard (these become the foci) and tie a piece of loose string (representing the constant sum) to the pins; then move a pencil within the string, keeping it taut.

With regard to an equation of an ellipse, we will limit ourselves to the cases in which the foci are symmetrically located on either coordinate axis. Thus, if $(-c, 0)$ and $(c, 0)$, $c > 0$, are the foci and $2a$ is the constant sum of the distances (note from Figure 20 that $2a > 2c$, hence $a > c$), then

$$d_1 + d_2 = 2a$$

$$\sqrt{(x + c)^2 + y^2} + \sqrt{(x - c)^2 + y^2} = 2a$$

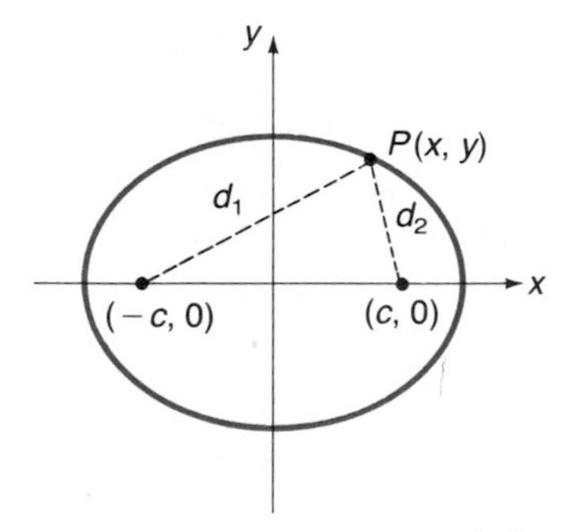

FIGURE 20

After eliminating radicals and simplifying—a good exercise for the reader—we eventually obtain

$$(a^2 - c^2)(x^2) + a^2y^2 = a^2(a^2 - c^2)$$

or

$$\frac{x^2}{a^2} + \frac{y^2}{a^2 - c^2} = 1$$

Since $a > c$, then $a^2 - c^2 > 0$. To simplify the equation further, we choose to let $b^2 = a^2 - c^2$, $b > 0$. Thus,

$$\frac{x^2}{a^2} + \frac{y^2}{b^2} = 1$$

Proceeding similarly with the foci on the vertical axis, we arrive at

$$\frac{x^2}{b^2} + \frac{y^2}{a^2} = 1$$

Combining these results, we can write

$$\frac{x^2}{m^2} + \frac{y^2}{n^2} = 1$$

as a standard form for an equation of an ellipse located as described above. We shift over to m and n to simplify our approach and because a and b have special significance in a more advanced treatment of the subject. It can be shown that if $m > n > 0$, then the foci are on the x axis; and if $n > m > 0$, then the foci are on the y axis. The two cases are summarized as follows:

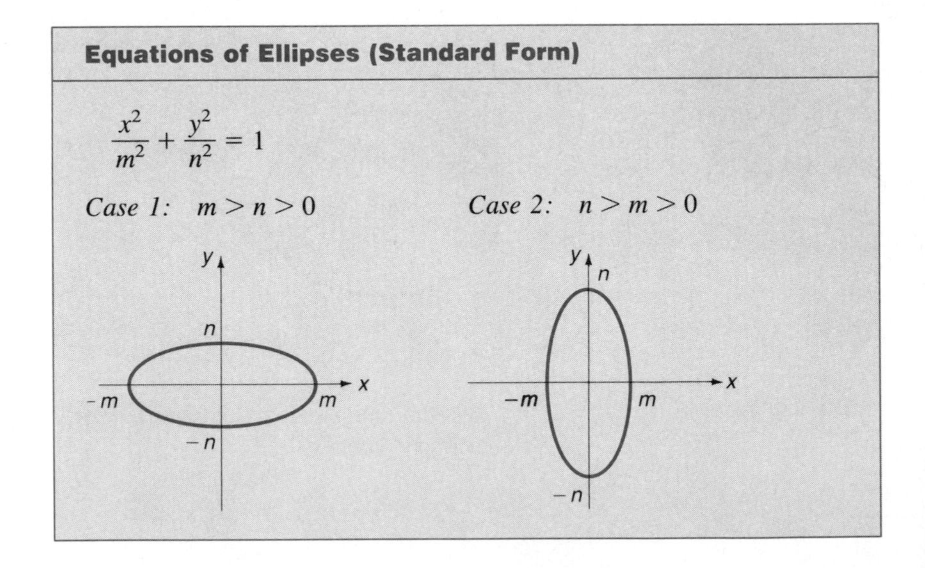

We now show how particular examples of these two cases can be sketched rather quickly.

Rapid Sketching of Ellipses

To graph $\frac{x^2}{m^2} + \frac{y^2}{n^2} = 1$:

Step 1: Find the x intercepts by letting $y = 0$ and solving for x.

Step 2: Find the y intercepts by letting $x = 0$ and solving for y.

Step 3: Sketch an ellipse passing through these intercepts.

Example 24 Graph $\frac{x^2}{16} + \frac{y^2}{9} = 1$.

Solution *Step 1:* Find x intercepts:

$$\frac{x^2}{16} + \frac{0}{9} = 1$$
$$x^2 = 16$$
$$x = \pm\sqrt{16} = \pm 4$$

Step 2: Find y intercepts:

$$\frac{0}{16} + \frac{y^2}{9} = 1$$
$$y^2 = 9$$
$$y = \pm\sqrt{9} = \pm 3$$

Step 3: Plot the intercepts and sketch in the ellipse.

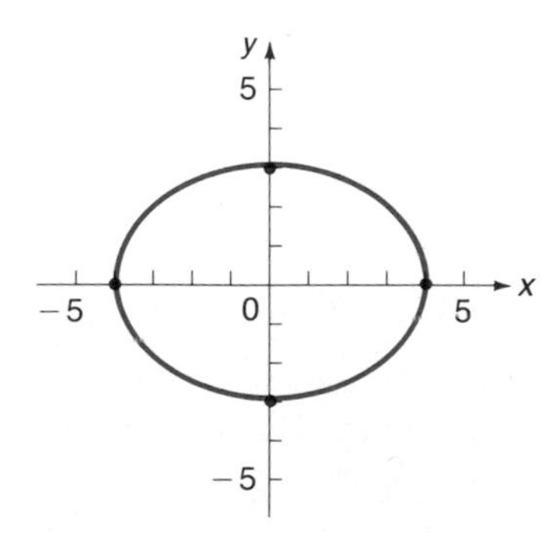

Problem 24 Graph $\frac{x^2}{9} + \frac{y^2}{16} = 1$.

HYPERBOLAS

Definition of a Hyperbola

A **hyperbola** is the set of all points such that the absolute value of the difference of the distances of each of two fixed points is constant. The two fixed points are called **foci**.

As with the ellipse, we will limit our investigation to cases in which the foci are symmetrically located on either coordinate axis. Thus, if $(-c, 0)$ and $(c, 0)$ are the foci and $2a$ is the constant difference (Figure 21), then

$$|d_1 - d_2| = 2a$$

$$|\sqrt{(x+c)^2 + y^2} - \sqrt{(x-c)^2 + y^2}| = 2a$$

After eliminating radicals and absolute-value signs (by appropriate use of squaring) and simplifying—another good exercise for the reader—we eventually obtain

$$\frac{x^2}{a^2} + \frac{y^2}{a^2 - c^2} = 1$$

which looks like the equation we obtained for the ellipse. However, from Figure 21 we see that $2a < 2c$; hence, $a^2 - c^2 < 0$. To simplify the equation further, we let $-b^2 = a^2 - c^2$. Thus,

$$\frac{x^2}{a^2} - \frac{y^2}{b^2} = 1$$

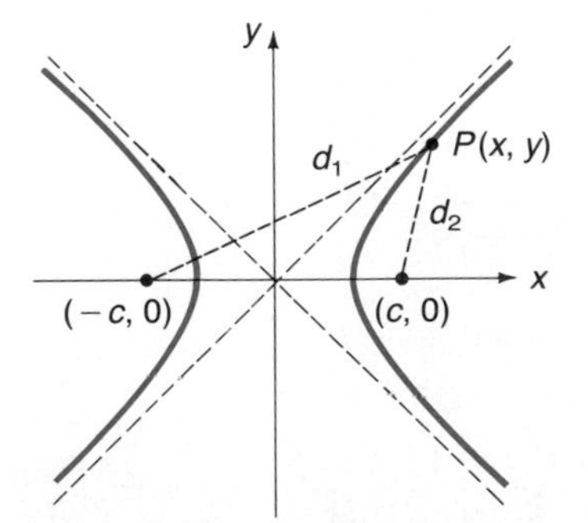

FIGURE 21

Proceeding similarly with the foci on the vertical axis, we obtain

$$\frac{y^2}{a^2} - \frac{x^2}{b^2} = 1$$

Combining these results, we can write

$$\frac{x^2}{m^2} - \frac{y^2}{n^2} = 1 \qquad \frac{y^2}{n^2} - \frac{x^2}{m^2} = 1$$

as standard forms for equations of hyperbolas located as described above. (Again, we shift over to m and n to simplify our approach and because a and b have special significance in a more advanced treatment of the subject.) In summary:

Equations of Hyperbolas (Standard Forms)

$$\frac{x^2}{m^2} - \frac{y^2}{n^2} = 1 \qquad \frac{y^2}{n^2} - \frac{x^2}{m^2} = 1 \qquad (1)$$

Opens left and right — Opens up and down

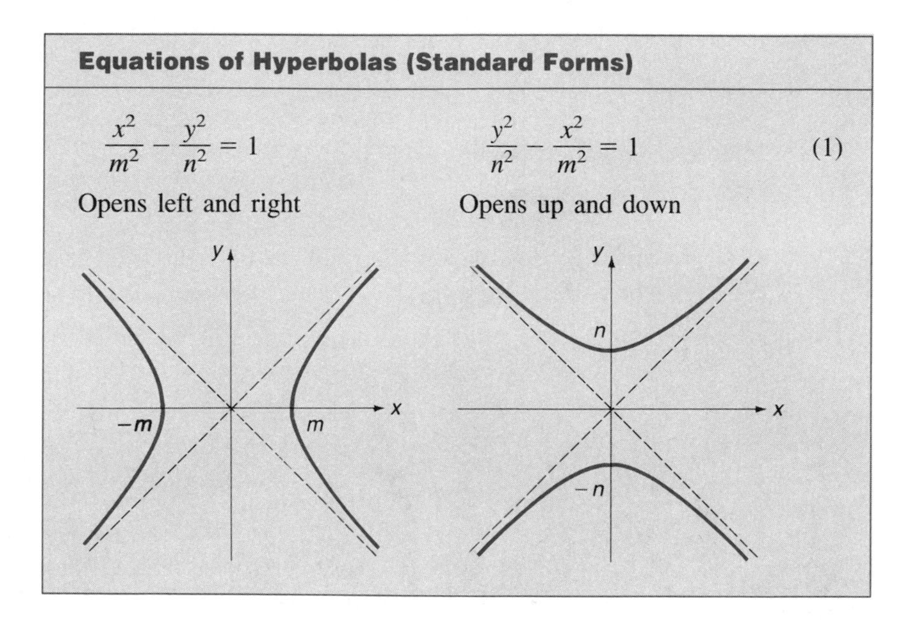

As an aid to graphing Equations (1) of hyperbolas in standard form, we solve each equation for y in terms of x. From the first equation, we obtain

$$y = \pm \frac{nx}{m}\sqrt{1 - \frac{m^2}{x^2}}$$

and from the second we obtain

$$y = \pm \frac{nx}{m}\sqrt{1 + \frac{m^2}{x^2}}$$

As x gets large, the radicals approach 1; hence, the equations (for large x) behave very much as if:

Asymptotes

$$y = \pm\frac{n}{m}x \qquad (2)$$

The two straight lines in Equation (2) are guidelines called **asymptotes**. The graphs of Equations (1) will approach these guidelines, but never touch them, as the graph moves farther and farther away from the origin. Quick sketches of hyperbolas can be made rather easily by following these steps:

Rapid Sketching of Hyperbolas

To graph $\frac{x^2}{m^2} - \frac{y^2}{n^2} = 1$ and $\frac{y^2}{n^2} - \frac{x^2}{m^2} = 1$:

Step 1: Draw a dashed rectangle with intercepts $x = \pm m$ and $y = \pm n$.

Step 2: Draw dashed diagonals of the rectangle and extend to form asymptotes. (These are the graphs of Equation 2.)

Step 3: Determine the true intercepts of the hyperbola (be particularly careful in this step); then sketch in the hyperbola (both branches).

Example 25 Graph: **(A)** $\frac{x^2}{25} - \frac{y^2}{16} = 1$ **(B)** $\frac{y^2}{16} - \frac{x^2}{25} = 1$

Solution **(A)** *Step 1:* Draw a dashed rectangle with intercepts $x = \pm\sqrt{25} = \pm 5$ and $y = \pm\sqrt{16} = \pm 4$.

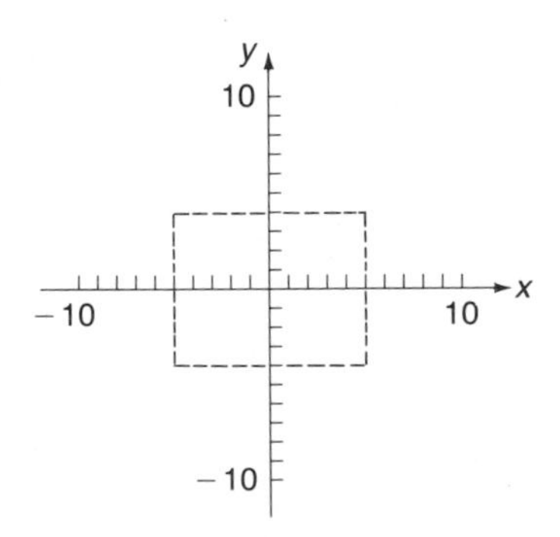

Step 2: Draw in asymptotes (extended diagonals of the rectangle).

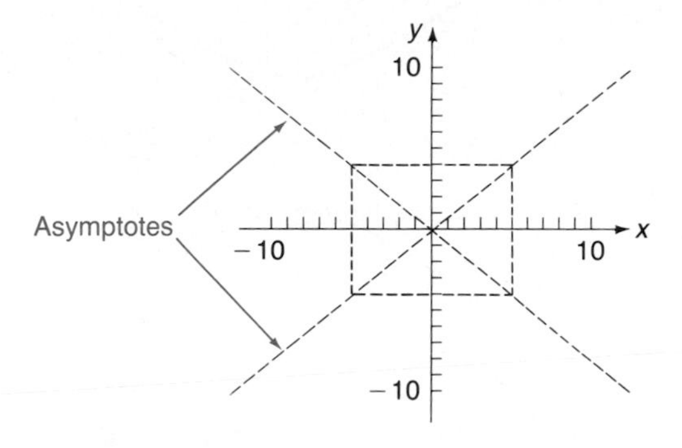

Step 3: Determine the true intercepts for the hyperbola; then sketch it in. Let $y = 0$; then

$$\frac{x^2}{25} - \frac{0}{16} = 1$$
$$x^2 = 25$$
$$x = \pm\sqrt{25} = \pm 5$$

If we let $x = 0$, then

$$\frac{0}{25} - \frac{y^2}{16} = 1$$
$$y^2 = -16$$
$$y = \pm\sqrt{-16} = \pm 4i$$

These are complex numbers and do not represent real intercepts. We conclude that the only real intercepts are $x = \pm 5$, and the hyperbola opens left and right.

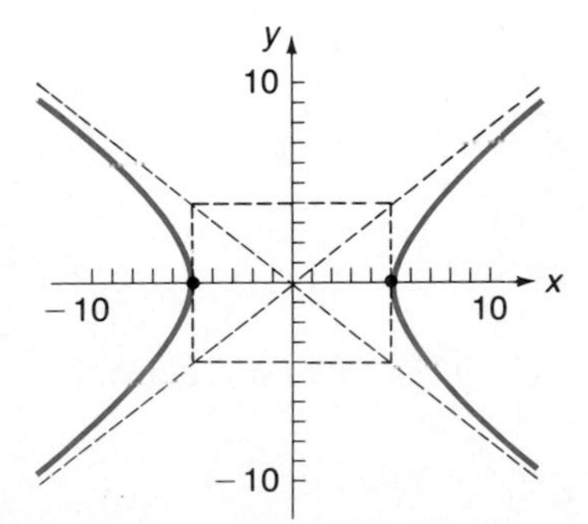

(B)

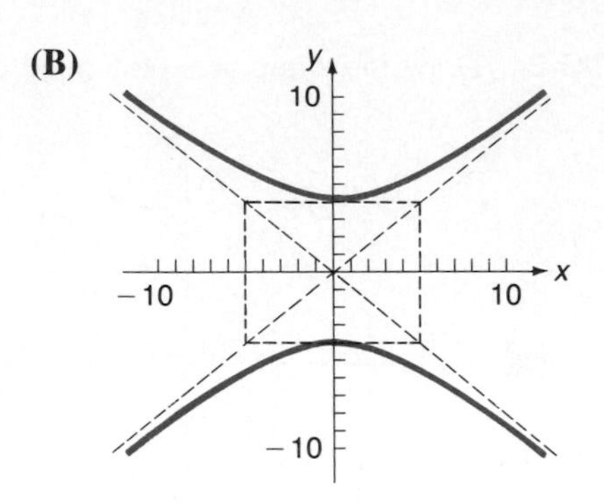

Problem 25 Graph: **(A)** $\frac{x^2}{4} - \frac{y^2}{9} = 1$ **(B)** $\frac{y^2}{9} - \frac{x^2}{4} = 1$

Hyperbolic forms are encountered in the study of comets, the loran system of navigation for ships and aircraft, some modern architectural structures, and optics, to name but a few examples among many.

ANSWERS TO MATCHED PROBLEMS

24.

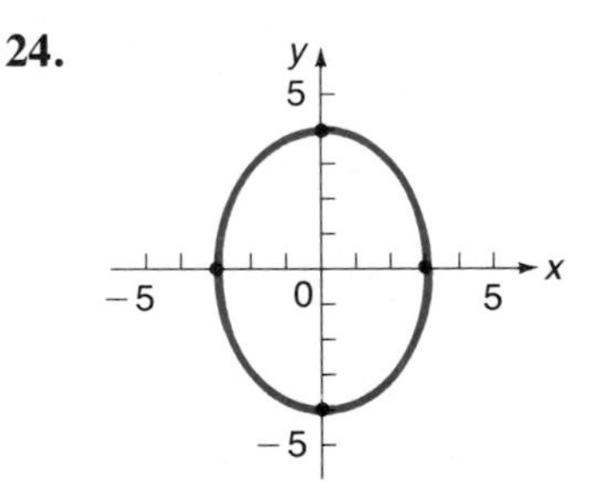

25. **(A)**

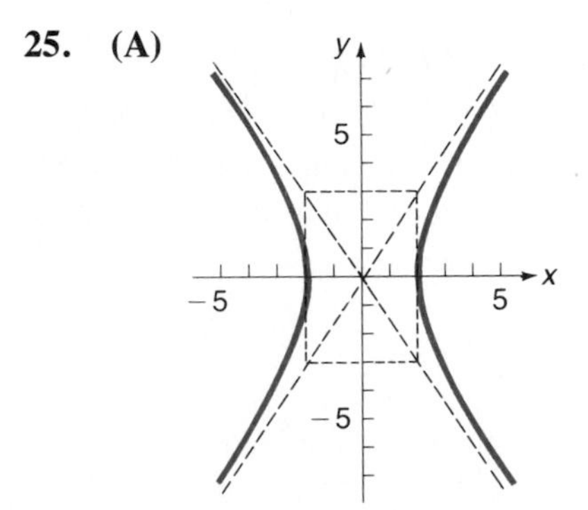

(B)

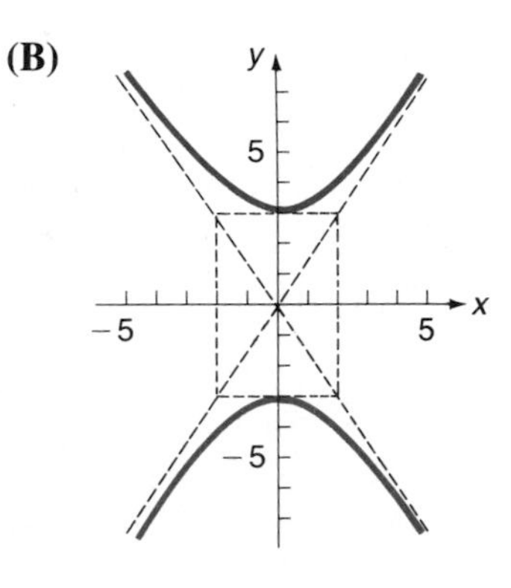

EXERCISE 7-6

A *Graph each of the following ellipses:*

1. $\frac{x^2}{25} + \frac{y^2}{4} = 1$

2. $\frac{x^2}{9} + \frac{y^2}{4} = 1$

3. $\frac{x^2}{4} + \frac{y^2}{25} = 1$

4. $\frac{x^2}{4} + \frac{y^2}{9} = 1$

Graph each of the following hyperbolas:

5. $\frac{x^2}{4} - \frac{y^2}{25} = 1$

6. $\frac{x^2}{16} - \frac{y^2}{9} = 1$

7. $\frac{y^2}{25} - \frac{x^2}{4} = 1$

8. $\frac{y^2}{9} - \frac{x^2}{16} = 1$

B *Graph each of the following equations after first writing the equation in one of the standard forms discussed in this section. For example,*

$$9x^2 + 4y^2 = 36$$

can be written in the form

$$\frac{x^2}{4} + \frac{y^2}{9} = 1$$

by dividing through by 36.

9. $4x^2 + 9y^2 = 36$

10. $4x^2 + 25y^2 = 100$

11. $4x^2 - 9y^2 = 36$

12. $25y^2 - 4x^2 = 100$

C **13.** Find the coordinates of the foci for the ellipse in Problem 3.

14. Find the coordinates of the foci for the ellipse in Problem 2.

15. Find the coordinates of the foci for the hyperbola in Problem 6.

16. Find the coordinates of the foci for the hyperbola in Problem 7.

7-7 CHAPTER REVIEW

A **cartesian coordinate system** is formed with two real number lines—one as a **horizontal axis**, the other as a **vertical axis**—intersecting at their origins. These axes divide the plane into four **quadrants**. Every point in the plane corresponds to its **coordinates**—a pair (a, b) where the **abscissa** a is the coordinate of the point projected to the horizontal axis and the **ordinate** b is the coordinate of the point projected to the vertical axis. The point $(0, 0)$ is the **origin**. *(7-1)*

The **graph of an equation** in two variables x and y is the set of all points whose coordinates (x, y) are solutions of the equation. The graph of a first-degree equation in x and y is a line. The point or value where a graph crosses the y axis is called the ***y* intercept**; where it crosses the x axis, it is called the ***x* intercept**. *(7-1)*

The **slope** of a nonvertical line is given by

$$\frac{y_1 - y_2}{x_1 - x_2}$$

where (x_1, y_1) and (x_2, y_2) are any two points on the line. The equation $y = mx + b$ represents a line in **slope–intercept form**; m is the slope of the line and b is the y intercept. The equation $y - y_1 = m(x - x_1)$ represents a line in **point–slope form**; m is the slope and (x_1, y_1) is any point on the line. A **vertical line** has equation of the form $x = c$ and has no slope. A **horizontal line** has equation of the form $y = c$ and slope 0. Two nonvertical lines are parallel when their slopes are equal, perpendicular when the product of their slopes is -1. *(7-2)*

A nonvertical line divides the plane into two **half-planes**; the **upper half-plane** above the line and the **lower half-plane** below. The graph of a linear inequality in two variables is a half-plane, possibly including the boundary line. *(7-3)*

The equation $y = ax^2 + bx + c$ has a graph similar to $y = x^2$, but it is stretched ($|a| > 1$) or flattened ($|a| < 1$), possibly turned over the x axis ($a < 0$), and shifted $\left|-\frac{b}{2a}\right|$ units right or left and $\left|-\frac{b^2 - 4ac}{4a}\right|$ units up or down. The graph is a **parabola**. *(7-4)*

The distance between two points (x_1, y_1) and (x_2, y_2) in the plane is given by

$$d = \sqrt{(x_2 - x_1)^2 + (y_2 - y_1)^2}$$

A **circle** is the set of all points a fixed distance (**radius**) from a fixed point (**center**). The equation of a circle of radius R with center (h, k) is $(x - h)^2 + (y - k)^2 = R^2$; if the center is the origin, this becomes $x^2 + y^2 = R^2$. A **parabola** is the set of all points equidistant from a fixed point (**focus**) and a fixed line (**directrix**). A parabola in standard form has equation

$$x = ky^2 \quad \text{Opens horizontally}$$

or

$$y = kx^2 \quad \text{Opens vertically} \quad \textit{(7-5)}$$

An **ellipse** is the set of all points such that the sum of the distances of each to two fixed points (**foci**; each separately is a **focus**) is constant. An ellipse in standard form has equation

$$\frac{x^2}{m^2} + \frac{y^2}{n^2} = 1$$

A **hyperbola** is the set of all points such that the absolute value of the difference of the distances of each to two fixed points (**foci**) is constant. A hyperbola in standard form has equation

$$\frac{x^2}{m^2} - \frac{y^2}{n^2} = 1 \quad \text{Opens horizontally}$$

or

$$\frac{y^2}{n^2} - \frac{x^2}{m^2} = 1 \quad \text{Opens vertically}$$

The lines

$$y = \pm\frac{n}{m}x$$

are **asymptotes** for the hyperbola. Circles, parabolas, ellipses, and hyperbolas are the principal **conic sections**—geometric figures obtained by intersecting a plane and a cone. *(7-6)*

REVIEW EXERCISE 7-7

Work through all the problems in this chapter review and check answers in the back of the book. (Answers to all problems are there, and following each answer is a number in italics indicating the section in which that type of problem is discussed.) Where weaknesses show up, review appropriate sections in the text.

A *Graph each in a rectangular coordinate system.*

1. $y = 2x - 3$ **2.** $2x + y = 6$ **3.** $x - y \geq 6$

4. $y > x - 1$ **5.** $x^2 + y^2 = 36$ **6.** $y = x^2 + 3x - 4$

7. What is the slope and y intercept for the graph of $y = -2x - 3$?

8. Write an equation of a line that passes through $(2, 4)$ with slope 2. Write the answer in the form $Ax + By = C$, $A > 0$.

9. What is the slope of the line that passes through $(1, 3)$ and $(3, 7)$?

10. What is an equation of a line that passes through $(1, 3)$ and $(3, 7)$? Write the answer in the form $y = mx + b$.

11. What is the distance between the two points $(1, 3)$ and $(3, 7)$?

12. What is an equation of a circle with radius 5 and center at the origin?

B *Graph each in a rectangular coordinate system.*

13. $3x - 2y = 9$

14. $y = \frac{1}{3}x - 2$

15. $x = -3$

16. $4x - 5y \leq 20$

17. $y = -x^2 + x + 5$

18. $y < \frac{x}{2} + 1$

19. $x \geq -3$

20. $-4 \leq y < 3$

21. $x^2 + y^2 = 49$

22. $(x - 2)^2 + (y + 3)^2 = 16$

23. $y^2 = -2x$

24. $\frac{x^2}{9} + \frac{y^2}{16} = 1$

25. $\frac{y^2}{16} - \frac{x^2}{9} = 1$

26. What is the slope and y intercept for the graph of $x + 2y = -6$?

27. Write an equation of a line that passes through $(-3, 2)$ with slope $-\frac{1}{3}$. Write the final answer in the form $y = mx + b$.

28. Write the equation of a line that passes through $(-3, 2)$ and $(3, -2)$. Write the final answer in the form $Ax + By = C$, $A > 0$.

29. Find the equation of a line that passes through $(3, -4)$ and is perpendicular to $x + 2y = -6$. Write the final answer in the form $y = mx + b$.

30. Write the equation of the vertical and horizontal lines that pass through $(5, -2)$.

31. Write an equation of a circle in the form $(x - h)^2 + (y - k)^2 = R^2$ if its center is at $(-3, 4)$ and it has a radius of 7.

C **32.** Find an equation of a circle with center at the origin that passes through $(12, -5)$.

33. Write the equation of a line that passes through $(-6, 2)$ and is parallel to $3x - 2y = 5$. Write your final answer in the form $y = mx + b$.

34. Graph $y = 3x^2 + 5x - 2$ and find all values where the graph crosses the x axis.

35. Transform the equation

$$x^2 + y^2 + 6x - 8y = 0$$

into the form

$$(x - h)^2 + (y - k)^2 = R^2$$

Since the graph is a circle, what is its radius and what are the coordinates of its center?

36. Graph the set $\{(x, y) \mid y \geq 0,\ 1 \leq x \leq 5,\ 2x + 3y \leq 18\}$.

8

SYSTEMS OF EQUATIONS AND INEQUALITIES

Up to now we have dealt with equations or inequalities individually. That is, we have solved a given single equation or a given inequality. In this chapter, we will turn our attention to *systems* of equations or inequalities consisting of more than one equation or inequality, such as

(1) $\begin{aligned} x + 3y &= 7 \\ 5x - y &= 3 \end{aligned}$ **(2)** $\begin{aligned} x + 3y &\le 7 \\ 5x - y &< 3 \end{aligned}$ **(3)** $\begin{aligned} x^2 + y^2 &= 25 \\ x + 2y &= 1 \end{aligned}$

To solve such a system means to find all pairs of real numbers (x, y) that satisfy each equation or inequality in the system.

Systems like (1) are studied in Sections 8-1 to 8-4. Those like (2) are considered in Section 8-5 and those like (3) in Section 8-6.

8-1 SYSTEMS OF LINEAR EQUATIONS IN TWO VARIABLES

- Solution by Graphing
- Solution by Substitution
- Solution by Elimination Using Addition

Many practical problems can be solved conveniently using two-equation–two-unknown methods. For example, if a 12-foot board is cut in two pieces so that one piece is 2 feet longer than the other piece, how long is each piece? We could solve this problem using one-equation–one unknown methods studied earlier, but we can also proceed as follows, using two variables. Let

x = Length of the longer piece

y = Length of the shorter piece

Then

$$x + y = 12$$
$$x - y = 2$$

To **solve** this system is to find all the ordered pairs of real numbers that satisfy both equations at the same time. In general, we are interested in solving linear systems of the type:

Linear System (Standard Form)

$$\begin{aligned} ax + by &= m \\ cx + dy &= n \end{aligned}$$

a, b, c, d, m, and n are constants
x and y are variables
a, b, c, d not all 0

There are several methods of solving systems of this type. We will consider three that are widely used: solution by graphing, solution by substitution, and solution by elimination using addition.

SOLUTION BY GRAPHING

We will proceed by graphing both equations on the same coordinate system. The coordinates of any points that the graphs have in common must be solutions to the system since they must satisfy both equations.

Example 1 Solve by graphing: $x + y = 12$
$x - y = 2$

Solution Graph each equation and find coordinates of points of intersection, if they exist.

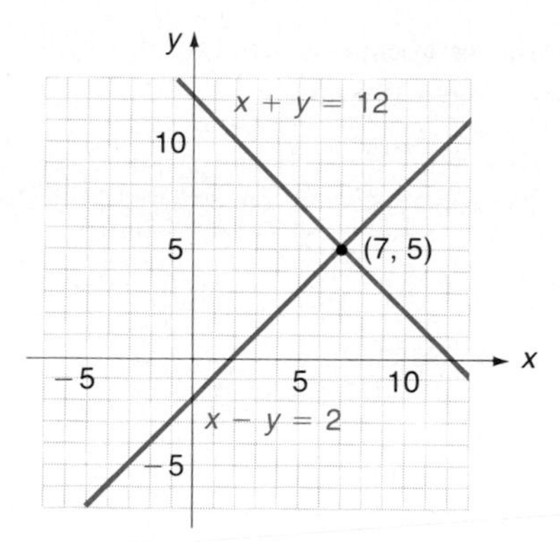

Solution:

$x = 7$ or $(x, y) = (7, 5)$

$y = 5$

Check $x + y = 12$ $\quad$ $x - y = 2$

$7 + 5 \stackrel{\checkmark}{=} 12$ $\quad$ $7 - 5 \stackrel{\checkmark}{=} 2$

Problem 1 Solve by graphing: $x + y = 10$

$x - y = 6$

It is clear that the system in Example 1 has exactly one solution since the lines have exactly one point of intersection. In general, two lines in the same rectangular coordinate system must be related to each other in one of three ways: (1) they intersect at one and only one point, (2) they are parallel, or (3) they coincide (see Example 2).

Example 2 Solve each of the following systems by graphing:

(A) $2x - 3y = 2$, $x + 2y = 8$ $\qquad$ **(B)** $4x + 6y = 12$, $2x + 3y = -6$ $\qquad$ **(C)** $2x - 3y = -6$, $-x + \frac{3}{2}y = 3$

Solution

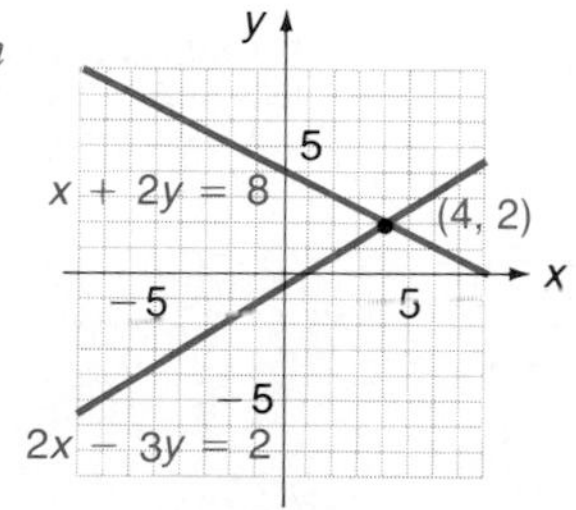

(A) Lines intersect at one point only: *Exactly one solution* $x = 4$, $y = 2$

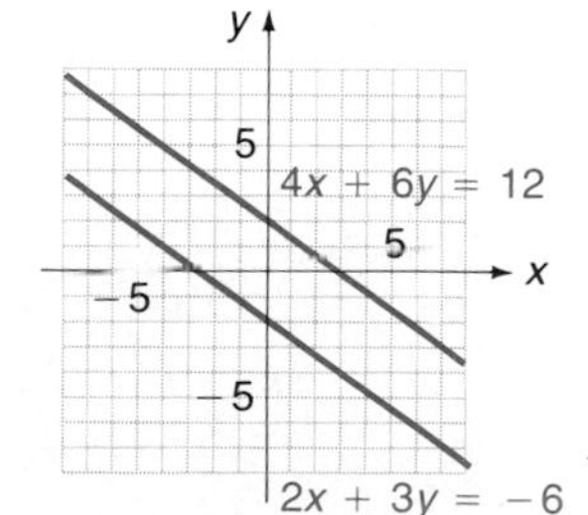

(B) Lines are parallel: *No solution* Such a system will be called inconsistent.

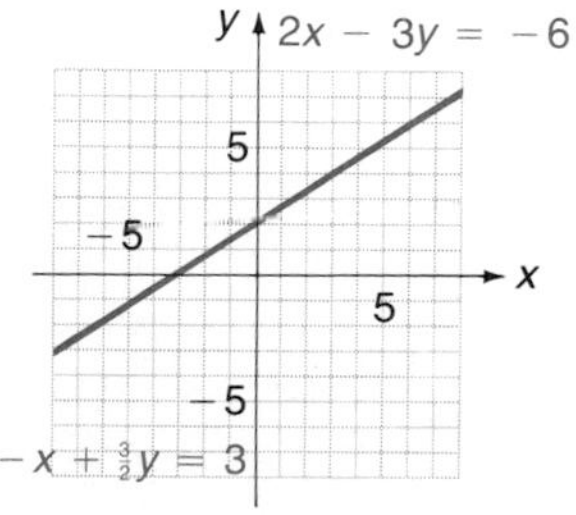

(C) Lines coincide: *Infinite number of solutions* Such a system will be called dependent.

Now we know exactly what to expect when solving a system of two linear equations in two unknowns:

Possible Solutions to a Linear System

1. Exactly one pair of numbers
2. No solution
3. Infinitely many solutions

Problem 2 Solve each of the following systems by graphing:

(A) $\begin{aligned} 2x + 3y &= 12 \\ x - 3y &= -3 \end{aligned}$ **(B)** $\begin{aligned} x - 3y &= -3 \\ -2x + 6y &= 12 \end{aligned}$ **(C)** $\begin{aligned} 2x - 3y &= 12 \\ -x + \tfrac{3}{2}y &= -6 \end{aligned}$

Generally, graphic methods give us only rough approximations of solutions. The methods of substitution and elimination using addition to be considered next will yield results to any decimal accuracy desired—assuming solutions exist.

SOLUTION BY SUBSTITUTION

Choose one of the two equations in a system and solve for one variable in terms of the other. (Choose an equation that avoids getting involved with fractions, if possible.) Then substitute the result into the other equation and solve the resulting linear equation in one variable. Now substitute this result back into either of the original equations to find the second variable. An example should make the process clear.

Example 3 Solve by substitution:

$$2x - 3y = 7 \qquad (1)$$

$$-3x + y = -7 \qquad (2)$$

Solution

$$y = 3x - 7$$

Solve Equation (2), the simplest choice, for y in terms of x. Then substitute the result in the other equation.

$$\begin{aligned} 2x - 3y &= 7 \\ 2x - 3(\mathbf{3x - 7}) &= 7 \\ 2x - 9x + 21 &= 7 \\ -7x &= -14 \\ \mathbf{x} &= \mathbf{2} \end{aligned}$$

Substitute $x = 2$ into one of the original equations, or more easily into Equation (2) rewritten as $y = 3x - 7$ and solve for y.

$$y = 3 \cdot \mathbf{2} - 7$$

$$\mathbf{y = -1}$$

Thus, $(2, -1)$ is a solution to the original system, as we can readily check:

Check $(2, -1)$ must satisfy *both* equations.

$$\begin{aligned} 2x - 3y &\overset{?}{=} 7 \\ 2(2) - 3(-1) &\overset{?}{=} 7 \\ 4 + 3 &\overset{\checkmark}{=} 7 \end{aligned} \qquad \begin{aligned} -3x + y &\overset{?}{=} -7 \\ -3(2) + (-1) &\overset{?}{=} -7 \\ -6 - 1 &\overset{\checkmark}{=} -7 \end{aligned}$$

Problem 3 Solve by substitution and check: $3x - 4y = 18$
$2x + y = 1$

SOLUTION BY ELIMINATION USING ADDITION

Now we turn to elimination using addition. This is probably the most important method of solution, since it is readily generalized to higher-order systems. The method involves the replacement of systems of equations with simpler *equivalent systems* (by performing appropriate operations) until we obtain a system with an obvious solution. **Equivalent systems** of equations are, as you would expect, systems that have exactly the same solution set. The following result lists operations that produce equivalent systems.

> Equivalent systems of equations result if:
>
> **(A)** Two equations are interchanged.
> **(B)** An equation is multiplied by a nonzero constant.
> **(C)** A constant multiple of another equation is added to a given equation.

Solving systems of equations by use of this result is best illustrated by examples.

Example 4 Solve by elimination using addition:

$$3x - 2y = 8 \qquad (3)$$
$$2x + 5y = -1 \qquad (4)$$

Solution We use the result above to eliminate one of the variables and thus obtain a system with an obvious solution.

$$\begin{aligned} 15x - 10y &= 40 \\ 4x + 10y &= -2 \\ \hline 19x \qquad &= 38 \\ x &= 2 \end{aligned}$$

If we multiply Equation (3) by 5 and Equation (4) by 2, we make the coefficients of y opposites. If we then add, we can eliminate y.

Now substitute $x = 2$ back into either of the original equations, say Equation (4), and solve for y:

$$
\begin{aligned}
2(\ ) + 5y &= -1 \\
5y &= -5 \\
y &= -1
\end{aligned}
$$

Thus, $(2, -1)$ is a solution to the original system. The check is completed as in Example 3.

Problem 4 Solve by elimination using addition and check your answer. Eliminate the variable x first.

$$
\begin{aligned}
2x + 5y &= -9 \\
3x - 4y &= -2
\end{aligned}
$$

Example 5 Solve by elimination using addition:

$$
\begin{aligned}
x + 3y &= 2 && (5) \\
2x + 6y &= -3 && (6)
\end{aligned}
$$

Solution

$$
\begin{aligned}
-2x - 6y &= -4 && \text{Multiply Equation (5) by } -2 \text{ and add.} \\
2x + 6y &= -3 \\
\hline
0 &= -7 && \text{A contradiction!}
\end{aligned}
$$

Our assumption that there are values for x and y that satisfy Equations (5) and (6) simultaneously must be false. (Otherwise, we have proved that $0 = -7$.) If you check the slope of each line, you will find that the two slopes are the same (but the y intercepts are different); hence, the lines are parallel and the system has no solution. Systems of this type are called **inconsistent**—conditions have been placed on the variables x and y that are impossible to meet. The equations represent parallel lines.

Problem 5 Solve by elimination using addition:

$$
\begin{aligned}
3x - 4y &= -2 \\
-6x + 8y &= 1
\end{aligned}
$$

Example 6 Solve by elimination using addition:

$$
\begin{aligned}
-2x + y &= -8 && (7) \\
x - \tfrac{1}{2}y &= 4 && (8)
\end{aligned}
$$

Solution

$$
\begin{aligned}
-2x + y &= -8 && \text{Multiply Equation (8) by 2 and add.} \\
2x - y &= 8 \\
\hline
0 &= 0
\end{aligned}
$$

Both sides have been eliminated. Actually, if we had multiplied Equation (8) by -2, we would have obtained Equation (7). When one equation is a constant multiple of the other, the system is said to be **dependent**, and their graphs will coincide. There are infinitely many solutions to the system—any solution of one equation will be a solution of the other. One way of expressing all solutions is to solve one of the equations for y in terms of x, say the first equation,

$$y = 2x - 8$$

Then

$$(x, 2x - 8)$$

is a solution to the system for any real number x, as can easily be checked. If $x = 1$, for example, then $y = 2x - 8 = 2 - 8 = -6$ and $(1, -6)$ is a solution. We could have chosen to solve for x in terms of y; the solution would then be described as the set of all pairs $(\frac{1}{2}y + 4, y)$.

Problem 6 Solve by elimination using addition:

$$\begin{aligned} 6x - 3y &= -2 \\ -2x + y &= \tfrac{2}{3} \end{aligned}$$

ANSWERS TO MATCHED PROBLEMS

1. $x = 8$, $y = 2$
2. **(A)** Graphs cross at $x = 3$ and $y = 2$
   (B) Graphs do not cross; no solution
   (C) Graphs coincide; infinitely many solutions
3. $(2, -3)$ **4.** $(-2, -1)$ **5.** No solution
6. $(x, 2x + \frac{2}{3})$ is a solution for x any real number.

EXERCISE 8-1

A *Solve by graphing.*

1. $\begin{aligned} 3x - 2y &= 12 \\ 7x + 2y &= 8 \end{aligned}$

2. $\begin{aligned} x + 5y &= -10 \\ -5x + y &= 24 \end{aligned}$

3. $\begin{aligned} 3x + 5y &= 15 \\ 6x + 10y &= -5 \end{aligned}$

4. $\begin{aligned} 3x - 5y &= 15 \\ x - \tfrac{5}{3}y &= 5 \end{aligned}$

Solve by substitution.

5. $\begin{aligned} 2x + y &= 6 \\ y &= x + 3 \end{aligned}$

6. $\begin{aligned} m - 2n &= 0 \\ -3m + 6n &= 8 \end{aligned}$

7. $\begin{aligned} 3x - y &= -3 \\ 5x + 3y &= -19 \end{aligned}$

8. $\begin{aligned} 2m - 3n &= 9 \\ m + 2n &= -13 \end{aligned}$

Solve by elimination using addition.

9. $3p + 8q = 4$
$15p + 10q = -10$

10. $3x - y = -3$
$5x + 3y = -19$

11. $6x - 2y = 18$
$-3x + y = -9$

12. $4m + 6n = 2$
$6m - 9n = 15$

B *Solve each system by graphing, by substitution, and by elimination using addition.*

13. $x - 3y = -11$
$2x + 5y = 11$

14. $5x + y = 4$
$x - 2y = 3$

15. $11x + 2y = 1$
$9x - 3y = 24$

16. $2x + y = 0$
$3x + y = 2$

Use any of the methods discussed in this section to solve each system.

17. $y = 3x - 3$
$6x = 8 + 3y$

18. $3m = 2n$
$n = -7 - 2m$

19. $\frac{1}{2}x - y = -3$
$-x + 2y = 6$

20. $y = 2x - 1$
$6x - 3y = -1$

21. $2x + 3y = 2y - 2$
$3x + 2y = 2x + 2$

22. $2u - 3v = 1 - 3u$
$4v = 7u - 2$

C **23.** $0.2x - 0.5y = 0.07$
$0.8x - 0.3y = 0.79$

24. $0.5m + 0.2n = 0.54$
$0.3m - 0.6n = 0.18$

25. $\frac{1}{4}x - \frac{2}{3}y = -2$
$\frac{1}{2}x - y = -2$

26. $\frac{2}{3}a + \frac{1}{2}b = 2$
$\frac{1}{2}a + \frac{1}{3}b = 1$

8-2 APPLICATION: MIXTURE PROBLEMS

Many of the applications already considered, using one-equation–one-unknown methods, can also be set up as a system of equations using two variables. This is also true of a type of application known as a mixture problem. The examples in this section will illustrate both the single-equation and system-of-equations approaches. Some problems lend themselves more naturally to one approach than the other. Experience will help you recognize this, as well as to decide which approach you prefer when both approaches work equally well.

Example 7 A change machine changes dollar bills into quarters and nickels. If you receive 12 coins after inserting a \$1 bill, how many of each type of coin did you receive?

Solution Let

x = Number of quarters
y = Number of nickels

Then

$$x + y = 12 \quad \text{Number of coins}$$
$$25x + 5y = 100 \quad \text{Value of coins in cents}$$

$$\begin{array}{rl} -5x - 5y &= -60 \\ 25x + 5y &= 100 \\ \hline 20x &= 40 \\ x &= 2 \quad \text{Quarters} \end{array}$$

$$x + y = 12$$
$$2 + y = 12$$
$$y = 10 \quad \text{Nickels}$$

Check $2 + 10 = 12$ coins in all $\quad 25 \cdot 2 + 5 \cdot 10 = 50 + 50$
$= 100$ cents or \$1

We could also have set this up using only one variable: x for the number of quarters, $12 - x$ for the number of nickels. The equation to solve would then be

$$25x + 5(12 - x) = 100$$

Problem 7 Repeat Example 7 with eight coins received in change from a \$1 bill.

Example 8 A concert brought in \$27,200 on the sale of 4,000 tickets. If tickets sold for \$5 and \$8, how many of each were sold?

Solution We use a one-variable approach. Let

x = Number of \$5 tickets sold

Then

$4,000 - x$ = Number of \$8 tickets sold

We now form an equation using the value of the tickets before and after combining the two kinds of tickets:

$$\text{Value before combining} = \text{Value after combining}$$

$$\begin{pmatrix}\text{Value of}\\ \$5 \text{ tickets}\\ \text{sold}\end{pmatrix} + \begin{pmatrix}\text{Value of}\\ \$8 \text{ tickets}\\ \text{sold}\end{pmatrix} = \begin{pmatrix}\text{Total value}\\ \text{of all}\\ \text{tickets sold}\end{pmatrix}$$

$$\begin{aligned} 5x \quad + 8(4{,}000 - x) &= \quad 27{,}200 \\ 5x + 32{,}000 - 8x &= 27{,}200 \\ -3x &= -4{,}800 \\ x &= 1{,}600 \quad \text{\$5 tickets} \\ 4{,}000 - x &= 2{,}400 \quad \text{\$8 tickets} \end{aligned}$$

Check $(\$5)(1{,}600) + (\$8)(2{,}400) = \$27{,}200$

Had we chosen a two-variable approach, we could have set

x = The number of \$5 tickets sold
y = The number of \$8 tickets sold

and obtained the system of equations

$$\begin{aligned} x + y &= 400 \\ 5x + 8y &= 27{,}200 \end{aligned}$$

You should try this approach.

Problem 8 Suppose you receive 40 nickels and quarters in change worth \$4. How many of each type of coin do you have?

Let us now consider some mixture problems involving percent. Recall that 23% in decimal form is 0.23, 6.5% is 0.065, and so on.

Example 9 A zoologist wishes to prepare a special diet that contains, among other things, 120 grams of protein and 17 grams of fat. Two available food mixes specify the following percentages of protein and fat:

MIX	PROTEIN (%)	FAT (%)
A	30	1
B	20	5

How many grams of each mix should be used to prepare the diet mix?

Solution Let

x = Number of grams of mix A used

y = Number of grams of mix B used

Set up one equation for the protein requirements and one equation for the fat requirements:

$$0.3x + 0.2y = 120 \quad \text{Protein requirements}$$
$$0.01x + 0.05y = 17 \quad \text{Fat requirements}$$

Multiply the top equation by 10 and the bottom equation by 100 to clear decimals (not necessary, but helpful):

$$\begin{aligned} 3x + 2y &= 1{,}200 \\ x + 5y &= 1{,}700 \end{aligned} \quad \text{Multiply bottom equation by } -3\text{; then add to eliminate } x.$$

$$\begin{aligned} 3x + 2y &= 1{,}200 \\ -3x - 15y &= -5{,}100 \\ \hline -13y &= -3{,}900 \\ y &= 300 \text{ grams} \quad \text{Mix } B \end{aligned}$$

$$\begin{aligned} x + 5y &= 1{,}700 \\ x + 5(300) &= 1{,}700 \\ x &= 200 \text{ grams} \quad \text{Mix } A \end{aligned}$$

The zoologist should use 200 grams of mix A and 300 grams of mix B to meet the diet requirements.

Check *Protein requirement*

Protein from mix A + Protein from mix $B \stackrel{?}{=} 120$ grams

30% of 200 grams + 20% of 300 grams $= 60 + 60 \stackrel{\checkmark}{=} 120$ grams

Fat requirement

Fat from mix A + Fat from mix $B \stackrel{?}{=} 17$ grams

1% of 200 grams + 5% of 300 grams $= 2 + 15 \stackrel{\checkmark}{=} 17$ grams

Problem 9 Repeat Example 9 for a diet mixture that is to contain 110 grams of protein and 8 grams of fat.

Example 10 A jeweler has two bars of gold alloy in stock, one 12-carat and the other 18-carat. (Note that 24-carat gold is pure gold, 12-carat gold is $\frac{12}{24}$ pure, 18-carat gold is $\frac{18}{24}$ pure, and so on.) How many grams of each alloy must be mixed to obtain 10 grams of 14-carat gold?

Solution Let

x = Number of grams of 12-carat gold used
y = Number of grams of 18-carat gold used

Then

$$x + y = 10$$ Amount of new alloy.

$$\tfrac{12}{24}x + \tfrac{18}{24}y = \tfrac{14}{24}(10)$$ Pure gold present before mixing equals pure gold present after mixing.

$$\begin{aligned} x + y &= 10 \\ 6x + 9y &= 70 \end{aligned}$$ Multiply the second equation by $\frac{24}{2}$ to simplify; then solve using the methods described above. (We use elimination here.)

$$\begin{aligned} -6x - 6y &= -60 \\ 6x + 9y &= 70 \\ \hline 3y &= 10 \end{aligned}$$

$$y = 3\tfrac{1}{3} \text{ grams}$$ 18-carat alloy

$$x + 3\tfrac{1}{3} = 10$$

$$x = 6\tfrac{2}{3} \text{ grams}$$ 12-carat alloy

The checking of the solution is left to you.

Problem 10 Repeat Example 10, but suppose that the jeweler has only 10-carat and pure gold in stock.

Example 11 How many milliliters of distilled water must be added to 60 milliliters of 70% acid solution to obtain a 60% solution? [*Note:* A 70% acid solution is 70% pure acid and 30% distilled water.]

Solution There is only one variable apparent in this problem, namely, the amount of distilled water added. Let x = Number of milliliters of distilled water added. We illustrate the situation before and after mixing, keeping in mind that the amount of acid present before mixing must equal the amount of acid present after mixing.

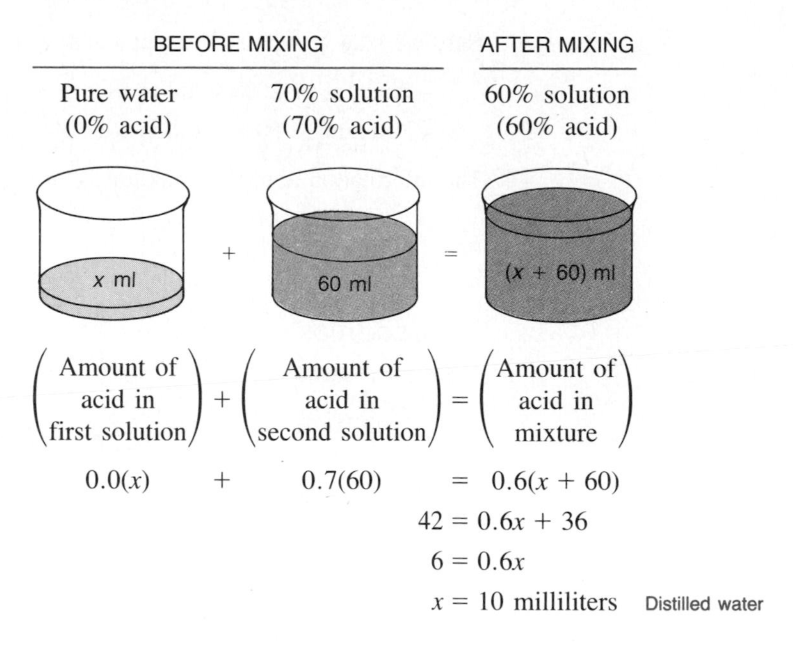

$$0.0(x) + 0.7(60) = 0.6(x + 60)$$
$$42 = 0.6x + 36$$
$$6 = 0.6x$$
$$x = 10 \text{ milliliters}$$ Distilled water

Check $\quad 0.0(10) + 0.7(60) \stackrel{?}{=} 0.6(10 + 60)$

$$42 \stackrel{\checkmark}{=} 42$$

The information contained in the diagrams in this problem can also be presented in a table:

	AMOUNT	% ACID	AMOUNT OF ACID
FIRST SOLUTION	x	0	0
SECOND SOLUTION	60 ml	70	0.7(60)
MIXTURE	x + 60 ml	60	0.7(60)

Since the mixture is to be a 60% solution, the amount of acid 0.7(60) must be 60% of the amount $x + 60$ in the mixture. That is,

$$0.7(60) = 0.6(x + 60)$$

Problem 11 How many centiliters of pure alcohol must be added to 35 centiliters of a 20% solution to obtain a 30% solution?

Example 12 A chemical storeroom has a 20% alcohol solution and a 50% alcohol solution. How many centiliters must be taken from each to obtain 24 centiliters of a 30% solution?

Solution Here we take a two-variable approach. Let

x = Amount of 20% solution used
y = Amount of 50% solution used

The information about the mixture is summarized in tabular form:

	AMOUNT	% ALCOHOL	AMOUNT OF ALCOHOL
20% SOLUTION	x	20	$0.2x$
50% SOLUTION	y	50	$0.5y$
MIXTURE	24	30	$0.3(24) = 7.2$

The two amounts we add together must equal what is in the mixture. Thus we get the system

$$x + y = 24 \quad \text{Amount of mixture}$$
$$0.2x + 0.5y = 7.2 \quad \text{Amount of alcohol}$$

Substitution yields the solution:

$$y = 24 - x$$
$$0.2x + 0.5(24 - x) = 7.2$$
$$0.2x + 12 - 0.5x = 7.2$$
$$-0.3x = -4.8$$
$$x = 16 \text{ centiliters} \quad 20\% \text{ solution}$$
$$24 - x = 8 \text{ centiliters} \quad 50\% \text{ solution}$$

Check $0.2(16) + 0.5(8) \stackrel{?}{=} 0.3(24)$
$7.2 \stackrel{\checkmark}{=} 7.2$

The information in the table can also be shown in diagrams as in the previous example:

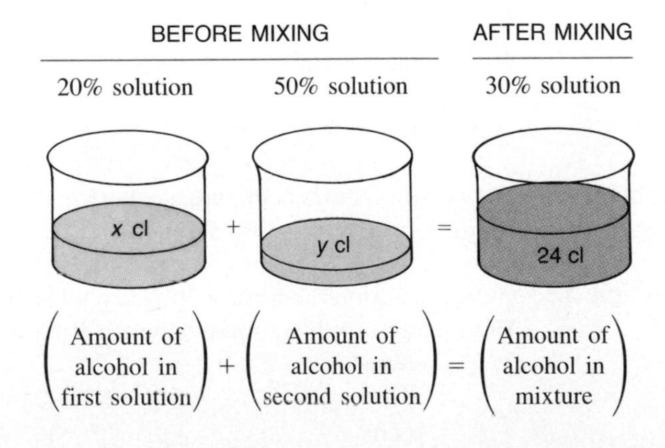

Problem 12 Repeat Example 12 using 10% and 40% stockroom solutions.

Example 13 A coffee shop wishes to blend \$3-per-pound coffee with \$4.25-per-pound coffee to produce a blend selling for \$3.50 per pound. How much of each should be used to produce 50 pounds of the new blend?

Solution This problem is mathematically very close to Example 12. We solve it by one-variable methods for comparison.

$$x = \text{Amount of \$3-per-pound coffee used}$$

Then

$$50 - x = \text{Amount of \$4.25-per-pound coffee used}$$

$$\text{Value before blending} = \text{Value after blending}$$

$$\begin{pmatrix}\text{Value of}\\ \text{\$3-per-pound}\\ \text{coffee used}\end{pmatrix} + \begin{pmatrix}\text{Value of}\\ \text{\$4.25-per-pound}\\ \text{coffee used}\end{pmatrix} = \begin{pmatrix}\text{Total value}\\ \text{of 50 pounds}\\ \text{of the blend}\end{pmatrix}$$

$$3x + 4.25(50 - x) = 3.50(50)$$

$$3x + 212.5 - 4.25x = 175$$

$$-1.25x = -37.5$$

$$x = 30 \text{ pounds} \quad \text{\$3 coffee}$$

$$50 - x = 20 \text{ pounds} \quad \text{\$4.25 coffee}$$

Check

$$(\$3)(30) + (\$4.25)(20) \stackrel{?}{=} (\$3.50)(50)$$

$$\$175 \stackrel{\checkmark}{=} \$175$$

A two-variable approach to this problem with

$$x = \text{Amount of \$3.00-per-pound coffee used}$$

$$y = \text{Amount of \$4.25-per-pound coffee used}$$

leads to the system

$$\begin{aligned} x + \quad\ \ y &= 50 \\ 3x + 4.25y &= 3.5(50) \end{aligned}$$

Problem 13 Repeat Example 13 using \$2.75-per-pound coffee and \$4-per-pound coffee.

ANSWERS TO MATCHED PROBLEMS

7. 3 quarters, 5 nickels
8. 30 nickels and 10 quarters
9. 300 grams of mix A, 100 grams of mix B

10. $2\frac{6}{7}$ grams of pure gold and $7\frac{1}{7}$ grams of 10-carat gold
11. 5 centiliters
12. 8 centiliters of 10% solution, 16 centiliters of 40% solution
13. 20 pounds of $2.75 coffee, 30 pounds of $4 coffee

EXERCISE 8-2

This set of exercises contains a variety of mixture problems classified according to subject area. The more difficult problems are marked with two stars (★★), the moderately difficult with one star (★), and the easier problems are not marked. Solve using either a one- or two-variable approach.

1. ***Puzzle*** A vending machine takes only dimes and quarters. If it contains 100 coins with a total value of $14.50, how many of each type of coin are in the machine?

2. ***Puzzle*** A parking meter contains only nickels and dimes. If it contains 50 coins at a total value of $3.50, how many of each type of coin are in the meter?

3. ***Business*** A school musical production brought in $12,600 on the sale of 3,500 tickets. If the tickets sold for $2 and $4, how many of each type were sold?

4. ***Business*** A concert brought in $60,000 on the sale of 8,000 tickets. If tickets sold for $6 and $10, how many of each type of ticket were sold?

5. ***Chemistry*** How many deciliters of alcohol must be added to 100 deciliters of a 40% alcohol solution to obtain a 50% solution?

6. ***Chemistry*** How many milliliters of hydrochloric acid must be added to 12 milliliters of a 30% solution to obtain a 40% solution?

7. ***Puzzle*** A bank gave you $1.50 in change consisting of only nickels and dimes. If there were 22 coins in all, how many of each type of coin did you receive?

8. ***Puzzle*** Your friend came out of a post office having spent $1.32 on thirty 4-cent and 5-cent stamps. How many of each type did he buy?

9. ***Geometry*** If the sum of two angles in a right triangle is 90° and their difference is 14°, find the two angles.

10. ***Geometry*** Find the dimensions of a rectangle with perimeter 72 inches if its length is 25% greater than its width.

11. ***Chemistry*** How many liters of distilled water must be added to 140 liters of an 80% alcohol solution to obtain a 70% solution?

12. ***Chemistry*** How many centiliters of distilled water must be added to 500 centiliters of a 60% acid solution to obtain a 50% solution?

13. ***Chemistry*** A chemical stockroom has a 20% alcohol solution and a 50% solution. How many deciliters of each should be used to obtain 90 deciliters of a 30% solution?

14. ***Chemistry*** A chemical supply company has a 30% sulfuric acid solution and a 70% sulfuric acid solution. How many liters of each should be used to obtain 100 liters of a 40% solution?

15. ***Business*** A tea shop wishes to blend a $5-per-kilogram tea with a $6.50-per-kilogram tea to produce a blend selling for $6 per kilogram. How much of each should be used to obtain 75 kilograms of the new blend?

16. ***Business*** A gourmet food store wishes to blend a $7-per-kilogram coffee with a $9.50-per-kilogram coffee to produce a blend selling for $8 per kilogram. How much of each should be used to obtain 100 kilograms of the new blend?

17. ***Finance*** You have inherited $20,000 and wish to invest it. If part is invested in a low-risk investment at 10% and the rest in a higher-risk investment at 15%, how much should you invest at each rate to produce the same yield as if all had been invested at 13%?

18. ***Finance*** An investor has $10,000 to invest. If part is invested in a low-risk investment at 11% and the rest in a higher-risk investment at 16%, how much should be invested at each rate to produce the same yield as if all had been invested at 12%?

⋆19. ***Business*** A packing carton contains 144 small packages, some weighing $\frac{1}{4}$ pound each and the others $\frac{1}{2}$ pound each. How many of each type are in the carton if the total contents of the carton weigh 51 pounds?

⋆20. ***Biology*** A biologist, in a nutrition experiment, wants to prepare a special diet for experimental animals. He requires a food mixture that contains, among other things, 20 ounces of protein and 6 ounces of fat. He is able to purchase food mixes of the following composition:

MIX	PROTEIN (%)	FAT (%)
A	20	2
B	10	6

How many ounces of each mix should he use to prepare the diet mix? Solve graphically and algebraically.

⋆21. ***Chemistry*** A chemist has two concentrations of hydrochloric acid in stock, a 50% solution and an 80% solution. How much of each should she mix to obtain 100 milliliters of a 68% solution?

★22. ***Business*** A newspaper printing plant has two folding machines for the final assembling of the evening newspaper—circulation, 29,000. The slower machine can fold papers at the rate of 6,000 per hour, and the faster machine at the rate of 10,000 per hour. If the use of the slower machine is delayed $\frac{1}{2}$ hour because of a minor breakdown, how much total time is required to fold all the papers? How much time does each machine spend on the job?

★23. ***Earth science*** A ship using sound-sensing devices above and below water recorded a surface explosion 6 seconds sooner by its underwater device than its abovewater device. Sound travels in air at about 1,100 feet per second and in seawater at about 5,000 feet per second.

(A) How long did it take each sound wave to reach the ship?

(B) How far was the explosion from the ship?

★24. ***Earth science*** An earthquake emits a primary wave and a secondary wave. Near the surface of the earth the primary wave travels at about 5 miles per second, and the secondary wave at about 3 miles per second. From the time lag between the two waves arriving at a given station, it is possible to estimate the distance to the quake. (The epicenter can be located by obtaining distance bearings at three or more stations.) Suppose a station measured a time difference of 16 seconds between the arrival of the two waves. How long did each wave travel, and how far was the earthquake from the station?

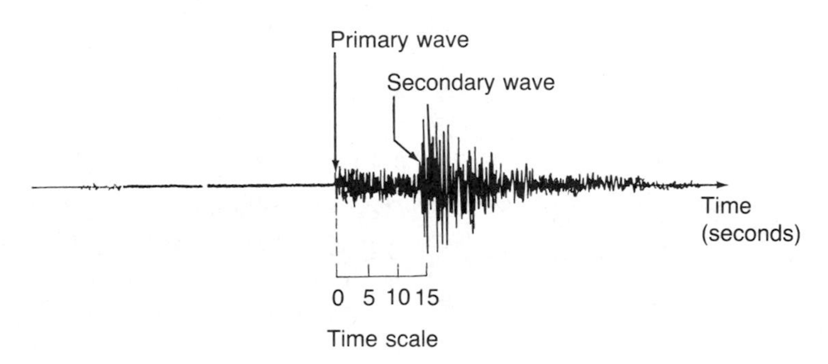

★★25. ***Domestic*** A 9-liter radiator contains a 50% solution of antifreeze in distilled water. How much should be drained and replaced with pure antifreeze to obtain a 70% solution?

★★26. ***Domestic*** A 12-liter radiator contains a 60% solution of antifreeze and distilled water. How much should be drained and replaced with pure antifreeze to obtain an 80% solution?

★★27. ***Business*** Two companies have offered you a sales position. Both jobs are essentially the same, but one company pays a straight 8% commission and the other pays \$51 per week plus a 5% commission. The best sales representatives with either company rarely have sales greater than

\$4,000 in any one week. Before accepting either offer, it would be helpful to know at what point both companies pay the same, and which of the companies pays more on either side of this point. Solve graphically and algebraically.

★★28. ***Business*** Solve Problem 27 with the straight-commission company paying 7% commission and the salary-plus-commission company paying \$75 per week plus 4% commission.

★★29. ***Economics*** In a particular city the weekly supply and demand for popular stereo records, relative to average price per record, p, are given by the equations

$$d = 5{,}000 - 1{,}000p \qquad \$1 \le p \le \$4$$
$$s = -3{,}000 + 3{,}000p \qquad \$1 \le p \le \$4$$

At what price does the supply equal the demand (equilibrium point)?
(A) Solve graphically. **(B)** Solve algebraically.

8-3 SYSTEMS OF LINEAR EQUATIONS IN THREE VARIABLES

- Solving Systems of Three Equations in Three Variables
- Application

Having learned how to solve systems of linear equations in two variables, we proceed to systems with three variables. Systems of the form

$$\begin{aligned} 3x - 2y + 4z &= 6 \\ 2x + 3y - 5z &= -8 \\ 5x - 4y + 3z &= 7 \end{aligned} \tag{1}$$

as well as systems with more variables are encountered frequently and are worth studying. In fact, systems of equations in many variables are so important in solving real-world problems that there are whole courses on this one topic. A triplet of numbers $x = 0$, $y = -1$, and $z = 1$ [also written as an ordered triplet $(0, -1, 1)$] is a **solution** of system (1) above since each equation is satisfied by this triplet. The set of all such ordered triplets of numbers is called the **solution set** of the system. Two systems are said to be **equivalent** if they have the same solution set.

SOLVING SYSTEMS OF THREE EQUATIONS IN THREE VARIABLES

We will use an extension of the method of elimination by addition discussed in Section 8-1 to solve systems in the form of (1). The basic result in Section 8-1 is behind the process.

Example 14 Solve:

$$3x - 2y + 4z = 6 \qquad (2)$$
$$2x + 3y - 5z = -8 \qquad (3)$$
$$5x - 4y + 3z = 7 \qquad (4)$$

Solution *Step 1* We look at the coefficients of the variables and choose to eliminate y from Equations (2) and (4) because of the convenient coefficients -2 and -4. Multiply Equation (2) by -2 and add to Equation (4):

$$\begin{array}{rl} -6x + 4y - 8z = -12 & \text{−2[Equation (2)]} \\ \underline{5x - 4y + 3z = 7} & \text{Equation (4)} \\ -x \qquad - 5z = -5 & \end{array}$$

Step 2 Now let us eliminate y (the same variable) from Equations (2) and (3). Multiply Equation (2) by 3 and Equation (3) by 2 and add:

$$\begin{array}{rl} 9x - 6y + 12z = 18 & \text{3[Equation (2)]} \\ \underline{4x + 6y - 10z = -16} & \text{2[Equation (3)]} \\ 13x \qquad + 2z = 2 & \end{array}$$

Step 3 From steps 1 and 2 we obtain the system

$$-x - 5z = -5 \qquad (5)$$
$$13x + 2z = 2 \qquad (6)$$

We solve this system as in Section 8-1. Multiply Equation (5) by 13 and add to Equation (6) to eliminate x:

$$\begin{array}{rl} -13x - 65z = -65 & \text{13[Equation (5)]} \\ \underline{13x + 2z = 2} & \\ -63z = -63 & \\ z = 1 & \end{array}$$

Substitute $z = 1$ back into either Equation (5) or (6)—we choose Equation (5)—to find x:

$$\begin{aligned} -x - 5z &= -5 \\ -x - 5 \cdot \mathbf{1} &= -5 \\ -x &= 0 \\ x &= 0 \end{aligned}$$

Step 4 Substitute $x = 0$ and $z = 1$ back into any of the three original equations—we choose Equation (2)—to find y:

$$\begin{aligned} 3x - 2y + 4z &= 6 \\ 3 \cdot \mathbf{0} - 2y + 4 \cdot \mathbf{1} &= 6 \\ -2y + 4 &= 6 \\ -2y &= 2 \\ y &= -1 \end{aligned}$$

Thus, the solution to the original system is $(0, -1, 1)$ or $x = 0$, $y = -1$, $z = 1$.

Check To check the solution, we must check *each* equation in the original system:

$$\begin{aligned} 3x - 2y + 4z &= 6 \\ 3 \cdot 0 - 2(-1) + 4 \cdot 1 &\stackrel{?}{=} 6 \\ 6 &\stackrel{\checkmark}{=} 6 \end{aligned} \qquad \begin{aligned} 2x + 3y - 5z &= -8 \\ 2 \cdot 0 + 3(-1) - 5 \cdot 1 &\stackrel{?}{=} -8 \\ -8 &\stackrel{\checkmark}{=} -8 \end{aligned}$$

$$\begin{aligned} 5x - 4y + 3z &= 7 \\ 5 \cdot 0 - 4(-1) + 3 \cdot 1 &\stackrel{?}{=} 7 \\ 7 &\stackrel{\checkmark}{=} 7 \end{aligned}$$

Steps in Solving Systems of Three Equations in Three Variables

Step 1 Choose two equations from the system and eliminate one of the three variables, using elimination by addition or subtraction. The result is generally one equation in two variables.

Step 2 Now eliminate the same variable from the unused equation and one of those used in step 1. We (generally) obtain another equation in two variables.

Step 3 The two equations from steps 1 and 2 form a system of two equations and two variables. Solve as in the preceding section.

Step 4 Substitute the solution from step 3 into any of the three original equations and solve for the third variable to complete the solution of the original system.

Problem 14 Solve:

$$\begin{aligned} 2x + 3y - 5z &= -12 \\ 3x - 2y + 2z &= 1 \\ 4x - 5y - 4z &= -12 \end{aligned}$$

If we encounter, in the process described above, an equation that states a contradiction, such as $0 = -2$, then we must conclude that the system has no solution (that is, the system is **inconsistent**). If, on the other hand, one of the equations turns out to be $0 = 0$, the system either has infinitely many solutions or it has none. We must proceed further to determine which. Notice how this last result differs from the two-equation–two-unknown case. There, when we obtained $0 = 0$, we *knew* that there were infinitely many solutions. If a system has infinitely many solutions, then it is said to be **dependent**.

For completeness, let us look at a system that turns out to be dependent to see how the solution set can be represented. Consider the system

$$x + y - z = 2 \tag{7}$$
$$3x + 2y - z = 5 \tag{8}$$
$$5x + 2y + z = 7 \tag{9}$$

We choose to eliminate z from two equations by adding Equation (9) to Equation (7) and by adding Equation (9) to Equation (8). Doing this we obtain the system

$$6x + 3y = 9$$
$$8x + 4y = 12$$

By multiplying the top equation by $\frac{1}{3}$ and the bottom equation by $\frac{1}{4}$, we obtain the simpler system

$$2x + y = 3 \tag{10}$$
$$2x + y = 3 \tag{11}$$

Since these two equations are the same, the original system must be dependent. [If we multiply either Equation (10) or (11) by -1 and add the result to the other we will obtain $0 = 0$.] To represent the solution set of the original system, we proceed as follows. Solve Equation (10) for y in terms of x:

$$y = 3 - 2x \tag{12}$$

Now replace y by $3 - 2x$ in any of the original equations and solve for z. We use Equation (9):

$$\begin{aligned} 5x + 2y + z &= 7 \\ 5x + 2(3 - 2x) + z &= 7 \\ 5x + 6 - 4x + z &= 7 \\ z &= 1 - x \end{aligned}$$

Thus, for *any* real number x, the ordered triplet (x, y, z) becomes

$$(x, 3 - 2x, 1 - x)$$

and is a solution of the original system. For example:

If $x = 1$, then $(\mathbf{1}, 3 - 2 \cdot \mathbf{1}, 1 - \mathbf{1}) = (1, 1, 0)$ is a solution.
If $x = -3$, then $(\mathbf{-3}, 3 - 2(\mathbf{-3}), 1 - (\mathbf{-3})) = (-3, 9, 4)$ is a solution.
And so on.

Other approaches to solving systems of three equations in three unknowns are given in the next section and in Appendix C.

APPLICATION

We now consider a real-world problem that leads to a system of equations in three variables.

Example 15 *Production scheduling* A small manufacturing plant makes three types of inflatable boats: one-person, two-person, and four-person models. Each boat requires the services of three departments, as listed in the table. The cutting, assembly, and packaging departments have available a maximum of 380, 330, and 120 workhours per week, respectively. How many boats of each type must be produced each week for the plant to operate at full capacity?

	ONE-PERSON BOAT	TWO-PERSON BOAT	FOUR-PERSON BOAT
CUTTING DEPARTMENT	0.6 hour	1.0 hour	1.5 hours
ASSEMBLY DEPARTMENT	0.6 hour	0.9 hour	1.2 hours
PACKAGING DEPARTMENT	0.2 hour	0.3 hour	0.5 hour

Solution Let

x = Number of one-person boats produced per week
y = Number of two-person boats produced per week
z = Number of four-person boats produced per week

The 380 workhours available in the cutting department will be used up by $0.6x$ workhours on the x one-person boats, $1.0y$ on the y two-person boats, and $1.5z$ on the z four-person boats, so that $380 = 0.6x + 1.0y + 1.5z$. Similar equations are obtained for the other departments. Thus,

$$
\begin{aligned}
0.6x + 1.0y + 1.5z &= 380 \quad \text{Cutting department}\\
0.6x + 0.9y + 1.2z &= 330 \quad \text{Assembly department}\\
0.2x + 0.3y + 0.5z &= 120 \quad \text{Packaging department}
\end{aligned}
$$

We can clear the system of decimals, if desired, by multiplying each side of each equation by 10. Thus,

$$6x + 10y + 15z = 3{,}800 \tag{13}$$
$$6x + 9y + 12z = 3{,}300 \tag{14}$$
$$2x + 3y + 5z = 1{,}200 \tag{15}$$

Let us start by eliminating x from Equations (13) and (14):

$$
\text{Add}\begin{cases}
6x + 10y + 15z = 3{,}800 & \text{Equation (13)}\\
-6x - 9y - 12z = -3{,}000 & -1[\text{Equation (14)}]
\end{cases}
$$
$$y + 3z = 500$$

Now we eliminate x from Equations (13) and (15):

$$
\text{Add}\begin{cases}
6x + 10y + 15z = 3{,}800 & \text{Equation (13)}\\
-6x - 9y - 15z = -3{,}600 & -3[\text{Equation (15)}]
\end{cases}
$$
$$y = 200$$

Substituting $y = 200$ into $y + 3z = 500$, we can solve for z:

$$
\begin{aligned}
200 + 3z &= 500\\
3z &= 300\\
z &= 100
\end{aligned}
$$

Now use Equation (13), (14), or (15) to find x (we use Equation 15):

$$
\begin{aligned}
2x + 3y + 5z &= 1{,}200\\
2x + 3(200) + 5(100) &= 1{,}200\\
2x &= 100\\
x &= 50
\end{aligned}
$$

Thus, each week the company should produce 50 one-person boats, 200 two-person boats, and 100 four-person boats to operate at full capacity. The check of the solution is left to you.

Problem 15 Repeat Example 15 assuming the cutting, assembly, and packaging departments have available a maximum of 260, 234, and 82 workhours per week, respectively.

ANSWERS TO MATCHED PROBLEMS

14. $(-1, 0, 2)$

15. 100 one-person boats, 140 two-person boats, and 40 four-person boats

EXERCISE 8-3

Solve and check each system.

A

1.
$$\begin{aligned} -2x \phantom{{}+2y+3z} &= 2 \\ x - 3y \phantom{{}+3z} &= 2 \\ -x + 2y + 3z &= -7 \end{aligned}$$

2.
$$\begin{aligned} 2y + z &= -4 \\ x - 3y + 2z &= 9 \\ -y &= 3 \end{aligned}$$

3.
$$\begin{aligned} 4y - z &= -13 \\ 3y + 2z &= 4 \\ 6x - 5y - 2z &= 0 \end{aligned}$$

4.
$$\begin{aligned} 2x + z &= -5 \\ x - 3z &= -6 \\ 4x + 2y - z &= -9 \end{aligned}$$

B

5.
$$\begin{aligned} 2x + y - z &= 5 \\ x - 2y - 2z &= 4 \\ 3x + 4y + 3z &= 3 \end{aligned}$$

6.
$$\begin{aligned} x - 3y + z &= 4 \\ -x + 4y - 4z &= 1 \\ 2x - y + 5z &= -3 \end{aligned}$$

7.
$$\begin{aligned} 2a + 4b + 3c &= 6 \\ a - 3b + 2c &= -7 \\ -a + 2b - c &= 5 \end{aligned}$$

8.
$$\begin{aligned} 3u - 2v + 3w &= 11 \\ 2u + 3v - 2w &= -5 \\ u + 4v - w &= -5 \end{aligned}$$

9.
$$\begin{aligned} 2x - 3y + 3z &= -15 \\ 3x + 2y - 5z &= 19 \\ 5x - 4y - 2z &= -2 \end{aligned}$$

10.
$$\begin{aligned} 3x - 2y - 4z &= -8 \\ 4x + 3y - 5z &= -5 \\ 6x - 5y + 2z &= -17 \end{aligned}$$

11.
$$\begin{aligned} 5x - 3y + 2z &= 13 \\ 2x + 4y - 3z &= -9 \\ 4x - 2y + 5z &= 13 \end{aligned}$$

12.
$$\begin{aligned} 4x - 2y + 3z &= 0 \\ 3x - 5y - 2z &= -12 \\ 2x + 4y - 3z &= -4 \end{aligned}$$

C

13.
$$\begin{aligned} x - 8y + 2z &= -1 \\ x - 3y + z &= 1 \\ 2x - 11y + 3z &= 2 \end{aligned}$$

14.
$$\begin{aligned} -x + 2y - z &= -4 \\ 4x + y - 2z &= 1 \\ x + y - z &= -1 \end{aligned}$$

15.
$$\begin{aligned} 4w - x &= 5 \\ -3w + 2x - y &= -5 \\ 2w - 5x + 4y + 3z &= 13 \\ 2w + 2x - 2y - z &= -2 \end{aligned}$$

16.
$$\begin{aligned} 2r - s + 2t - u &= 5 \\ r - 2s + t + u &= 1 \\ -r + s - 3t - u &= -1 \\ -r - 2s + t + 2u &= -4 \end{aligned}$$

APPLICATIONS

17. ***Geometry*** A circle in a rectangular coordinate system can be written in the form $x^2 + y^2 + Dx + Ey + F = 0$. Find D, E, and F so that the circle passes through $(-2, -1)$, $(-1, -2)$, and $(6, -1)$.

18. ***Geometry*** Repeat Problem 17 with the circle passing through $(6, -8)$, $(6, 0)$, and $(0, -8)$.

★19. ***Production scheduling*** A garment company manufactures three shirt styles. Each style of shirt requires the services of three departments, as listed in the table. The cutting, sewing, and packaging departments have available a maximum of 1,160, 1,560, and 480 workhours per week, respectively. How many of each style shirt must be produced each week for the plant to operate at full capacity?

	STYLE A	STYLE B	STYLE C
CUTTING DEPARTMENT	0.2 hour	0.4 hour	0.3 hour
SEWING DEPARTMENT	0.3 hour	0.5 hour	0.4 hour
PACKAGING DEPARTMENT	0.1 hour	0.2 hour	0.1 hour

★20. ***Production scheduling*** Repeat Problem 19 with the cutting, sewing, and packaging departments having available a maximum of 1,180, 1,560, and 510 workhours per week, respectively.

★21. ***Diet*** In an experiment involving guinea pigs, a zoologist finds she needs a food mix that contains, among other things, 23 grams of protein, 6.2 grams of fat, and 16 grams of moisture. She has on hand mixes with the compositions shown in the table. How many grams of each mix should she use to get the desired diet mix?

MIX	PROTEIN (%)	FAT (%)	MOISTURE (%)
A	20	2	15
B	10	6	10
C	15	5	5

★22. ***Diet*** Repeat Problem 21 assuming the diet mix is to contain 18.5 grams of protein, 4.9 grams of fat, and 13 grams of moisture.

★★23. ***Business*** A newspaper firm uses three printing presses, of different ages and capacities, to print the evening paper. With all three presses running, the paper can be printed in 2 hours. If the newest press breaks down, the older two presses can print the paper in 4 hours; if the middle press breaks down, the newest and oldest together can print the paper in 3 hours. How long would it take each press alone to print the paper? [*Hint:* Use $(2/x) + (2/y) + (2/z) = 1$ as one of the equations.]

8-4 SYSTEMS OF EQUATIONS AND MATRICES (OPTIONAL)

- Matrices and Augmented Matrices
- Solving Linear Systems Using Augmented Matrix Methods

In solving systems of equations by elimination in the preceding sections, the coefficients of the variables and constant terms played a central role. The process can be made more efficient for generalization and computer work by the introduction of a mathematical form called a matrix (plural: matrices).

MATRICES AND AUGMENTED MATRICES

A **matrix** is a rectangular array of numbers written within brackets. Some examples are

$$\begin{bmatrix} 3 & 5 \\ 0 & -2 \end{bmatrix}$$

$$\begin{bmatrix} 2 \\ -3 \\ 0 \end{bmatrix}$$

$$\begin{bmatrix} 1 & -1 & 0 & 5 \end{bmatrix}$$

$$\begin{bmatrix} -1 & 2 & -5 & 0 \\ 0 & 3 & 2 & 1 \end{bmatrix}$$

$$\begin{bmatrix} 1 & 0 & 0 \\ 0 & 1 & 0 \\ 0 & 0 & 1 \end{bmatrix}$$

Each number in a matrix is called an **element** of the matrix.

Each linear system of the form

$$\begin{aligned} ax + by &= m \\ cx + dy &= n \end{aligned} \tag{1}$$

where x and y are variables, is associated with a matrix called the **augmented matrix** of the system:

$$\left[\begin{array}{cc|c} a & b & m \\ c & d & n \end{array}\right] \tag{2}$$

This matrix contains the essential parts of system (1). The vertical bar is included only to separate the coefficients of the variables from the constant terms. Our objective is to learn how to manipulate augmented matrices in such a way that a solution to system (1) will result, if a solution exists. The manipulative process is a direct outgrowth of the elimination process discussed in Sections 8-1 and 8-3.

Recall that two linear systems are said to be **equivalent** if they have exactly the same solution set. How did we transform linear systems into equivalent linear systems? We used a result, which we restate here as Theorem 1 for convenient reference:

Theorem 1

Producing Equivalent Systems

A system of linear equations is transformed into an equivalent system if:

1. Two equations are interchanged.
2. An equation is multiplied by a nonzero constant.
3. A constant multiple of another equation is added to a given equation.

Paralleling the previous discussion, we say that two augmented matrices are **row-equivalent**, denoted by the symbol ~ between the two matrices, if they are augmented matrices of equivalent systems of equations. (Think about this.) How do we transform augmented matrices into row-equivalent matrices? We use Theorem 2 below, which is a direct consequence of Theorem 1.

Theorem 2

Producing Row-Equivalent Matrices

An augmented matrix is transformed into a row-equivalent matrix if:

1. Two rows are interchanged ($R_i \leftrightarrow R_j$).
2. A row is multiplied by a nonzero constant ($kR_i \rightarrow R_i$).
3. A constant multiple of another row is added to a given row ($R_i + kR_j \rightarrow R_i$).

[*Note:* The arrow $\rightarrow$ means "replaces."]
These three operations are called **row operations**.

SOLVING LINEAR SYSTEMS USING AUGMENTED MATRIX METHODS

The use of Theorem 2 in solving systems in the form of system (1) is best illustrated by examples.

Example 16 Solve using augmented matrix methods:

$$\begin{aligned} 3x + 4y &= 1 \\ x - 2y &= 7 \end{aligned} \tag{3}$$

Solution We start by writing the augmented matrix corresponding to system (3):

$$\left[\begin{array}{cc|c} 3 & 4 & 1 \\ 1 & -2 & 7 \end{array}\right] \tag{4}$$

Our objective is to use row operations from Theorem 2 to try to transform system (4) into the form

$$\left[\begin{array}{cc|c} 1 & k & m \\ 0 & 1 & n \end{array}\right] \tag{5}$$

where k, m, and n are real numbers. The solution to system (3) will then be easy since matrix (5) will be the augmented matrix of the following system, which can be solved by substitution:

$$\begin{aligned} x + ky &= m \\ y &= n \end{aligned}$$

We now proceed to use row operations to transform (4) into form (5).

Step 1 To get a 1 in the upper left corner, we interchange rows 1 and 2 (Theorem 2, part 1).

$$\left[\begin{array}{cc|c} 3 & 4 & 1 \\ 1 & -2 & 7 \end{array}\right] \overset{R_1 \leftrightarrow R_2}{\sim} \left[\begin{array}{cc|c} 1 & -2 & 7 \\ 3 & 4 & 1 \end{array}\right]$$

Step 2 To get a 0 in the lower left corner, we multiply R_1 by (-3) and add the result to R_2 (Theorem 2, part 3). The operation here is to replace R_2 by $R_2 + (-3)R_1$ so that this changes R_2 but not R_1. Some people find it useful to write $(-3)R_1$ outside the matrix to help reduce errors in arithmetic, as shown.

$$\begin{array}{ccc} -3 & 6 & -21 \end{array}$$

$$\left[\begin{array}{cc|c} 1 & -2 & 7 \\ 3 & 4 & 1 \end{array}\right] \overset{R_2 + (-3)R_1 \to R_2}{\sim} \left[\begin{array}{cc|c} 1 & -2 & 7 \\ 0 & 10 & -20 \end{array}\right]$$

Step 3 To get a 1 in the second row, second column, we multiply R_2 by $\frac{1}{10}$ (Theorem 2, part 2).

$$\left[\begin{array}{cc|c} 1 & -2 & 7 \\ 0 & 10 & -20 \end{array}\right] \overset{\frac{1}{10}R_2 \to R_2}{\sim} \left[\begin{array}{cc|c} 1 & -2 & 7 \\ 0 & 1 & -2 \end{array}\right]$$

We have accomplished our objective! The last matrix is the augmented matrix for the system

$$\begin{aligned} x - 2y &= 7 \\ y &= -2 \end{aligned} \tag{6}$$

which can be solved by substitution. Substituting $y = -2$ into the first equation, we obtain $x + 4 = 7$ or $x = 3$. Since system (6) is equivalent to system (3), our starting system, we have solved (3); that is, $x = 3$ and $y = -2$.

Check

$$\begin{aligned} 3x + 4y &= 1 \\ 3(3) + 4(-2) &\overset{?}{=} 1 \\ 9 - 8 &\overset{\checkmark}{=} 1 \end{aligned} \qquad \begin{aligned} x - 2y &= 7 \\ 3 - 2(-2) &\overset{?}{=} 7 \\ 3 + 4 &\overset{\checkmark}{=} 7 \end{aligned}$$

The solution process above is written more compactly as follows:

Step 1: Need a 1 here

$$\left[\begin{array}{cc|c} 3 & 4 & 1 \\ 1 & -2 & 7 \end{array}\right] \quad R_1 \leftrightarrow R_2$$

Step 2: Need a 0 here

$$\sim \left[\begin{array}{cc|c} 1 & -2 & 7 \\ 3 & 4 & 1 \end{array}\right] \quad R_2 + (-3)R_1 \to R_2$$

$$-3 \quad 6 \quad -21$$

Step 3: Need a 1 here

$$\sim \left[\begin{array}{cc|c} 1 & -2 & 7 \\ 0 & 10 & -20 \end{array}\right] \quad \tfrac{1}{10}R_2 \to R_2$$

$$\sim \left[\begin{array}{cc|c} 1 & -2 & 7 \\ 0 & 1 & -2 \end{array}\right]$$

Problem 16 Solve using augmented matrix methods:

$$\begin{aligned} 2x - y &= -7 \\ x + 2y &= 4 \end{aligned}$$

Example 17 Solve using augmented matrix methods:

$$\begin{aligned} 2x - 3y &= 7 \\ 3x + 4y &= 2 \end{aligned}$$

Solution

Step 1: Need a 1 here

$$\begin{bmatrix} 2 & -3 & | & 7 \\ 3 & 4 & | & 2 \end{bmatrix} \quad \tfrac{1}{2}R_1 \to R_1$$

Step 2: Need a 0 here

$$\sim \begin{bmatrix} 1 & -\frac{3}{2} & | & \frac{7}{2} \\ 3 & 4 & | & 2 \end{bmatrix} \quad R_2 + (-3)R_1 \to R_2$$

$$-3 \quad \tfrac{9}{2} \quad -\tfrac{21}{2}$$

Step 3: Need a 1 here

$$\sim \begin{bmatrix} 1 & -\frac{3}{2} & | & \frac{7}{2} \\ 0 & \frac{17}{2} & | & -\frac{17}{2} \end{bmatrix} \quad \tfrac{2}{17}R_2 \to R_2$$

$$\sim \begin{bmatrix} 1 & -\frac{3}{2} & | & \frac{7}{2} \\ 0 & 1 & | & -1 \end{bmatrix}$$

The original system is therefore equivalent to

$$x - \tfrac{3}{2}y = \tfrac{7}{2}$$
$$y = -1$$

Thus, $y = -1$ and $x + \frac{3}{2} = \frac{7}{2}$ so $x = 2$. You should check the solution $(2, -1)$.

Problem 17 Solve using augmented matrix methods:

$$5x - 2y = 12$$
$$2x + 3y = 1$$

Example 18 Solve using augmented matrix methods:

$$2x - y = 4$$
$$-6x + 3y = -12$$

Solution

$$\begin{bmatrix} 2 & -1 & | & 4 \\ -6 & 3 & | & -12 \end{bmatrix} \quad \begin{matrix} \tfrac{1}{2}R_1 \to R_1 \text{ (This produces a 1 in the upper left corner.)} \\ \tfrac{1}{3}R_2 \to R_2 \text{ (This simplifies } R_2\text{.)} \end{matrix}$$

$$\sim \begin{bmatrix} 1 & -\frac{1}{2} & | & 2 \\ -2 & 1 & | & -4 \end{bmatrix} \quad R_2 + 2R_1 \to R_2 \text{ (This produces a 0 in the lower left corner.)}$$

$$2 \quad -1 \quad 4$$

$$\sim \begin{bmatrix} 1 & -\frac{1}{2} & | & 2 \\ 0 & 0 & | & 0 \end{bmatrix}$$

The last matrix corresponds to the system

$$x - \tfrac{1}{2}y = 2$$
$$0x + 0y = 0$$

Thus, $x = \frac{1}{2}y + 2$. Hence, for any real number y, $(\frac{1}{2}y + 2, y)$ is a solution. If $y = 6$, for example, then (5, 6) is a solution; if $y = -2$, then $(1, -2)$ is a solution; and so on. Geometrically, the graphs of the two original equations coincide and there are infinitely many solutions. In general, if we end up with a row of 0's in an augmented matrix for a two-equation–two-variable system, the system is dependent and there are infinitely many solutions.

Problem 18 Solve using augmented matrix methods:
$$\begin{aligned} -2x + 6y &= 6 \\ 3x - 9y &= -9 \end{aligned}$$

Example 19 Solve using augmented matrix methods:

$$\begin{aligned} 2x + 6y &= -3 \\ x + 3y &= 2 \end{aligned}$$

Solution

$$\left[\begin{array}{cc|c} 2 & 6 & -3 \\ 1 & 3 & 2 \end{array}\right] \quad R_1 \leftrightarrow R_2$$

$$\sim \left[\begin{array}{cc|c} 1 & 3 & 2 \\ 2 & 6 & -3 \end{array}\right] \quad R_2 + (-2)R_1 \to R_2$$

$$\begin{array}{ccc} -2 & -6 & -4 \end{array}$$

$$\sim \left[\begin{array}{cc|c} 1 & 3 & 2 \\ 0 & 0 & -7 \end{array}\right] \quad R_2 \text{ implies the contradiction } 0 = -7.$$

The system is inconsistent and has no solution—otherwise, we have proved that $0 = -7$! Thus, if in a row of an augmented matrix we obtain all 0's to the left of the vertical bar and a nonzero number to the right of the bar, the system is inconsistent and there are no solutions.

Problem 19 Solve using augmented matrix methods:
$$\begin{aligned} 2x - y &= 3 \\ 4x - 2y &= -1 \end{aligned}$$

Summary

FORM 1 A UNIQUE SOLUTION	FORM 2 INFINITELY MANY SOLUTIONS (DEPENDENT)	FORM 3 NO SOLUTION (INCONSISTENT)
$\left[\begin{array}{cc\|c} 1 & k & m \\ 0 & 1 & n \end{array}\right]$	$\left[\begin{array}{cc\|c} 1 & m & n \\ 0 & 0 & 0 \end{array}\right]$	$\left[\begin{array}{cc\|c} 1 & m & n \\ 0 & 0 & p \end{array}\right]$

k, m, n, p real numbers; $p \neq 0$

The augmented matrix method is readily applied to systems with three (or more) variables.

Example 20 Solve using augmented matrix methods:

$$\begin{aligned} 2x - 2y + z &= 3 \\ 3x + y - z &= 7 \\ x - 3y + 2z &= 0 \end{aligned}$$

Solution We change to equivalent matrices, aiming at a matrix of the form

$$\left[\begin{array}{ccc|c} 1 & m & n & p \\ 0 & 1 & q & r \\ 0 & 0 & 1 & s \end{array}\right]$$

as our final result.

Need a 1 here

$$\left[\begin{array}{ccc|c} 2 & -2 & 1 & 3 \\ 3 & 1 & -1 & 7 \\ 1 & -3 & 2 & 0 \end{array}\right] \quad R_1 \leftrightarrow R_3$$

Need 0's here

$$\sim \left[\begin{array}{ccc|c} 1 & -3 & 2 & 0 \\ 3 & 1 & -1 & 7 \\ 2 & -2 & 1 & 3 \end{array}\right] \quad \begin{array}{l} R_2 + (-3)R_1 \to R_2 \\ R_3 + (-2)R_1 \to R_3 \end{array}$$

Need a 1 here

$$\sim \left[\begin{array}{ccc|c} 1 & -3 & 2 & 0 \\ 0 & 10 & -7 & 7 \\ 0 & 4 & -3 & 3 \end{array}\right] \quad \tfrac{1}{10}R_2 \to R_2$$

Need a 0 here

$$\sim \left[\begin{array}{ccc|c} 1 & -3 & 2 & 0 \\ 0 & 1 & -\frac{7}{10} & \frac{7}{10} \\ 0 & 4 & -3 & 3 \end{array}\right] \quad R_3 + (-4)R_2 \to R_3$$

Need a 1 here

$$\sim \left[\begin{array}{ccc|c} 1 & -3 & 2 & 0 \\ 0 & 1 & -\frac{7}{10} & \frac{7}{10} \\ 0 & 0 & -\frac{1}{5} & \frac{1}{5} \end{array}\right] \quad (-5)R_3 \to R_3$$

$$\sim \left[\begin{array}{ccc|c} 1 & -3 & 2 & 0 \\ 0 & 1 & -\frac{7}{10} & \frac{7}{10} \\ 0 & 0 & 1 & -1 \end{array}\right]$$

The resulting equivalent system

$$\begin{aligned} x - 3y + 2z &= 0 \\ y - \tfrac{7}{10}z &= \tfrac{7}{10} \\ z &= -1 \end{aligned}$$

is solved by substitution:

$$y - \tfrac{7}{10}z = y + \tfrac{7}{10} = \tfrac{7}{10} \qquad \text{so} \qquad y = 0$$
$$x - 3y + 2z = x - 0 - 2 = 0 \qquad \text{so} \qquad x = 2$$

You should check the solution (2, 0, −1).

Problem 20 Solve using augmented matrix methods:

$$\begin{aligned} 3x + y - 2z &= 2 \\ x - 2y + z &= 3 \\ 2x - y - 3z &= 3 \end{aligned}$$

The process of solving systems of equations described in this section is known as **gaussian elimination**. It is a powerful method that can be used to solve large-scale systems on a computer. The concept of a matrix is also a powerful mathematical tool with applications far beyond solving systems of equations. Matrices are discussed further in subsequent courses.

ANSWERS TO MATCHED PROBLEMS

16. $x = -2$, $y = 3$ **17.** $x = 2$, $y = -1$
18. The system is dependent. For y any real number, $(3y - 3, y)$ is a solution.
19. Inconsistent—no solution. **20.** $x = 1$, $y = -1$, $z = 0$

EXERCISE 8-4

A *Perform each of the indicated row operations on the following matrix:*

$$\left[\begin{array}{cc|c} 1 & -3 & 2 \\ 4 & -6 & -8 \end{array}\right]$$

1. $R_1 \leftrightarrow R_2$ **2.** $\tfrac{1}{2}R_2 \to R_2$ **3.** $-4R_1 \to R_1$

4. $-2R_1 \to R_1$ **5.** $2R_2 \to R_2$ **6.** $-1R_2 \to R_2$

7. $R_2 + (-4)R_1 \to R_2$ **8.** $R_1 + (-\tfrac{1}{2})R_2 \to R_1$

9. $R_2 + (-2)R_1 \to R_2$ **10.** $R_2 + (-3)R_1 \to R_2$

11. $R_2 + (-1)R_1 \to R_2$ **12.** $R_2 + (1)R_1 \to R_2$

B *Solve using augmented matrix methods.*

13. $x + y = 5$
$x - y = 1$

14. $x - y = 2$
$x + y = 6$

15. $x - 2y = 1$
$2x - y = 5$

16. $x + 3y = 1$
$3x - 2y = 14$

17. $x - 4y = -2$
$-2x + y = -3$

18. $x - 3y = -5$
$-3x - y = 5$

19. $3x - y = 2$
$x + 2y = 10$

20. $2x + y = 0$
$x - 2y = -5$

21. $x + 2y = 4$
$2x + 4y = -8$

22. $2x - 3y = -2$
$-4x + 6y = 7$

23. $2x + y = 6$
$x - y = -3$

24. $3x - y = -5$
$x + 3y = 5$

25. $3x - 6y = -9$
$-2x + 4y = 6$

26. $2x - 4y = -2$
$-3x + 6y = 3$

27. $4x - 2y = 2$
$-6x + 3y = -3$

28. $-6x + 2y = 4$
$3x - y = -2$

C *Solve using augmented matrix methods.*

29. $3x - y = 7$
$2x + 3y = 1$

30. $2x - 3y = -8$
$5x + 3y = 1$

31. $3x + 2y = 4$
$2x - y = 5$

32. $4x + 3y = 26$
$3x - 11y = -7$

33. $0.2x - 0.5y = 0.07$
$0.8x - 0.3y = 0.79$

34. $0.3x - 0.6y = 0.18$
$0.5x - 0.2y = 0.54$

35. $2x + 4y - 10z = -2$
$3x + 9y - 21z = 0$
$x + 5y - 12z = 1$

36. $3x + 5y - z = -7$
$x + y + z = -1$
$2x + y + 11z = 7$

37. $3x + 8y - z = -18$
$2x + y + 5z = 8$
$2x + 4y + 2z = -4$

38. $2x + 7y + 15z = -12$
$4x + 7y + 13z = -10$
$3x + 6y + 12z = -9$

Augmented matrix methods extend naturally to any number of variables. Solve using augmented matrix methods.

39. $x - 2y + w = 1$
$-2x + 5y + 3z - w = -5$
$3x - 6y + z + 7w = 2$
$x - y + 4z + 7w = -3$

40. $x - y + z - w = 8$
$2x - y + 4z + w = 20$
$x + 4z + 8w = 20$
$-2x + 3y + z + 12w = -3$

8-5 SYSTEMS OF LINEAR INEQUALITIES

- Systems of Inequality Statements
- Application

As in systems of linear equations in two variables, we say that the ordered pair of numbers (x_0, y_0) is a solution of a system of linear inequalities in two variables if the ordered pair satisfies each inequality in the system. Thus, **the graph of a system of linear inequalities** is the intersection of the graphs of each inequality in the system; that is, the set of all points that satisfy each inequality in the system. In this book we will limit our investigation of solutions of systems of inequalities to graphic methods.

SYSTEMS OF INEQUALITY STATEMENTS

Examples will illustrate the process for solving a system of linear inequalities. Recall from Section 7-3 that a linear equation divides the plane into two half-planes. The solution of a linear inequality is one of the half-planes, and a test point will indicate which. For example, $x + 3y \leq 6$ is the half-plane indicated by the direction of the arrows:

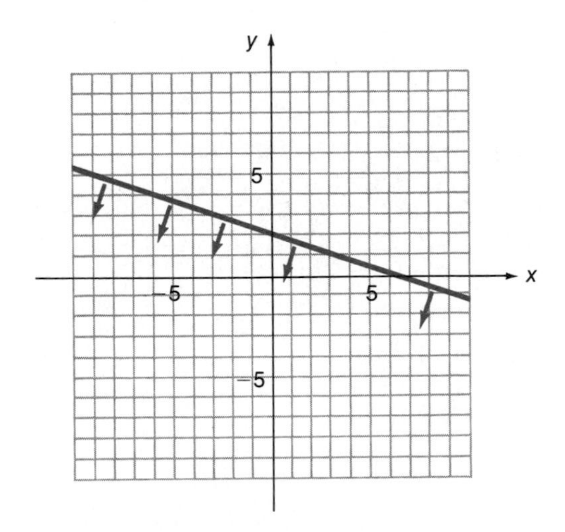

Example 21 Solve the following linear system graphically:

$$0 \leq x \leq 8$$
$$0 \leq y \leq 4$$

Solution This system is actually equivalent to the system

$$\left.\begin{aligned} x &\geq 0 \\ x &\leq 8 \\ y &\geq 0 \\ y &\leq 4 \end{aligned}\right\}$$ The solution to the system is the intersection of all four solution sets.

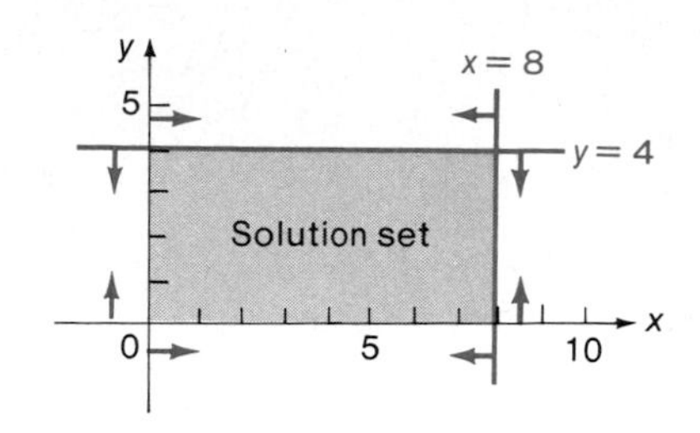

Problem 21 Solve graphically: $2 \leq x \leq 6$
$1 \leq y \leq 3$

Example 22 Solve graphically:

$$\begin{aligned} 3x + 5y &\leq 60 \\ 4x + 2y &\leq 40 \\ x &\geq 0 \\ y &\geq 0 \end{aligned}$$

Solution

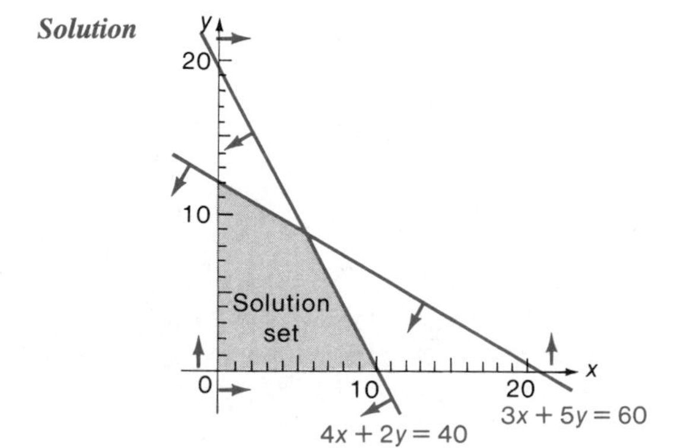

Problem 22 Solve graphically:

$$\begin{aligned} x + 2y &\geq 12 \\ 3x + 2y &\geq 24 \\ x &\geq 0 \\ y &\geq 0 \end{aligned}$$

Example 23 Solve graphically:

$$\begin{aligned} x + y &\leq 1 \\ y - x &\leq 1 \\ y &\geq -1 \end{aligned}$$

Solution

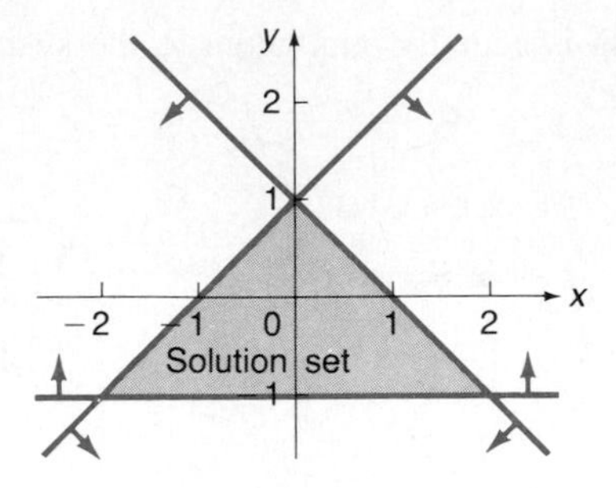

Problem 23 Solve graphically:

$$x + 3y \le 6$$
$$x - 2y \le 2$$
$$x \ge -2$$

APPLICATION

Example 24 *Production scheduling* A manufacturer of sailboards makes a standard model and a competition model. The relevant manufacturing data are shown in the table. What combinations of boards can be produced each week so as not to exceed the number of workhours available in each department per week?

	STANDARD MODEL (WORKHOURS PER BOARD)	COMPETITION MODEL (WORKHOURS PER BOARD)	MAXIMUM WORKHOURS AVAILABLE PER WEEK
FABRICATING	6	8	120
FINISHING	1	3	30

Solution Let x and y be the respective number of standard and competition boards produced per week. These variables are restricted as follows:

$$6x + 8y \le 120 \quad \text{Fabricating}$$
$$x + 3y \le 30 \quad \text{Finishing}$$
$$x \ge 0$$
$$y \ge 0$$

The solution set of this system of inequalities is the shaded area in the figure on the next page and is referred to as the **feasible region**. Any point within the shaded area would represent a possible production schedule. Any point outside the shaded area would represent an impossible schedule. For example, it would be possible to produce 10 standard boards and 5 competition boards per week, but it would not be possible to produce 13 standard boards and 6 competition boards per week.

SYST
S

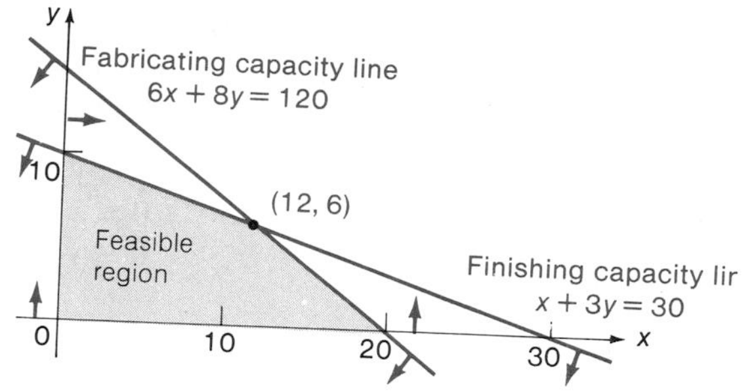

Problem 24 Repeat Example 24 using 5 hours for fabricating a standard board in place of 6 hours and a maximum of 27 workhours for the finishing department.

ANSWERS TO MATCHED PROBLEMS

21.

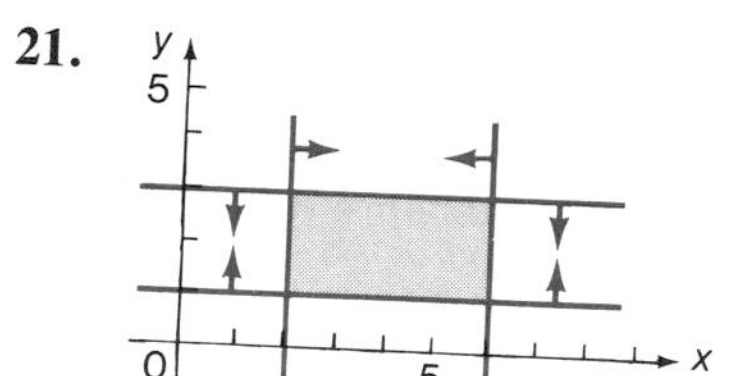

22.

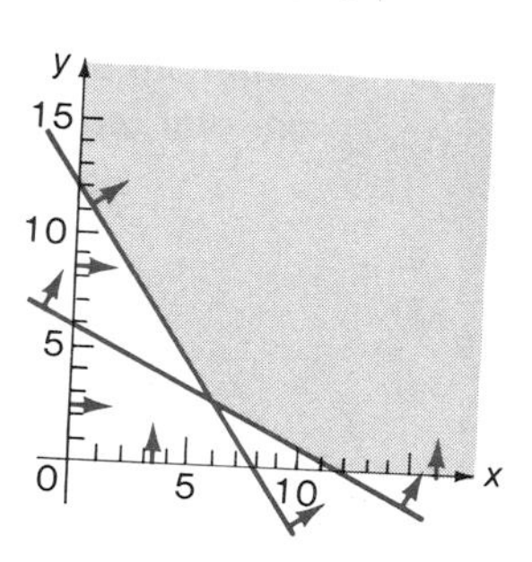

23.

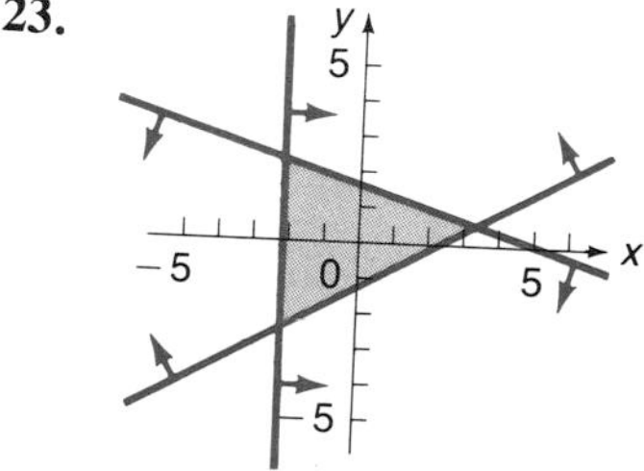

24.

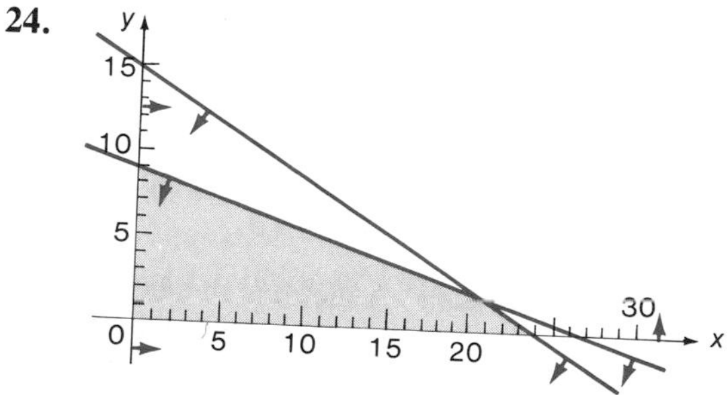

EXERCISE 8-5 *Find the solution set of each system graphically.*

A

1. $-2 \leq x < 2$
$-1 < y \leq 6$

2. $-4 \leq x < -1$
$-2 < y \leq 5$

3. $-1 \leq y \leq 2$
$x \geq -3$
$x + y \leq 1$

19. $x^2 + y^2 = 20$, $x^2 = y$

20. $x^2 - y^2 = 2$, $y^2 = x$

21. $x^2 + y^2 = 16$, $y^2 = 4 - x$

22. $x^2 + y^2 = 5$, $x^2 = 4(2 - y)$

23. Find the dimensions of a rectangle with area 32 square feet and perimeter 36 feet.

24. Find two numbers such that their sum is 1 and their product is 1.

C **25.** $2x^2 + y^2 = 18$, $xy = 4$

26. $x^2 - y^2 = 3$, $xy = 2$

27. $x^2 + 2xy + y^2 = 36$, $x^2 - xy = 0$

28. $2x^2 - xy + y^2 = 8$, $(x - y)(x + y) = 0$

29. $x^2 - 2xy + 2y^2 = 16$, $x^2 - y^2 = 0$

30. $x^2 + xy - 3y^2 = 3$, $x^2 + 4xy + 3y^2 = 0$

8-7 CHAPTER REVIEW

The standard form of a system of two linear equations in two unknowns is

$$ax + by = m$$
$$cx + dy = n$$

The **solutions of the system** are all ordered pairs (x, y) that satisfy both equations. Solutions can be estimated by **solving graphically**—that is, by graphing each equation and estimating the intersection point, if any, from the graph. Solutions can be found by **substitution** by solving one equation for one variable in terms of the other, substituting the result in the other equation, and then solving. **Equivalent systems** are those with exactly the same solution set. These operations produce equivalent systems:

1. Two equations are interchanged.
2. An equation is multiplied by a nonzero constant.
3. A constant multiple of another equation is added to a given equation.

These operations can be used to solve the system by **elimination**—that is, by eliminating one variable. A system that has no solution is called **inconsistent**; one that has an infinite number of solutions is called **dependent**. *(8-1)*

Applied problems can often be solved using systems of equations. *(8-2)*

A system of three equations in three unknowns can be solved by elimination. **Solutions**, **equivalent systems**, and **dependent systems** mean the same as for the two-equation–two-unknown case. *(8-3)*

A **matrix** is a rectangular array of numbers (**elements** of the matrix) written within brackets. The **augmented matrix** of the standard system of two equations in two unknowns is

$$\left[\begin{array}{cc|c} a & b & m \\ c & d & n \end{array}\right]$$

The operations that are performed on the system need only be done on the coefficients and constants in the augmented matrix to obtain a **row-equivalent matrix**—that is, an augmented matrix of an equivalent system. These operations are

1. Two rows are interchanged.
2. A row is multiplied by a nonzero constant.
3. A constant multiple of another row is added to a given row.

The **augmented matrix method** is a simple, concise way to solve systems and can be extended to large-scale systems. *(8-4)*

The **graph of a system of linear inequalities** is the intersection of half-planes representing the graph of each inequality. *(8-5)*

Systems involving second-degree equations can sometimes be solved by substitution. *(8-6)*

REVIEW EXERCISE 8-7

Work through all the problems in this chapter review and check answers in the back of the book. (Answers to all problems are there, and following each answer is a number in italics indicating the section in which that type of problem is discussed.) Where weaknesses show up, review appropriate sections in the text.

A

1. Solve graphically: $\begin{aligned} x - y &= 5 \\ x + y &= 7 \end{aligned}$

2. Solve graphically: $\begin{aligned} 2x + y &\le 8 \\ 2x + 3y &\le 12 \\ x &\ge 0 \\ y &\ge 0 \end{aligned}$

3. Solve by substitution: $\begin{aligned} 2x + 3y &= 7 \\ 3x - y &= 5 \end{aligned}$

4. Solve by elimination using addition: $\begin{aligned} 2x + 3y &= 7 \\ 3x - y &= 5 \end{aligned}$

Solve each system in Problems 5–8.

5. $y + 2z = 4$
$x - z = -2$
$x + y = 1$

6. $x^2 + y^2 = 2$
$2x - y = 3$

7. $x^2 - y^2 = 7$
$x^2 + y^2 = 25$

8. $x + 2y + z = 3$
$2x + 3y + 4z = 3$
$x + 2y + 3z = 1$

B

9. Solve graphically: $2x + y \geq 8$
$x + 3y \geq 12$
$x \geq 0$
$y \geq 0$

Solve each system.

10. $3x - 2y = -1$
$-6x + 4y = 3$

11. $3x - 2y - 7z = -6$
$-x + 3y + 2z = -1$
$x + 5y + 3z = 3$

12. $3x^2 - y^2 = -6$
$2x^2 + 3y^2 = 29$

13. $x^2 = y$
$y = 2x - 2$

C

14. $2x - 6y = -3$
$-\frac{2}{3}x + 2y = 1$

15. $x^2 + 2xy - y^2 = -4$
$x^2 - xy = 0$

Solve using augmented matrix methods.

16. $3x + 2y = 3$
$x + 3y = 8$

17. $x + y = 1$
$x - z = -2$
$y + 2z = 4$

18. $x + 2y + 3z = 1$
$2x + 3y + 4z = 3$
$x + 2y + z = 3$

19. $x + 2y - z = 2$
$2x + 3y + z = -3$
$3x + 5y + z = -1$

APPLICATIONS

20. ***Puzzle*** If you have 30 nickels and dimes worth \$2.30 in your pocket, how many of each do you have?

21. ***Business*** If \$6,000 is to be invested, part at 10% and the rest at 6%, how much should be invested at each rate so that the total annual return from both investments is \$440?

22. ***Geometry*** The perimeter of a rectangle is 22 centimeters. If its area is 30 square centimeters, find the length of each side.

★**23.** ***Chemistry*** A chemist has one 40% and one 70% solution of acid in stock. How much of each should she take to get 100 grams of a 49% solution?

★**24.** ***Business*** A container contains 120 packages. Some of the packages weigh $\frac{1}{2}$ pound each, and the rest weigh $\frac{1}{3}$ pound each. If the total contents of the container weigh 48 pounds, how many are there of each type of package?

★**25.** ***Diet*** A lab assistant wishes to obtain a food mix that contains, among other things, 27 grams of protein, 5.4 grams of fat, and 19 grams of moisture. He has available mixes of the compositions as listed in the table. How many grams of each mix should be used to get the desired diet mix? Set up a system of equations and solve using augmented matrix methods.

MIX	PROTEIN (%)	FAT (%)	MOISTURE (%)
A	30	3	10
B	20	5	20
C	10	4	10

9

FUNCTIONS

Correspondence rules among various sets of objects abound in our daily activities. For example:

To each person there corresponds an age.

To each item in a drugstore there corresponds a price.

To each automobile there corresponds a vehicle identification number.

To each circle there corresponds an area.

To each number there corresponds its cube.

To each nonzero real number there corresponds two square roots.

One of the most important aspects of science is establishing relationships among various types of phenomena. Once a relationship is known, predictions can be made. A chemist can use a gas law to predict the pressure of an enclosed gas given its temperature; an engineer can use a formula to predict the deflections of a beam subject to different loads; an economist would like to be able to predict interest rates given the rate of change of the money supply; and so on. In each case some rule is sought that relates the various phenomena studied, as for example, how interest rates are related to or depend on the rate of change in the money supply.

Establishing and working with such relationships are so fundamental to both pure and applied science that people have found it necessary to describe them in the precise language of mathematics. Special rules called *functions* represent one of the most important concepts in all of mathematics. Your efforts to understand and use this concept correctly right from the beginning will be rewarded many times. This chapter introduces the concept of function and relates it to several of the algebraic topics we have studied thus far.

9-1 FUNCTIONS

- Functions
- Common Ways of Specifying Functions
- Functions Specified by Equations
- A Brief History of Function

What do all the examples given on page 400 have in common? Each deals with the matching of elements from a first set, called the **domain**, with elements in a second set, called the **range**, by some correspondence rule. We will be most interested in those correspondence rules whereby each element in the domain is matched to a *unique* corresponding element in the range. Such rules are called *functions*. For example, the correspondence of automobile to vehicle identification number will be a function since the number is unique. On the other hand, the correspondence of a number to its square roots will not be a function since nonzero numbers have two square roots. We begin our study of functions in this section.

FUNCTIONS

The important term **function** is defined as follows:

> **Definition of a Function**
>
> A **function** is a rule (process or method) that produces a correspondence between a first set of elements called the **domain** and a second set of elements called the **range** such that to each element in the domain there corresponds *one and only one* element in the range.

Consider Table 1 showing three rules involving the cube, square, and square root. (The choice of small domains enables us to introduce concepts in

TABLE 1

RULE 1		RULE 2		RULE 3	
DOMAIN (NUMBER)	RANGE (CUBE)	DOMAIN (NUMBER)	RANGE (SQUARE)	DOMAIN (NUMBER)	RANGE (SQUARE ROOT)
$0 \to 0$		$-2 \to 4$		$0 \to 0$	
$1 \to 1$		$-1 \to 1$		$1 \to 1$, $1 \to -1$	
$2 \to 8$		$0 \to 0$		$4 \to 2$, $4 \to -2$	
		$1 \to 1$		$9 \to 3$, $9 \to -3$	
		$2 \to 4$			

a relatively simple setting. Shortly, we will consider rules with infinite domains.) Rules 1 and 2 are functions since to each domain value there corresponds exactly one range value. (For example, the square of -2 is 4 and no other number.) On the other hand, rule 3 is not a function, since to at least one domain value there corresponds more than one range value. (For example, to the domain value 9 there corresponds -3 and 3, both square roots of 9.)

Example 1 Indicate which rules are functions.

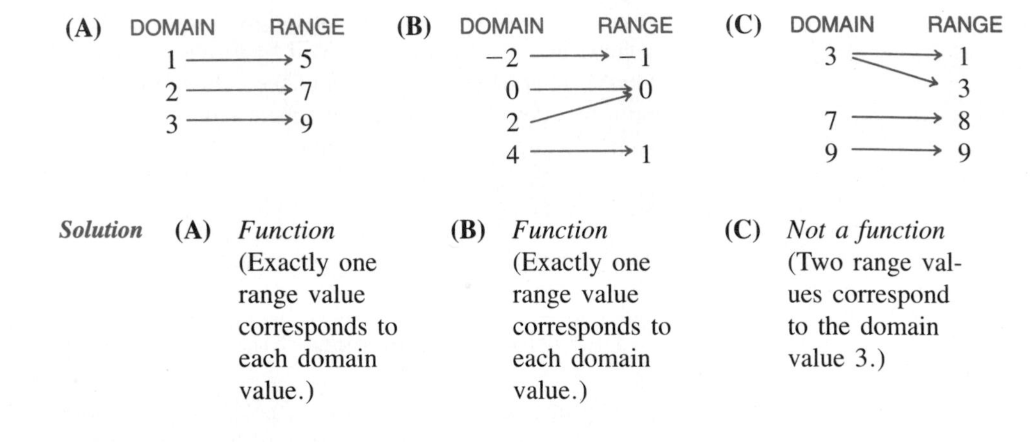

Solution

(A) *Function* (Exactly one range value corresponds to each domain value.)

(B) *Function* (Exactly one range value corresponds to each domain value.)

(C) *Not a function* (Two range values correspond to the domain value 3.)

Problem 1 Indicate which rules are functions:

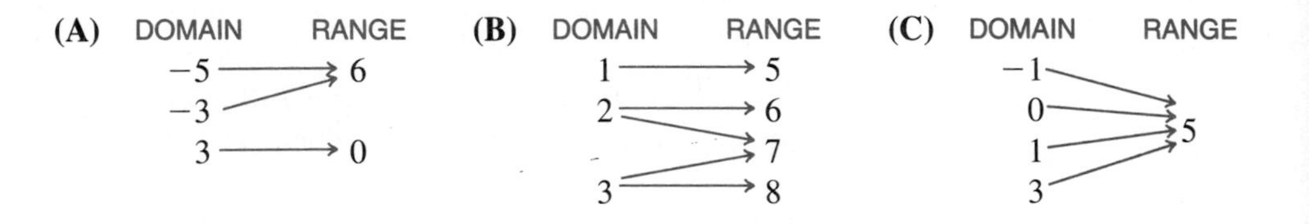

Since in a function elements in the range are paired with elements in the domain by some rule or process, this correspondence (pairing) can be illustrated using ordered pairs of elements where the first component represents a domain element and the second component a corresponding range element. Thus, we can write Rules 1–3 in Table 1 as

$$\text{Rule } 1 = \{(0, 0), (1, 1), (2, 8)\}$$
$$\text{Rule } 2 = \{(-2, 4), (-1, 1), (0, 0), (1, 1), (2, 4)\}$$
$$\text{Rule } 3 = \{(0, 0), (1, 1), (1, -1), (4, 2), (4, -2), (9, 3), (9, -3)\}$$

This suggests an alternative but equivalent way of defining functions that provides additional insight into this concept:

Definition of a Function: Set Form

A **function** is a set of ordered pairs of elements, with the restriction that no two distinct ordered pairs can have the same first component.

The set of first components in a function is called the **domain** of the function, and the set of second components is called the **range**.

According to this definition, we see (as before) that Rule 3 is not a function, since there are two distinct ordered pairs [(1, 1) and (1, −1), for example] that have the same first component. (More than one range element is associated with a given domain element.)

The rule form of the definition of a function suggests a formula or a "machine" operating on domain values to produce range values—a dynamic process. On the other hand, the set definition of this concept is closely related to graphs in a cartesian coordinate system—a static form. Each approach has its advantages in certain situations.

COMMON WAYS OF SPECIFYING FUNCTIONS

One of the main objectives of this section is to introduce you to the more common ways in which functions are specified and to provide you with experience in determining whether a given rule is or is not a function.

As a consequence of the definitions of a function, we find that a function can be specified in many different ways: by an equation, by a table, by a set of ordered pairs of elements, and by a graph, to name a few of the more common ways (see Table 2). All that matters is that we are given a set of elements called the domain and a rule (method or process) for obtaining corresponding range values for each domain value.

TABLE 2 COMMON WAYS OF SPECIFYING FUNCTIONS

METHOD	ILLUSTRATION	EXAMPLE
Equation	$y = x^2 - x,\ x \in R$†	$x = -1$ corresponds to $y = 2$
Table	<table><tr><td>m</td><td>1</td><td>2</td><td>3</td></tr><tr><td>n</td><td>1</td><td>8</td><td>27</td></tr></table>	$m = 2$ corresponds to $n = 8$
Sets of ordered pairs of elements	**(A)** $\{(1, 1), (2, 8), (3, 27)\}$ **(B)** $\{(x, y) \mid y = x^3,\ x \in R\}$	3 corresponds to 27 $x = -2$ corresponds to $y = -8$

†Recall that R is the set of real numbers.

TABLE 2 COMMON WAYS OF SPECIFYING FUNCTIONS *(Continued)*

METHOD	ILLUSTRATION	EXAMPLE
Graph		$x = 0$ corresponds to $y = -4$, $x = 2$ to $y = 0$

For a given function it is often convenient to be able to shift from one representation to another or to use more than one representation. For example, the following all specify the same function with domain $\{1, 2, 3, 4\}$:

(A) Correspondence rule:

$$\begin{aligned} 1 &\longrightarrow -2 \\ 2 &\longrightarrow -2 \\ 3 &\longrightarrow 0 \\ 4 &\longrightarrow 4 \end{aligned}$$

(B) Ordered pairs: $\{(1, -2), (2, -2), (3, 0), (4, 4)\}$

(C) Equation: $y = x^2 - 3x$

(D) Table:

x	y
1	−2
2	−2
3	0
4	4

(E) Graph:

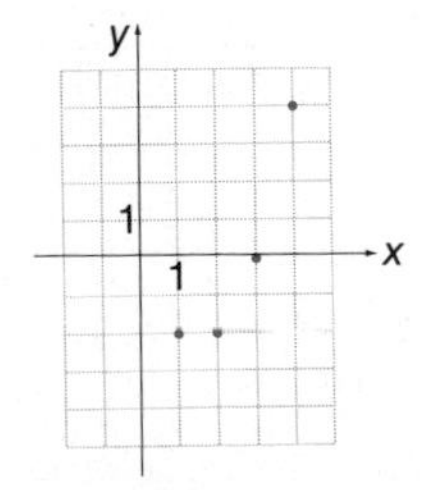

It is particularly useful to note how a function can be specified by a table and a graph. For example, a laboratory experiment may yield the table and graph shown below:

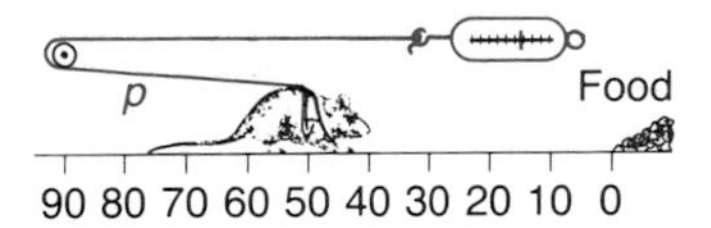

DISTANCE d (in centimeters)	PULL TOWARD FOOD p (in grams)
30	64
50	60
70	56
90	52
110	48
130	44
150	40
170	36

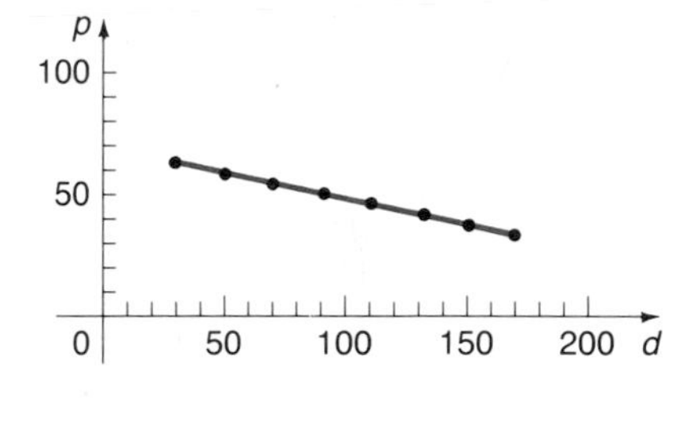

Both the table and the graph establish the same correspondence between domain values d and range values p; hence, both specify the same rule and we call the graph the graph of the function. More generally, for any function with domain and range real numbers, the graph of all ordered pairs (x, y) belonging to the function is called the **graph of the function**.

> *Note:* It is the usual practice to associate domain values with the horizontal axis and range values with the vertical axis. Thus, the first coordinate (abscissa) of the coordinates of a point on the graph is a domain value and the second coordinate (ordinate) is a range value.

We have graphed other relationships between variables x and y, where the relationship does not give y as a function of x. For example, in the graphs of $x^2 + y^2 = 16$ and $x = y^2$ on the next page, we see that x may correspond to more than one y, so neither relation is a function.

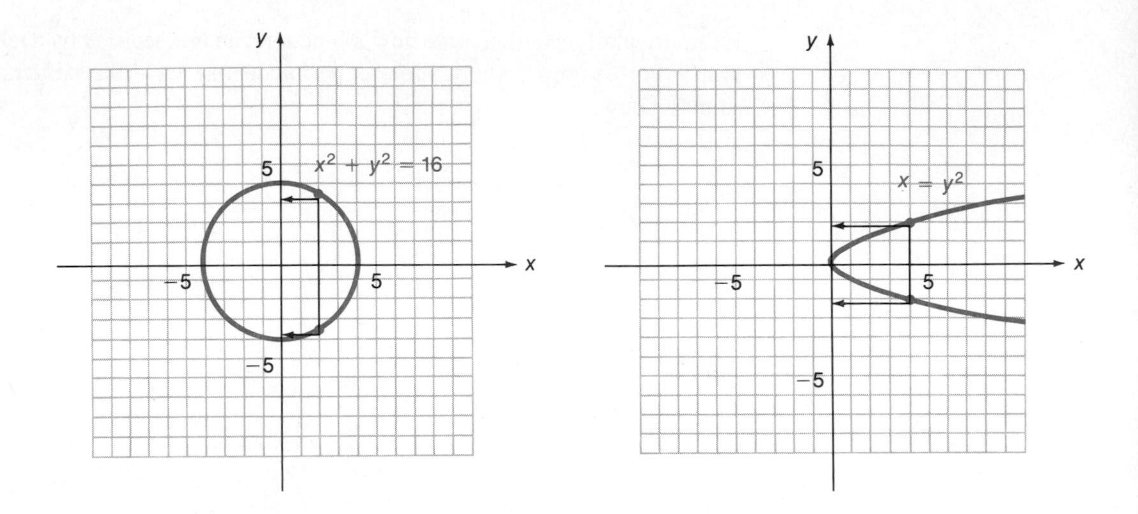

It is very easy to determine whether a relationship is a function if you have its graph.

Vertical-Line Test for a Function

A relationship between real numbers where x corresponds to y is a function if each vertical line in the coordinate system passes through *at most* one point on the graph of the relationship. (If a vertical line passes through two or more points on the graph of a relationship, then the relationship is not a function.)

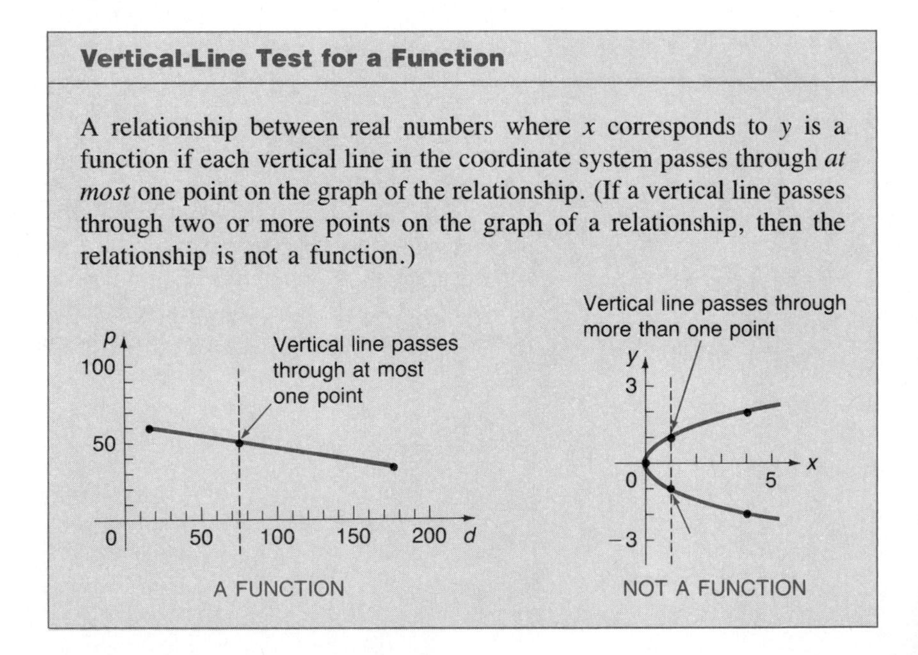

FUNCTIONS SPECIFIED BY EQUATIONS

Most of the domains and ranges included in this text will be sets of numbers, and the rules associating range values with domain values will be equations in two variables.

Consider the equation

$$y = x^2 - x$$

For each **input** x we obtain one **output** y. For example:

If $x = 3$, then $y = 3^2 - 3 = 6$.
If $x = -\frac{1}{2}$, then $y = (-\frac{1}{2})^2 - (-\frac{1}{2}) = \frac{1}{4} + \frac{1}{2} = \frac{3}{4}$.

The input values are domain values and the output values are range values. The equation (a rule) assigns each domain value x a range value y. The variable x is called an independent variable (since values are independently assigned to x from the domain), and y is called a dependent variable (since y's value depends on the value assigned to x). In general, any variable used for domain values is called an **independent variable**; any variable used for range values is called a **dependent variable**.

Unless stated to the contrary, we will adhere to the following convention regarding domains and ranges for functions specified by equations:

Agreement on Domains and Ranges

If a function is specified by an equation and the domain is not indicated, then we will assume that the domain is the set of all real number replacements of the independent variable (inputs) that produce real values for the dependent variable (outputs). The range is the set of all outputs corresponding to input values.

The domains of functions are often restricted due to physical constraints, for instance, a measurement of a length or weight that must be at least zero. There may also be mathematical restrictions on the function rule that limit the domain. Two common mathematical restrictions are that we cannot divide by zero and we cannot take an even root of a negative number.

Example 2 Assuming x is an independent variable, find the domain of the relation specified by the indicated equation:

(A) $y = 2x - 1$

(B) $y = \sqrt{x - 2}$

(C) $y = \sqrt{x^2 - 4}$

(D) $y = \dfrac{x + 1}{x - 1}$

Solution **(A)** For each real x, y is defined and is real. Thus,

Domain: R (the set of real numbers)

(B) For y to be real, $x - 2$ cannot be negative; that is,

$$x - 2 \geq 0$$
$$x \geq 2$$

Thus,

Domain: $x \geq 2$ or $[2, \infty)$

(C) For y to be real, $x^2 - 4$ cannot be negative; that is,

$$x^2 - 4 \geq 0$$
$$(x - 2)(x + 2) \geq 0$$

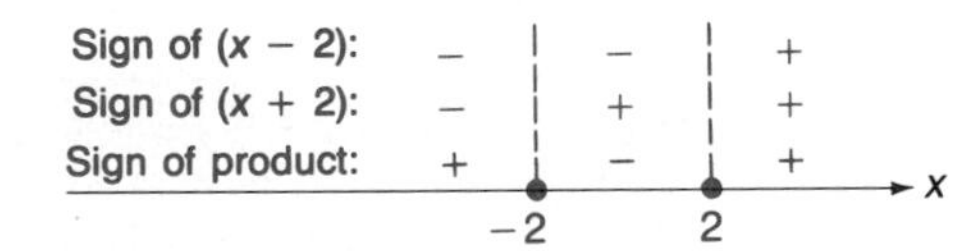

Thus,

Domain: $x \leq -2$ or $x \geq 2$ or $(-\infty, -2] \cup [2, \infty)$

(D) For y to be defined, $x - 1$ cannot be 0; that is, $x \neq 1$. Thus,

Domain: $x \neq 1$ or $(-\infty, 1) \cup (1, \infty)$

Problem 2 Assuming x is an independent variable, find the domain of the relation specified by the indicated equation:

(A) $y = \dfrac{1}{x^2 - 4}$ **(B)** $y = \sqrt{2 - x}$ **(C)** $y = \sqrt{4 - x^2}$

(D) $y = x^2 - 4$

By considering their graphs, we have already noted that the equations $x^2 + y^2 = 1$ and $x = y^2$ do not determine a function rule with x as the independent variable. An equation will specify a function with x as the independent variable when each value of x corresponds to a *unique* value of y. We can test this in several ways:

1. Graph the equation and apply the vertical-line test.
2. Solve the equation for y in terms of x and check whether each x produces a unique y.
3. Look for a particular x that has two or more corresponding y values; this determines that the equation does not specify a function.

[Remember, however, that it is permissible for two elements in the domain of a function to correspond to the same element in the range. For example, for $y = x^2$, both 2 and -2 correspond to 4.]

Example 3 Determine whether or not the equation specifies a function with x as the independent variable.

(A) $3x + 4y = 6$ **(B)** $3x^2 + 4y^2 = 1$

Solution **(A)** Solve for y, obtaining

$$y = \tfrac{1}{4}(6 - 3x)$$

to see that every value for x produces only one corresponding y value. Thus, the equation does specify the function. We could also have seen this from the graph of the equation.

(B) Solve for y:

$$4y^2 = 1 - 3x^2$$

$$y^2 = \frac{1 - 3x^2}{4}$$

$$y = \pm\frac{\sqrt{1 - 3x^2}}{2}$$

The $\pm$ indicates that any value of x for which the radical is not 0 will have two corresponding values for y. Thus, the equation does not specify a function.

We could also have observed from the original equation that for $y = 0$,

$$x = \pm\frac{\sqrt{3}}{3}$$

or we could have graphed the equation (the resulting ellipse fails the vertical-line test).

Problem 3 Determine whether or not the equation specifies a function with x as the independent variable.

(A) $3x^2 - 4y^2 = 1$ **(B)** $3x^2 + 4y = 6$

A BRIEF HISTORY OF FUNCTION

In reviewing the history of function, we are made aware of the tendency of mathematicians to extend and generalize a concept. The word "function" appears to have been first used by Leibniz in 1694 to stand for any quantity associated with a curve. By 1718, Johann Bernoulli considered a function any expression made up of constants and a variable. Later in the same century, Euler came to regard a function as any equation made up of constants and variables. Euler made extensive use of the extremely important notation $f(x)$, which we will consider in the next section, although its origin is generally attributed to Clairaut (1734).

The form of the definition of function that was used until well into this century (many texts still contain this definition) was formulated by Dirichlet (1805–1859). He stated that if two variables x and y are so related that for each value of x there corresponds exactly one value of y, then y is said to be a (single-valued) function of x. He called x, the variable to which values are assigned at will, the independent variable, and y, the variable whose values depend on the values assigned to x, the dependent variable. He called the values assumed by x the domain of the function, and the corresponding values assumed by y he termed the range of the function.

Now, since set concepts permeate almost all mathematics, we have the more general definitions of function presented in this section in terms of sets of ordered pairs of elements. The function is one of the most important concepts in mathematics, and as such it plays a central role as a guide for the selection and development of material in many mathematics courses. (Look at the section titles in this chapter and the next.)

ANSWERS TO MATCHED PROBLEMS

1. **(A)** Function **(B)** Not a function **(C)** Function
2. **(A)** All real numbers, except ± 2 **(B)** $x \le 2$ or $(-\infty, 2]$ **(C)** $-2 \le x \le 2$ or $[-2, 2]$ **(D)** All real numbers
3. **(A)** No **(B)** Yes

EXERCISE 9-1

A *Indicate whether each rule in Problems 1–6 is or is not a function.*

1.

DOMAIN		RANGE
3	⟶	0
5	⟶	1
7	⟶	2

2.

DOMAIN		RANGE
−1	⟶	5
−2	⟶	7
−3	⟶	9

3. DOMAIN RANGE

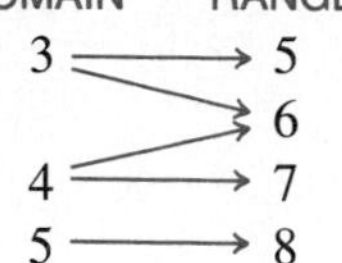

4. DOMAIN RANGE

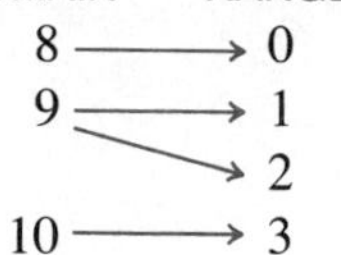

5. DOMAIN RANGE

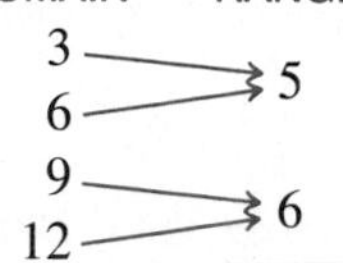

6. DOMAIN RANGE

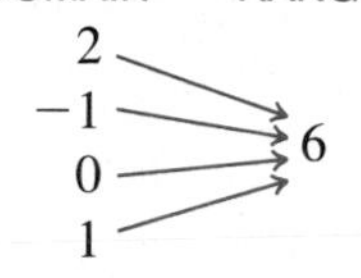

For each graph in Problems 7–12, determine whether it is the graph of a function with x the independent variable.

7.

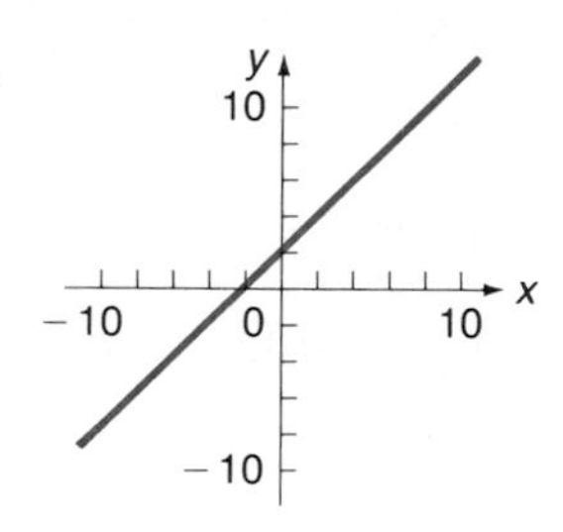

8.

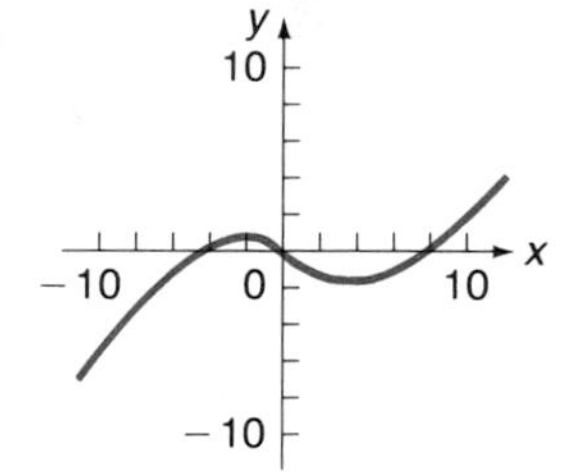

9.

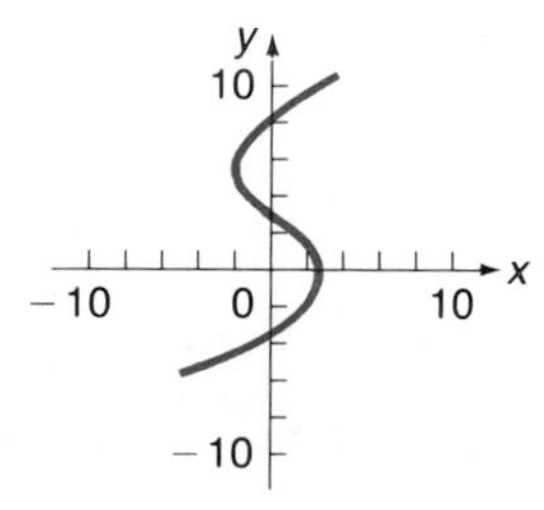

10.

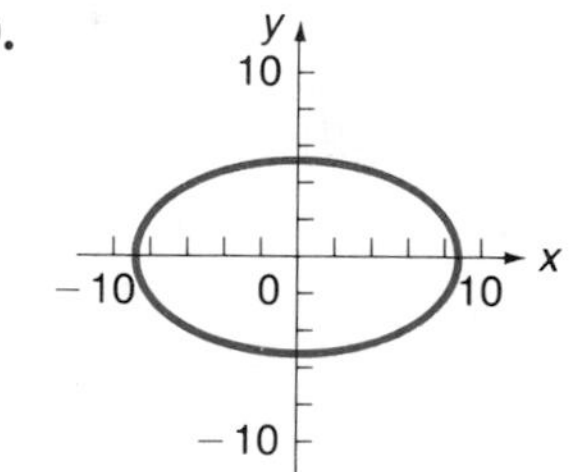

11.

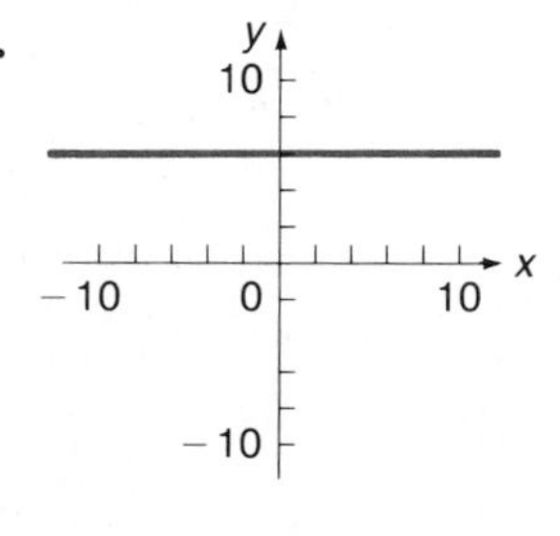

12.

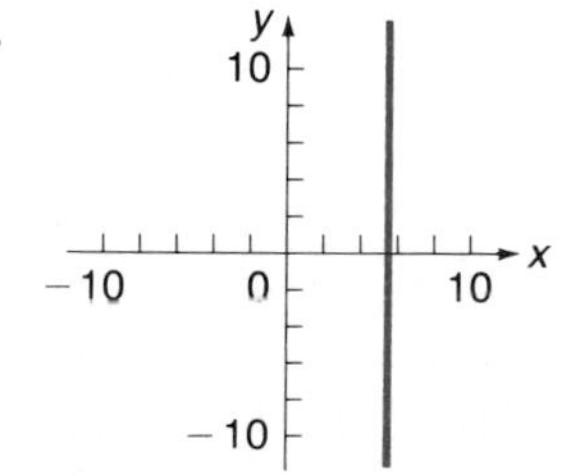

B *Determine whether or not the equation specifies a function with x as the independent variable.*

13. $y = 3x - 1$

14. $y = \frac{x}{2} - 1$

15. $y = x^2 - 3x + 1$

16. $y = x^3$

17. $y^2 = x$

18. $x^2 + y^2 = 25$

19. $x = y^2 - y$

20. $x = (y - 1)(y + 2)$

21. $y = x^4 - 3x^2$

22. $2x - 3y = 5$

23. $y = \dfrac{x + 1}{x - 1}$

24. $y = \dfrac{x^2}{1 - x}$

Graph each set of ordered pairs. Determine whether the rule so represented is a function.

25. $F = \{(1, 1), (2, 1), (3, 2), (3, 3)\}$

26. $f = \{(2, 4), (4, 2), (2, 0), (4, -2)\}$

27. $G = \{(-1, -2), (0, -1), (1, 0), (2, 1), (3, 2), (4, 1)\}$

28. $g = \{(-2, 0), (0, 2), (2, 0)\}$

Graph all pairs (x, y) satisfying the given equation. Determine whether the rule so represented is a function.

29. $y = 6 - 2x,\ x \in \{0, 1, 2, 3, 4\}$

30. $y = \dfrac{x}{2} - 4,\ x \in \{0, 1, 2, 3, 4\}$

31. $y^2 = x,\ x \in \{0, 1, 4\}$

32. $y = x^2,\ x \in \{-2, 0, 2\}$

33. $x^2 + y^2 = 4,\ x \in \{-2, 0, 2\}$

34. $x^2 + y^2 = 9,\ x \in \{-3, 0, 3\}$

35. $y = |x|,\ x \in \{-2, 0, 2\}$

36. $|y| = x,\ x \in \{0, 1, 4\}$

The equations in Problems 37–50 do specify a function. Determine the domain.

37. $y = 5 - x$

38. $y = 5x + 2$

39. $y = 3x^2 - 2x + 1$

40. $y = (5 - x)^2$

41. $y = \dfrac{1}{x}$

42. $y = \dfrac{1}{x - 1}$

43. $y = \dfrac{x - 1}{(x + 2)(x - 3)}$

44. $y = \dfrac{(x + 5)}{(x - 4)(x + 3)}$

45. $y = \dfrac{3x}{x^2 + x - 12}$

46. $y = \dfrac{x + 1}{x^2 - 7x + 12}$

47. $y = \sqrt{4 - x}$

48. $y = \sqrt{x - 5}$

49. $y = \sqrt{\frac{x - 1}{x + 3}}$

50. $y = \sqrt{x^2 + 3x - 10}$

C *Graph the set of ordered pairs in Problems 51–56. Determine whether the rule so represented is a function.*

51. $H = \left\{(x, y) \mid y = \frac{x}{2}, x \in \{-4, -2, 0, 2, 4\}\right\}$

52. $h = \{(x, y) \mid y = x + 3, x \in \{-3, -1, 0, 2\}\}$

53. $F = \{(x, y) \mid 0 \le y \le x, 0 \le x \le 3; x, y \in J\}$

54. $G = \{(x, y) \mid 0 \le y < |x|, -2 \le x \le 2; x, y \in J\}$

55. $f = \left\{(x, y) \mid y = \frac{|x|}{x}, x \in \{-2, -1, 1, 2, 3\}\right\}$

56. $g = \{(x, y) \mid y = (-1)^x, x \in \{-2, -1, 0, 1, 2\}\}$

APPLICATIONS

57. If an arrow is shot straight upward from the ground with an initial velocity of 160 feet per second, its distance d in feet above the ground at the end of t seconds (neglecting air resistance) is given by

$$d = 160t - 16t^2 \qquad 0 \le t \le 10$$

(A) Graph this rule (t is the independent variable).
(B) What are its domain and range?
(C) Is it a function?

58. The distance s that an object falls (neglecting air resistance) in t seconds is given by

$$s = 16t^2 \qquad t \ge 0$$

(A) Graph this rule (t is the independent variable).
(B) What are its domain and range?
(C) Is it a function?

9-2 FUNCTION NOTATION

- The Function Symbol $f(x)$
- Use of the Function Symbol $f(x)$

In this section we introduce a notation that allows us to name functions and describe the function rule.

THE FUNCTION SYMBOL $f(x)$

We have just seen that a function involves two sets of elements, a domain and a range, and a rule of correspondence that enables us to assign each element in the domain to exactly one element in the range. We use different letters to denote names for numbers; in essentially the same way, we will now use different letters to denote names for functions. For example, f and g may be used to name the two functions

$$f: \quad y = 2x + 1$$
$$g: \quad y = x^2 + 2x - 3$$

If x represents an element in the domain of a function f, then we will often use the symbol

$$f(x)$$

in place of y to designate the number in the range of the function f to which x is paired (Figure 1).

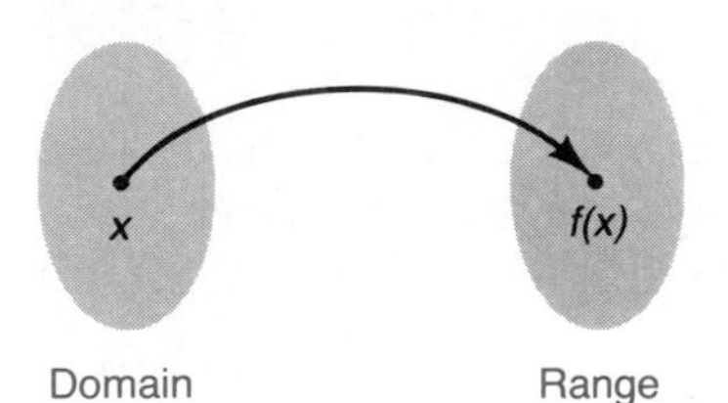

The function f "maps" the domain value x into the range value $f(x)$.

FIGURE 1

Do not confuse this new function symbol and think of it as the product of f and x. The symbol $f(x)$ is read "f of x" or "the value of f at x." The variable x is an independent variable; both y and $f(x)$ are dependent variables.

USE OF THE FUNCTION SYMBOL $f(x)$

This new function notation is extremely important, and its correct use should be mastered as early as possible. For example, in place of the more formal representation of the functions f and g given above, we can now write

$$f(x) = 2x + 1 \qquad \text{and} \qquad g(x) = x^2 + 2x - 3$$

The function symbols $f(x)$ and $g(x)$ have advantages over the variable y in certain situations. For example, if we write $f(3)$ and $g(5)$, then each symbol indicates in a concise way that these are range or output values of particular functions associated with particular domain or input values. Let us find $f(3)$ and $g(5)$.

To find $f(3)$, we replace x by 3 wherever x occurs in

$$f(x) = 2x + 1$$

and evaluate the right side:

$$
\begin{aligned}
f(3) &= 2 \cdot 3 + 1 \\
&= 6 + 1 \\
&= 7
\end{aligned}
$$

Thus,

$f(3) = 7$ The function f assigns the range value 7 to the domain value 3; the ordered pair (3, 7) belongs to f.

To find $g(5)$, we replace x by 5 wherever x occurs in

$$g(x) = x^2 + 2x - 3$$

and evaluate the right side:

$$
\begin{aligned}
g(5) &= 5^2 + 2 \cdot 5 - 3 \\
&= 25 + 10 - 3 \\
&= 32
\end{aligned}
$$

Thus,

$g(5) = 32$ The function g assigns the range value 32 to the domain value 5; the ordered pair (5, 32) belongs to g.

It is very important to understand and remember the definition of the symbol $f(x)$:

The Function Symbol $f(x)$

For any element x in the domain of the function f, the function symbol

$$f(x)$$

represents the element in the range of f corresponding to x in the domain of f. [If x is an input value, then $f(x)$ is an output value; or, symbolically, $f\colon\ x \to f(x)$.] The ordered pair $(x, f(x))$ belongs to the function f.

Figure 2 on thc next page, illustrating a ''function machine,'' may give you additional insight into the nature of function and the new function symbol $f(x)$. We can think of a function machine as a device that produces exactly one output (range) value for each input (domain) value.

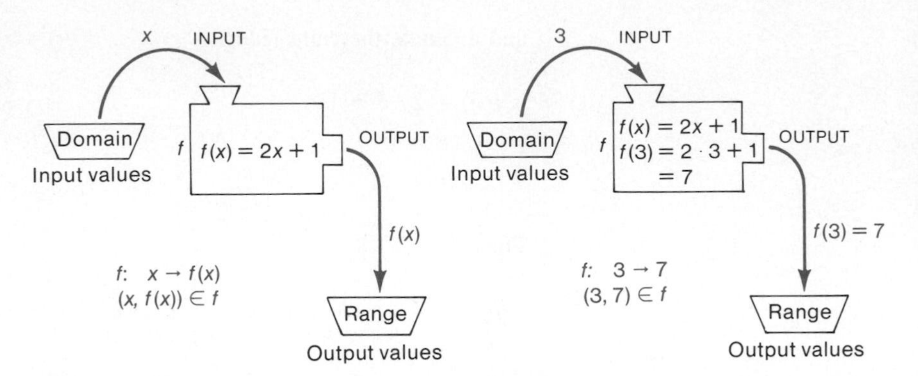

FIGURE 2 "Function machine"—exactly one output for each input

For the function $f(x) = 2x + 1$, the machine takes each domain value (input), multiplies it by 2, then adds 1 to the result to produce the range value (output). Different rules inside the machine result in different functions.

Example 4 If $f(x) = \dfrac{x}{2} + 1$ and $g(x) = 1 - x^2$, find:

(A) $f(6)$ **(B)** $g(-2)$ **(C)** $f(4) + g(0)$ **(D)** $\dfrac{2g(-3) + 6}{f(8)}$

Solution **(A)** $f(6) = \frac{6}{2} + 1 = 3 + 1 = 4$

(B) $g(-2) = 1 - (-2)^2 = 1 - 4 = -3$

(C) $f(4) + g(0) = \underbrace{(\tfrac{4}{2} + 1)}_{f(4)} + \underbrace{(1 - 0^2)}_{g(0)} = 3 + 1 = 4$

(D) $\dfrac{2g(-3) + 6}{f(8)} = \dfrac{2[1 - (-3)^2] + 6}{\frac{8}{2} + 1} = \dfrac{2(-8) + 6}{5} = \dfrac{-10}{5} = -2$

(with $g(-3)$ labelling $[1 - (-3)^2]$ and $f(8)$ labelling $\frac{8}{2} + 1$)

Problem 4 If $f(x) = \dfrac{x}{3} - 2$ and $g(x) = 4 - x^2$, find:

(A) $f(9)$ **(B)** $g(-2)$ **(C)** $f(0) + g(2)$ **(D)** $\dfrac{4g(-1) - 4}{f(12)}$

Example 5 For $f(x) = 5x + 3$, find: **(A)** $f(z)$ **(B)** $f(z + 1)$ **(C)** $f(2q)$

Solution **(A)** $f(z) = 5z + 3$ Replace x in $f(x) = 5x + 3$ with z.

(B) $f(z + 1) = 5(z + 1) + 3$ Replace x in $f(x) = 5x + 3$ with $z + 1$ and simplify.

$= 5z + 5 + 3$

$= 5z + 8$

(C) $f(2q) = 5(2q) + 3$ Replace x in $f(x) = 5x + 3$ with $2q$ and simplify.
$= 10q + 3$

Problem 5 For $f(x) = 2 - 3x$, find: **(A)** $f(m)$ **(B)** $f(m - 1)$ **(C)** $f(5p)$

Example 6 For $f(x) = \frac{x}{2} + 1$ and $g(x) = 1 - x^2$, find:

(A) $g(2 + h)$ **(B)** $\frac{g(2 + h) - g(2)}{h}$ **(C)** $f[g(3)]$

Solution **(A)** $g(2 + h) = 1 - (2 + h)^2 = 1 - (4 + 4h + h^2)$ Replace x in $g(x) = 1 - x^2$ with $(2 + h)$.
$= -3 - 4h - h^2$

(B) $\frac{g(2 + h) - g(2)}{h} = \frac{[1 - (2 + h)^2] - (1 - 2^2)}{h}$ Be careful here! The brackets and parentheses are important.

$= \frac{-3 - 4h - h^2 + 3}{h}$

$= \frac{-4h - h^2}{h} = -4 - h$

(C) $f[g(3)] = f(1 - 3^2)$ Evaluate $g(3)$ first; then evaluate f for this value.

$= f(-8) = \frac{-8}{2} + 1 = -3$

Problem 6 For $f(x) = \frac{x}{3} - 2$ and $g(x) = 4 - x^2$, find:

(A) $g(3 + h)$ **(B)** $\frac{g(3 + h) - g(3)}{h}$ **(C)** $g[f(3)]$

ANSWERS TO MATCHED PROBLEMS

4. **(A)** 1 **(B)** 0 **(C)** -2 **(D)** 4
5. **(A)** $2 - 3m$ **(B)** $5 - 3m$ **(C)** $2 - 15p$
6. **(A)** $-5 - 6h - h^2$ **(B)** $-6 - h$ **(C)** 3

EXERCISE 9-2

A

If $f(x) = 3x - 2$, *find the following:*

1. $f(2)$ **2.** $f(1)$ **3.** $f(-2)$
4. $f(-1)$ **5.** $f(0)$ **6.** $f(4)$

If $g(x) = x - x^2$, *find the following:*

7. $g(2)$ **8.** $g(1)$ **9.** $g(4)$

10. $g(5)$ **11.** $g(-2)$ **12.** $g(-1)$

B *For* $f(x) = 10x - 7$, $g(t) = 6 - 2t$, $F(u) = 3u^2$, *and* $G(v) = v - v^2$, *find the following:*

13. $f(-2)$ **14.** $F(-1)$ **15.** $g(2)$

16. $G(-3)$ **17.** $g(0)$ **18.** $G(0)$

19. $f(3) + g(2)$ **20.** $F(2) + G(3)$

21. $2g(-1) - 3G(-1)$ **22.** $4G(-2) - g(-3)$

23. $\dfrac{f(2) \cdot g(-4)}{G(-1)}$ **24.** $\dfrac{F(-1) \cdot G(2)}{g(-1)}$

25. $g(u - 2)$ **26.** $f(v + 1)$ **27.** $G(3a)$

28. $F(2c)$ **29.** $g(2 + h)$ **30.** $F(2 + h)$

31. $\dfrac{g(2 + h) - g(2)}{h}$ **32.** $\dfrac{F(2 + h) - F(2)}{h}$

33. $\dfrac{f(3 + h) - f(3)}{h}$ **34.** $\dfrac{G(2 + h) - G(2)}{h}$

35. $F[g(1)]$ **36.** $G[F(1)]$ **37.** $g[f(1)]$

38. $g[G(0)]$ **39.** $f[G(1)]$ **40.** $G[g(2)]$

41. If $A(w) = \dfrac{w - 3}{w + 5}$, find $A(5)$, $A(0)$, and $A(-5)$.

42. If $h(s) = \dfrac{s}{s - 2}$, find $h(3)$, $h(0)$, and $h(2)$.

C *For* $f(x) = 10x - 7$ *and* $g(t) = 6 - 2t$, *find the following:*

43. $\dfrac{f(x + h) - f(x)}{h}$ **44.** $\dfrac{g(t + h) - g(t)}{h}$

45. For $f(x) = 5x$:
(A) Does $f(at) = af(t)$? **(B)** Does $f(a + b) = f(a) + f(b)$?
(C) Does $f(ab) = f(a) \cdot f(b)$?

46. For $g(x) = x^2$:
(A) Does $g(at) = ag(t)$? **(B)** Does $g(a + b) = g(a) + g(b)$?
(C) Does $g(ab) = g(a) \cdot g(b)$?

APPLICATIONS *Each of the statements in Problems 47–50 can be described by a function. Write an equation that specifies the function.*

47. ***Cost function*** The cost $C(x)$ of x records at \$8.60 per record. (The cost depends on the number of records purchased.)

48. ***Cost function*** The cost $C(x)$ of manufacturing x pairs of skis if fixed costs are \$800 per day and the variable costs are \$60 per pair of skis. (The cost per day depends on the number of skis manufactured per day.)

49. ***Temperature conversion*** The temperature in Celsius degrees C(F) can be found from the temperature in Fahrenheit degrees F by subtracting 32 from the Fahrenheit temperature and multiplying the difference by $\frac{5}{9}$.

50. ***Earth science*** The pressure $P(d)$ in the ocean in pounds per square inch depends on the depth d. To find the pressure, divide the depth by 33, add 1 to the quotient, then multiply the result by 15.

51. ***Distance–rate–time*** Let the distance that a car travels at 30 miles per hour in t hours be given by $d(t) = 30t$. Find:

(A) $d(1)$, $d(10)$ **(B)** $\dfrac{d(2 + h) - d(2)}{h}$

★52. ***Physics*** The distance in feet that an object falls in t seconds in a vacuum is given by $s(t) = 16t^2$. Find:

(A) $s(0)$, $s(1)$, $s(2)$, and $s(3)$ **(B)** $\dfrac{s(2 + h) - s(2)}{h}$

What happens as h tends to 0? Interpret physically.

9-3 GRAPHING POLYNOMIAL FUNCTIONS

- Linear Functions
- Quadratic Functions
- Higher-Degree Polynomial Functions
- Application

In Chapters 6 and 7 we studied linear and quadratic forms in some detail. These forms with certain restrictions define **linear** and **quadratic functions**. Linear and quadratic functions are important special cases of a larger class of functions called **polynomial functions**. In this section we will look at the graphs of linear and quadratic functions (and some of their properties) and at an effective technique for graphing polynomial functions in general. The **graph of a function** is the graph of all ordered pairs of numbers that constitute the function.

LINEAR FUNCTIONS

Any nonvertical line in a rectangular coordinate system defines a linear function. (A vertical line does not define a function. Why?) Thus, any function defined by an equation of the form

$$f(x) = ax + b \quad \text{Linear function}$$

where a and b are constants and x is a variable, is called a **linear function**. We know from Section 7-2 that the graph of this equation is a straight line (nonvertical) with slope a and y intercept b.

Example 7 Graph the linear function defined by $f(x) = \dfrac{x}{3} + 1$, and indicate its slope and y intercept.

Solution

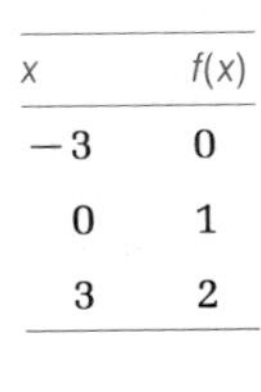

x	$f(x)$
-3	0
0	1
3	2

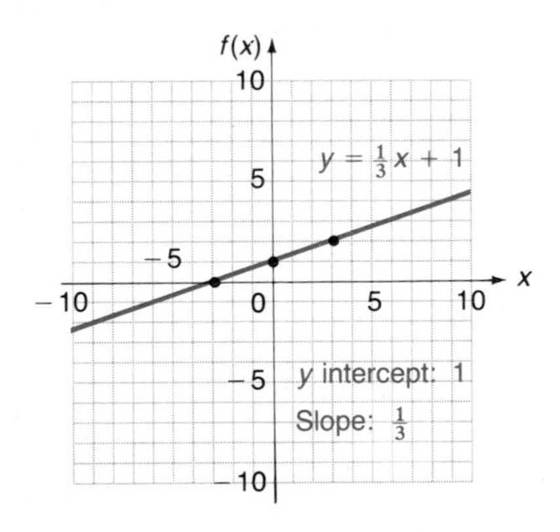

Problem 7 Graph the linear function defined by $f(x) = -\dfrac{x}{2} + 3$, and indicate its slope and y intercept.

QUADRATIC FUNCTIONS

Any function defined by an equation of the form

$$f(x) = ax^2 + bx + c \qquad a \neq 0 \quad \text{Quadratic function}$$

where a, b, and c are constants and x is a variable, is called a **quadratic function**. We know from Section 7-4 that the graph of this equation is a parabola with vertex at $x = \dfrac{-b}{2a}$ and y intercept c. The x intercepts, if any, are

$$\frac{-b \pm \sqrt{b^2 - 4ac}}{2a}$$

The parabola opens up if $a > 0$, down if $a < 0$.

Example 8 Graph $f(x) = -2x^2 + 10x + 5$. Find the coordinates of the vertex and any intercepts.

Solution Since $a = -2$, the parabola opens down. The vertex is at

$$x = -\frac{b}{2a} = \frac{-10}{-4} = \frac{5}{2} = 2\tfrac{1}{2}$$

At this value of x, $f(x) = -2(\tfrac{5}{2})^2 + 10(\tfrac{5}{2}) + 5 = 17\tfrac{1}{2}$. The y intercept is 5. The x intercepts are at

$$x = \frac{-10 \pm \sqrt{100 + 40}}{-4} = \frac{-10 \pm \sqrt{140}}{-4} \approx -0.46,\ 5.46.$$

With this information we can sketch a graph of the parabola:

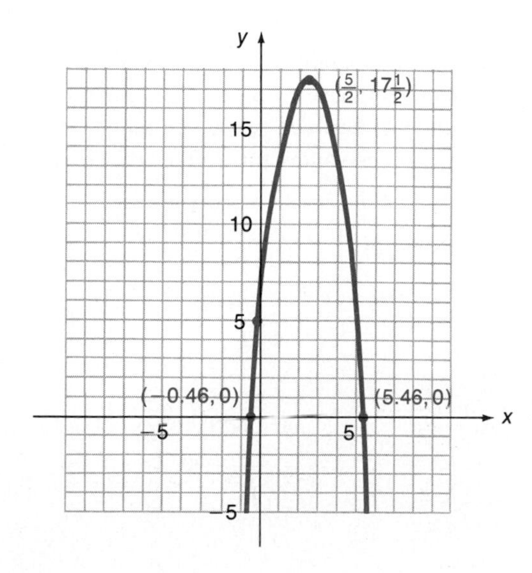

Problem 8 Graph $f(x) = 3x^2 - 9x + 8$. Find the coordinates of the vertex and any intercepts.

HIGHER-DEGREE POLYNOMIAL FUNCTIONS

As we indicated at the start of this section, linear and quadratic functions are special cases of a general class of functions called **polynomial functions** (functions whose range values are determined by use of a polynomial). In general, a function f defined by an equation of the form

$$f(x) = a_n x^n + a_{n-1}x^{n-1} + \cdots + a_1 x + a_0 \qquad a_n \neq 0$$

where the coefficients a_i are constants and n is a nonnegative integer, is called an ***n*th-degree polynomial function**. The following equations define polynomial functions of various degrees:

$f(x) = 2x - 3$	First-degree (linear)
$g(x) = 2x^2 - 3x + 2$	Second-degree (quadratic)
$P(x) = x^3 - 2x^2 + x - 1$	Third-degree
$Q(x) = x^4 - 5$	Fourth-degree

Unless otherwise stated, the domain of a polynomial function is the set of all real numbers. In advanced courses it can be shown that graphs of polynomial functions have no holes or breaks in them—the graphs are continuous smooth curves. There are a number of other properties of polynomial functions that are also studied in later courses. We will graph polynomial functions by plotting enough points to be fairly confident of its shape.

Example 9 Graph $P(x) = x^3 + 3x^2 - x - 3, \ -4 \le x \le 2$.

Solution We construct a table of ordered pairs of numbers belonging to the function P. We then plot these points and join them with a smooth curve. The more points we compute between two given points, the greater the accuracy of the graph. Here, integer values for x give us a reasonable picture:

x	$P(x)$
−4	−15
−3	0
−2	3
−1	0
0	−3
1	0
2	15

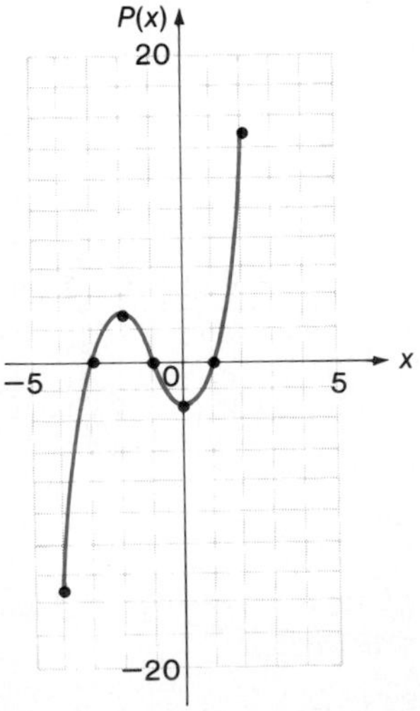

Problem 9 Graph $P(x) = x^3 - 4x^2 - 4x + 16, \; -3 \le x \le 5$.

APPLICATION

Example 10 A rectangular dog pen is to be made with 160 feet of fencing.

(A) If x represents the width of the pen, express its area $A(x)$ in terms of x.
(B) What is the domain of the function A (determined by the physical restrictions)?

Solution **(A)** Draw a figure and label the sides:

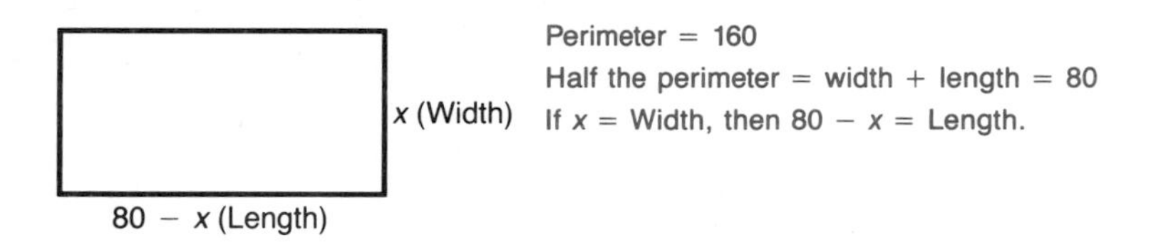

$$A(x) = (\text{Length})(\text{Width}) = (80 - x)x \quad \text{Area depends on width } x.$$

(B) The area cannot be negative; hence, x cannot be negative and x cannot be greater than 80. [Look at $A(x) = (80 - x)x$.] Thus,

Domain: $0 \le x \le 80$ Inequality notation
$[0, 80]$ Interval notation

Problem 10 Work Example 10 with the added assumption that a large barn is to be used as one side of the pen.

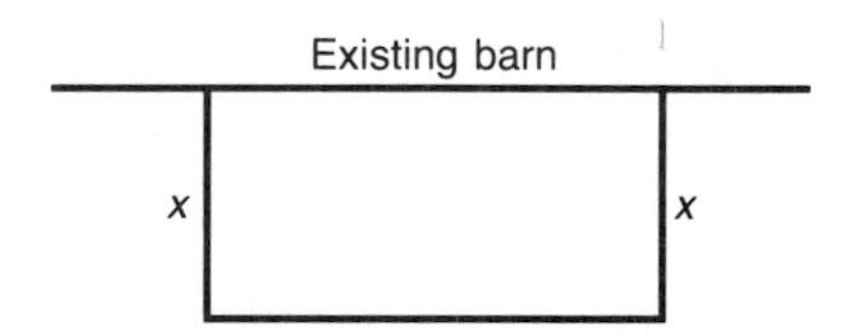

ANSWERS TO MATCHED PROBLEMS

7.

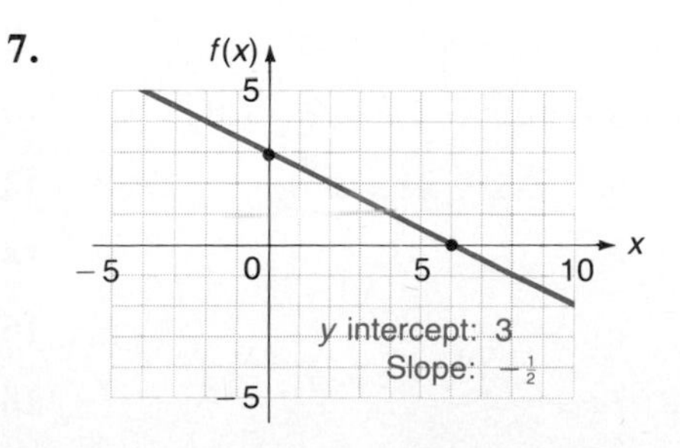

8. y intercept: 8
No x intercept

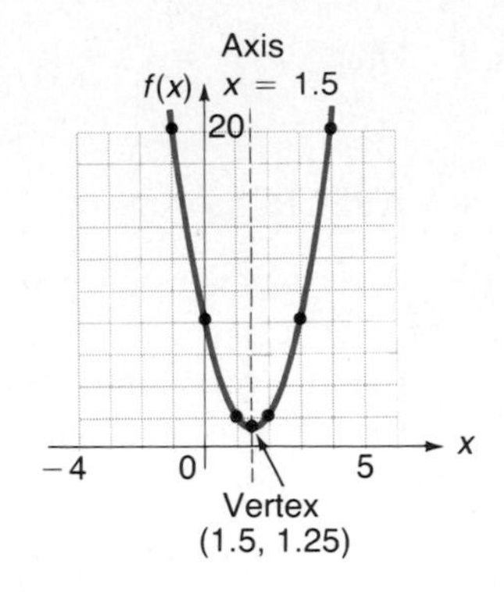

9.

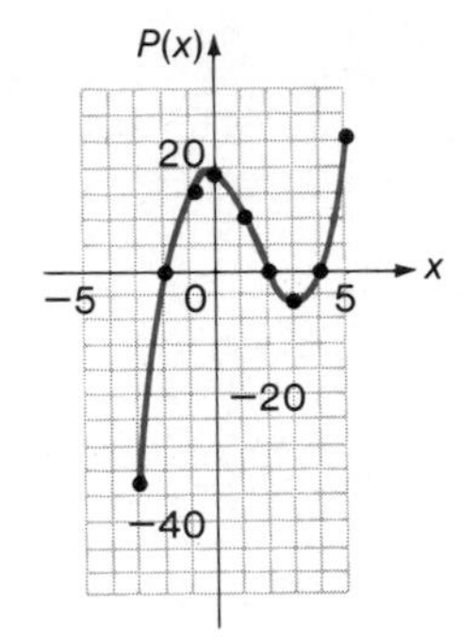

10. **(A)** $A(x) = (160 - 2x)x$ **(B)** Domain: $0 \le x \le 80$ or $[0, 80]$

EXERCISE 9-3

A *Graph the following linear functions. Indicate the slope and y intercept for each.*

1. $f(x) = 2x - 4$

2. $g(x) = \dfrac{x}{2}$

3. $h(x) = 4 - 2x$

4. $f(x) = -\dfrac{x}{2} + 3$

5. $g(x) = -\frac{2}{3}x + 4$

6. $f(x) = 3$

Graph each of the following quadratic functions. Find the vertex and any intercepts.

7. $f(x) = x^2 + 8x + 16$

8. $h(x) = x^2 - 2x - 3$

9. $f(u) = u^2 - 2u + 4$

10. $f(x) = x^2 - 10x + 25$

11. $h(x) = 2 + 4x - x^2$

12. $g(x) = -x^2 - 6x - 4$

13. $f(x) = 6x - x^2$

14. $G(x) = 16x - 2x^2$

15. $F(s) = s^2 - 4$

16. $g(t) = t^2 + 4$

17. $F(x) = 4 - x^2$

18. $G(x) = 9 - x^2$

B **19.** $f(x) = x^2 - 7x + 10$

20. $g(t) = t^2 - 5t + 2$

21. $g(t) = 4 + 3t - t^2$

22. $h(x) = 2 - 5x - x^2$

23. $f(x) = \frac{1}{2}x^2 + 2x$

24. $f(x) = 2x^2 - 12x + 14$

25. $f(x) = -2x^2 - 8x - 2$

26. $f(x) = -\frac{1}{2}x^2 + 4x - 4$

Graph each polynomial function.

27. $P(x) = x^3 - 5x^2 + 2x + 8; \quad -2 \le x \le 5$

28. $P(x) = x^3 + 2x^2 - 5x - 6; \quad -4 \le x \le 3$

29. $P(x) = x^3 + 4x^2 - x - 4; \quad -5 \le x \le 2$

30. $P(x) = x^3 - 2x^2 - 5x + 6; \quad -3 \le x \le 4$

C **31.** $P(x) = x^4 - 2x^3 - 2x^2 + 8x - 8$

32. $P(x) = x^4 - 2x^2 + 16x - 15$

33. $P(x) = x^4 + 4x^3 - x^2 - 20x - 20$

34. $P(x) = x^4 - 4x^2 - 4x - 1$

APPLICATIONS

35. ***Cost equation*** The cost equation for a particular company to produce stereos is found to be

$$C = g(n) = 96{,}000 + 80n$$

where \$96,000 is fixed costs (tooling and overhead) and \$80 is the variable cost per unit (material, labor, and so on). Graph this function for $0 \le n \le 1{,}000$.

36. ***Demand equation*** After extensive surveys the research department in a stereo company produced the demand equation

$$n = f(p) = 8{,}000 - 40p \qquad 100 \le p \le 200$$

where n is the number of units that retailers are likely to purchase per week at a price of p dollars per unit. Graph the function for the indicated domain.

37. ***Construction*** A rectangular feeding pen for cattle is to be made with 100 meters of fencing.

(A) If x represents the width of the pen, express its area $A(x)$ in terms of x.

(B) What is the domain of the function A (determined by the physical restrictions)?

(C) Graph the function for this domain.

(D) What dimension pen will produce the largest area? What is the largest area?

38. ***Construction*** Work the preceding problem with the added assumption that a large straight river is to be used as one side of the pen.

39.** ***Packaging A candy box is to be made out of a rectangular piece of cardboard that measures 8 by 12 inches. Equal-sized squares (x by x inches) will be cut out of each corner, and then the ends and sides will be folded up to form a rectangular box.

(A) Write the volume of the box $V(x)$ in terms of x.
(B) Considering the physical limitations, what is the domain of the function V?
(C) Graph the function for this domain.
(D) From the graph, estimate to the nearest half-inch the size square that must be cut from each corner to yield a box with the largest volume. What is the largest volume?

****40.** ***Packaging*** A parcel delivery service will deliver only packages with length plus girth (distance around) not exceeding 108 inches. A packaging company wishes to design a box with a square base (x by x inches) that will have a maximum volume but meet the delivery service's restrictions.

(A) Write the volume of the box $V(x)$ in terms of x.
(B) Considering the physical limitations imposed by the delivery service, what is the domain of the function V?
(C) Graph the function for this domain.
(D) From the graph, estimate to the nearest inch the dimensions of the box with the largest volume. What is the largest volume?

9-4 INVERSE FUNCTIONS

- Inverses
- One-to-One Correspondence and Inverses

In this section we are going to discuss an important method for obtaining new functions from old functions. In particular, we will try to "reverse" the input–output rule for a function to obtain a new function called the *inverse function*. This will not always be possible but can be done for those functions that meet the additional condition that each output comes from a unique input. Such functions will be called *one-to-one*. We will use this method in Chapter 10 to obtain the logarithmic functions from the exponential functions.

INVERSES

A function f associates an output value $f(x)$ with an input value x. We can ask if the process can be reversed. For example, if

$$y = f(x) = 2x - 1$$

we can ask what input x yields output 7—that as, if $f(x) = 7$, what is x? This amounts to solving $7 = 2x - 1$ for x:

$$7 = 2x - 1$$
$$8 = 2x$$
$$x = 4$$

Thus, for the given range value 7 we have found the corresponding domain value 4. Proceeding in the same way, we can find the corresponding domain value for each range value for the function f.

The process just described leads to a new function, the **inverse of f**, which is denoted by f^{-1}. The range for f becomes the domain for f^{-1}, and the domain for f becomes the range for f^{-1}. In terms of ordered pairs, if (a, b) belongs to f, then (b, a) belongs to f^{-1}. When f is specified by an equation, we may be able to find an equation that specifies f^{-1} by interchanging the variables and solving for y in terms of x as follows:

f: $\quad y = 2x - 1$ — Interchange variables to form an inverse rule.

f^{-1}: $\quad x = 2y - 1$ — Solve for y in terms of x.

$$x + 1 = 2y$$

$$y = \frac{x + 1}{2}$$ — Inverse rule.

Thus,

$$f^{-1}(x) = \frac{x + 1}{2}$$

Note that $(4, 7)$ is an element of function f and $(7, 4)$ is an element of the inverse function f^{-1}; that is, $f(4) = 7$ and $f^{-1}(7) = 4$.

Example 11 The function $f(x) = \sqrt{x}$ has an inverse. Find the inverse rule and its domain and range.

Solution For our given function f we have

Rule: $y = f(x) = \sqrt{x}$

Domain: $x \geq 0$

Range: $y \geq 0$

For the inverse rule, the domain and range will be reversed; that is, the inputs for the inverse will be the outputs from the original function. We find the inverse rule by interchanging variables and solving for y in terms of x:

$y = f(x) = \sqrt{x}$ Interchange the variables

$x = f(y) = \sqrt{y}$ Solve for y

$x^2 = y$ This is the inverse rule

Thus, we have for the rule for f^{-1}:

Rule: $y = f^{-1}(x) = x^2$

Domain: $x \geq 0$

Range: $y \geq 0$

Problem 11 The function $f(x) = x^3$ has an inverse. Find the inverse rule and its domain.

If a function $f(x)$ has an inverse, there is a relationship between the graphs of $f(x)$ and $f^{-1}(x)$. A point (a, b) is on one graph if and only if (b, a) is on the other. The points (a, b) and (b, a) are symmetric about the 45° line through 0, that is, about the line $y = x$.

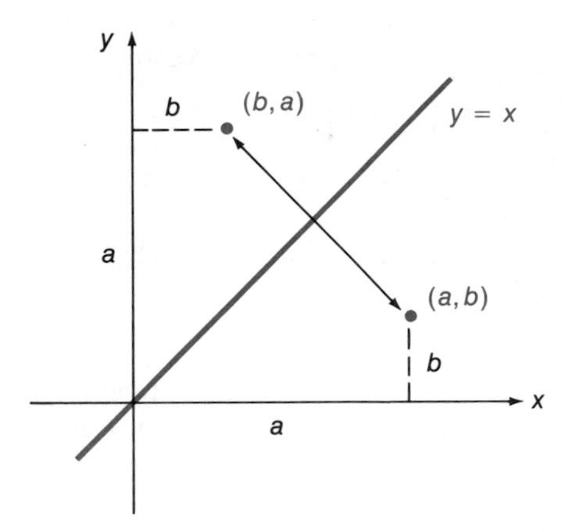

The graph of f^{-1} is therefore the graph of f reflected about the 45° line as a mirror image.

Example 12 Graph the function and its inverse.

(A) $f(x) = 2x - 1$ **(B)** $f(x) = \sqrt{x}$.

Solution **(A)** We already know

$$f^{-1}(x) = \frac{x+1}{2}$$

We can rewrite $\frac{x+1}{2}$ as $\frac{1}{2}x + \frac{1}{2}$ to see that the graph is a straight line. Thus, we get the graphs

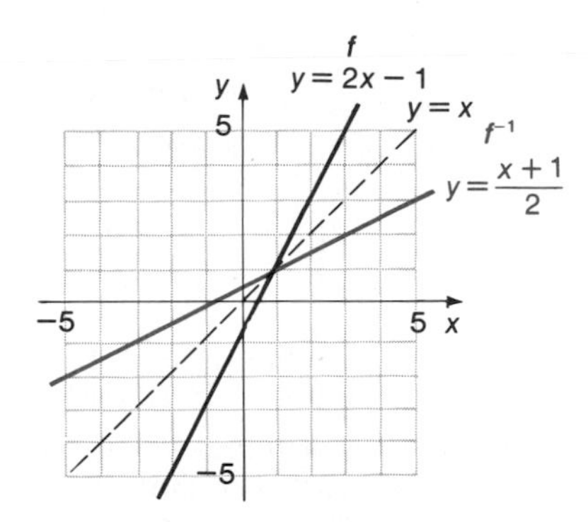

(B) We already know $f^{-1}(x) = x^2$ for $x \geq 0$. The two graphs are as shown.

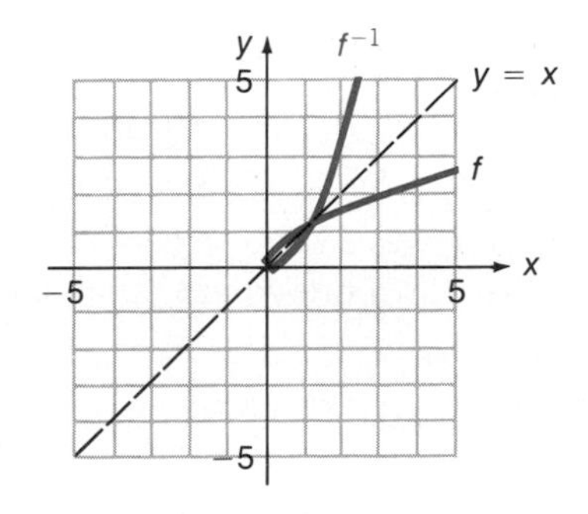

Problem 12 Graph the function $f(x) = x^3$ and its inverse.

ONE-TO-ONE CORRESPONDENCE AND INVERSES

Not all functions have inverses. If we are given a function f, how can we tell in advance whether it has an inverse function f^{-1}? The answer is contained in the concept of one-to-one correspondence. A **one-to-one correspondence** exists between two sets if each element in the first set corresponds to exactly one element in the second set, and each element in the second set corresponds to exactly one element in the first set. Consider the two functions f and g and their inverse rules:

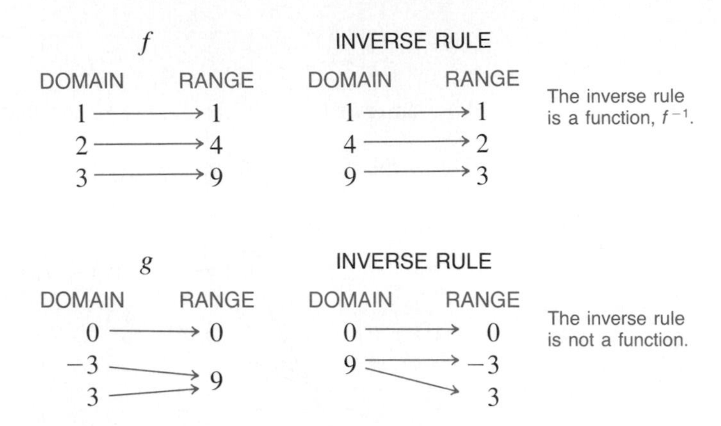

Function f has a one-to-one correspondence between domain and range values. (Notice that f^{-1} is also a function.) Function g does not have a one-to-one correspondence between domain and range values. (Notice that the inverse rule is not a function.) More generally we have this result:

Inverses

A function f has an inverse function if and only if there exists a one-to-one correspondence between domain and range values of f. In this case, we say that f is a **one-to-one function** and note that

$$f[f^{-1}(y)] = y \qquad \text{and} \qquad f^{-1}[f(x)] = x$$

This result is interpreted schematically in Figure 3.

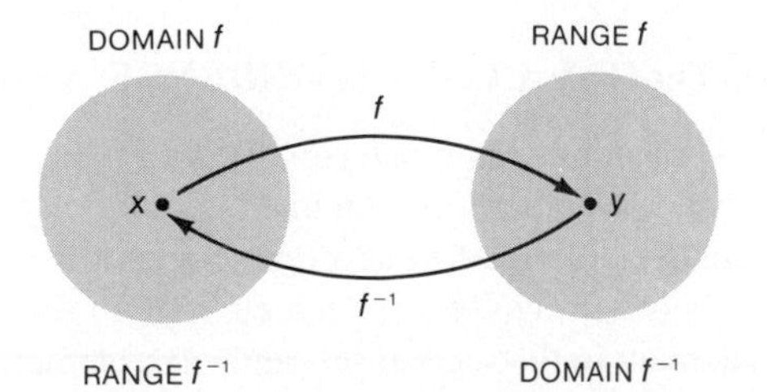

If the function f is one-to-one and if f maps x into y, then f^{-1} will map y back into x.

FIGURE 3

Figure 4 illustrates some functions that are one-to-one, and Figure 5 illustrates some that are not. In Figure 4, each domain value corresponds to exactly one range value, and each range value corresponds to exactly one domain value. In Figure 5, each domain value corresponds to exactly one range value, but some range values correspond to more than one domain value. (In Figure 5a, y_1 corresponds to x_1 and x_2; in Figure 5b, y_1 corresponds to x_1, x_2, and x_3.)

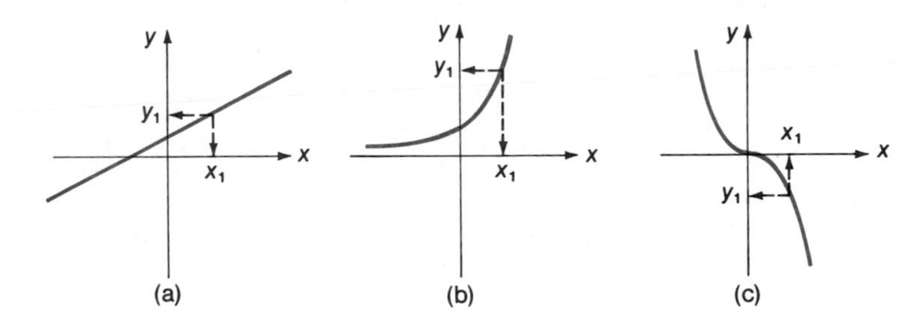

FIGURE 4 Functions that are one-to-one (each has an inverse that is a function)

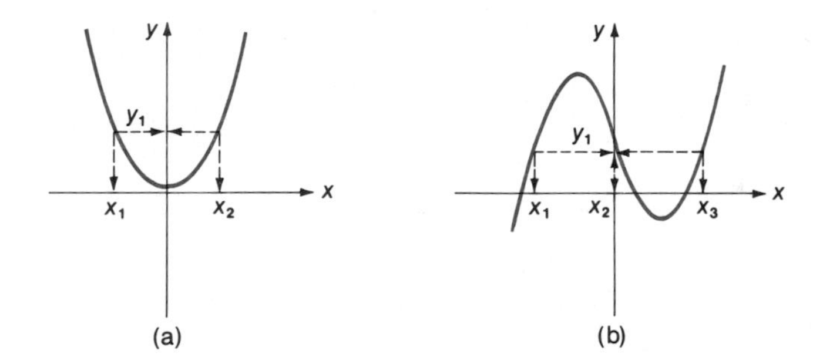

FIGURE 5 Functions that are not one-to-one (neither has an inverse that is a function)

Given the graph of a function, it is easy to see whether it has an inverse which is a function.

Horizontal-Line Test for an Inverse Function

A function has an inverse if each horizontal line in the coordinate system crosses the graph of the function at most once.

Notice how this test corresponds to the results in Figures 4 and 5.

Example 13 Determine whether the given function has an inverse:

(A) $g(x) = x^2 - 4$ **(B)** $f(x) = 3x + 2$

Solution **(A)** From the graph, $g(x)$ has no inverse.

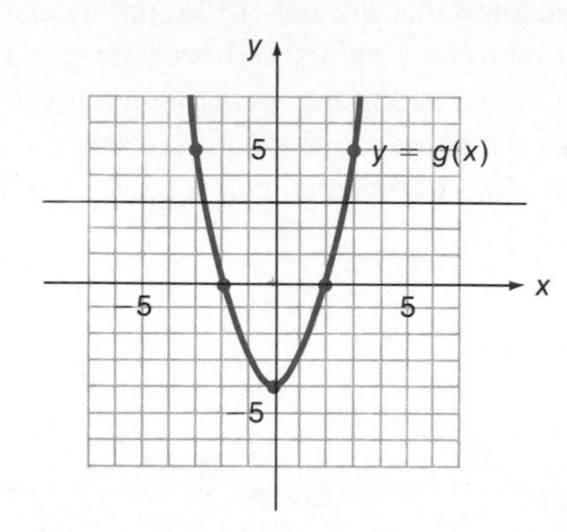

Horizontal line cross $f(x)$ twice.

(B) From the graph, $f(x)$ has an inverse.

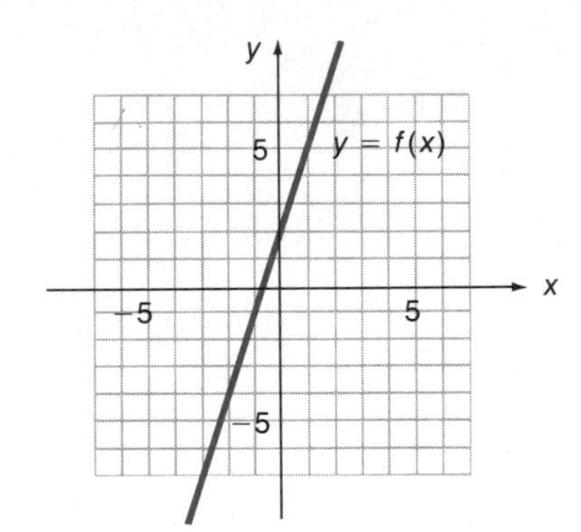

No horizontal line will cross $f(x)$ twice.

Problem 13 Determine whether the given function has an inverse:

(A) $g(x) = 3 - 4x$ **(B)** $f(x) = 3 - x^2$

Example 14 Given $f(x) = 3x + 2$, find:

(A) $f^{-1}(x)$ **(B)** $f^{-1}(5)$ **(C)** $f^{-1}[f(5)]$ **(D)** $f^{-1}[f(x)]$

Solution We know from Example 13(B) that $f(x)$ has an inverse.

(A)

$$f\colon\quad y = 3x + 2$$ Replace $f(x)$ with y in $f(x) = 3x + 2$.

$$f^{-1}\colon\quad x = 3y + 2$$ Interchange variables x and y to obtain f^{-1}.

$$y = \frac{x - 2}{3}$$ Solve for y in terms of x.

Thus,

$$f^{-1}(x) = \frac{x-2}{3} \quad \text{Replace } y \text{ with } f^{-1}(x).$$

(B) $f^{-1}(5) = \frac{5-2}{3} = \frac{3}{3} = 1$

(C) $f^{-1}[f(5)] = \frac{f(5)-2}{3}$ We are just verifying the basic result about inverses.

$$= \frac{17-2}{3}$$

$$= \frac{15}{3} = 5$$

(D) $f^{-1}[f(x)] = \frac{f(x)-2}{3}$ See comment in part (C).

$$= \frac{(3x+2)-2}{3} = x$$

Problem 14 Given $g(x) = \frac{x}{3} - 2$, find:

(A) $g^{-1}(x)$ **(B)** $g^{-1}(-2)$ **(C)** $g^{-1}(g(3)]$ **(D)** $g^{-1}[g(x)]$

ANSWERS TO MATCHED PROBLEMS

11. $f^{-1}(x) = \sqrt[3]{x}$; Domain: all real numbers

12.

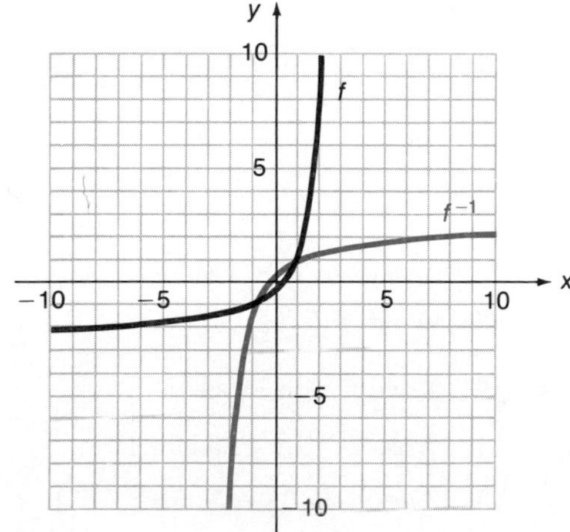

13. **(A)** Yes **(B)** No

14. **(A)** $3x + 6$ **(B)** 0 **(C)** 3 **(D)** x

EXERCISE 9-4

A *Determine whether or not the function has an inverse.*

1. $R = \{(-2, 0), (-1, 1), (0, 2), (1, 3), (2, 4)\}$
2. $S = \{(-2, -2), (-1, -1), (0, 0), (1, 1), (2, 2)\}$
3. $T = \{(-2, 0), (-1, 1), (0, 2), (1, 1), (2, 0)\}$
4. $U = \{(-2, 2), (-1, 1), (0, 0), (1, 1), (2, 2)\}$

Find the inverse rule and its domain.

5. $f = \{(-2, 1), (-1, 2), (0, 3), (1, 4), (2, 5)\}$
6. $g = \{(-2, -2), (-1, -4), (0, -6), (1, -8), (2, -10)\}$
7. $F = \{(-2, \frac{1}{5}), (-1, \frac{1}{4}), (0, \frac{1}{3}), (1, \frac{1}{2}), (2, 1)\}$
8. $G = \{(-2, \frac{1}{9}), (-1, \frac{1}{3}), (0, 1), (1, 3), (2, 9)\}$

Graph the function, its inverse, and $y = x$ on the same coordinate system.

9. f in Problem 5
10. g in Problem 6
11. F in Problem 7
12. G in Problem 8

B *Find the inverse for each of the following functions in the form of an equation.*

13. f: $y = 3x - 2$
14. g: $y = 2x + 3$
15. F: $y = \frac{x}{3} - 2$
16. G: $y = \frac{x}{2} + 5$
17. h: $y = \frac{x^2}{2}, x \geq 0$
18. H: $y = -\sqrt{x}$

In Problems 19–22 graph each pair of functions in the same coordinate system, along with $y = x$.

19. f and f^{-1} in Problem 13
20. g and g^{-1} in Problem 14
21. h and h^{-1} in Problem 17
22. H and H^{-1} in Problem 18
23. For $f(x) = 3x - 2$, find: **(A)** $f^{-1}(x)$ **(B)** $f^{-1}(2)$ **(C)** $f[f^{-1}(3)]$
24. For $g(x) = 2x + 3$, find: **(A)** $g^{-1}(x)$ **(B)** $g^{-1}(5)$ **(C)** $g[g^{-1}(4)]$

25. For $F(x) = \frac{x}{3} - 2$; find:

(A) $F^{-1}(x)$ **(B)** $F^{-1}(-1)$ **(C)** $F^{-1}[F(4)]$

26. For $G(x) = \frac{x}{2} + 5$, find:

(A) $G^{-1}(x)$ **(B)** $G^{-1}(8)$ **(C)** $G^{-1}[G(-4)]$

C **27.** For $f(x) = \frac{x}{3} + 2$, find: **(A)** $f^{-1}(x)$ **(B)** $f[f^{-1}(a)]$

28. For $g(x) = 4x + 2$, find: **(A)** $g^{-1}(x)$ **(B)** $g^{-1}[g(a)]$

29. Find $F[F^{-1}(x)]$ for Problem 15.

30. Find $G^{-1}[G(x)]$ for Problem 16.

31. Let $G(x) = 3^x$, $x \in \{-2, -1, 0, 1, 2\}$.
(A) Find the inverse rule for G.
(B) Graph G, G^{-1}, and $y = x$ in the same coordinate system.

32. Let $H(x) = \frac{1}{x}$.
(A) Find $H^{-1}(x)$.
(B) Find $H^{-1}(3)$ and $H^{-1}(H(x))$.

9-5 VARIATION

- Direct Variation
- Inverse Variation
- Joint Variation
- Combined Variation

In reading scientific material, one is likely to come across statements such as "the pressure of an enclosed gas varies directly as the absolute temperature," or "the frequency of vibration of air in an organ pipe varies inversely as the length of the pipe," or even more complicated statements such as "the force of attraction between two bodies varies jointly as their masses and inversely as the square of the distance between the two bodies." These statements have precise mathematical meaning in that they represent particular types of functions. The purpose of this section is to investigate these special functions.

The statement **y varies directly as x** means†

$$y = kx \qquad k \neq 0$$

where k is a constant called the **constant of variation**. Similarly, the statement "y varies directly as the square of x" means

$$y = kx^2 \qquad k \neq 0$$

and so on. The first equation defines a linear function, and the second a quadratic function.

Direct variation is illustrated by the familiar formulas

$$C = \pi D \qquad \text{and} \qquad A = \pi r^2$$

where the first formula asserts that the circumference of a circle varies directly as the diameter and the second states that the area of a circle varies directly as the square of the radius. In both cases, π is the constant of variation.

Example 15 Translate each statement into an appropriate equation, and find the constant of variation if $y = 16$ when $x = 4$.

(A) y varies directly as x.
(B) y varies directly as the cube of x.

Solution **(A)** y varies directly as x means

$$y = kx \quad \text{Do not forget } k.$$

To find the constant of variation k, substitute $x = 4$ and $y = 16$ and solve for k:

$$16 = k \cdot 4$$
$$k = \tfrac{16}{4} = 4$$

Thus, $k = 4$ and the equation of variation is

$$y = 4x$$

† You will sometimes hear "y is proportional to x" in place of "y varies directly as x." Both mean the same thing.

(B) y varies directly as the cube of x means

$$y = kx^3 \quad \text{Do not forget } k.$$

To find k, substitute $x = 4$ and $y = 16$:

$$16 = k \cdot 4^3$$
$$k = \tfrac{16}{64} = \tfrac{1}{4}$$

Thus, the equation of variation is

$$y = \tfrac{1}{4}x^3$$

Problem 15 If $y = 4$ when $x = 8$, find the equation of variation for each statement:

(A) y varies directly as x.
(B) y varies directly as the cube root of x.

INVERSE VARIATION

The statement ***y* varies inversely as *x*** means†

$$y = \frac{k}{x} \qquad k \neq 0$$

where k is a constant (the constant of variation). As in the case of direct variation, we also discuss y varying inversely as the square of x and so on.

An illustration of inverse variation is given in the distance–rate–time formula $d = rt$ in the form $t = d/r$ for a fixed distance d. In driving a fixed distance, say $d = 400$ miles, time varies inversely as the rate; that is,

$$t = \frac{400}{r}$$

where 400 is the constant of variation—as the rate increases, the time decreases, and vice versa.

Example 16 Translate each statement into an appropriate equation, and find the constant of variation if $y = 16$ when $x = 4$.

† You will sometimes hear ''y is inversely proportional to x'' in place of ''y varies inversely as x.'' Both mean the same thing.

(A) y varies inversely as x.
(B) y varies inversely as the square root of x.

Solution **(A)** y varies inversely as x means

$$y = \frac{k}{x} \quad \text{Do not forget } k.$$

To find k, substitute $x = 4$ and $y = 16$:

$$16 = \frac{k}{4}$$
$$k = 64$$

Thus, the equation of variation is

$$y = \frac{64}{x}$$

(B) y varies inversely as the square root of x means

$$y = \frac{k}{\sqrt{x}}$$

To find k, substitute $x = 4$ and $y = 16$:

$$16 = \frac{k}{\sqrt{4}}$$
$$k = 32$$

Thus, the equation of variation is

$$y = \frac{32}{\sqrt{x}}$$

Problem 16 If $y = 4$ when $x = 8$, find the equation of variation for each statement:

(A) y varies inversely as x.
(B) y varies inversely as the square of x.

JOINT VARIATION

The statement **w varies jointly as x and y** means

$$w = kxy \qquad k \neq 0$$

where k is a constant (the constant of variation). Similarly, if

$$w = kxyz^2 \qquad k \neq 0$$

we would say that "w varies jointly as x, y, and the square of z," and so on. For example, the area of a rectangle varies jointly as its length and width (recall $A = lw$), and the volume of a right circular cylinder varies jointly as the square of its radius and its height (recall $V = \pi r^2 h$). What is the constant of variation in each case?

In the joint variation $w = kxy$, w is a function of either x or y if the other is held fixed (w is also a function of both x and y, but we will not consider this kind of function in this text).

COMBINED VARIATION

The basic types of variation introduced above are often combined. For example, the statement "w varies jointly as x and y and inversely as the square of z" means

$$w = k\frac{xy}{z^2} \qquad k \neq 0$$

We do not write $w = \frac{kxy}{kz^2}$. This is wrong because it eliminates the proportionality constant k.

Thus, the statement "the force of attraction F between two bodies varies jointly as their masses m_1 and m_2 and inversely as the square of the distance d between the two bodies" means

$$F = k\frac{m_1 m_2}{d^2} \qquad k \neq 0$$

If (assuming k is positive) either of the two masses is increased, the force of attraction increases; on the other hand, if the distance is increased, the force of attraction decreases.

Example 17 The pressure P of an enclosed gas varies directly as the absolute temperature T and inversely as the volume V. If 500 cubic feet of gas yields a pressure of 10 pounds per square foot at a temperature of 300 K (absolute temperature†), what will be the pressure of the same gas if the volume is decreased to 300 cubic feet and the temperature increased to 360 K?

† A kelvin (absolute) and a Celsius degree are the same size, but 0 on the kelvin scale is −273° on the Celsius scale. This is the point at which molecular action is supposed to stop and is called *absolute zero*.

Solution *Method 1* Write the equation of variation $P = k(T/V)$, and find k using the first set of values:

$$10 = k(\tfrac{300}{500})$$
$$k = \tfrac{50}{3}$$

Hence, the equation of variation for this particular gas is $P = \frac{50}{3}(T/V)$.

Now find the new pressure P using the second set of values:

$$P = \tfrac{50}{3}(\tfrac{360}{300}) = 20 \text{ pounds per square foot}$$

Method 2 (generally faster than Method 1) Write the equation of variation $P = k(T/V)$; then convert to the equivalent form:

$$\frac{PV}{T} = k$$

If P_1, V_1, and T_1 are the first set of values for the gas and P_2, V_2, and T_2 are the second set, then

$$\frac{P_1V_1}{T_1} = k \qquad \text{and} \qquad \frac{P_2V_2}{T_2} = k$$

Hence,

$$\frac{P_1V_1}{T_1} = \frac{P_2V_2}{T_2}$$

Since all values are known except P_2, substitute and solve. Thus,

$$\frac{(10)(500)}{300} = \frac{P_2(300)}{360}$$
$$P_2 = 20 \text{ pounds per square foot}$$

Problem 17 The length L of skid marks of a car's tires (when brakes are applied) varies directly as the square of the speed v of the car. If skid marks of 20 feet are produced at 30 miles per hour, how fast would the same car be going if it produced skid marks of 80 feet? Solve in two ways (see Example 17).

Example 18 The frequency of pitch f of a given musical string varies directly as the square root of the tension T and inversely as the length L. What is the effect on the frequency if the tension is increased by a factor of 4 and the length is cut in half?

Solution Write the equation of variation:

$$f = \frac{k\sqrt{T}}{L} \qquad \text{or equivalently} \qquad \frac{f_2L_2}{\sqrt{T_2}} = \frac{f_1L_1}{\sqrt{T_1}}$$

We are given that $T_2 = 4T_1$ and $L_2 = 0.5L_1$. Substituting in the second equation, we have

$$\frac{f_2 0.5L_1}{\sqrt{4T_1}} = \frac{f_1L_1}{\sqrt{T_1}} \qquad \text{Solve for } f_2.$$

$$\frac{f_2 0.5L_1}{2\sqrt{T_1}} = \frac{f_1L_1}{\sqrt{T_1}}$$

$$f_2 = \frac{2\sqrt{T_1}f_1L_1}{0.5L_1\sqrt{T_1}} = 4f_1$$

Thus, the frequency of pitch is increased by a factor of 4.

Problem 18 The weight w of an object on or above the surface of the earth varies inversely as the square of the distance d between the object and the center of the earth. If an object on the surface of the earth is moved into space so as to double its distance from the earth's center, what effect will this move have on its weight?

ANSWERS TO MATCHED PROBLEMS

15. **(A)** $y = \frac{1}{2}x$ **(B)** $y = 2\sqrt[3]{x}$
16. **(A)** $y = 32/x$ **(B)** $y = 256/x^2$
17. $v = 60$ miles per hour
18. It will be one-fourth as heavy.

EXERCISE 9-5

A *Translate each problem into an equation using k as the constant of variation.*

1. F varies directly as the square of v.
2. u varies directly as v.
3. The pitch or frequency f of a guitar string of a given length varies directly as the square root of the tension T of the string.
4. Geologists have found in studies of earth erosion that the erosive force (sediment-carrying power) P of a swiftly flowing stream varies directly as the sixth power of the velocity v of the water.
5. y varies inversely as the square root of x.
6. I varies inversely as t.

7. The biologist Reaumur suggested in 1735 that the length of time t that it takes fruit to ripen during the growing season varies inversely as the sum T of the average daily temperatures during the growing season.

8. In a study on urban concentration, F. Auerbach discovered an interesting law. After arranging all the cities of a given country according to their population size, starting with the largest, he found that the population P of a city varied inversely as the number n indicating its position in the ordering.

9. R varies jointly as S, T, and V.

10. g varies jointly as x and the square of y.

11. The volume of a cone V varies jointly as its height h and the square of the radius r of its base.

12. The amount of heat put out by an electrical appliance (in calories) varies jointly as time t, resistance R in the circuit, and the square of the current I.

Solve using either of the two methods illustrated in Example 17.

13. u varies directly as the square root of v. If $u = 2$ when $v = 2$, find u when $v = 8$.

14. y varies directly as the square of x. If $y = 20$ when $x = 2$, find y when $x = 5$.

15. L varies inversely as the square root of M. If $L = 9$ when $M = 9$, find L when $M = 3$.

16. I varies inversely as the cube of t. If $I = 4$ when $t = 2$, find I when $t = 4$.

B *Translate each problem into an equation using k as the constant of variation.*

17. U varies jointly as a and b and inversely as the cube of c.

18. w varies directly as the square of x and inversely as the square root of y.

19. The maximum safe load L for a horizontal beam varies jointly as its width w and the square of its height h and varies inversely as its length l.

20. Joseph Cavanaugh, a sociologist, found that the number of long-distance phone calls n between two cities in a given time period varied (approximately) jointly as the populations P_1 and P_2 of the two cities and inversely as the distance d between the two cities.

Solve using either of the two methods illustrated in Example 17.

21. Q varies jointly as m and the square of n and inversely as P. If $Q = -4$ when $m = 6$, $n = 2$, and $P = 12$, find Q when $m = 4$, $n = 3$, and $P = 6$.

22. w varies jointly as x, y, and z and inversely as the square of t. If $w = 2$ when $x = 2$, $y = 3$, $z = 6$, and $t = 3$, find w when $x = 3$, $y = 4$, $z = 2$, and $t = 2$.

23. The weight w of an object on or above the surface of the earth varies inversely as the square of the distance d between the object and the center of the earth. If a girl weighs 100 pounds on the surface of the earth, how much would she weigh (to the nearest pound) 400 miles above the earth's surface? (Assume that the radius of the earth is 4,000 miles.)

24. A child was struck by a car in a crosswalk. The driver of the car had slammed on his brakes and left skid marks 160 feet long. He told the police he had been driving at 30 miles per hour. The police know that the length of skid marks L (when brakes are applied) varies directly as the square of the speed of the car v and that at 30 miles per hour (under ideal conditions) skid marks would be 40 feet long. How fast was the driver actually going before he applied his brakes?

25. Ohm's law states that the current I in a wire varies directly as the electromotive force E and inversely as the resistance R. If $I = 22$ amperes when $E = 110$ volts and $R = 5$ ohms, find I if $E = 220$ volts and $R = 11$ ohms.

26. Anthropologists, in their study of race and human genetic groupings, often use an index called the *cephalic index*. The cephalic index C varies directly as the width w of the head and inversely as the length l of the head (both when viewed from the top). If an Indian in Baja California (Mexico) has measurements of $C = 75$, $w = 6$ inches, and $l = 8$ inches, what is C for an Indian in northern California with $w = 8.1$ inches and $l = 9$ inches?

C **27.** If the horsepower P required to drive a speedboat through water varies directly as the cube of the speed v of the boat, what change in horsepower is required to double the speed of the boat?

28. The intensity of illumination E on a surface varies inversely as the square of its distance d from a light source. What is the effect on the total illumination on a book if the distance between the light source and the book is doubled?

29. The frequency of vibration f of a musical string varies directly as the square root of the tension T and inversely as the length L of the string. If the tension of the string is increased by a factor of 4 and the length of the string is doubled, what is the effect on the frequency?

30. In an automobile accident the destructive force F of a car varies (approximately) jointly as the weight w of the car and the square of the speed v of the car. (This is why accidents at high speed are generally so serious.) What would be the effect on the destructive force of a car if its weight were doubled and its speed were doubled?

ADDITIONAL APPLICATIONS

The following problems include significant applications from many different areas and are arranged according to subject area. The more difficult problems are marked with two stars (★★), the moderately difficult problems are marked with one star (★), and the easier problems are not marked.

31. ***Astronomy*** The square of the time t required for a planet to make one orbit around the sun varies directly as the cube of its mean (average) distance d from the sun. Write the equation of variation, using k as the constant of variation.

★**32.** ***Astronomy*** The centripetal force F of a body moving in a circular path at constant speed varies inversely as the radius r of the path. What happens to F if r is doubled?

33. ***Astronomy*** The length of time t a satellite takes to complete a circular orbit of the earth varies directly as the radius r of the orbit and inversely as the orbital velocity v of the satellite. If $t = 1.42$ hours when $r = 4{,}050$ miles and $v = 18{,}000$ miles per hour *(Sputnik I),* find t for $r = 4{,}300$ miles and $v = 18{,}500$ miles per hour.

34. ***Life Science*** The number N of gene mutations resulting from x-ray exposure varies directly as the size of the x-ray dose r. What is the effect on N if r is quadrupled?

35. ***Life Science*** In biology there is an approximate rule, called the *bioclimatic rule* for temperate climates, which states that the difference d in time for fruit to ripen (or insects to appear) varies directly as the change in altitude h. If $d = 4$ days when $h = 500$ feet, find d when $h = 2{,}500$ feet.

36. ***Physics and Engineering*** Over a fixed distance d, speed r varies inversely as time t. Police use this relationship to set up speed traps. (The graph of the resulting function is a hyperbola.) If in a given speed trap $r = 30$ miles per hour when $t = 6$ seconds, what would be the speed of a car if $t = 4$ seconds?

★**37.** ***Physics and Engineering*** The length L of skid marks of a car's tires (when the brakes are applied) varies directly as the square of the speed v of the car. How is the length of skid marks affected by doubling the speed?

38. ***Physics and Engineering*** The time t required for an elevator to lift a weight varies jointly as the weight w and the distance d through which it is lifted and inversely as the power P of the motor. Write the equation of variation, using k as the constant of variation.

39. ***Physics and Engineering*** The total pressure P of the wind on a wall varies jointly as the area of the wall A and the square of the velocity of the wind v. If $P = 120$ pounds when $A = 100$ square feet and $v = 20$ miles per hour, find P if $A = 200$ square feet and $v = 30$ miles per hour.

★★**40.** ***Physics and Engineering*** The thrust T of a given type of propeller varies jointly as the fourth power of its diameter d and the square of the number of revolutions per minute n it is turning. What happens to the thrust if the diameter is doubled and the number of revolutions per minute is cut in half?

41. ***Psychology*** In early psychological studies on sensory perception (hearing, seeing, feeling, and so on), the question was asked: "Given a certain level of stimulation S, what is the minimum amount of added stimulation ΔS that can be detected?" A German physiologist, E. H. Weber (1795–1878) formulated, after many experiments, the famous law that now bears his name: "The amount of change ΔS that will be just noticed varies directly as the magnitude S of the stimulus."

(A) Write the law as an equation of variation.

(B) If a person lifting weights can just notice a difference of 1 ounce at the 50-ounce level, what will be the least difference she will be able to notice at the 500-ounce level?

(C) Determine the just noticeable difference in illumination a person is able to perceive at 480 candlepowers if he is just able to perceive a difference of 1 candlepower at the 60-candlepower level.

42. ***Psychology*** Psychologists in their study of intelligence often use an index called IQ. IQ varies directly as mental age MA and inversely as chronological age CA (up to the age of 15). If a 12-year-old boy with a mental age of 14.4 has an IQ of 120, what will be the IQ of an 11-year-old girl with a mental age of 15.4?

43. ***Music*** The frequency of vibration of air in an open organ pipe varies inversely as the length of the pipe. If the air column in an open 32-foot pipe vibrates 16 times per second (low C), how fast would the air vibrate in a 16-foot pipe?

44. ***Music*** The frequency of pitch f of a musical string varies directly as the square root of the tension T and inversely as the length l and the diameter d. Write the equation of variation using k as the constant of variation. (It is interesting to note that if pitch depended on only length, then pianos would have to have strings varying from 3 inches to 38 feet.)

45. ***Photography*** The f-stop numbers N on a camera, known as *focal ratios,* vary directly as the focal length F of the lens and inversely as the diameter d of the diaphragm opening (effective lens opening). Write the equation of variation using k as the constant of variation.

★46. ***Photography*** In taking pictures using flashbulbs, the lens opening (f-stop number) N varies inversely as the distance d from the object being photographed. What adjustment should you make on the f-stop number if the distance between the camera and the object is doubled?

★47. ***Chemistry*** Atoms and molecules that make up the air constantly fly about like microscopic missiles. The velocity v of a particular particle at a fixed temperature varies inversely as the square root of its molecular weight w. If an oxygen molecule in air at room temperature has an average velocity of 0.3 mile per second, what will be the average velocity of a hydrogen molecule, given that the hydrogen molecule is one-sixteenth as heavy as the oxygen molecule?

48. ***Chemistry*** The Maxwell–Boltzmann equation says that the average velocity v of a molecule varies directly as the square root of the absolute temperature T and inversely as the square root of its molecular weight w. Write the equation of variation using k as the constant of variation.

49. ***Business*** The amount of work A completed varies jointly as the number of workers W used and the time t they spend. If 10 workers can finish a job in 8 days, how long will it take 4 workers to do the same job?

50. ***Business*** The simple interest I earned in a given time varies jointly as the principal p and the interest rate r. If \$100 at 4% interest earns \$8, how much will \$150 at 3% interest earn in the same period?

★51. ***Geometry*** The volume of a sphere varies directly as the cube of its radius r. What happens to the volume if the radius is doubled?

★52. ***Geometry*** The surface area S of a sphere varies directly as the square of its radius r. What happens to the area if the radius is cut in half?

9-6 CHAPTER REVIEW

A **function** is a rule that produces a correspondence between a first set of elements (**domain**) and a second set (**range**) such that to each element in the domain there corresponds exactly one element in the range. A function can also be thought of as a set of ordered pairs. Functions may be specified by diagrams, tables, equations, sets of ordered pairs, or graphs. The values in the domain can be thought of as **input** values, those in the range as **output**. A function with domain and range sets of real numbers can be graphed by graphing all pairs (x, y), where x represents the input and y the corresponding output. A graph represents a function if no vertical line crosses it more than once (**vertical-line test**). A variable representing the input of a function is called the **independent variable**; a variable for values in the range is called the **dependent variable**. *(9-1)*

If a function is denoted by f, the notation $f(x)$ denotes the value in the range corresponding to x in the domain. *(9-2)*

A function $f(x) = ax + b$ is called a **linear function**; $f(x) = ax^2 + bx + c$ is a **quadratic function** for $a \neq 0$. The graph of a linear function is a nonvertical straight line; that of a quadratic function is a parabola. If $f(x)$ is specified by a polynomial in x, the function is called a **polynomial function**. Polynomial functions are graphed by plotting a sufficient number of points and joining these points with a smooth curve. *(9-3)*

A function f is **one-to-one** if every range element corresponds to exactly one domain element. The reverse rule for such a function f is called the **inverse of f** and is denoted f^{-1}. The inverse of a function specified by an equation $y = f(x)$ may, in some cases, be found by interchanging variables, obtaining $x = f(y)$, and solving for y. The graph of the inverse function f^{-1} is the graph of f reflected about the line $y = x$. A function will have an inverse when no horizontal line crosses its graph more than once (**horizontal-line test**). *(9-4)*

Direct, **inverse**, **joint**, and **combined variations** are relations that occur often in applications:

Direct:	y varies directly as x	$y = kx$
Inverse:	y varies inversely as x	$y = \frac{k}{x}$
Joint:	w varies jointly as x and y	$w = kxy$
Combined:	w varies directly as x and inversely as y	$w = \frac{kx}{y}$

In each case, k is called the **constant of variation**. *(9-5)*

REVIEW EXERCISE 9-6

Work through all the problems in this chapter review and check answers in the back of the book. (Answers to all problems are there, and following each answer is a number in italics indicating the section in which that type of problem is discussed.) Where weaknesses show up, review appropriate sections in the text.

A *Which of the rules in Problems 1–12 are functions? The variable x is independent.*

1. 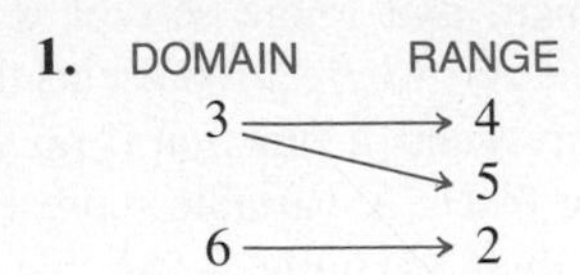

2. 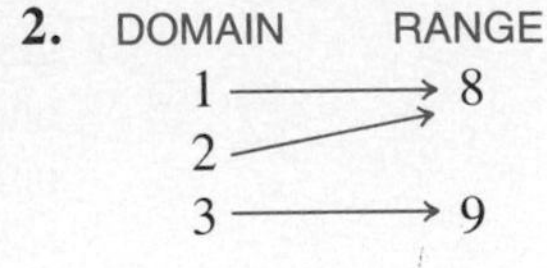

3. 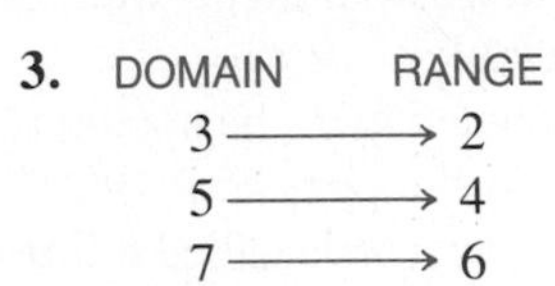

4. 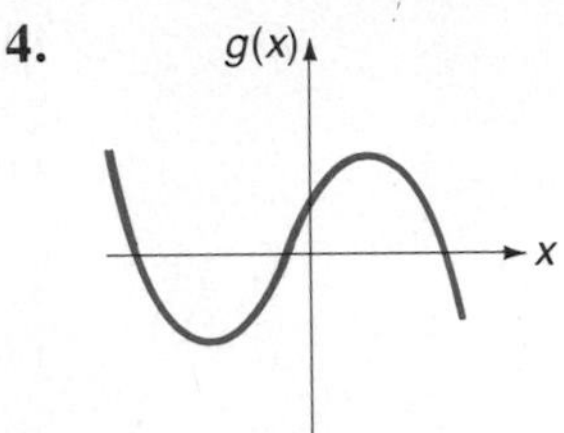

5. 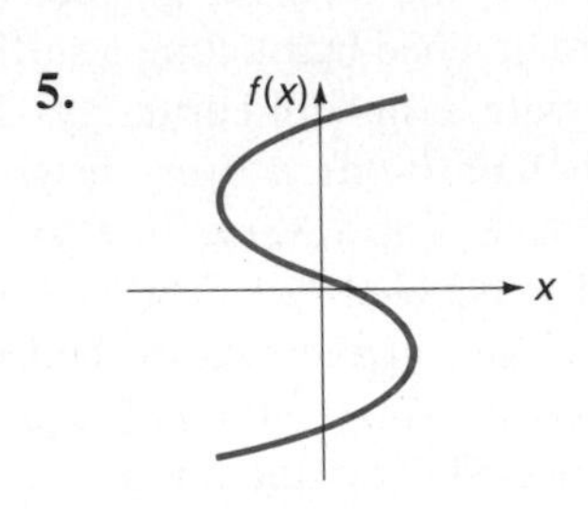

6. 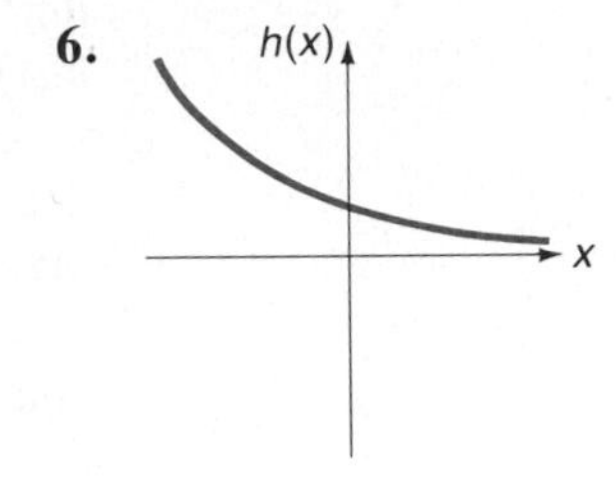

7. $y = x^3 - 2x$

8. $y^2 = x$

9. $x^2 + y^2 = 25$

10. $\{(1, 2), (1, -2), (0, 3)\}$

11. $\{(-1, 2), (1, 3), (2, 4)\}$

12. $\{(-2, 3), (0, 3), (2, 3)\}$

Find the domain of the function:

13. $\{(1, 2), (3, 4), (5, 6)\}$

14. $f(x) = \sqrt{x - 1}$

15. $f(x) = x - 1$

16. If $f(x) = 6 - x$, find:
(A) $f(6)$ **(B)** $f(0)$ **(C)** $f(-3)$ **(D)** $f(m)$

17. If $G(z) = z - 2z^2$, find:
(A) $G(2)$ **(B)** $G(0)$ **(C)** $G(-1)$ **(D)** $G(c)$

18. Graph $f(x) = 2x - 4$. Indicate its slope and y intercept.

19. Graph $g(x) = \dfrac{x^2}{2}$.

20. Graph the function $M = \{(0, 5), (2, 7), (4, 9)\}$, its inverse M^{-1}, and $y = x$, all on the same coordinate system.

21. What are the domain and range of M^{-1} in the preceding problem?

Translate each statement into an equation using k as the constant of variation.

22. m varies directly as the square of n.

23. P varies inversely as the cube of Q.

24. A varies jointly as a and b.

25. y varies directly as the cube of x and inversely as the square root of z.

B **26.** Does every function have an inverse? Explain.

27. If $f(t) = 4 - t^2$ and $g(t) = t - 3$, find:

(A) $f(0) - g(0)$ **(B)** $\dfrac{g(6)}{f(-1)}$ **(C)** $g(x) - f(x)$

(D) $f[g(2)]$

28. If $f(x) = 2x - 3$, find:

(A) $f(3 + h)$ **(B)** $\dfrac{f(3 + h) - f(3)}{h}$

29. Graph $g(t) = -\frac{3}{2}t + 6$ and indicate its slope and y intercept.

30. Graph $f(x) = x^2 - 4x + 5$. Find its vertex and intercepts.

31. Graph $P(x) = x^3 - 2x^2 - 5x + 6$ for $-3 \le x \le 4$.

Indicate which functions are one-to-one in the problems indicated. (In Problems 1–12, you identified which were functions.)

32. Problems 1–3

33. Problems 4–6

34. Problems 7–9

35. Problems 10–12

36. Which of the functions in Problems 4 and 6 have inverses?

37. Which of the functions in Problems 2 and 3 have inverses?

38. Determine the domain of the function specified by the equation

$$y = \sqrt{\frac{x + 2}{x - 5}}$$

39. Let $M(x) = \dfrac{x + 3}{2}$.

(A) Find $M^{-1}(x)$.
(B) Find $M^{-1}(2)$.
(C) Find $M^{-1}[M(3)]$.

40. If y varies directly as x and inversely as z:
(A) Write the equation of variation.
(B) If $y = 4$ when $x = 6$ and $z = 2$, find y when $x = 4$ and $z = 4$.

C **41.** If $g(t) = 1 - t^2$, find: **(A)** $g(2 + h)$ **(B)** $\dfrac{g(2 + h) - g(2)}{h}$

42. Graph $g(t) = 96t - 16t^2$. Indicate its vertex and intercepts.

43. Let $f(x) = x^2$, $x \geq 0$.
(A) Find $f^{-1}(x)$.
(B) Graph f and f^{-1} on the same coordinate system along with $y = x$.
(C) Find $f^{-1}(9)$ and $f^{-1}[f(x)]$.

44. Let $E(x) = 2^x$, $x \in \{-2, -1, 0, 1, 2\}$. Graph E, E^{-1}, and $y = x$ on the same coordinate system.

APPLICATIONS

45. ***Cost function*** The cost $C(x)$ for renting a business copying machine is \$200 for 1 month plus 5 cents a copy for x copies. Express this functional relationship in terms of an equation and graph it for $0 \leq x \leq 3{,}000$.

46. ***Revenue function*** The revenue function for a company producing stereo radios (for a particular model) is

$$R = f(p) = 6{,}000p - 30p^2 \qquad 0 \leq p \leq 200$$

where p is the price per unit.
(A) Graph f.
(B) At what price p will the revenue be largest? What is the largest revenue?

47. ***Variation*** The time t required for an elevator to lift a weight varies jointly as the weight w and the distance d through which it is lifted and inversely as the power P of the motor. Write the equation of variation using k as the constant of variation. If it takes a 400-horsepower motor 4 seconds to lift an 800-kilogram elevator 8 meters, how long will it take the same motor to lift a 1,600-kilogram elevator 24 meters?

★48. ***Variation*** The total force F of a wind on a wall varies jointly as the area of the wall A and the square of the velocity of the wind v. How is the total force on a wall affected if the area is cut in half and the wind velocity is doubled?

10

EXPONENTIAL AND LOGARITHMIC FUNCTIONS

For a fixed $b > 0$, $b \neq 1$, the function

$$f(x) = b^x$$

is called an *exponential function*. This function has an inverse, called a *logarithm* or *logarithmic function*. In this chapter, we will explore these two new classes of functions.

10-1 EXPONENTIAL FUNCTIONS

- Exponential Functions
- Graphing an Exponential Function
- Base *e*
- Basic Exponential Properties

In this section and the next we will consider two new kinds of functions that use variable exponents in their definitions. To start, note that

$$f(x) = 2^x \qquad \text{and} \qquad g(x) = x^2$$

are not the same function. The function g has a variable base x with a fixed exponent 2; for the function f, the base is fixed and the exponent varies. The function g is a quadratic function, which we have already discussed; the function f is a new function called an *exponential function*.

EXPONENTIAL FUNCTIONS

An **exponential function** is a function defined by an equation of the form:

Exponential Function

$$f(x) = b^x \qquad b > 0, \quad b \neq 1$$

where b is a constant, called the **base**, and the exponent x is a variable. The replacement set for the exponent, the **domain of f**, is the set of real numbers R. The **range of f** is the set of positive real numbers. We require b to be positive to avoid complex numbers such as $(-2)^{1/2}$, and we insist that $b \neq 1$ since $1^x = 1$ would simply be a constant function.

GRAPHING AN EXPONENTIAL FUNCTION

Many students, if asked to graph an exponential function such as $f(x) = 2^x$, would not hesitate at all. They would likely make up a table by assigning integers to x, plot the resulting points, and then join these points with a smooth curve (Figure 1).

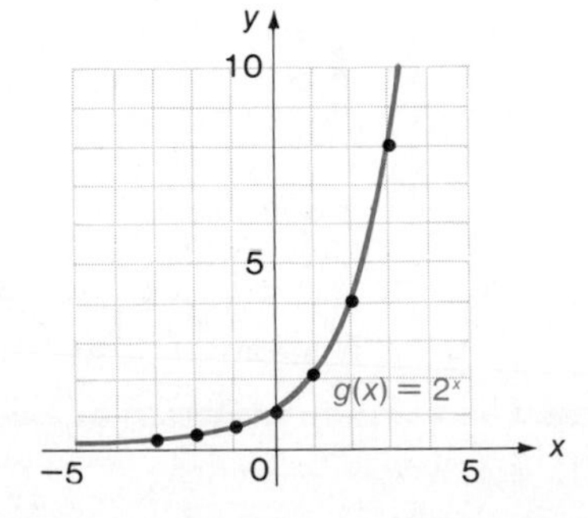

x	$g(x)$
-3	$\frac{1}{8}$
-2	$\frac{1}{4}$
-1	$\frac{1}{2}$
0	1
1	2
2	4
3	8

FIGURE 1

The only catch is that 2^x has not been defined at this point for all real numbers. We know what 2^5, 2^{-3}, $2^{2/3}$, $2^{-3/5}$, $2^{1.4}$, and $2^{-3.15}$ all mean (that is, 2^p, where p is a rational number), but what does

$$2^{\sqrt{2}}$$

mean? The question is not easy to answer at this time. In fact, a precise definition of $2^{\sqrt{2}}$ must wait for more advanced courses, where we can show that

$$b^x$$

names a real number for b a positive real number and x any real number and that the graph of $g(x) = 2^x$ is as indicated in Figure 1.

It is useful to compare the graphs of $y = 2^x$ and $y = (\frac{1}{2})^x = (2^{-1})^x = 2^{-x}$ by plotting both on the same coordinate system (Figure 2a).

FIGURE 2

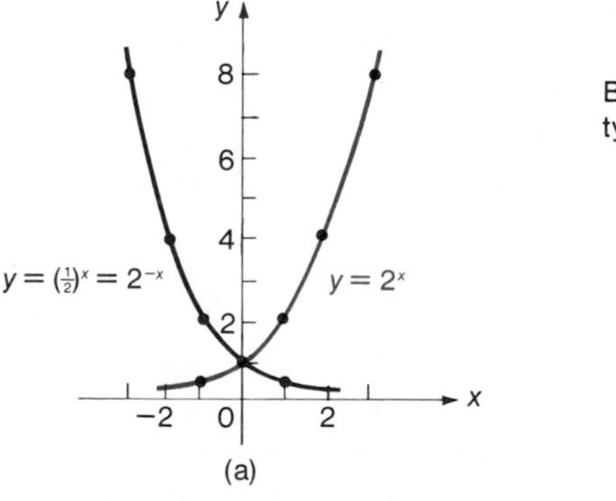

(a)

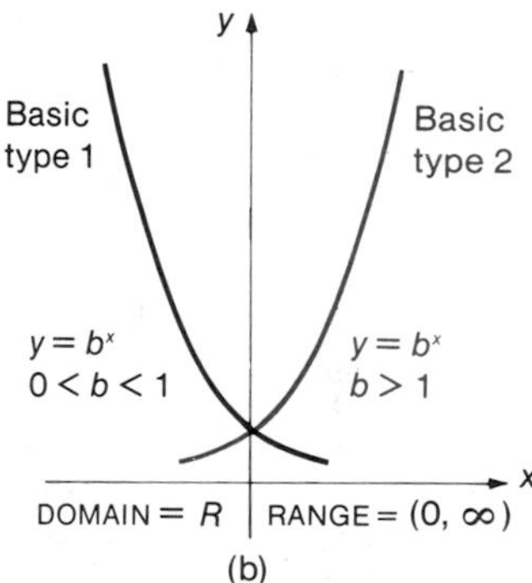

(b)

The graph of

$$f(x) = b^x \qquad b > 1 \text{ (Figure 2b)}$$

will look very much like the graph of $y = 2^x$, and the graph of

$$f(x) = b^x \qquad 0 < b < 1 \text{ (Figure 2b)}$$

will look very much like the graph of $y = (\frac{1}{2})^x$. Note in both cases that the x axis is a horizontal asymptote† and the graphs will never touch it.

We can see from the graphs that an exponential function is either increasing or decreasing and hence is one-to-one and has an inverse that is a function. This fact will be important to us in the next section when we define a logarithmic function as an inverse of an exponential function.

† Asymptotes were discussed in Section 7-6.

Example 1 Graph $y = \frac{1}{2}(4^x)$ for $-3 \le x \le 3$.

Solution

x	y
−3	0.01
−2	0.03
−1	0.13
0	0.50
1	2.00
2	8.00
3	32.00

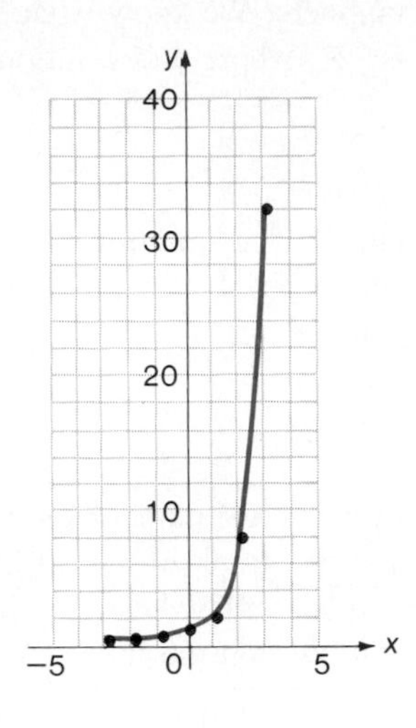

Problem 1 Graph $y = \frac{1}{2}(4^{-x})$ for $-3 \le x \le 3$.

Exponential functions are often referred to as *growth functions* because of their widespread use in describing different kinds of growth. These functions are used to describe population growth of people, wildlife, and bacteria; radioactive decay (negative growth); increasing concentration of a new chemical substance in a chemical reaction; increase or decline in the temperature of a substance being heated or cooled; growth of money at compound interest; light absorption (negative growth) as it passes through air, water, or glass; decline of atmospheric pressure as altitude is increased; and growth of learning a skill such as swimming or typing relative to practice.

BASE *e*

For introductory purposes, the bases 2 and $\frac{1}{2}$ were convenient choices; however, a certain irrational number, denoted by e, is by far the most frequently used exponential base for both theoretical and practical purposes. In fact,

$$f(x) = e^x$$

is often referred to as *the* exponential function because of its widespread use. The reasons for the preference for e as a base are made clear in advanced courses. And at that time, it is shown that e is approximated by $(1 + 1/n)^n$ to any decimal accuracy desired by making n (an integer) sufficiently large. The irrational number e to eight decimal places is

$$e \approx 2.718\ 281\ 83$$

Similarly, e^x can be approximated by using $(1 + 1/n)^{nx}$ for sufficiently large n. Because of the importance of e^x and e^{-x}, tables for their evaluation are readily available and many hand calculators can evaluate these functions directly. We will rely on hand calculators to obtain values for e^x. The important constant e along with two other important constants $\sqrt{2}$ and π are shown on the number line in Figure 3a, and the graph of $y = e^x$ is shown in Figure 3b.

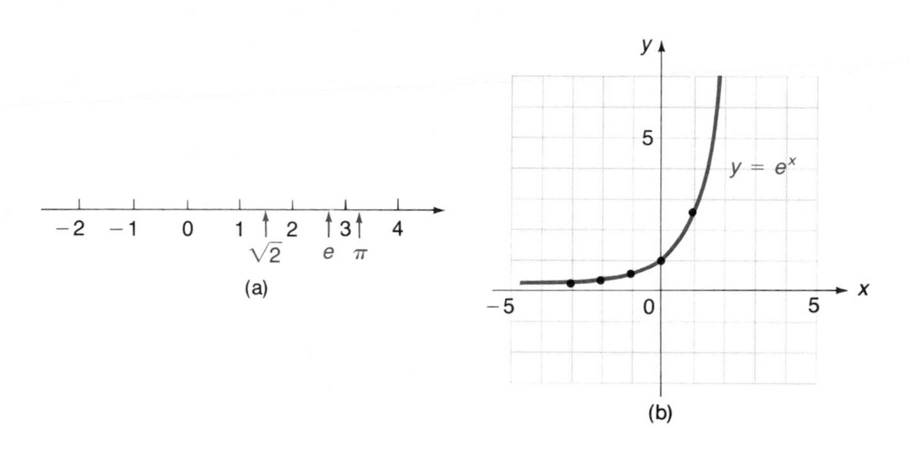

FIGURE 3

Example 2 Graph $y = 10e^{-0.5x}$, $-3 \le x \le 3$, using a hand calculator.

Solution

x	y
−3	44.82
−2	27.18
−1	16.49
0	10.00
1	6.07
2	3.68
3	2.23

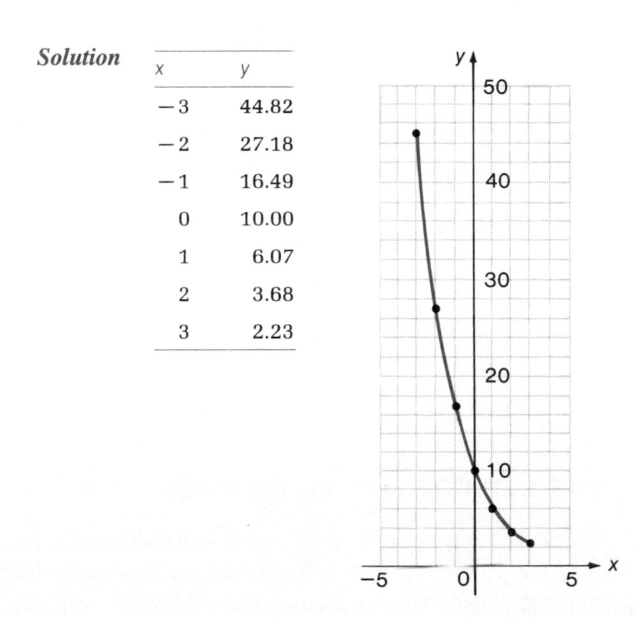

Problem 2 Graph $y = 10e^{0.5x}$, $-3 \le x \le 3$, using a hand calculator.

Example 3 If an initial amount, or principal, P is invested at $100r\%$ compounded continuously, then the amount A in the account at the end of t years is given by (from the mathematics of finance):

$$A = Pe^{rt}$$

If \$100 is invested at 12% compounded continuously, graph the amount in the account relative to time for a period of 10 years.

Solution We wish to graph

$$A = 100e^{0.12t} \qquad 0 \le t \le 10$$

We make up a table of values using a calculator, graph the points from the table, and then join the points with a smooth curve, as follows:

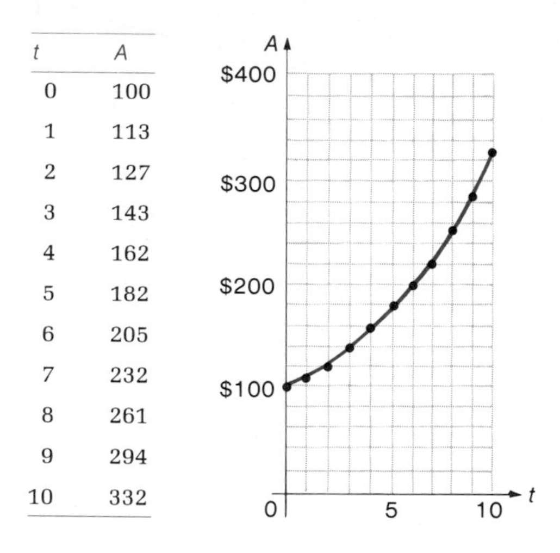

t	A
0	100
1	113
2	127
3	143
4	162
5	182
6	205
7	232
8	261
9	294
10	332

Problem 3 Repeat Example 3 with \$5,000 being invested at 20% compounded continuously.

BASIC EXPONENTIAL PROPERTIES

Earlier (Sections 2-1, 5-1, 5-2, and 5-4) we discussed five laws of exponents for rational exponents. It can be shown that these same laws hold for irrational exponents. Thus, we now assume that all five laws of exponents hold for any real exponents as long as the bases involved are positive.

As a consequence of an exponential function being either increasing or decreasing and thus one-to-one, we have

> For a, $b > 0$; a, $b \neq 1$,
>
> $a^m = a^n$ if and only if $m = n$
> $a^m = b^m$ $(m \neq 0)$ if and only if $a = b$

Thus, if $2^{15} = 2^{3x}$, then $3x = 15$ and $x = 5$. Also, if $x^{15} = 2^{15}$, then $x = 2$.

ANSWERS TO MATCHED PROBLEMS

1. $y = \frac{1}{2}(4^{-x})$

x	y
−3	32.00
−2	8.00
−1	2.00
0	0.50
1	0.13
2	0.03
3	0.01

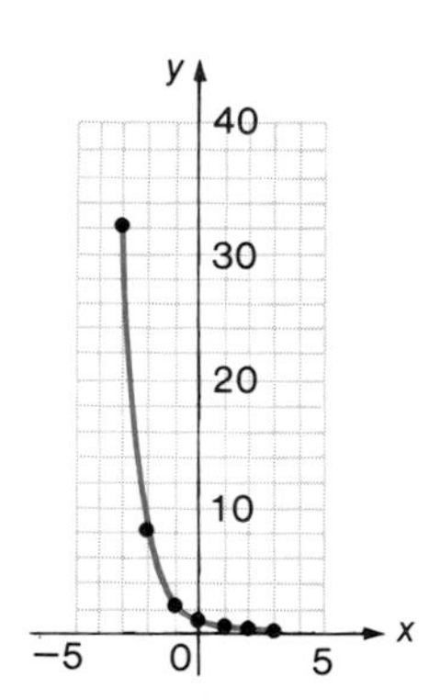

2. $y = 10e^{0.5x}$

x	y
−3	2.23
−2	3.68
−1	6.07
0	10.00
1	16.49
2	27.18
3	44.82

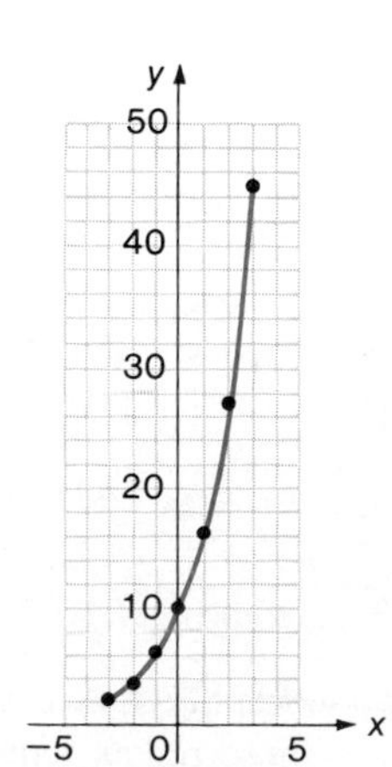

3. $A = 5{,}000e^{0.2t}$

t	A
0	5,000
1	6,107
2	7,459
3	9,111
4	11,128
5	13,591
6	16,601
7	20,276
8	24,765
9	30,248
10	36,945

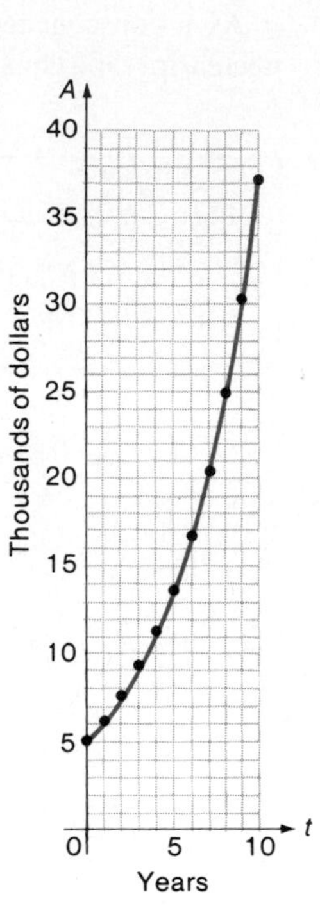

EXERCISE 10-1

A *Graph each exponential function for* $-3 \le x \le 3$ *using a calculator. (Plot points using integers for x; then join the points with a smooth curve.)*

1. $y = 3^x$

2. $y = 2^x$

3. $y = (\frac{1}{3})^x = 3^{-x}$

4. $y = (\frac{1}{2})^x = 2^{-x}$

5. $y = 4 \cdot 3^x$ [*Note:* $4 \cdot 3^x \neq 12^x$]

6. $y = 5 \cdot 2^x$

B **7.** $y = 2^{x+3}$

8. $y = 3^{x+1}$

9. $y = 7(\frac{1}{2})^{2x} = 7 \cdot 2^{-2x}$

10. $y = 11 \cdot 2^{-2x}$

Graph Problems 11–14 for $-3 \le x \le 3$. *Use a calculator.*

11. $y = e^x$

12. $y = e^{-x}$

13. $y = 10e^{-0.12x}$

14. $y = 100e^{0.25x}$

C **15.** Graph $y = 10 \cdot e^{-x^2}$ for $-2 \le x \le 2$.

16. Graph $y = e^{-x^2}$ for $x = -1.5, -1.0, -0.5, 0, 0.5, 1.0, 1.5$, and join these points with a smooth curve.

17. Graph $y = y_0 2^x$, where y_0 is the value of y when $x = 0$. (Express the vertical scale in terms of y_0.)

18. Graph $y = y_0 e^{-0.22x}$, where y_0 is the value of y when $x = 0$. (Express the vertical scale in terms of y_0.)

19. Graph $y = 2^x$ and $x = 2^y$ on the same coordinate system.

20. Graph $f(x) = 10^x$ and $y = f^{-1}(x)$ on the same coordinate system.

APPLICATIONS

21. If we start with 2 cents and double the amount each day, at the end of n days we will have 2^n cents. Graph $f(n) = 2^n$ for $1 \leq n \leq 10$. (Pick the scale on the vertical axis so that the graph will not go off the paper.)

22. ***Compound interest*** If a certain amount of money P, called the *principal,* is invested at $100r\%$ interest compounded annually, the amount of money A after t years is given by

$$A = P(1 + r)^t$$

Graph this equation for $P = \$10$, $r = 0.10$, and $0 \leq t \leq 10$.

★23. ***Earth science*** The atmospheric pressure P, in pounds per square inch, can be calculated approximately using the formula

$$P = 14.7e^{-0.21x}$$

where x is altitude relative to sea level in miles. Graph the equation for $-1 \leq x \leq 5$.

★24. ***Bacterial growth*** If bacteria in a certain culture double every hour, write a formula that gives the number of bacteria N in the culture after n hours, assuming the culture has N_0 bacteria to start with.

25. ***Radioactive decay*** Radioactive strontium-90 has a half-life of 28 years; that is, in 28 years one-half of any amount of strontium-90 will change to another substance because of radioactive decay. If we place a bar containing 100 milligrams of strontium-90 in a nuclear reactor, the amount of strontium-90 that will be left after t years is given by $A = 100(\frac{1}{2})^{t/28}$. Graph this exponential function for $t = 0$, 28, 2(28), 3(28), 4(28), 5(28), and 6(28), and join these points with a smooth curve.

26. ***Radioactive decay*** Radioactive argon-39 has a half-life of 4 minutes; that is, in 4 minutes one-half of any amount of argon-39 will change to another substance because of radioactive decay. If we start with A_0 milligrams of argon-39, the amount left after t minutes is given by $A = A_0(\frac{1}{2})^{t/4}$. Graph this exponential function for $A_0 = 100$ and $t = 0$, 4, 8, 12, 16, and 20, and join these points with a smooth curve.

★27. ***Sociology—small-group analysis*** Two sociologists, Stephan and Mischler, found that when the members of a discussion group of 10 were ranked according to the number of times each participated, the number of times $N(i)$ the ith-ranked person participated was given approximately by the exponential function

$$N(i) = N_1 e^{-0.11(i-1)} \qquad 1 \le i \le 10$$

where N_1 is the number of times the top-ranked person participated in the discussion. Graph the exponential function, using $N_1 = 100$.

10-2 LOGARITHMIC FUNCTIONS

- Logarithmic Functions
- From Logarithmic to Exponential, and Vice Versa
- Finding x, b, or y in $y = \log_b x$
- Logarithmic–Exponential Identities

Since an exponential function is one-to-one, it has an inverse. The inverse will be called a **logarithmic function**. Here you will see why we placed special emphasis on the general concept of inverse functions in Section 9-4. If you know quite a bit about a function, then (knowing about inverses in general) you will automatically know quite a bit about its inverse. For example, the graph of f^{-1} is the graph of f reflected across the line $y = x$, and the domain and range of f^{-1} are, respectively, the range and domain of f.

LOGARITHMIC FUNCTIONS

If we start with the exponential function

$$f\colon \quad y = 2^x$$

and interchange the variables x and y, we obtain the inverse of f:

$$f^{-1}\colon \quad x = 2^y$$

The graphs of f and f^{-1} (along with $y = x$) are shown in Figure 4. This new function is given the name **logarithmic function with base 2** and is symbolized as follows (since we cannot "algebraically" solve $x = 2^y$ for y):

$$y = \log_2 x$$ This is read "y equals the log of x to the base 2" or "y equals log to the base 2 of x."

Thus,

$$y = \log_2 x \quad \text{is equivalent to} \quad x = 2^y$$

That is, $\log_2 x$ is the power to which 2 must be raised to obtain x. (Symbolically, $x = 2^y = 2^{\log_2 x}$.) For example, $\log_2 32 = 5$ since $2^5 = 32$.

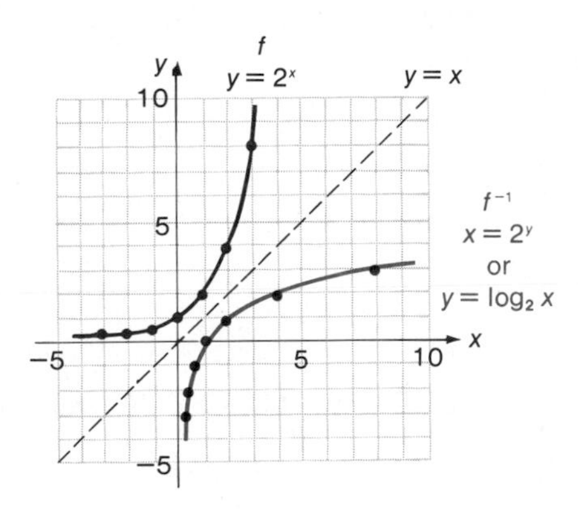

f

x	$y = 2^x$
-3	$\frac{1}{8}$
-2	$\frac{1}{4}$
-1	$\frac{1}{2}$
0	1
1	2
2	4
3	8

f^{-1}

$x = 2^y$	y
$\frac{1}{8}$	-3
$\frac{1}{4}$	-2
$\frac{1}{2}$	-1
1	0
2	1
4	2
8	3

Ordered pairs reversed

FIGURE 4

DOMAIN OF $f = R =$ RANGE OF f^{-1}
RANGE OF $f = (0, \infty) =$ DOMAIN OF f^{-1}

In general, we define the **logarithmic function with base *b*** to be the inverse of the exponential function with base b ($b > 0$, $b \neq 1$).

Definition of Logarithmic Function

For $b > 0$ and $b \neq 1$:

$y = \log_b x$ is equivalent to $x = b^y$

(The log to the base b of x is the power to which b must be raised to obtain x.)

$y = \log_{10} x$ is equivalent to $x = 10^y$
$y = \log_e x$ is equivalent to $x = e^y$

Remember that $y = \log_b x$ and $x = b^y$ define the same function, and as such can be used interchangeably.

Since the domain of an exponential function includes all real numbers and its range is the set of positive real numbers, the **domain** of a logarithmic function is the set of all positive real numbers and its **range** is the set of all real numbers. Thus, $\log_{10} 3$ is defined, but $\log_{10} 0$ and $\log_{10}(-5)$ are not defined (3 is a logarithmic domain value, but 0 and -5 are not). Typical logarithmic curves are shown in Figure 5 on page 462.

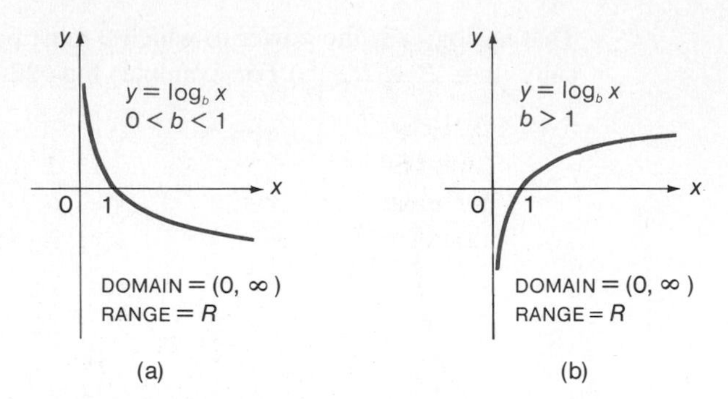

FIGURE 5 Typical logarithmic graphs

FROM LOGARITHMIC TO EXPONENTIAL, AND VICE VERSA

We now look into the matter of converting logarithmic forms to equivalent exponential forms, and vice versa.

Example 4 Change to an equivalent exponential form:

(A) $\log_2 8 = 3$ **(B)** $\log_{25} 5 = \frac{1}{2}$ **(C)** $\log_2 \frac{1}{4} = -2$

Solution **(A)** $\log_2 8 = 3$ is equivalent to $8 = 2^3$
(B) $\log_{25} 5 = \frac{1}{2}$ is equivalent to $5 = 25^{1/2}$
(C) $\log_2 \frac{1}{4} = -2$ is equivalent to $\frac{1}{4} = 2^{-2}$

Problem 4 Change to an equivalent exponential form:

(A) $\log_3 27 = 3$ **(B)** $\log_{36} 6 = \frac{1}{2}$ **(C)** $\log_3 \frac{1}{9} = -2$

Example 5 Change to an equivalent logarithmic form:

(A) $49 = 7^2$ **(B)** $3 = \sqrt{9}$ **(C)** $\frac{1}{5} = 5^{-1}$

Solution **(A)** $49 = 7^2$ is equivalent to $\log_7 49 = 2$
(B) $3 = \sqrt{9}$ is equivalent to $\log_9 3 = \frac{1}{2}$ $\sqrt{9} = 9^{1/2}$
(C) $\frac{1}{5} = 5^{-1}$ is equivalent to $\log_5 \frac{1}{5} = -1$

Problem 5 Change to equivalent logarithmic form:

(A) $64 = 4^3$ **(B)** $2 = \sqrt[3]{8}$ **(C)** $\frac{1}{16}=4^{-2}$

FINDING x, b, OR y IN $y = \log_b x$

To gain a little deeper understanding of logarithmic functions and their relationship to the exponential functions, we will look at a few problems where we are to find x, b, or y in $y = \log_b x$, given the other two values. All values were chosen so that the problems can be solved without tables or a calculator.

Example 6 Find x, b, or y as indicated.

(A) Find y: $y = \log_4 8$ **(B)** Find x: $\log_3 x = -2$
(C) Find b: $\log_b 1{,}000 = 3$

Solution **(A)** Write $y = \log_4 8$ in equivalent exponential form:

$$8 = 4^y \quad \text{Write each number to the same base 2.}$$

$$\boxed{2^3 = (2^2)^y}$$

$$2^3 = 2^{2y} \quad \text{Recall that } b^m = b^n \text{ if and only if } m = n.$$

$$2y = 3$$

$$y = \tfrac{3}{2}$$

Thus, $\frac{3}{2} = \log_4 8$.

(B) Write $\log_3 x = -2$ in equivalent exponential form:

$$x = 3^{-2}$$

$$= \frac{1}{3^2} = \frac{1}{9}$$

Thus, $\log_3 \frac{1}{9} = -2$.

(C) Write $\log_b 1{,}000 = 3$ in equivalent exponential form:

$$1{,}000 = b^3 \quad \text{Write 1,000 as a third power.}$$

$$10^3 = b^3 \quad \text{Recall that for } a, b > 0, \; a^m = b^m \text{ if and only if } a = b.$$

$$b = 10$$

Thus, $\log_{10} 1{,}000 = 3$.

Problem 6 Find x, b, or y as indicated:

(A) Find y: $y = \log_9 27$ **(B)** Find x: $\log_2 x = -3$
(C) Find b: $\log_b 100 = 2$

LOGARITHMIC–EXPONENTIAL IDENTITIES

Recall from Section 9-4 that

$$f^{-1}[f(x)] = x \qquad \text{and} \qquad f[f^{-1}(x)] = x$$

Applying these general properties to $f(x) = b^x$ and $f^{-1}(x) = \log_b x$, we see that

$$\begin{aligned} f^{-1}[f(x)] &= x & f[f^{-1}(x)] &= x \\ \log_b[f(x)] &= x & b^{f^{-1}(x)} &= x \\ \log_b b^x &= x & b^{\log_b x} &= x \end{aligned}$$

Thus, we have the useful logarithmic–exponential identities:

Logarithmic–Exponential Identities

For $b > 0$, $b \neq 1$:

1. $\log_b b^x = x$ — That is, the power to which b must be raised to obtain b^x is x.
2. $b^{\log_b x} = x \qquad x > 0$ — That is, the power to which b must be raised to obtain x is $\log_b x$.

Example 7 Find each of the following:

(A) $\log_{10} 10^5$ **(B)** $\log_{10} 0.01$ **(C)** $\log_e e^{2x+1}$
(D) $\log_4 1$ **(E)** $10^{\log_{10} 7}$ **(F)** $e^{\log_e x^2}$

Solution **(A)** $\log_{10} 10^5 = 5$ **(B)** $\log_{10} 0.01 = \log_{10} 10^{-2} = -2$
(C) $\log_e e^{2x+1} = 2x + 1$ **(D)** $\log_4 1 = \log_4 4^0 = 0$
(E) $10^{\log_{10} 7} = 7$ **(F)** $e^{\log_e x^2} = x^2$

Problem 7 Find each of the following:

(A) $\log_{10} 10^{-5}$ **(B)** $\log_5 25$ **(C)** $\log_{10} 1$
(D) $\log_e e^{m+n}$ **(E)** $10^{\log_{10} 4}$ **(F)** $e^{\log_e(x^4+1)}$

ANSWERS TO MATCHED PROBLEMS

4. **(A)** $27 = 3^3$ **(B)** $6 = 36^{1/2}$ **(C)** $\frac{1}{9} = 3^{-2}$
5. **(A)** $\log_4 64 = 3$ **(B)** $\log_8 2 = \frac{1}{3}$ **(C)** $\log_4 \frac{1}{16} = -2$
6. **(A)** $y = \frac{3}{2}$ **(B)** $x = \frac{1}{8}$ **(C)** $b = 10$
7. **(A)** -5 **(B)** 2 **(C)** 0 **(D)** $m + n$ **(E)** 4
(F) $x^4 + 1$

EXERCISE 10-2

A *Rewrite in exponential form.*

1. $\log_3 9 = 2$ **2.** $\log_2 4 = 2$ **3.** $\log_3 81 = 4$

4. $\log_5 125 = 3$ **5.** $\log_{10} 1{,}000 = 3$ **6.** $\log_{10} 100 = 2$

7. $\log_e 1 = 0$ **8.** $\log_8 1 = 0$

Rewrite in logarithmic form.

9. $64 = 8^2$ **10.** $25 = 5^2$ **11.** $10{,}000 = 10^4$

12. $1{,}000 = 10^3$ **13.** $u = v^x$ **14.** $a = b^c$

15. $9 = 27^{2/3}$ **16.** $8 = 4^{3/2}$

Find each of the following:

17. $\log_{10} 10^5$ **18.** $\log_5 5^3$ **19.** $\log_2 2^{-4}$

20. $\log_{10} 10^{-7}$ **21.** $\log_6 36$ **22.** $\log_3 9$

23. $\log_{10} 1{,}000$ **24.** $\log_{10} 0.001$

Find x, y, or b as indicated.

25. $\log_2 x = 2$ **26.** $\log_3 x = 2$

27. $\log_4 16 = y$ **28.** $\log_8 64 = y$

29. $\log_b 16 = 2$ **30.** $\log_b 10^{-3} = -3$

B *Rewrite in exponential form.*

31. $\log_{10} 0.001 = -3$ **32.** $\log_{10} 0.01 = -2$

33. $\log_{81} 3 = \frac{1}{4}$ **34.** $\log_4 2 = \frac{1}{2}$

35. $\log_{1/2} 16 = -4$ **36.** $\log_{1/3} 27 = -3$

37. $\log_a N = e$ **38.** $\log_k u = v$

Rewrite in logarithmic form.

39. $0.01 = 10^{-2}$ **40.** $0.001 = 10^{-3}$ **41.** $1 = e^0$

42. $1 = (\frac{1}{2})^0$ **43.** $\frac{1}{8} = 2^{-3}$ **44.** $\frac{1}{8} = (\frac{1}{2})^3$

45. $\frac{1}{3} = 81^{-1/4}$ **46.** $\frac{1}{2} = 32^{-1/5}$ **47.** $7 = \sqrt{49}$

48. $11 = \sqrt{121}$

Find each of the following:

49. $\log_b b^u$ **50.** $\log_b b^{uv}$ **51.** $\log_e e^{1/2}$

52. $\log_e e^{-3}$ **53.** $\log_2 \sqrt{8}$ **54.** $\log_5 \sqrt[3]{5}$

55. $\log_{23} 1$ **56.** $\log_{17} 1$ **57.** $\log_4 8$

58. $\log_4 \frac{1}{4}$

Find x, y, or b as indicated.

59. $\log_4 x = \frac{1}{2}$ **60.** $\log_{25} x = \frac{1}{2}$ **61.** $\log_{1/3} 9 = y$

62. $\log_{49} \frac{1}{7} = y$ **63.** $\log_b 1{,}000 = \frac{3}{2}$ **64.** $\log_b 4 = \frac{2}{3}$

C **65.** $\log_b 1 = 0$ **66.** $\log_b b = 1$

67. For $f = \{(x, y) \mid y = 1^x\}$, discuss the domain and range for f. Does f have an inverse?

68. Why is 1 not a suitable logarithmic base? [*Hint:* Try to find $\log_1 5.1$.]

69. **(A)** For $f = \{(x, y) \mid y = 10^x\}$, graph f and f^{-1} using the same coordinate axes.
(B) Discuss the domain and range of f and f^{-1}.
(C) What other name could you use for the inverse of f?

70. Prove that $\log_b(1/x) = -\log_b x$.

71. If $\log_b x = 3$, find $\log_b(1/x)$.

10-3 PROPERTIES OF LOGARITHMIC FUNCTIONS

- Basic Logarithmic Properties
- Use of the Logarithmic Properties

Since logarithms are exponents—remember, $\log_b x$ is the power to which b must be raised to obtain x—they have properties related to the laws of exponents. These properties will enable us to convert multiplication problems into addition problems, division problems into subtraction problems, and power and root problems into multiplication problems. Moreover, we will be able to solve exponential equations such as $2 = 10^x$.

BASIC LOGARITHMIC PROPERTIES

Properties of Logarithmic Functions

If b, M, and N are positive real numbers, $b \neq 1$, and p is a real number, then:

1. $\log_b b^u = u$ Logarithmic–exponential identity
2. $\log_b MN = \log_b M + \log_b N$
3. $\log_b \frac{M}{N} = \log_b M - \log_b N$
4. $\log_b M^p = p \log_b M$
5. $\log_b 1 = 0$

The first property, the logarithmic–exponential identity from Section 10-2, follows from the definition of the logarithm function. The last property is a restatement of $b^0 = 1$. The other three properties are logarithmic versions of the laws of exponents:

2. $b^m b^n = b^{m+n}$
3. $\frac{b^m}{b^n} = b^{m-n}$
4. $(b^m)^p = b^{pm}$

Consider the second property. Let

$$\log_b M = m \qquad \log_b N = n$$

Convert to exponential form:

$$b^m = M \qquad b^n = N$$

Then

$$\begin{aligned} \log_b MN &= \log_b b^m b^n \\ &= \log_b b^{m+n} \\ &= m + n && \text{Property 1} \\ &= \log_b M + \log_b N \end{aligned}$$

The other two properties are established in a similar manner.

USE OF THE LOGARITHMIC PROPERTIES

We will now see how logarithmic properties can be used to convert multiplication problems into addition problems, division problems into subtraction problems, and power and root problems into multiplication problems.

Example 8 Rewrite using the properties of logarithms:

(A) $\log_{10} 10^5$ **(B)** $\log_b 3x$ **(C)** $\log_b \dfrac{x}{5}$ **(D)** $\log_b x^7$

(E) $\log_b \dfrac{mn}{pq}$ **(F)** $\log_b(mn)^{2/3}$ **(G)** $\log_b \dfrac{x^8}{y^{1/5}}$

Solution

(A) $\log_{10} 10^5 = 5$ — $\log_b b^u = u$

(B) $\log_b 3x = \log_b 3 + \log_b x$ — $\log_b MN = \log_b M + \log_b N$

(C) $\log_b \dfrac{x}{5} = \log_b x - \log_b 5$ — $\log_b \dfrac{M}{N} = \log_b M - \log_b N$

(D) $\log_b x^7 = 7 \log_b x$ — $\log_b M^p = p \log_b M$

(E) $\log_b \dfrac{mn}{pq} = \log_b mn - \log_b pq$ — $\log_b \dfrac{M}{N} = \log_b M - \log_b N$

$= \log_b m + \log_b n - (\log_b p + \log_b q)$ — $\log_b MN = \log_b M + \log_b N$

$= \log_b m + \log_b n - \log_b p - \log_b q$

(F) $\log_b(mn)^{2/3} = \frac{2}{3} \log_b mn$ — $\log_b M^p = p \log_b M$

$= \frac{2}{3}(\log_b m + \log_b n)$ — $\log_b MN = \log_b M + \log_b N$

(G) $\log_b \dfrac{x^8}{y^{1/5}} = \log_b x^8 - \log_b y^{1/5}$ — $\log_b \dfrac{M}{N} = \log_b M - \log_b N$

$= 8 \log_b x - \frac{1}{5} \log_b y$ — $\log_b M^p = p \log_b M$

Problem 8 Rewrite using the properties of logarithms:

(A) $\log_b \dfrac{r}{uv}$ **(B)** $\log_b \left(\dfrac{m}{n}\right)^{3/5}$ **(C)** $\log_b \dfrac{u^{1/3}}{v^5}$

Example 9 If $\log_e 3 = 1.10$ and $\log_e 7 = 1.95$, find:

(A) $\log_e \frac{7}{3}$ **(B)** $\log_e \sqrt[3]{21}$

Solution **(A)** $\log_e \frac{7}{3} = \log_e 7 - \log_e 3 = 1.95 - 1.10 = 0.85$

(B) $\log_e \sqrt[3]{21} = \log_e (21)^{1/3} = \frac{1}{3}\log_e(3 \cdot 7) = \frac{1}{3}(\log_e 3 + \log_e 7)$
$= \frac{1}{3}(1.10 + 1.95) = 1.02$

Problem 9 If $\log_e 5 = 1.609$ and $\log_e 8 = 2.079$, find:

(A) $\log_e \dfrac{5^{10}}{8}$ **(B)** $\log_e \sqrt[4]{\dfrac{8}{5}}$

Finally, we note that since logarithmic functions are one-to-one:

$$\log_b m = \log_b n \quad \text{if and only if} \quad m = n$$

Thus, if $\log_{10} x = \log_{10} 32.15$, then $x = 32.15$.

The following example and problem, though somewhat artificial, will give you additional practice in using the properties of logarithms.

Example 10 Find x so that $\log_b x = \frac{2}{3}\log_b 27 + 2\log_b 2 - \log_b 3$ without using a calculator or table.

Solution

$\log_b x = \frac{2}{3}\log_b 27 + 2\log_b 2 - \log_b 3$ — Express the right side in terms of a single log.

$= \log_b 27^{2/3} + \log_b 2^2 - \log_b 3$ — Property 4: $p \log_b M = \log_b M^p$

$= \log_b 9 + \log_b 4 - \log_b 3$ — $27^{2/3} = 9$, $2^2 = 4$

$= \log_b \dfrac{9 \cdot 4}{3} = \log_b 12$ — Properties 2 and 3: $\log_b M + \log_b N = \log_b M \cdot N$ and $\log_b M - \log_b N = \log_b \dfrac{M}{N}$

Thus,

$$\log_b x = \log_b 12$$

Hence,

$$x = 12$$

Problem 10 Find x so that $\log_b x = \frac{2}{3}\log_b 8 + \frac{1}{2}\log_b 9 - \log_b 6$ without using a calculator or table.

ANSWERS TO MATCHED PROBLEMS

8. (A) $\log_b r - \log_b u - \log_b v$ (B) $\frac{3}{5}(\log_b m - \log_b n)$
(C) $\frac{1}{3} \log_b u - 5 \log_b v$

9. (A) 14.01 (to four significant digits)†
(B) 0.1175 (to four significant digits)

10. $x = 2$

EXERCISE 10-3

A *Write in terms of simpler logarithmic forms (going as far as you can with logarithmic properties—see Example 8).*

1. $\log_b uv$
2. $\log_b rt$
3. $\log_b(A/B)$
4. $\log_b(p/q)$
5. $\log_b u^5$
6. $\log_b w^{25}$
7. $\log_b N^{3/5}$
8. $\log_b u^{-2/3}$
9. $\log_b \sqrt{Q}$
10. $\log_b \sqrt[5]{M}$
11. $\log_b uvw$
12. $\log_b(u/vw)$

Write each expression in terms of a single logarithm with a coefficient of 1. *[Example:* $\log_b u^2 - \log_b v = \log_b(u^2/v)$]

13. $\log_b A + \log_b B$
14. $\log_b P + \log_b Q + \log_b R$
15. $\log_b X - \log_b Y$
16. $\log_b x^2 - \log_b y^3$
17. $\log_b w + \log_b x - \log_b y$
18. $\log_b w - \log_b x - \log_b y$

If $\log_b 0.69$, $\log_b 3 = 1.10$, *and* $\log_b 5 = 1.61$, *find the logarithm to the base b of each of the following numbers:*

19. $\log_b 30$
20. $\log_b 6$
21. $\log_b \frac{2}{5}$
22. $\log_b \frac{5}{3}$
23. $\log_b 27$
24. $\log_b 16$

B *Write in terms of simpler logarithmic forms (going as far as you can with logarithmic properties—see Example 8).*

25. $\log_b u^2v^7$
26. $\log_b u^{1/2}v^{1/3}$
27. $\log_b \dfrac{1}{a}$
28. $\log_b \dfrac{1}{M^3}$
29. $\log_b \dfrac{\sqrt[3]{N}}{p^2q^3}$
30. $\log_b \dfrac{m^5n^3}{\sqrt{p}}$
31. $\log_b \sqrt[4]{\dfrac{x^2y^3}{\sqrt{z}}}$
32. $\log_b \sqrt[5]{\left(\dfrac{x}{y^4z^9}\right)^3}$

† Significant digits are discussed in Appendix B.

Write each expression in terms of a single logarithm with a coefficient of 1.

33. $2 \log_b x - \log_b y$

34. $\log_b m - \frac{1}{2} \log_b n$

35. $3 \log_b x + 2 \log_b y - 4 \log_b z$

36. $\frac{1}{3} \log_b w - 3 \log_b x - 5 \log_b y$

37. $\frac{1}{5}(2 \log_b x + 3 \log_b y)$

38. $\frac{1}{3}(\log_b x - \log_b y)$

If $\log_b 2 = 0.69$, $\log_b 3 = 1.10$, *and* $\log_b 5 = 1.61$, *find the logarithm to the base b of the following numbers:*

39. $\log_b 7.5$

40. $\log_b 1.5$

41. $\log_b \sqrt[3]{2}$

42. $\log_b \sqrt{3}$

43. $\log_b \sqrt{0.9}$

44. $\log_b \sqrt[3]{\frac{3}{2}}$

C **45.** Find x so that $\frac{3}{2} \log_4 4 - \frac{2}{3} \log_b 8 + 2 \log_b 2 = \log_b x$.

46. Find x so that $3 \log_b 2 + \frac{1}{2} \log_b 25 - \log_b 20 = \log_b x$.

47. Write $\log_b y - \log_b c + kt = 0$ in exponential form free of logarithms.

48. Write $\log_e x - \log_e 100 = -0.08t$ in exponential form free of logarithms.

49. Prove that $\log_b(M/N) = \log_b M - \log_b N$ for $b, M, N > 0$.

50. Prove that $\log_b M^p = p \log_b M$ for $b, M > 0$.

51. Prove that $\log_b MN = \log_b M + \log_b N$ by starting with $M = b^{\log_b M}$ and $N = b^{\log_b N}$.

52. Prove that $\log_b(M/N) = \log_b M - \log_b N$ by starting with $M = b^{\log_b M}$ and $N = b^{\log_b N}$.

10-4 LOGARITHMS TO VARIOUS BASES

- Common and Natural Logarithms—Calculator Evaluation
- Change-of-Base Formula
- Graphing Logarithmic Functions

John Napier (1550–1617) is credited with the invention of logarithms. They evolved out of an interest in reducing the computational strain in astronomy research. This new computational tool was immediately accepted by the scientific world. Now, with the availability of inexpensive hand calculators, logarithms have lost most of their importance as a computational device. However, the logarithmic concept has been greatly generalized since its conception, and logarithmic functions are used widely in both theoretical and applied sciences. For example, even with a very good scientific hand calculator, we still need logarithmic functions to solve the simple-looking exponential equation from popluation growth studies and the mathematics of finance:

$$2 = 1.08^x$$

Of all possible logarithmic bases, the base e and the base 10 are used almost exclusively. Before we can use logarithms in certain practical problems, we need to be able to approximate the logarithm of any positive number to either base 10 or base e. And conversely, if we are given the logarithm of a number to base 10 or base e, we need to be able to approximate the number. Historically, tables such as Tables II and III in the back of the book were used for this purpose, but now with inexpensive scientific hand calculators readily available, most people will choose a calculator, since it is faster and far more accurate than almost any table you might use.

COMMON AND NATURAL LOGARITHMS—CALCULATOR EVALUATION

Common logarithms (also called **Briggsian logarithms**) are logarithms with base 10. **Natural logarithms** (also called **Napierian logarithms**) are logarithms with base e. Most scientific calculators have a button labeled "log" (or "LOG") and a button labeled "ln" (or "LN"). The former represents a common (base 10) logarithm and the latter a natural (base e) logarithm. In fact, "log" and "ln" are both used extensively in mathematical literature, and whenever you see either used in this book without a base indicated, they will be interpreted as follows:

Logarithmic Notation

$\log x = \log_{10} x \qquad \ln x = \log_e x$

To find the common or natural logarithm using a scientific calculator is very easy. You simply enter a number from the domain of the function and push the log or ln button. Tables for evaluating logarithms are also readily available. Their use is not discussed in this book. We will rely on the calculator to obtain values for logarithms.

Example 11 Use a scientific calculator to find each to six decimal places:

(A) log 3,184 **(B)** ln 0.000 349 **(C)** log(−3.24)

Solution

ENTER	PRESS	DISPLAY
(A) 3,184	log	3.502973
(B) 0.000 349	ln	−7.960439
(C) −3.24	log	Error

Why is an error indicated in part (C)? Because -3.24 is not in the domain of the log function.

Problem 11 Use a scientific calculator to find each to six decimal places:

(A) $\log 0.013\,529$ **(B)** $\ln 28.693\,28$ **(C)** $\ln(-0.438)$

Example 12 Use a scientific calculator to evaluate each to three decimal places:

(A) $n = \dfrac{\log 2}{\log 1.1}$ **(B)** $n = \dfrac{\ln 3}{\ln 1.08}$

Solution **(A)** First note that $(\log 2)/(\log 1.1) \neq \log 2 - \log 1.1$. Recall (see Section 10-3) that $\log_b(M/N) = \log_b M - \log_b N$, which is not the same as $(\log_b M)/(\log_b N)$.

$$n = \frac{\log 2}{\log 1.1} = \frac{0.301\,03}{0.041\,39} = 7.272\,540\,9 \approx 7.273$$

(B) $$n = \frac{\ln 3}{\ln 1.08} = \frac{1.098\,612\,3}{0.076\,961} = 14.274\,915 \approx 14.275$$

Problem 12 Use a scientific calculator to evaluate each to two decimal places:

(A) $n = \dfrac{\ln 2}{\ln 1.1}$ **(B)** $n = \dfrac{\log 3}{\log 1.08}$

We will now turn to the second problem: given the logarithm of a number, find the number. We can make direct use of the logarithmic–exponential relationships we discussed in Section 10-2.

Logarithmic–Exponential Relationships

$\log x = y$	is equivalent to	$x = 10^y$
$\ln x = y$	is equivalent to	$x = e^y$

Example 13 Find x to three significant digits, given the indicated logarithms:

(A) $\log x = -9.315$ **(B)** $\ln x = 2.386$

Solution **(A)** $\log x = -9.315$ Change to equivalent exponential form.

$x = 10^{-9.315}$

$= 4.84 \times 10^{-10}$ Notice the answer is displayed in scientific notation in the calculator.

(B) $\ln x = 2.386$ Change to equivalent exponential form.

$$x = e^{2.386}$$
$$= 10.9$$

Problem 13 Find x to four significant figures, given the indicated logarithms:

(A) $\ln x = -5.062$ **(B)** $\log x = 12.0821$

CHANGE-OF-BASE FORMULA

If we have a means (through either a calculator or table) of finding logarithms of numbers to one base, then by means of the change-of-base formula we can find the logarithm of a number to any other base.

Change-of-Base Formula

$$\log_b N = \frac{\log_a N}{\log_a b} = \frac{\log N}{\log b} = \frac{\ln N}{\ln b}$$

The derivation of the change-of-base formula is as follows:

$$y = \log_b N$$

$$N = b^y \quad \text{Definition of logarithm.}$$

$$\log_a N = \log_a b^y \quad \text{Apply log to the base } a \text{ to each side.}$$

$$\log_a N = y \log_a b \quad \text{Property 4}$$

$$y = \frac{\log_a N}{\log_a b} \quad \text{Divide}$$

$$\log_b N = \frac{\log_a N}{\log_a b} \quad \text{Since } y = \log_b N$$

Example 14 Fing $\log_5 14$ to three decimal places using common logarithms.

Solution

$$\log_5 14 = \frac{\log_{10} 14}{\log_{10} 5} = \frac{1.146\ 128}{0.698\ 97} = 1.639\ 738\ 5 \ldots \approx 1.640$$

Problem 14 Find $\log_7 729$ to four decimal places.

GRAPHING LOGARITHMIC FUNCTIONS

With the aid of a scientific calculator to find values, logarithmic functions can be graphed by plotting points.

Example 15 Graph $y = 4 \log x$.

Solution

x	y
0.01	−8
0.1	−4
1	0
2	1.20
5	2.80
10	4

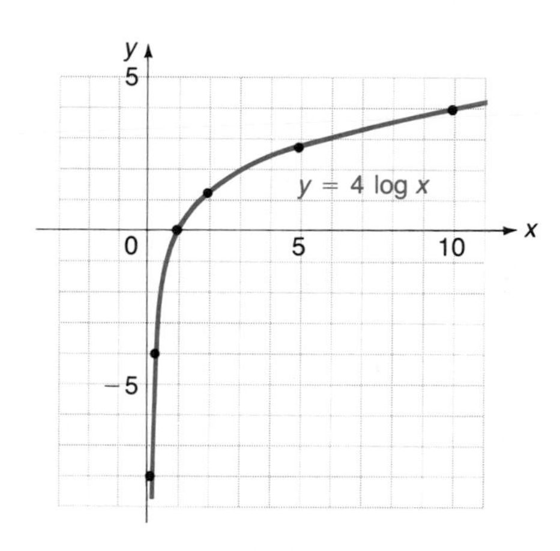

Problem 15 Graph $y = 2 \log 10x$.

Example 16 Graph $y = 3 \ln 2x$.

Solution

x	y
0.1	−4.8
0.5	0
1	2.08
2	4.16
3	5.37
4	6.23
6	7.45
8	8.32

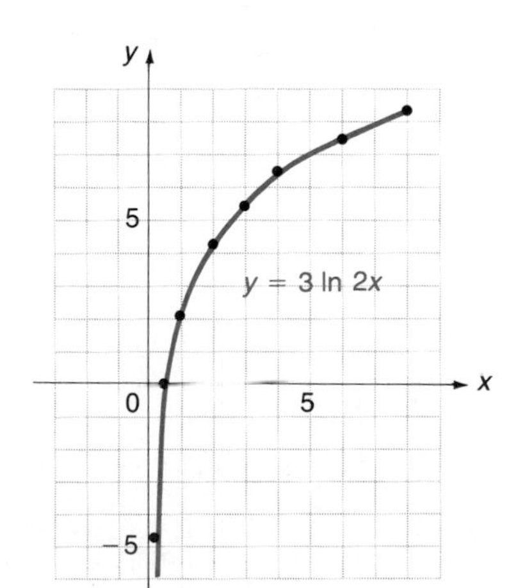

Problem 16 Graph $y = \frac{1}{2} \ln 6x$.

Example 17 Graph $y = \log_{0.5} x$.

Solution To find values for y, use $\log_{0.5} x = \dfrac{\ln x}{\ln 0.5}$.

x	y
0.1	3.32
0.5	1
1	0
2	−1
3	−1.58
4	−2
6	−2.58

Problem 17 Graph $y = 2 \log_{1/4} x$.

ANSWERS TO MATCHED PROBLEMS

11. **(A)** −1.868 734 **(B)** 3.356 663 **(C)** Not possible

12. **(A)** 7.27 **(B)** 14.27

13. **(A)** 0.006 333 **(B)** 1.21×10^{12}

14. 3.3875

15.

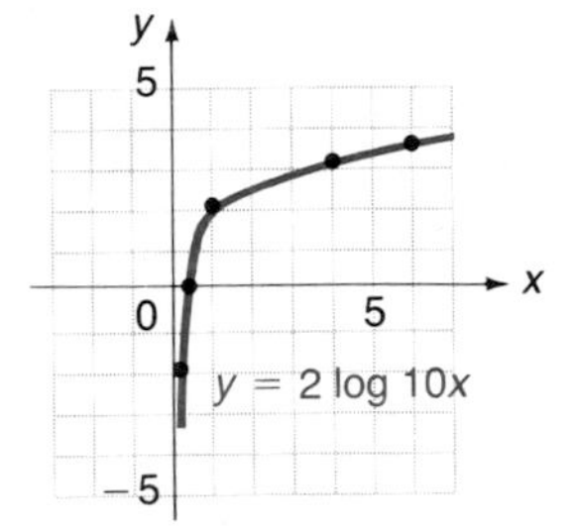

16.

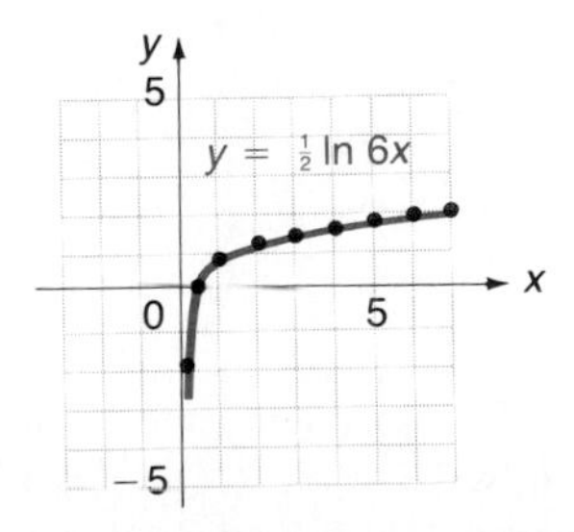

17.

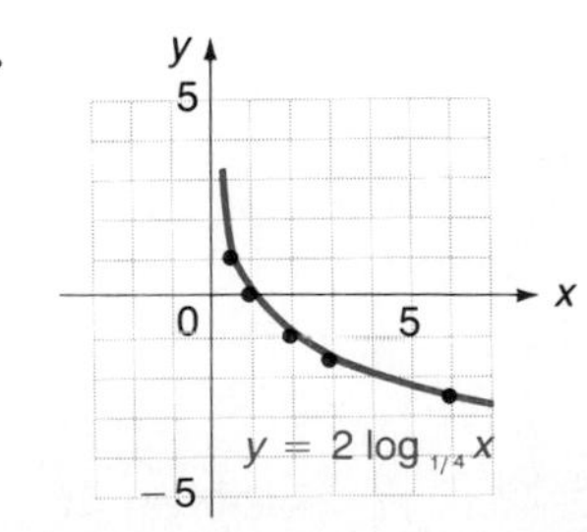

EXERCISE 10–4

A *Use a calculator to find each to four decimal places.*

1. $\log 82{,}734$ **2.** $\log 843{,}250$ **3.** $\log 10.001\,439$

4. $\log 0.035\,604$ **5.** $\ln 43.046$ **6.** $\ln 2{,}843{,}100$

7. $\ln 0.081\,043$ **8.** $\ln 0.000\,032\,4$

B *Use a calculator to find x to four significant digits given the following:*

9. $\log x = 5.3027$ **10.** $\log x = 1.9168$

11. $\log x = -3.1773$ **12.** $\log x = -2.0411$

13. $\ln x = 3.8655$ **14.** $\ln x = 5.0884$

15. $\ln x = -0.3916$ **16.** $\ln x = -4.1083$

Evaluate each of the following to three decimal places using a calculator:

17. $n = \dfrac{\log 2}{\log 1.15}$ **18.** $n = \dfrac{\log 2}{\log 1.12}$ **19.** $n = \dfrac{\ln 3}{\ln 1.15}$

20. $n = \dfrac{\ln 4}{\ln 1.2}$ **21.** $x = \dfrac{\ln 0.5}{-0.21}$ **22.** $t = \dfrac{\log 200}{2 \log 2}$

Use the change-of-base formula and a calculator with either log or ln to find each to four decimal places.

23. $\log_5 372$ **24.** $\log_4 23$ **25.** $\log_8 0.0352$

26. $\log_2 0.005\,439$ **27.** $\log_3 0.1483$ **28.** $\log_{12} 435.62$

Graph.

29. $y = \frac{1}{2} \ln 5x$ **30.** $y = 3 \ln \dfrac{x}{4}$

31. $y = 2 \log x^2$ **32.** $y = 4 \log(\frac{1}{2}x)$

33. $y = 4 \log_5 x$ **34.** $y = 10 \log_{0.10} x$

C *Find x to five significant digits using a calculator.*

35. $x = \log(5.3147 \times 10^{12})$ **36.** $x = \log(2.0991 \times 10^{17})$

37. $x = \ln(6.7917 \times 10^{-12})$ **38.** $x = \ln(4.0304 \times 10^{-8})$

39. $\log x = 32.068\,523$ **40.** $\log x = -12.731\,64$

41. $\ln x = -14.667\,13$ **42.** $\ln x = 18.891\,143$

10-5 EXPONENTIAL AND LOGARITHMIC EQUATIONS

- Exponential Equations
- Logarithmic Equations

Equations involving exponential and logarithmic functions such as

$$2^{3x-2} = 5 \qquad \text{and} \qquad \log(x + 3) + \log x = 1$$

are called **exponential** and **logarithmic equations**, respectively. Logarithmic properties play a central role in their solution.

EXPONENTIAL EQUATIONS

The following examples illustrate the use of logarithmic properties in solving exponential equations:

Example 18 Solve $2^{3x-2} = 5$ for x to four decimal places.

Solution

$$2^{3x-2} = 5$$

How can we get x out of the exponent? Use logs! If two positive quantities are equal, their logs are equal.

$$\log 2^{3x-2} = \log 5$$

Use $\log_b N^p = p \log_b N$ to get $3x - 2$ out of the exponent position.

$$(3x - 2)\log 2 = \log 5$$

$$3x - 2 = \frac{\log 5}{\log 2}$$

Remember: $\frac{\log 5}{\log 2} \neq \log 5 - \log 2$

$$x = \frac{1}{3}\left(2 + \frac{\log 5}{\log 2}\right)$$

$$= \frac{1}{3}\left(2 + \frac{0.698\ 97}{0.301\ 03}\right)$$

$$= 1.4406$$

To four decimal places

Problem 18 Solve $35^{1-2x} = 7$ for x to four decimal places.

Example 19 If a certain amount of money P (principal) is invested at $100r\%$ interest compounded annually, then the amount of money A in the account after n years, assuming no withdrawals, is given by

$$A = P(1 + r)^n$$

How long will it take the money to double if it is invested at 6% compounded annually?

Solution To find the doubling time, we replace A in $A = P(1.06)^n$ with $2P$ and solve for n:

$$2P = P(1.06)^n \quad \text{Divide both sides by } P.$$

$$2 = 1.06^n \quad \text{Take the common or natural log of both sides.}$$

$$\log 2 = \log 1.06^n$$

$$\log 2 = n \log 1.06 \quad \text{Note how log properties are used to get } n \text{ out of the exponent position.}$$

$$n = \frac{\log 2}{\log 1.06}$$

$$= \frac{0.301\ 03}{0.025\ 30} \approx 11.89$$

$$= 12 \text{ years} \quad \text{To the next whole year}$$

Problem 19 Repeat Example 19, changing the interest rate from 6% compounded annually to 9% compounded annually.

Example 20 The atmospheric pressure P (in pounds per square inch) at x miles above sea level is given approximately by

$$P = 14.7e^{-0.21x}$$

At what height will the atmospheric pressure be half of the sea-level pressure? Compute the answer to two significant digits.

Solution Sea-level pressure is the pressure at $x = 0$. Thus,

$$P = 14.7e^0 = 14.7$$

One-half of sea-level pressure is $\frac{14.7}{2} = 7.35$. Now our problem is to find x so that $P = 7.35$; that is, we solve $7.35 = 14.7e^{-0.21x}$ for x:

$$7.35 = 14.7e^{-0.21x} \quad \text{Divide both sides by 14.7 to simplify.}$$

$$0.5 = e^{-0.21x} \quad \text{Take the natural log of both sides.}$$

$$\ln 0.5 = \ln e^{-0.21x} \quad \text{Why use natural logs? Compare with the common log to see why.}$$

$$\ln 0.5 = -0.21x$$

$$x = \frac{\ln 0.5}{-0.21} \quad \text{Use a hand calculator.}$$

$$= 3.3 \text{ miles} \quad \text{To two significant digits}$$

Problem 20 Using the formula in Example 20, find the altitude in miles to two significant digits so that the atmospheric pressue will be one-eighth that at sea level.

LOGARITHMIC EQUATIONS

The next two examples illustrate approaches to solving some types of logarithmic equations.

Example 21 Solve $\log(x + 3) + \log x = 1$ and check.

Solution

$$\log(x + 3) + \log x = 1 \quad \text{Combine the left side using } \log M + \log N = \log MN.$$
$$\log[x(x + 3)] = 1 \quad \text{Change to the equivalent exponential form.}$$
$$x(x + 3) = 10^1 \quad \text{Write in } ax^2 + bx + c = 0 \text{ form.}$$
$$x^2 + 3x - 10 = 0 \quad \text{Solve.}$$
$$(x + 5)(x - 2) = 0$$
$$x = -5, 2$$

Check $x = -5$: $\log(-5 + 3) + \log(-5)$ is not defined   Why?

Thus, -5 must be discarded.

$x = 2$: $\log(2 + 3) + \log 2 = \log 5 + \log 2$
$= \log(5 \cdot 2) = \log 10 \overset{\checkmark}{=} 1$

Thus, 2 is the only solution.

Remember, answers should be checked in the original equation to see whether any should be discarded. An extraneous solution can be introduced in shifting from logarithmic to exponential form because the logarithm has a restricted domain.

Problem 21 Solve $\log(x - 15) = 2 - \log x$ and check.

Example 22 Solve $(\ln x)^2 = \ln x^2$.

Solution

$$(\ln x)^2 = \ln x^2$$
$$(\ln x)^2 = 2 \ln x \quad \text{This is a quadratic equation in } \ln x. \text{ Move all nonzero terms to the left and factor.}$$
$$(\ln x)^2 - 2 \ln x = 0$$
$$(\ln x)(\ln x - 2) = 0$$

$\ln x = 0$ or $\ln x - 2 = 0$

$x = e^0$ $\ln x = 2$

$= 1$ $x = e^2$ Check these results.

Problem 22 Solve $\log x^2 = (\log x)^2$.

ANSWERS TO MATCHED PROBLEMS

18. 0.2263
19. More than double in 9 years, but not quite double in 8 years
20. 9.9 miles **21.** 20 **22.** 100, 1

EXERCISE 10-5

A *Solve to three significant digits.*

1. $10^{-x} = 0.0347$ **2.** $10^x = 14.3$ **3.** $10^{3x+1} = 92$

4. $10^{5x-2} = 348$ **5.** $e^x = 3.65$ **6.** $e^{-x} = 0.0142$

7. $e^{2x-1} = 405$ **8.** $e^{3x+5} = 23.8$ **9.** $5^x = 18$

10. $3^x = 4$ **11.** $2^{-x} = 0.238$ **12.** $3^{-x} = 0.074$

Solve exactly.

13. $\log 5 + \log x = 2$ **14.** $\log x - \log 8 = 1$

15. $\log x + \log(x - 3) = 1$ **16.** $\log(x - 9) + \log 100x = 3$

B *Solve to three significant digits.*

17. $2 = 1.05^x$ **18.** $3 = 1.06^x$

19. $e^{-1.4x} = 13$ **20.** $e^{0.32x} = 632$

21. $123 = 500e^{-0.12x}$ **22.** $438 = 200e^{0.25x}$

Solve exactly.

23. $\log x - \log 5 = \log 2 - \log(x - 3)$

24. $\log(6x + 5) - \log 3 = \log 2 - \log x$

25. $(\ln x)^3 = \ln x^4$ **26.** $(\log x)^3 = \log x^4$

27. $\ln(\ln x) = 1$ **28.** $\log(\log x) = 1$

29. $x^{\log x} = 100x$ **30.** $3^{\log x} = 3x$

C *In Problems 31–36 solve for the indicated letter in terms of all others using common or natural logs, whichever produces the simplest results.*

31. $I = I_0e^{-kx}$ for x *X-ray absorption*

32. $A = P(1 + i)^n$ for n *Compound interest*

33. $N = 10 \log \frac{I}{I_0}$ for I *Sound intensity—decibels*

34. $t = \frac{-1}{k}(\ln A - \ln A_0)$ for A *Radioactive decay*

35. $I = \frac{E}{R}(1 - e^{-Rt/L})$ for t *Electric circuits*

36. $S = R\frac{(1+i)^n - 1}{i}$ for n *Future value of an annuity*

37. Find the fallacy:

$$3 > 2$$
$$(\log \tfrac{1}{2})3 > (\log \tfrac{1}{2})2$$
$$3 \log \tfrac{1}{2} > 2 \log \tfrac{1}{2}$$
$$\log(\tfrac{1}{2})^3 > \log(\tfrac{1}{2})^2$$
$$(\tfrac{1}{2})^3 > (\tfrac{1}{2})^2$$
$$\tfrac{1}{8} > \tfrac{1}{4}$$

38. Find the fallacy:

$$-2 < -1$$
$$\ln e^{-2} < \ln e^{-1}$$
$$2 \ln e^{-1} < \ln e^{-1}$$
$$2 < 1$$

APPLICATIONS

39. ***Compound interest*** How long will it take a sum of money to double if it is invested at 15% interest compounded annually (see Example 19)?

40. ***Compound interest*** How long will it take money to quadruple if it is invested at 20% interest compounded annually (see Example 19)?

41. ***Bacterial growth*** A single cholera bacterium divides every $\frac{1}{2}$ hour to produce two complete cholera bacteria. If we start with a colony of 5,000 bacteria, then after t hours we will have

$$A = 5{,}000 \cdot 2^{2t}$$

bacteria. How long will it take for A to equal 1,000,000?

42. ***Astronomy*** An optical instrument is required to observe stars beyond the sixth magnitude, the limit of ordinary vision. However, even optical instruments have their limitations. The limiting magnitude L of an optical telescope with lens diameter D in inches is given by

$$L = 8.8 + 5.1 \log D$$

(A) Find the limiting magnitude for a homemade 6-inch reflecting telescope.
(B) Find the diameter of a lens that would have a limiting magnitude of 20.6.

★43. ***World populations*** A mathematical model for world population growth over short periods of time is given by

$$P = P_0 e^{rt}$$

where P_0 = Population at $t = 0$, r = Rate compounded continuously, t = Time in years, and P = Population at time t. How long will it take the earth's population to double if it continues to grow at its current rate of 2% per year (compounded continuously)? [*Hint:* Given $r = 0.02$, find t so that $P = 2P_0$.]

44. ★★ ***World population*** If the world population is now 4 billion people and if it continues to grow at 2% per year (compounded continuously), how long will it be before there is only 1 square yard of land per person? Use the formula in Problem 43 and the fact that there are 1.7×10^{14} square yards of land on earth.

45. ★ ***Nuclear reactors—strontium-90*** Radioactive strontium-90 is used in nuclear reactors and decays according to

$$A = Pe^{-0.0248t}$$

where P is the amount present at $t = 0$ and A is the amount remaining after t years. Find the half-life of strontium-90; that is, find t so that $A = 0.5P$.

46. ★ ***Archaeology—carbon-14 dating*** Cosmic-ray bombardment of the atmosphere produces neutrons, which in turn react with nitrogen to produce radioactive carbon-14. Radioactive carbon-14 enters all living tissues through carbon dioxide, which is first absorbed by plants. As long as a plant or animal is alive, carbon-14 is maintained in a constant amount on its tissues. Once dead, however, it ceases taking in carbon and, to the slow beat of time, the carbon-14 diminishes by radioactive decay according to the equation

$$A = A_0e^{-0.000124t}$$

where t is time in years. Estimate the age of a skull uncovered in an archaeological site if 10% of the original amount of carbon-14 is still present. [*Hint:* Find t such that $A = 0.1A_0$.]

47. ★ ***Sound intensity—decibels*** Because of the extraordinary range of sensitivity of the human ear (a range of over 1,000 million million to 1), it is helpful to use a logarithmic scale to measure sound intensity over this range rather than an absolute scale. The unit of measure is called the *decibel,* after the inventor of the telephone, Alexander Graham Bell. If we let N by the number of decibels, I the power of the sound in question in watts per cubic centimeter, and I_0 the power of sound just below the threshold of hearing (approximately 10^{-16} watt per square centimeter), then

$$I = I_0 10^{N/10}$$

Show that this formula can be written in the form

$$N = 10 \log \frac{I}{I_0}$$

48. ***Sound intensity—decibels*** Use the formula in Problem 47 (with $I_0 = 10^{-16}$ watt per square centimeter) to find the decibel ratings of the following sounds:

(A) Whisper (10^{-13} watt per square centimeter)
(B) Normal conversation (3.16×10^{-10} watt per square centimeter)
(C) Heavy traffic (10^{-8} watt per square centimeter)
(D) Jet plane with afterburner (10^{-1} watt per square centimeter)

★49. ***Earth science*** For relatively clear bodies of fresh water or salt water, light intensity is reduced according to the exponential function

$$I = I_0 e^{-kd}$$

where I is the intensity at d feet below the surface and I_0 is the intensity at the surface; k is called the coefficient of extinction. Two of the clearest bodies of water in the world are the freshwater Crystal Lake in Wisconsin ($k = 0.0485$) and the saltwater Sargasso Sea off the West Indies ($k = 0.009\ 42$). Find the depths (to the nearest foot) in these two bodies of water at which the light is reduced to 1% of that at the surface.

50. ***Psychology—learning*** In learning a particular task, such as typing or swimming, one progresses faster at the beginning and then levels off. If you plot the level of performance against time, you will obtain a curve of the type shown in the figure. This is called a learning curve and can be very closely approximated by an exponential equation of the form $y = a(1 - e^{-cx})$, where a and c are positive constants. Curves of this type have applications in psychology, education, and industry. Suppose a particular person's history of learning to type is given by the exponential equation $N = 80(1 - e^{-0.08n})$, where N is the number of words per minute typed after n weeks of instruction. Approximately how many weeks did it take the person to learn to type 60 words per minute?

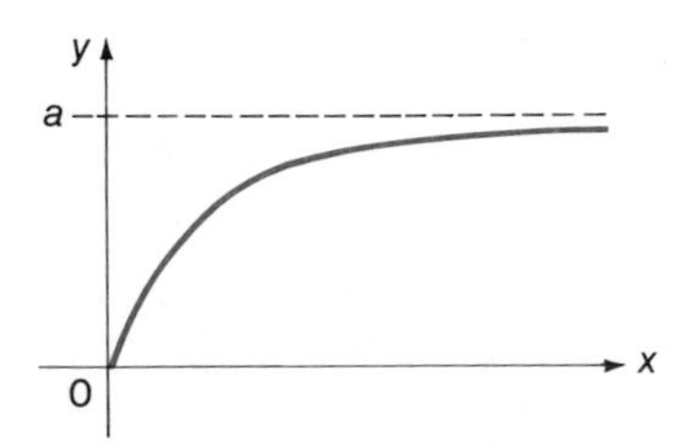

10-6 CHAPTER REVIEW

An **exponential function** is a function defined by $f(x) = b^x$, where the **base** $b > 0$ and $b \neq 1$. The domain of an exponential function is R; the range is the set of positive real numbers. The number $\boldsymbol{e} \approx 2.718$ is the base used for *the* exponential function $f(x) = e^x$. The basic properties of exponents continue to hold for b^x. In addition, $b^m = b^n$ if and only if $m = n$ ($b > 0, b \neq 1$), and for $a, b > 0$, $a^m = b^m$ if and only if $a = b$. *(10-1)*

The inverse of an exponential function b^x is a **logarithmic function**, $y = \log_b x$; thus, $y = \log_b x$ means $b^y = x$. The domain of a logarithmic function is the set of positive real numbers; the range is the set of all real numbers. The logarithmic–exponential identities are

$\log_b b^x = x$ and $b^{\log_b x} = x$ *(10-2)*

Logarithms satisfy these basic properties:

1. $\log_b b^M = M$
2. $\log_b MN = \log_b M + \log_b N$
3. $\log_b \dfrac{M}{N} = \log_b M - \log_b N$
4. $\log_b M^p = p \log_b M$
5. $\log_b 1 = 0$

Also $\log_b M = \log_b N$ if and only if $M = N$. *(10-3)*

Logarithms using base 10 are called **common logarithms**; those with base e are **natural logarithms**. These two types are denoted by $\log_{10} x =$ **log *x*** and $\log_e x =$ **ln *x***. The **change-of-base** formula

$$\log_b N = \frac{\log_a N}{\log_a b}$$

converts logarithms using base a to logarithms using base b. *(10-4)*

Equations involving an exponential function are called **exponential equations**; those involving logarithms are called **logarithmic equations**. *(10-5)*

REVIEW EXERCISE 10-6

Work through all the problems in this chapter review and check answers in the back of the book. (Answers to all problems are there, and following each answer is a number in italics indicating the section in which that type of problem is discussed.) Where weaknesses show up, review appropriate sections in the text.

You will find a scientific hand calculator useful in many of the problems in this exercise.

A

1. Write $m = 10^n$ in logarithmic form with base 10.
2. Write $\log x = y$ in exponential form.

Solve for x exactly. Do not use a calculator or table.

3. $\log_2 x = 3$ 4. $\log_x 25 = 2$ 5. $\log_3 27 = x$

Solve for x to three significant digits.

6. $10^x = 17.5$ **7.** $e^x = 143{,}000$

In Problems 8 and 9, solve for x exactly. Do not use a calculator or table.

8. $\log x - 2 \log 3 = 2$ **9.** $\log x + \log(x - 3) = 1$

B **10.** Write $\ln y = x$ in exponential form.

11. Write $x = e^y$ in logarithmic form with base e.

Solve for x exactly. Do not use a calculator or table.

12. $\log_{1/4} 16 = x$ **13.** $\log_x 9 = -2$ **14.** $\log_{16} x = \frac{3}{2}$

15. $\log_x e^5 = 5$ **16.** $10^{\log_{10} x} = 33$ **17.** $\ln x = 0$

Solve for x to three significant digits.

18. $25 = 5(2)^x$

19. $4{,}000 = 2{,}500e^{0.12x}$

20. $0.01 = e^{-0.05x}$

In Problems 21–24 solve for x exactly. Do not use a table or calculator.

21. $\log 3x^2 - \log 9x = 2$

22. $\log x - \log 3 = \log 4 - \log(x + 4)$

23. $(\log x)^3 = \log x^9$

24. $\ln(\log x) = 1$

25. Calculate $\log_5 23$ to three significant digits.

Graph.

26. $y = 3e^{-2x}$ **27.** $y = 10^{x/2}$

28. $y = 2 \log 3x$ **29.** $y = 2 \ln \dfrac{x}{3}$

C **30.** Write $\ln y = -5t + \ln c$ in an exponential form free of logarithms; then solve for y in terms of the other letters.

31. For $f = \{(x, y) \mid y = \log_2 x\}$, graph f and f^{-1} using the same coordinate system. What are the domains and ranges for f and f^{-1}?

32. Explain why 1 cannot be used as a logarithmic base.

33. Prove that $\log_b(M/N) = \log_b M - \log_b N$.

APPLICATIONS

34. ***Population growth*** Many countries in the world have a population growth rate of 3% (or more) per year. At this rate how long will it take a population to double? Use the population growth model

$$P = P_0(1.03)^t$$

which assumes annual compounding. Compute the answer to three significant digits.

35. ***Population growth*** Repeat Problem 34 using the continuous population growth model

$$P = P_0e^{0.03t}$$

which assumes continuous compounding. Compute the answer to three significant digits.

36. ***Carbon-14 dating*** Refer to Problem 46, Exercise 10-5. How long, to three significant digits, will it take for the carbon-14 to diminish to 1% of the original amount after the death of a plant or animal?

$$A = A_0e^{-0.000124t} \qquad \text{where } t \text{ is time in years}$$

37. ***X-ray absorption*** Solve $x = -(1/k)\ln(I/I_0)$ for I in terms of the other letters.

38. ***Amortization—time payments*** Solve $r = P\{i/[1 - (1 + i)^{-n}]\}$ for n in terms of the other letters.

11

SEQUENCES AND SERIES

Most of the functions we have considered thus far have had domain R, the set of real numbers, or intervals in this set. In this chapter, we consider functions where the domain is the set N of natural numbers, or particular subsets of N. The functions are then called *sequences*. The values of the function can be added to get a sum called a *series*.

11-1 SEQUENCES AND SERIES

- Sequences
- Series

Intuitively, a sequence of numbers is simply a list presented in order, such as

$$1, 2, 4, 8, 16, 32, \ldots$$

In this example, 4 is in the third place, 16 is in the fifth, etc. The rule by which each place corresponds to the number is a function. This observation allows a careful consideration of sequences in mathematical terms.

SEQUENCES

An **infinite sequence** is a function a whose domain is the set of all natural numbers $N = \{1, 2, 3, \ldots, n, \ldots\}$. The range of the function is $a(1)$, $a(2)$, $a(3)$, . . . , $a(n)$, . . . , which is usually written in the form:

Infinite Sequence

$$a_1, a_2, a_3, \ldots, a_n, \ldots \qquad \text{where } a_n = a(n)$$

The elements in the range are called the **terms of the sequence**; a_1 is the first term, a_2 the second term, and a_n the nth term. For example, if

$$a_n = \frac{n-1}{n} \qquad n \in N \tag{1}$$

the function a is a sequence with terms

$$0, \frac{1}{2}, \frac{2}{3}, \frac{3}{4}, \ldots, \frac{n-1}{n}, \ldots$$

Replace n in Equation (1) with 1, 2, 3, and so on.

If the domain of a function a is the set of positive integers $\{1, 2, 3, \ldots, n\}$ for some fixed n, then a is called a **finite sequence**. Thus, if

$$a_1 = 5$$

and

$$a_n = a_{n-1} + 2 \qquad n \in \{2, 3, 4\}$$

the function a is a finite sequence with terms

$$5, 7, 9, 11 \quad a_2 = a_{2-1} + 2 = a_1 + 2 = 5 + 2 = 7, \text{ and so on.}$$

The two examples presented above illustrate two common ways in which sequences are specified:

1. The nth term a_n is expressed by means of n.
2. One or more terms are given, and the nth term is expressed by means of preceding terms.

Example 1 Find the first four terms of a sequence whose nth term is

$$a_n = \frac{1}{2^n}$$

Solution

$$a_1 = \frac{1}{2^1} = \frac{1}{2} \qquad a_2 = \frac{1}{2^2} = \frac{1}{4} \qquad a_3 = \frac{1}{2^3} = \frac{1}{8} \qquad a_4 = \frac{1}{2^4} = \frac{1}{16}$$

Problem 1 Find the first four terms of a sequence whose nth term is

$$a_n = \frac{n}{n^2 + 1}$$

Example 2 Find the first five terms of a sequence specified by

$$a_1 = 5 \qquad a_n = \tfrac{1}{2}a_{n-1} \qquad n \ge 2$$

Solution

$$a_1 = 5$$
$$a_2 = \tfrac{1}{2}a_{2-1} = \tfrac{1}{2}a_1 = \tfrac{1}{2}(5) = \tfrac{5}{2}$$
$$a_3 = \tfrac{1}{2}a_{3-1} = \tfrac{1}{2}a_2 = \tfrac{1}{2}(\tfrac{5}{2}) = \tfrac{5}{4}$$
$$a_4 = \tfrac{1}{2}a_{4-1} = \tfrac{1}{2}a_3 = \tfrac{1}{2}(\tfrac{5}{4}) = \tfrac{5}{8}$$
$$a_5 = \tfrac{1}{2}a_{5-1} = \tfrac{1}{2}a_4 = \tfrac{1}{2}(\tfrac{5}{8}) = \tfrac{5}{16}$$

Problem 2 Find the first five terms of a sequence specified by

$$a_1 = 3 \qquad a_n = a_{n-1} + 4 \qquad n \ge 2$$

Now let us look at the problem in reverse. That is, given the first few terms of a sequence (and assuming the sequence continues in the indicated pattern), find a_n in terms of n.

Example 3 Find a_n in terms of n for the given sequences.

(A) 5, 6, 7, 8, . . . **(B)** 2, −4, 8, −16, . . .

Solution **(A)** $a_n = n + 4$

(B) $a_n = (-1)^{n+1}2^n$ Note how $(-1)^{n+1}$ functions as a sign alternator.

Problem 3 Find a_n in terms of n for:

(A) 3, 5, 7, 9, . . . **(B)** $1, -\frac{1}{2}, \frac{1}{4}, -\frac{1}{8}, \ldots$

SERIES

The sum of the terms of a sequence is called a **series**. Thus, if a_1, a_2, a_3, . . . , a_n are the terms of a sequence, then

$$S_n = a_1 + a_2 + a_3 + \cdots + a_n$$

is called a series. If the sequence is infinite, the corresponding series is called an **infinite series**. We will restrict our attention to finite series in this section.

A series is often represented in a compact form using **summation notation**, as follows:

Summation Notation

$$S_n = \sum_{k=1}^{n} a_k = a_1 + a_2 + a_3 + \cdots + a_n$$

where the terms of the series on the right are obtained from the middle expression by successively replacing the subscript k in a_k with integers, starting with 1 and ending with n. Thus, if a sequence is given by

$$\frac{1}{2}, \frac{1}{4}, \frac{1}{8}, \ldots, \frac{1}{2^n}$$

the corresponding series is given by

$$S_n = \frac{1}{2} + \frac{1}{4} + \frac{1}{8} + \cdots + \frac{1}{2^n}$$

or

$$S_n = \sum_{k=1}^{n} \frac{1}{2^k}$$

Example 4 Write $S_5 = \sum_{k=1}^{5} \frac{k-1}{k}$ without summation notation.

Solution

$$S_5 = \sum_{k=1}^{5} \frac{k-1}{k} \qquad \text{Replace } k \text{ in } \frac{k-1}{k} \text{ successively with 1, 2, 3, 4, and 5; then add.}$$

$$= \frac{1-1}{1} + \frac{2-1}{2} + \frac{3-1}{3} + \frac{4-1}{4} + \frac{5-1}{5}$$

$$= 0 + \tfrac{1}{2} + \tfrac{2}{3} + \tfrac{3}{4} + \tfrac{4}{5}$$

Problem 4 Write $S_6 = \sum_{k=1}^{6} \frac{(-1)^{k+1}}{2k-1}$ without summation notation.

Example 5 Write the following series using summation notation:

$$S_6 = 1 - \tfrac{1}{2} + \tfrac{1}{3} - \tfrac{1}{4} + \tfrac{1}{5} - \tfrac{1}{6}$$

Solution We first note that the nth term of the series is given by

$$a_n = (-1)^{n+1} \frac{1}{n}$$

Hence,

$$S_6 = \sum_{k=1}^{6} (-1)^{k+1} \frac{1}{k}$$

Problem 5 Write the following series using summation notation:

$$S_5 = 1 - \tfrac{2}{3} + \tfrac{4}{9} - \tfrac{8}{27} + \tfrac{16}{81}$$

ANSWERS TO MATCHED PROBLEMS

1. $a_1 = \frac{1}{2}$, $a_2 = \frac{2}{5}$, $a_3 = \frac{3}{10}$, $a_4 = \frac{4}{17}$
2. $a_1 = 3$, $a_2 = 7$, $a_3 = 11$, $a_4 = 15$, $a_5 = 19$
3. **(A)** $a_n = 2n + 1$
   (B) $a_n = (-1)^{n+1}/2^{n-1}$ or $a_n = (-\frac{1}{2})^{n-1}$ (two choices of many)
4. $S_6 = 1 - \frac{1}{3} + \frac{1}{5} - \frac{1}{7} + \frac{1}{9} - \frac{1}{11}$
5. $S_5 = \sum_{k=1}^{5} (-\frac{2}{3})^{k-1}$ or $S_5 = \sum_{k=1}^{5} (-1)^{k+1}(\frac{2}{3})^{k-1}$ (two choices of many)

EXERCISE 11-1

A *Write the first four terms for each sequence.*

1. $a_n = n - 2$

2. $a_n = n + 3$

3. $a_n = \dfrac{n-1}{n+1}$

4. $a_n = \left(1 + \dfrac{1}{n}\right)^n$

5. $a_n = (-2)^{n+1}$

6. $a_n = \dfrac{(-1)^{n+1}}{n^2}$

7. Write the 8th term of the sequence in Problem 1.

8. Write the 10th term of the sequence in Problem 2.

9. Write the 100th term of the sequence in Problem 3.

10. Write the 200th term of the sequence in Problem 4.

Write each series in expanded form without summation notation.

11. $S_5 = \sum_{k=1}^{5} k$

12. $S_4 = \sum_{k=1}^{4} k^2$

13. $S_3 = \sum_{k=1}^{3} \dfrac{1}{10^k}$

14. $S_5 = \sum_{k=1}^{5} (\frac{1}{3})^k$

15. $S_4 = \sum_{k=1}^{4} (-1)^k$

16. $S_6 = \sum_{k=1}^{6} (-1)^{k+1}k$

B *Write the first five terms of each sequence.*

17. $a_n = (-1)^{n+1}n^2$

18. $a_n = (-1)^{n+1}\dfrac{1}{2^n}$

19. $a_n = \dfrac{1}{3}\left(1 - \dfrac{1}{10^n}\right)$

20. $a_n = n[1 - (-1)^n]$

21. $a_1 = 7;\ a_n = a_{n-1} - 4,\ n \ge 2$

22. $a_1 = a_2 = 1;\ a_n = a_{n-1} + a_{n-2},\ n \ge 3$ (Fibonacci sequence)

23. $a_1 = 4;\ a_n = \frac{1}{4}a_{n-1},\ n \ge 2$

24. $a_1 = 2;\ a_n = 2a_{n-1},\ n \ge 2$

Find a_n in terms of n.

25. 4, 5, 6, 7, . . .

26. −2, −1, 0, 1, . . .

27. 3, 6, 9, 12, . . .

28. −2, −4, −6, −8, . . .

29. $\frac{1}{2}, \frac{2}{3}, \frac{3}{4}, \frac{4}{5}, \ldots$

30. $\frac{1}{2}, \frac{3}{4}, \frac{5}{6}, \frac{7}{8}, \ldots$

31. $1, -1, 1, -1, \ldots$

32. $1, -2, 3, -4, \ldots$

33. $-2, 4, -8, 16, \ldots$

34. $1, -3, 5, -7, \ldots$

35. $x, \frac{x^2}{2}, \frac{x^3}{3}, \frac{x^4}{4}, \ldots$

36. $x, -x^3, x^5, -x^7, \ldots$

Write each series in expanded form without summation notation.

37. $\sum_{k=1}^{4} \frac{(-2)^{k+1}}{k}$

38. $\sum_{k=1}^{5} (-1)^{k+1}(2k-1)^2$

39. $S_3 = \sum_{k=1}^{3} \frac{1}{k} x^{k+1}$

40. $S_5 = \sum_{k=1}^{5} x^{k-1}$

41. $\sum_{k=1}^{5} \frac{(-1)^{k+1}}{k} x^k$

42. $\sum_{k=0}^{4} \frac{(-1)^k x^{2k+1}}{2k+1}$

Write each series using summation notation.

43. $S_4 = 1^2 + 2^2 + 3^2 + 4^2$

44. $S_5 = 2 + 3 + 4 + 5 + 6$

45. $S_5 = \frac{1}{2} + \frac{1}{2^2} + \frac{1}{2^3} + \frac{1}{2^4} + \frac{1}{2^5}$

46. $S_4 = 1 - \frac{1}{2} + \frac{1}{3} - \frac{1}{4}$

47. $S_n = 1 + \frac{1}{2^2} + \frac{1}{3^2} + \cdots + \frac{1}{n^2}$

48. $S_n = 2 + \frac{3}{2} + \frac{4}{3} + \cdots + \frac{n+1}{n}$

49. $S_n = 1 - 4 + 9 + \cdots + (-1)^{n+1}n^2$

50. $S_n = \frac{1}{2} - \frac{1}{4} + \frac{1}{8} + \cdots + \frac{(-1)^{n+1}}{2^n}$

C **51.** Show that: $\sum_{k=1}^{n} ca_k = c \sum_{k=1}^{n} a_k$

52. Show that: $\sum_{k=1}^{n} (a_k + b_k) = \sum_{k=1}^{n} a_k + \sum_{k=1}^{n} b_k$

The sequence

$$a_n = \frac{a_{n-1}^2 + P}{2a_{n-1}} \qquad n \geq 2,\ P \text{ a positive real number}$$

can be used to find $\sqrt{P}$ to any decimal accuracy desired. To start the sequence, choose a_1 arbitrarily from the positive real numbers.

53. **(A)** Find the first four terms of the sequence to three decimal places:

$$a_1 = 3 \qquad a_n = \frac{a_{n-1}^2 + 2}{2a_{n-1}} \qquad n \geq 2$$

(A small hand calculator is useful, but not necessary.)

(B) Compare terms with decimal approximation of $\sqrt{2}$ from a table or a calculator.

(C) Repeat parts (A) and (B) by letting a_1 be any other positive number, say 1.

54. **(A)** Find the first four terms of the sequence to three decimal places:

$$a_1 = 2 \qquad a_n = \frac{a_{n-1}^2 + 5}{2a_{n-1}} \qquad n \geq 2$$

(B) Find $\sqrt{5}$ in a table or by a calculator and compare with part (A).

(C) Repeat parts (A) and (B) by letting a_1 be any other positive number, say 3.

11-2 ARITHMETIC SEQUENCES AND SERIES

- Arithmetic Sequences
- Arithmetic Series

Consider the sequence

5, 9, 13, 17, . . .

Can you guess what the fifth term is? If you guessed 21, you have observed that each term after the first can be obtained from the preceding one by adding 4 to it. This is an example of an arithmetic sequence.

ARITHMETIC SEQUENCES

In general, a sequence

$$a_1,\ a_2,\ a_3,\ \ldots,\ a_n,\ \ldots$$

is called an **arithmetic sequence** (or an **arithmetic progression**) if there exists a constant d, called the **common difference**, such that each term is d more than the preceding term. That is,

$a_n = a_{n-1} + d \quad$ for every $n > 1$	ARITHMETIC SEQUENCE

Example 6 Which sequence is an arithmetic sequence and what is its common difference?

(A) 1, 2, 3, 5, . . . **(B)** 3, 5, 7, 9, . . .

Solution Sequence (B) is an arithmetic sequence with $d = 2$.

Problem 6 Repeat Example 6 with: **(A)** $-4, -1, 2, 5, \ldots$ **(B)** 2, 4, 8, 16, . . .

Arithmetic sequences have several convenient properties. For example, it is easy to derive formulas for the nth term in terms of n and the sum of any number of consecutive terms. To obtain an nth-term formula, we note that if a is an arithmetic sequence, then

$$a_2 = a_1 + d$$
$$a_3 = a_2 + d = a_1 + 2d$$
$$a_4 = a_3 + d = a_1 + 3d$$

which suggests:

$a_n = a_1 + (n - 1)d \quad$ for every $n > 1$	nTH-TERM FORMULA

Example 7 If the 1st and 10th terms of an arithmetic sequence are 3 and 30, respectively, find the 50th term of the sequence.

Solution First find d:

$$a_n = a_1 + (n - 1)d$$
$$a_{10} = a_1 + (10 - 1)d$$
$$30 = 3 + 9d$$
$$d = 3$$

Now find a_{50}:

$$\begin{aligned} a_{50} &= 3 + (50 - 1)3 \quad \text{Use } a_n = a_1 + (n-1)d. \\ &= 3 + 49 \cdot 3 \\ &= 150 \end{aligned}$$

Problem 7 If the 1st and 15th terms of an arithmetic sequence are -5 and 23, respectively, find the 73d term of the sequence.

ARITHMETIC SERIES

The sum of the terms of an arithmetic sequence is called an **arithmetic series**. We will derive two simple and very useful formulas for finding the **sum of the first *n* terms of an arithmetic series**.

Consider first the sum

$$S = 1 + 2 + 3 + \cdots + (n-1) + n$$

We can rewrite the terms in reverse order without affecting S:

$$S = n + (n-1) + \cdots + 3 + 2 + 1$$

Now add the corresponding sides of these two equations:

$$\begin{aligned} S &= 1 + 2 + \cdots + (n-1) + n \\ S &= n + (n-1) + \cdots + 2 + 1 \\ \hline 2S &= (n+1) + (n+1) + \cdots + (n+1) + (n+1) \end{aligned}$$

Thus,

$$\begin{aligned} 2S &= n(n+1) \quad \text{There are } n \text{ terms on the right.} \\ S &= \frac{n(n+1)}{2} \end{aligned}$$

The same technique can be used to sum an arbitrary arithmetic series. Let

$$S_n = a_1 + (a_1 + d) + \cdots + [a_1 + (n-2)d] + [a_1 + (n-1)d]$$

Reversing the order of the sum, we obtain

$$S_n = [a_1 + (n-1)d] + [a_1 + (n-2)d] + \cdots + (a_1 + d) + a_1$$

Adding left-hand sides and corresponding elements of the right sides of the two equations, we see that

$$\begin{aligned} 2S_n &= [2a_1 + (n-1)d] + [2a_1 + (n-1)d] + \cdots + [2a_1 + (n-1)d] \\ &= n[2a_1 + (n-1)d] \end{aligned}$$

or

$$S_n = \frac{n}{2}[2a_1 + (n-1)d]$$ SUM FORMULA—FORM 1

Use this form when given the number of terms n, the first term a_1, and the common difference d.

By replacing $a_1 + (n-1)d$ with a_n, we obtain a second useful formula for the sum:

$$S_n = \frac{n}{2}(a_1 + a_n)$$ SUM FORMULA—FORM 2

Use this form when given the number of terms n, the first term a_1, and the last term a_n.

Example 8 Find the sum of the first 26 terms of an arithmetic series if the first term is -7 and $d = 3$.

Solution

$$S_n = \frac{n}{2}[2a_1 + (n-1)d]$$ Use sum formula—form 1.

$$\begin{aligned} S_{26} &= \tfrac{26}{2}[2(-7) + (26-1)3] \\ &= 793 \end{aligned}$$

Problem 8 Find the sum of the first 52 terms of an arithmetic series if the first term is 23 and $d = -2$.

Example 9 Find the sum of all the odd numbers from 51 to 99, inclusive.

Solution First find n:

$$\begin{aligned} a_n &= a_1 + (n-1)d \\ 99 &= 51 + (n-1)2 \\ n &= 25 \end{aligned}$$

Now find S_{25}:

$$S_n = \frac{n}{2}(a_1 + a_n) \quad \text{Use sum formula—form 2.}$$

$$S_{25} = \tfrac{25}{2}(51 + 99)$$
$$= 1{,}875$$

Problem 9 Find the sum of all the even numbers from -22 to 52, inclusive.

ANSWERS TO MATCHED PROBLEMS

6. Sequence (A) with $d = 3$ **7.** $a_{73} = 139$ **8.** $S_{52} = -1{,}456$
9. 570

EXERCISE 11-2

A

1. Determine which of the following are arithmetic sequences. Find d and the next two terms for those that are.
(A) 2, 4, 8, . . . **(B)** 7, 6.5, 6, . . .
(C) $-11, -16, -21, \ldots$ **(D)** $\frac{1}{2}, \frac{1}{6}, \frac{1}{18}, \ldots$

2. Repeat Problem 1 for:
(A) $5, -1, -7, \ldots$ **(B)** $12, 4, \frac{4}{3}, \ldots$
(C) $\frac{1}{2}, \frac{2}{3}, \frac{3}{4}, \ldots$ **(D)** 16, 48, 80, . . .

Let $a_1, a_2, a_3, \ldots, a_n, \ldots$ be an arithmetic sequence. In Problems 3–18 find the indicated quantities.

3. $a_1 = -5, d = 4, a_2 = ?, a_3 = ?, a_4 = ?$

4. $a_1 = -18, d = 3, a_2 = ?, a_3 = ?, a_4 = ?$

5. $a_1 = -3, d = 5, a_{15} = ?, S_{11} = ?$

6. $a_1 = 3, d = 4, a_{22} = ?, S_{21} = ?$

7. $a_1 = 1, a_2 = 5, S_{21} = ?$

8. $a_1 = 5, a_2 = 11, S_{11} = ?$

B

9. $a_1 = 7, a_2 = 5, a_{15} = ?$

10. $a_1 = -3, d = -4, a_{10} = ?$

11. $a_1 = 3, a_{20} = 117, d = ?, a_{101}?$

12. $a_1 = 7, a_8 = 28, d = ?, a_{25} = ?$

13. $a_1 = -12, a_{40} = 22, S_{40} = ?$

14. $a_1 = 24, a_{24} = -28; S_{24} = ?$

15. $a_1 = \frac{1}{3}, a_2 = \frac{1}{2}, a_{11} = ?, S_{11} = ?$

16. $a_1 = \frac{1}{6}, a_2 = \frac{1}{4}, a_{19} = ?, S_{19} = ?$

17. $a_3 = 13,\ a_{10} = 55,\ a_1 = ?$

18. $a_9 = -12,\ a_{13} = 3,\ a_1 = ?$

19. $S_{51} = \sum_{k=1}^{51} (3k + 3) = ?$

20. $S_{40} = \sum_{k=1}^{40} (2k - 3) = ?$

C

21. Find $g(1) + g(2) + g(3) + \cdots + g(51)$ if $g(t) = 5 - t$.

22. Find $f(1) + f(2) + f(3) + \cdots + f(20)$ if $f(x) = 2x - 5$.

23. Find the sum of all the even integers between 21 and 135.

24. Find the sum of all the odd integers between 100 and 500.

25. Show that the sum of the first n odd natural numbers is n^2, using appropriate formulas from this section.

26. Show that the sum of the first n even natural members is $n + n^2$, using appropriate formulas from this section.

27. If in a given sequence $a_1 = -3$ and $a_n = a_{n-1} + 3$, $n > 1$, find a_n in terms of n.

28. For the sequence in Problem 27 find $S_n = \sum_{k=1}^{n} a_k$ in terms of n.

APPLICATIONS

29. ***Earth science*** An object falling from rest in a vacuum near the surface of the earth falls 16 feet during the 1st second, 48 feet during the 2d second, 80 feet during the 3d second, and so on.
(A) How far will the object fall during the 11th second?
(B) How far will the object fall in 11 seconds?
(C) How far will the object fall in t seconds?

30. ***Business*** In investigating different job opportunities, you find that firm A will start you at \$10,000 per year and guarantee you a raise of \$500 each year, while firm B will start you at \$13,000 per year but will only guarantee you a raise of \$200 each year. Over a 15-year period, which firm will pay the greatest total amount?

11-3 GEOMETRIC SEQUENCES AND SERIES

- Geometric Sequences
- Geometric Series
- Infinite Geometric Series

Consider the sequence

$$2,\ -4,\ 8,\ -16,\ \ldots$$

Can you guess what the fifth and sixth terms are? If you guessed 32 and -64, respectively, you have observed that each term after the first can be obtained from the preceding one by multiplying it by -2. This is an example of a geometric sequence.

GEOMETRIC SEQUENCES

In general, a sequence

$$a_1, a_2, a_3, \ldots, a_n, \ldots$$

is called a **geometric sequence** (or a **geometric progression**) if there exists a nonzero constant r, called the **common ratio**, such that each term is r times the preceding term. That is,

$$a_n = ra_{n-1} \qquad \text{for every } n > 1 \qquad \text{GEOMETRIC SEQUENCE}$$

Example 10 Which sequence is a geometric sequence and what is its common ratio?

(A) 2, 6, 8, 10, . . . **(B)** $-1, 3, -9, 27, \ldots$

Solution Sequence (B) is a geometric sequence with $r = -3$.

Problem 10 Repeat Example 10 with: **(A)** $\frac{1}{4}, \frac{1}{2}, 1, 2, \ldots$ **(B)** $\frac{1}{2}, \frac{1}{4}, \frac{1}{16}, \frac{1}{256}, \ldots$

Just as with arithmetic sequences, geometric sequences have several convenient properties. It is easy to derive formulas for the nth term in terms of n and the sum of any number of consecutive terms. To obtain an nth-term formula, we note that if a is a geometric sequence, then

$$a_2 = ra_1$$
$$a_3 = ra_2 = r^2a_1$$
$$a_4 = ra_3 = r^3a_1$$

which suggests that

$$a_n = a_1r^{n-1} \qquad \text{for every } n > 1 \qquad n\text{TH-term formula}$$

Example 11 Find the seventh term of the geometric sequence $1, \frac{1}{2}, \frac{1}{4}, \ldots$.

Solution

$$r = \tfrac{1}{2}$$
$$a_n = a_1r^{n-1}$$
$$a_7 = 1\left(\frac{1}{2}\right)^{7-1} = \frac{1}{2^6} = \frac{1}{64}$$

Problem 11 Find the eighth term of the geometric sequence $\frac{1}{64}, -\frac{1}{32}, \frac{1}{16}, \ldots$.

Example 12 If the first and tenth terms of a geometric sequence are 1 and 2, respectively, find the common ratio r.

Solution

$$a_n = a_1 r^{n-1}$$
$$2 = 1r^{10-1}$$
$$r = 2^{1/9} \approx 1.08 \quad \text{Calculation by calculator.}$$

Problem 12 If the first and eighth terms of a geometric sequence are 2 and 16, respectively, find the common ratio r.

GEOMETRIC SERIES

A **geometric series** is any series whose terms form a geometric sequence. As was the case with an arithmetic series, we can derive two simple and very useful formulas for finding the **sum of the first *n* terms of a geometric series**.

For example, consider the geometric series

$$S = 1 + 3 + 9 + 27 + 81 + 243$$

Then, multiplying both sides by the common ratio 3, we obtain

$$3S = 3 + 9 + 27 + 81 + 243 + 729$$

Subtracting the first sum from the second,

$$2S = 729 - 1$$
$$S = \frac{729 - 1}{2} = 364$$

The same technique works for geometric series in general. Let

$$S_n = a_1 + a_1 r + a_1 r^2 + a_1 r^3 + \cdots + a_1 r^{n-1}$$

and multiply both members by r to obtain

$$rS_n = a_1 r + a_1 r^2 + a_1 r^3 + \cdots + a_1 r^{n-1} + a_1 r^n$$

Now subtract the left side of the second equation from the left side of the first, and the right side of the second equation from the right side of the first, to obtain

$$S_n - rS_n = a_1 - a_1 r^n \quad \text{Note how many terms dropped out on the right side.}$$
$$S_n(1 - r) = a_1 - a_1 r^n$$

Thus:

$$S_n = \frac{a_1 - a_1r^n}{1 - r} = \frac{a_1(1 - r^n)}{1 - r} \qquad r \neq 1$$

SUM FORMULA—FORM 1

Use this formula when given the number of terms n, the first term a_1, and the common ratio r.

Since $a_n = a_1r^{n-1}$, or $ra_n = a_1r^n$, the sum formula can also be written as

$$S_n = \frac{a_1 - ra_n}{1 - r} \qquad r \neq 1$$

SUM FORMULA—FORM 2

Use this formula when given the common ratio r, the first term a_1, and the last term a_n.

Example 13 Find the sum to three significant figures of the first 20 terms of a geometric series if the first term is 1 and $r = 2$.

Solution

$$S_n = \frac{a_1 - a_1r^n}{1 - r} \qquad \text{Use sum formula—form 1.}$$

$$= \frac{1 - 1 \cdot 2^{20}}{1 - 2} \approx 1{,}050{,}000 \qquad \text{Calculation by calculator.}$$

Problem 13 Find the sum to four significant figures of the first 14 terms of a geometric series if the first term is $\frac{1}{64}$ and $r = -2$.

INFINITE GEOMETRIC SERIES

Given a geometric series, what happens to the sum S_n as n increases? To answer this question, we first write the sum formula in the more convenient form

$$S_n = \frac{a_1 - a_1r^n}{1 - r} = \frac{a_1}{1 - r} - \frac{a_1r^n}{1 - r}$$

It is possible to show that if $|r| < 1$, that is, if $-1 < r < 1$, then r^n will tend to 0 as n increases. (See what happens, for example, if $r = \frac{1}{2}$ and n gets large.) Thus, S_n can be made as close to

$$\frac{a_1}{1 - r}$$

as we wish by taking n sufficiently large. Thus, we define

$$S_\infty = \frac{a_1}{1-r} \qquad |r| < 1$$

and call this the **sum of an infinite geometric series**. If $|r| \geq 1$, an infinite geometric series has no sum, since r^n can (in absolute value) be made as large as we like by making n sufficiently large. (What happens to r^n if $r = 3$ and n gets large?)

Example 14 Represent the repeating decimal $0.4545\overline{45}$ as the quotient of two integers. (The bar over the last two digits indicates that these digits repeat indefinitely.)

Solution

$$0.4545\overline{45} = 0.45 + 0.0045 + 0.000045 + \cdots$$

The right side of the equation is an infinite geometric series with $a_1 = 0.45$ and $r = 0.01$. Thus,

$$S_\infty = \frac{a_1}{1-r} = \frac{0.45}{1-0.01} = \frac{0.45}{0.99} = \frac{5}{11}$$

Hence, $0.4545\overline{45}$ and $\frac{5}{11}$ name the same rational number. Check the result by dividing 5 by 11.

Problem 14 Repeat Example 14 for $0.8181\overline{81}$.

ANSWERS TO MATCHED PROBLEMS

10. Sequence (A) with $r = 2$ **11.** -2 **12.** $r \approx 1.346$
13. -85.33 **14.** $\frac{9}{11}$

EXERCISE 11-3

A

1. Determine which of the following are geometric sequences. Find r and the next two terms for those that are.

(A) $2, -4, 8, \ldots$ **(B)** $7, 6.5, 6, \ldots$
(C) $-11, -16, -21, \ldots$ **(D)** $\frac{1}{2}, \frac{1}{6}, \frac{1}{18}, \ldots$

2. Repeat Problem 1 for:

(A) $5, -1, -7, \ldots$ **(B)** $12, 4, \frac{4}{3}, \ldots$
(C) $\frac{1}{2}, \frac{2}{3}, \frac{3}{4}, \ldots$ **(D)** $16, 48, 80, \ldots$

Let $a_1, a_2, a_3, \ldots, a_n, \ldots$ be a geometric sequence. Find each of the indicated quantities.

3. $a_1 = -6$, $r = -\frac{1}{2}$, $a_2 = ?$, $a_3 = ?$, $a_4 = ?$

4. $a_1 = 12$, $r = \frac{2}{3}$, $a_2 = ?$, $a_3 = ?$, $a_4 = ?$

5. $a_1 = 81,\ r = \frac{1}{3},\ a_{10} = ?$

6. $a_1 = 64,\ r = \frac{1}{2},\ a_{13} = ?$

7. $a_1 = 3,\ a_7 = 2{,}187,\ r = 3,\ S_7 = ?$

8. $a_1 = 1,\ a_7 = 729,\ r = -3,\ S_7 = ?$

B

9. $a_1 = 100,\ a_6 = 1,\ r = ?$

10. $a_1 = 10,\ a_{10} = 30,\ r = ?$

11. $a_1 = 5,\ r = -2,\ S_{10} = ?$

12. $a_1 = 3,\ r = 2,\ S_{10} = ?$

13. $a_1 = 9,\ a_4 = \frac{8}{3},\ a_2 = ?,\ a_3 = ?$

14. $a_1 = 12,\ a_4 = -\frac{4}{9},\ a_2 = ?,\ a_3 = ?$

15. $S_7 = \sum_{k=1}^{7} (-3)^{k-1} = ?$

16. $S_7 = \sum_{k=1}^{7} 3^k = ?$

17. Find $g(1) + g(2) + \cdots + g(10)$ if $g(x) = (\frac{1}{2})^x$.

18. Find $f(1) + f(2) + \cdots + f(10)$ if $f(x) = 2^x$.

19. Find a positive number x such that $-2 + x - 6$ is a geometric series with three terms.

20. Find a positive number x such that $6 + x + 8$ is a geometric series with three terms.

Find the sum of each infinite geometric series that has a sum.

21. $3 + 1 + \frac{1}{3} + \cdots$

22. $16 + 4 + 1 + \cdots$

23. $2 + 4 + 8 + \cdots$

24. $4 + 6 + 9 + \cdots$

25. $2 - \frac{1}{2} + \frac{1}{8} - \cdots$

26. $21 - 3 + \frac{3}{7} - \cdots$

C *Represent each repeating decimal fraction as the quotient of two integers.*

27. $0.777\overline{7}$

28. $0.555\overline{5}$

29. $0.5454\overline{54}$

30. $0.2727\overline{27}$

31. $3.216\overline{216}$

32. $5.63636\overline{3}$

APPLICATIONS

★33. ***Business*** If P dollars is invested at $r\%$ compounded annually, the amount A present after n years forms a geometric progression with a constant ratio $(1 + r)$. Write a formula for the amount present after n years. How long will it take for a sum of money P to double if invested at 6% interest compounded annually?

★**34.** ***Life science*** If a population of A_0 people grows at the constant rate of $r\%$ per year, the population after t years forms a geometric progression with a constant ratio $(1 + r)$. Write a formula for the total population after t years. If the world's population is increasing at the rate of 2% per year, how long will it take to double?

35. ***Engineering*** A rotating flywheel coming to rest rotates 300 revolutions the first minute. If in each subsequent minute it rotates two-thirds as many times as in the preceding minute, how many revolutions will the wheel make before coming to rest?

36. ***Physics*** The first swing of a bob on a pendulum is 10 inches. If on each subsequent swing it travels 0.9 as far as on the preceding swing, how far will the bob travel before coming to rest?

37. ***Economics*** The government, through a subsidy program, distributes \$1,000,000. If we assume that each individual or agency spends 0.8 of what is received, and 0.8 of this is spent, and so on, how much total increase in spending results from this government action? (Let $a_1 = \$800{,}000$.)

★**38.** ***Zeno's paradox*** Visualize a hypothetical 440-yard oval racetrack that has tapes stretched across the track at the halfway point and at each point that marks the halfway point of each remaining distance thereafter. A runner running around the track has to break the first tape before the second, the second before the third, and so on. From this point of view it appears that the runner will never finish the race. (This famous paradox is attributed to the Greek philosopher Zeno, 495–435 B.C.) If we assume the runner runs at 440 yards per minute, the times between tape breakings form an infinite geometric progression. What is the sum of this progression?

11-4 BINOMIAL FORMULA

- Factorial
- Binomial Formula

The binomial form

$$(a + b)^n$$

with n a natural number, appears in many mathematical contexts. The coefficients in its expansion play an important role in probability studies. In this section we will give an informal derivation of the famous binomial formula, which will enable us to expand $(a + b)^n$ directly for any natural number n, however large. First, we will introduce the concept of factorial which is useful in stating the formula.

FACTORIAL

For n a natural number, ***n* factorial**—denoted by $n!$—is the product of the first n natural numbers. **Zero factorial** is defined to be 1. Symbolically:

> **n Factorial**
>
> $n! = n(n-1) \cdot \cdots \cdot 2 \cdot 1$ $\quad 5! = 5 \cdot 4 \cdot 3 \cdot 2 \cdot 1 = 120$
>
> $1! = 1$
>
> $0! = 1$

It is also useful to note that

> $n! = n \cdot (n-1)!$ $\quad 8! = 8 \cdot 7!$

Example 15 Evaluate each: **(A)** $4!$ **(B)** $6!$ **(C)** $\frac{7!}{6!}$ **(D)** $\frac{8!}{5!}$

Solution

(A) $4! = 4 \cdot 3 \cdot 2 \cdot 1 = 24$

(B) $6! = 6 \cdot 5 \cdot 4 \cdot 3 \cdot 2 \cdot 1 = 720$

(C) $\frac{7!}{6!} = \frac{7 \cdot \cancel{6!}}{\cancel{6!}} = 7$

(D) $\frac{8!}{5!} = \frac{8 \cdot 7 \cdot 6 \cdot \cancel{5!}}{\cancel{5!}} = 336$

Problem 15 Evaluate: **(A)** $7!$ **(B)** $\frac{8!}{7!}$ **(C)** $\frac{6!}{3!}$

A form involving factorials that is very useful is given by the notation

> $$\binom{n}{r} = \frac{n!}{(n-r)!r!} \qquad 0 \le r \le n,\ n \in N$$

Example 16 Find: **(A)** $\binom{5}{2}$ **(B)** $\binom{4}{4}$

Solution **(A)** $\binom{5}{2} = \dfrac{5!}{(5-2)!2!} = \dfrac{5!}{3!2!} = 10$

(B) $\binom{4}{4} = \dfrac{4!}{(4-4)!4!} = \dfrac{4!}{0!4!} = 1$

Problem 16 Find: **(A)** $\binom{9}{2}$ **(B)** $\binom{5}{5}$

BINOMIAL FORMULA

Let us try to discover a formula for the expansion of $(a + b)^n$, for n a natural number:

$$(a + b)^1 = a + b$$
$$(a + b)^2 = a^2 + 2ab + b^2$$
$$(a + b)^3 = a^3 + 3a^2b + 3ab^2 + b^3$$
$$(a + b)^4 = a^4 + 4a^3b + 6a^2b^2 + 4ab^3 + b^4$$
$$(a + b)^5 = a^5 + 5a^4b + 10a^3b^2 + 10a^2b^3 + 5ab^4 + b^5$$

OBSERVATIONS

1. The expansion of $(a + b)^n$ has $n + 1$ terms.
2. The power of a starts at n and decreases by 1 for each term until it is 0 in the last term.
3. The power of b starts at 0 in the first term and increases by 1 for each term until it is n in the last term.
4. The sum of the powers of a and b in each term is the constant n.
5. The coefficient of any term after the first can be obtained from the preceding term as follows. In the preceding term multiply the coefficient by the exponent of a, and then divide this product by the number representing the position of the preceding term in the series. For example, in $(a + b)^5$ above, the coefficient 10 in the third term is obtained by:
   (A) Taking the coefficient 5 in the second term
   (B) Multiplying this coefficient by the exponent 4 of a in the second term to obtain 20
   (C) Dividing by the position (2) of the second term to obtain 10

We now postulate these same properties for the general case:

$$(a+b)^n = a^n + \frac{n}{1}a^{n-1}b + \frac{n(n-1)}{1 \cdot 2}a^{n-2}b^2 + \frac{n(n-1)(n-2)}{1 \cdot 2 \cdot 3}a^{n-3}b^3 + \cdots + b^n$$

$$= \frac{n!}{0!(n-0)!}a^n + \frac{n!}{1!(n-1)!}a^{n-1}b + \frac{n!}{2!(n-2)!}a^{n-2}b^2 + \frac{n!}{3!(n-3)!}a^{n-3}b^3 + \cdots + \frac{n!}{n!(n-n)!}b^n$$

$$= \binom{n}{0}a^n + \binom{n}{1}a^{n-1}b + \binom{n}{2}a^{n-2}b^2 + \binom{n}{3}a^{n-3} + \cdots + \binom{n}{n}b^n$$

Thus, it appears that:

Binomial Formula

$$(a+b)^n = \sum_{k=0}^{n} \binom{n}{k} a^{n-k}b^k \qquad n \geq 1$$

This result is known as the **binomial formula**, and its general proof requires a method of proof called mathematical induction, which is considered in more advanced courses.

Example 17 Use the binomial formula to expand $(x + y)^6$.

Solution

$$(x+y)^6 = \sum_{k=0}^{6} \binom{6}{k} x^{6-k}y^k$$

$$= \binom{6}{0}x^6 + \binom{6}{1}x^5y + \binom{6}{2}x^4y^2 + \binom{6}{3}x^3y^3 + \binom{6}{4}x^2y^4 + \binom{6}{5}xy^5 + \binom{6}{6}y^6$$

$$= x^6 + 6x^5y + 15x^4y^2 + 20x^3y^3 + 15x^2y^4 + 6xy^5 + y^6$$

Problem 17 Use the binomial formula to expand $(x + 1)^5$.

Example 18 Use the binomial formula to find the fourth term in the expansion of $(x - 2)^{20}$.

Solution

$$4\text{th term} = \binom{20}{3} x^{20-3}(-2)^3$$

Since the 1st term is given when $k = 0$, the 4th term is given when $k = 3$.

$$= \frac{20!}{(20 - 3)!\,3!} x^{17}(-2)^3$$

$$= \frac{20 \cdot 19 \cdot 18}{3 \cdot 2 \cdot 1} x^{17}(-8)$$

$$= -9{,}120x^{17}$$

Problem 18 Use the binomial formula to find the fifth term in the expansion of $(u - 1)^{18}$.

ANSWERS TO MATCHED PROBLEMS

15. **(A)** 5,040 **(B)** 8 **(C)** 120
16. **(A)** 36 **(B)** 1
17. $x^5 + 5x^4 + 10x^3 + 10x^2 + 5x + 1$
18. $3{,}060u^{14}$

EXERCISE 11-4

A *Evaluate.*

1. $6!$ **2.** $4!$ **3.** $\frac{20!}{19!}$ **4.** $\frac{5!}{4!}$

5. $\frac{10!}{7!}$ **6.** $\frac{9!}{6!}$ **7.** $\frac{6!}{4!2!}$ **8.** $\frac{5!}{2!3!}$

9. $\frac{9!}{0!(9 - 0)!}$ **10.** $\frac{8!}{8!(8 - 8)!}$ **11.** $\frac{8!}{2!(8 - 2)!}$ **12.** $\frac{7!}{3!(7! - 3)!}$

Write as the quotient of two factorials.

13. 9 **14.** 12

15. $6 \cdot 7 \cdot 8$ **16.** $9 \cdot 10 \cdot 11 \cdot 12$

B *Evaluate.*

17. $\binom{9}{5}$ **18.** $\binom{5}{2}$ **19.** $\binom{6}{5}$ **20.** $\binom{7}{1}$

21. $\binom{9}{9}$ **22.** $\binom{5}{0}$ **23.** $\binom{17}{13}$ **24.** $\binom{20}{16}$

Expand using the binomial formula.

25. $(u + v)^5$ **26.** $(x + y)^4$ **27.** $(y - 1)^4$

28. $(x - 2)^5$ **29.** $(2x - y)^5$ **30.** $(m + 2n)^6$

Find the indicated term in each expansion.

31. $(u + v)^{15}$; 7th term **32.** $(a + b)^{12}$; 5th term

33. $(2m + n)^{12}$; 11th term **34.** $(x + 2y)^{20}$; 3d term

35. $[(w/2) - 2]^{12}$; 7th term **36.** $(x - 3)^{10}$; 4th term

C **37.** Evaluate $(1.01)^{10}$ to four decimal places, using the binomial formula. [*Hint:* Let $1.01 = 1 + 0.01$.]

38. Evaluate $(0.99)^6$ to four decimal places, using the binomial formula.

39. Show that: $\binom{n}{r} = \binom{n}{n - r}$

40. Can you guess what the next two rows in **Pascal's triangle** are? Compare the numbers in the triangle with the binomial coefficients obtained with the binomial formula.

```
        1
      1   1
    1   2   1
  1   3   3   1
1   4   6   4   1
```

11-5 CHAPTER REVIEW

An **infinite sequence** is a function whose domain is the set of all natural numbers. A **finite sequence** is a function whose domain is a set $\{1, 2, 3, \ldots, n\}$. The element associated with the natural number n is called the ***n*th term** of the sequence and is usually denoted with a subscript as a_n. The sum of the terms of a sequence is called a **series**—an **infinite series** if the sequence is infinite. **Summation notation** is useful for representing a series:

$$S_n = \sum_{k=1}^{n} a_k = a_1 + a_2 + \cdots + a_n \quad (11\text{-}1)$$

An **arithmetic sequence** has nth term $a_n = a_{n-1} + d = a_1 + (n - 1)d$, where the constant d is called the **common difference**. The sum of the first n

terms of an arithmetic sequence is called an **arithmetic series** and is given by the formula

$$S_n = \frac{n}{2}[2a_1 + (n-1)d] = \frac{n}{2}(a_1 + a_n) \quad \textit{(11-2)}$$

A **geometric sequence** has nth term $a_n = ra_{n-1} = a_1 r^{n-1}$, where the constant $r \neq 0$ is called the **common ratio**. A **geometric series** is the sum of terms in a geometric sequence. For n terms the series sum is

$$S_n = \frac{a_1 - a_1 r^n}{1-r} = \frac{a_1(1-r^n)}{1-r} = \frac{a_1 - ra_n}{1-r} \qquad r \neq 1$$

For an infinite series with $|r| < 1$, the sum is

$$S_\infty = \frac{a_1}{1-r} \quad \textit{(11-3)}$$

The **binomial formula** is

$$(a+b)^n = \sum_{k=0}^{n} \binom{n}{k} a^{n-k}b^k$$

where

$$\binom{n}{k} = \frac{n!}{(n-r)!r!}$$

$n! = n(n-1) \cdot \cdots \cdot 2 \cdot 1$, $1! = 1$, and $0! = 1$. The number $n!$ is called ***n* factorial**. *(11-4)*

REVIEW EXERCISE 11-5

Work through all the problems in this chapter review and check answers in the back of the book. (Answers to all problems are there, and following each answer is a number in italics indicating the section in which that type of problem is discussed.) Where weaknesses show up, review appropriate sections in the text.

A

1. Identify all arithmetic and all geometric sequences from the following list of sequences:

(A) $16, -8, 4, \ldots$ **(B)** $5, 7, 9, \ldots$
(C) $-8, -5, -2, \ldots$ **(D)** $2, 3, 5, 8, \ldots$
(E) $-1, 2, -4, \ldots$

In Problems 2–5:

(A) *Write the first four terms of each sequence.*

(B) *Find* a_{10}.
(C) *Find* S_{10}.

2. $a_n = 2n + 3$

3. $a_n = 32(\frac{1}{2})^n$

4. $a_1 = -8;\ a_n = a_{n-1} + 3,\ n \geq 2$

5. $a_1 = -1;\ a_n = (-2)a_{n-1},\ n \geq 2$

6. Find S_∞ in Problem 3.

Evaluate.

7. $6!$ 8. $\dfrac{22!}{19!}$ 9. $\dfrac{7!}{2!(7-2)!}$

B *Write Problems 10 and 11 without summation notation and find the sums.*

10. $S_{10} = \sum_{k=1}^{10} (2k - 8)$ 11. $S_7 = \sum_{k=1}^{7} \dfrac{16}{2^k}$

12. $S_\infty = 27 - 18 + 12 - \cdots = ?$

13. Write $S_n = \dfrac{1}{3} - \dfrac{1}{9} + \dfrac{1}{27} + \cdots + \dfrac{(-1)^{n+1}}{3^n}$ using summation notation and find S_∞.

14. If in an arithmetic sequence $a_1 = 13$ and $a_7 = 31$, find the common difference d and the fifth term a_5.

Evaluate.

15. $\dfrac{20!}{18!(20-18)!}$ 16. $\dbinom{16}{12}$ 17. $\dbinom{11}{11}$

18. Expand $(x - y)^5$ using the binomial formula.

19. Find the tenth term in the expansion of $(2x - y)^{12}$.

C 20. Write $0.727\overline{272}$ as the quotient of two integers.

21. A free-falling body travels $g/2$ feet in the first second, $3g/2$ feet during the next second, $5g/2$ feet the next, and so on, where g is the gravitational constant. Find the distance fallen during the 25th second and the total distance fallen from the start to the end of the 25th second. Express answers in terms of g.

22. Expand $(x + i)^6$, i the complex unit, using the binomial formula.

APPENDIX A

A WORD PROBLEM TECHNIQUE

A strategy for solving word problems is given in Section 1-6:

A Strategy for Solving Word Problems

1. Read the problem carefully—several times if necessary—until you understand the problem, know what is to be found, and know what is given.
2. If appropriate, draw figures or diagrams and label known and unknown parts. Look for formulas connecting the known quantities with the unknown quantities.
3. Let one of the unknown quantities be represented by a variable, say x, and try to represent all other unknown quantities in terms of x. This is an important step and must be done carefully. Be sure you clearly understand what you are letting x represent.
4. Form an equation relating the unknown quantities with the known quantities. This step may involve the translation of an English sentence into an algebraic sentence, the use of relationships in a geometric figure, the use of certain formulas, and so on.
5. Solve the equation and write answers to *all* parts of the problem requested.
6. Check all solutions in the original problem.

As noted there, the key step and often the most difficult one is step 4, forming the equation. There is no general technique for doing this, but there is a procedure that works well for many students and can be used in a wide variety of word problems.

If you have difficulty finding the relationships that lead to an appropriate equation, guess a numerical answer and try it to see if it works. This then becomes an arithmetic problem, exactly the same as checking your answer after solving the problem. Most likely your guess will be incorrect, but in checking you will better see the structure of the problem. You can then replace your guess by a variable and follow your same logic to get the appropriate equation. Some examples will illustrate the process.

Example 1 Find three consecutive even numbers such that twice the first plus the third is 10 more than the second.

Solution Let us guess that the first number is 6 so that the numbers are 6, 8, and 10. Then

$$\text{Twice the first plus the third} = 2 \cdot 6 + 10 = 22$$
$$\text{10 more than the second} = 8 + 10 = 18$$

Our guess is incorrect. However, if we replace our initial guess of 6 by x, then 8 is replaced by $x + 2$ and 10 by $x + 4$:

$$\text{Twice the first plus the third} = 2x + (x + 4)$$
$$\text{10 more than the second} = (x + 2) + 10$$

The equation we seek is, therefore,

$$2x + (x + 4) = (x + 2) + 10$$

The solution is completed in Example 32, Section 1-6.

Example 2 The length of a rectangle is 1 inch more than twice its width. If the area is 21 square inches, find its dimensions.

Solution Suppose we guess that the width is 4 inches. Then

$$\text{Width} = 4$$
$$\text{Length} = 2 \cdot 4 + 1 = 9$$
$$\text{Area} = \text{Width} \cdot \text{Length} = 4 \cdot 9 = 36$$

Our guess is incorrect, so we replace our incorrect guess by x:

$$\text{Width} = x$$
$$\text{Length} = 2x + 1$$
$$\text{Area} = x(2x + 1) = 21$$

This last equation is what we sought:

$$x(2x + 1) = 21$$

The solution is completed in Example 25, Section 2-7.

Example 3 A jet plane leaves San Francisco and travels at 650 kilometers per hour toward Los Angeles. At the same time another plane leaves Los Angeles and travels at 800 kilometers per hour toward San Francisco. If the cities are 570 kilometers apart, how long will it take the jets to meet, and how far from San Francisco will they be at that time?

Solution We guess the planes will meet in $\frac{1}{2}$ hour. Then

Plane from SF to LA travels $650 \cdot \frac{1}{2} = 325$ km
Plane from LA to SF travels $800 \cdot \frac{1}{2} = 400$ km

The planes have thus traveled 725 km, so they have passed each other. (This also tells us that the guess of $\frac{1}{2}$ is too large.)

Now replace the guess of $\frac{1}{2}$ by x.

Plane from SF to LA travels $650x$ km
Plane from LA to SF travels $800x$ km

The total kilometers must be 650, so we get the equation

$$650x + 800x = 650$$

The solution is completed in Example 9, Section 4-2.

Example 4 A speedboat takes 1.5 times longer to go 120 miles up a river than to return. If the boat cruises at 25 miles per hour in still water, what is the rate of the current?

Solution Let us guess that the current is 3 miles per hour. Then the speed of the boat upstream is 22 (that is, $25 - 3$) and the speed downstream is 28 (that is, $25 + 3$). Since $D = RT$, $T = D/R$, so

Time upstream is $\frac{120}{22} = 5.4545\ldots$
Time downstream is $\frac{120}{28} = 4.2857\ldots$

Is the time up 1.5 times the time down? No, 1.5 times 4.2857 is more than 6, so it is not equal to the time downstream. Thus, our guess is incorrect and we replace it by x. Now the speed upstream is $25 - x$ and the speed downstream is $25 + x$. Thus,

Time upstream is $\dfrac{120}{25 - x}$

Time downstream is $\dfrac{120}{25 + x}$

and our equation is

$$\frac{120}{25 - x} = 1.5 \frac{120}{(25 + x)}$$

The solution is completed in Example 11, Section 4-2.

Example 5 Five people form a glider club and decide to share the cost of a glider equally. They find, however, that if they let three more join the club, the share for each of the original five will be reduced by \$480. What is the total cost of the glider?

Solution Suppose we guess the cost of the glider to be \$5,000. Then five shares would be \$1,000 each, that is, \$5,000/5. With three additional shares, each would be \$5,000/8 = \$625. Since the cost of each share has not been reduced by \$480, our guess is incorrect. Replace the guess by x:

For five shares: $\dfrac{x}{5}$ each

For eight shares: $\dfrac{x}{8}$ each

The cost per share for five members should be \$480 more than the cost per share for eight members, that is,

$$\frac{x}{5} = \frac{x}{8} + 480 \qquad \text{or} \qquad \frac{x}{8} = \frac{x}{5} - 480$$

The solution is completed in Example 13, Section 4-2.

Example 6 A change machine changes dollar bills into quarters and nickels. If you receive 12 coins after inserting a \$1 bill, how many of each type of coin did you receive?

Solution If we guess 3 quarters and 9 nickels (the total number of coins is 12), then the value of the coins would be

Quarters: $3 \cdot 25 = 75$ cents

Nickels: $9 \cdot 5 = 45$ cents

which totals more than 100 cents, more than \$1. If we replace our guesses of 3 and 9 by x and y, then

$$x + y = 12$$

and

$$x \cdot 25 + y \cdot 5 = 100$$

or

$$\begin{aligned} x + y &= 12 \\ 25x + 5y &= 100 \end{aligned}$$

Using this last system of equations, the solution is completed in Example 7, Section 8-2.

Example 7 A chemical storeroom has a 20% alcohol solution and a 50% alcohol solution. How many centiliters must be taken from each to obtain 24 centiliters of a 30% solution?

Solution Let us guess that we need 10 centiliters of the weak (20%) solution and, therefore, 14 centiliters of the strong. Then our mixture will be 24 centiliters. How much alcohol does it contain?

The 10 centiliters is 20% alcohol, so the amount is 20% of 10, that is, $0.2(10) = 2$. The 14 centiliters is 50% alcohol, so the amount there is $0.5(14) = 7$. The total amount of alcohol in the mixture is, therefore, $2 + 7 = 9$ centiliters. Is this 30% of 24?

$$30\% \text{ of } 24 = 0.3(24) = 7.2$$

Thus our mixture has too much alcohol (we have used too much of the stronger solution). Replace our guesses of 10 and 14 by x and y, respectively. Then

$$\begin{aligned} x + y &= 24 \\ 0.2x + 0.5y &= 7.2 \end{aligned}$$

The solution is completed in Example 12, Section 8-2.

There are no matched problems and no exercise set for this appendix. There are an ample number of word problems in the text—especially in Sections 1-6, 2-7, 4-1, 4-2, 6-3, 8-1, 8-2, and 8-3—upon which to practice this technique.

APPENDIX B

SIGNIFICANT DIGITS

Most calculations involving problems of the real world deal with figures that are only approximate. It would therefore seem reasonable to assume that a final answer could not be any more accurate than the least accurate figure used in the calculation. This is an important point, since calculators tend to give the impression that greater accuracy is achieved than is warranted.

Suppose we wish to compute the length of the diagonal of a rectangular field from measurements of its sides of 237.8 meters and 61.3 meters. Using the Pythagorean theorem and a calculator, we find

$$\begin{aligned} d &= \sqrt{237.8^2 + 61.3^2} \\ &= 245.573\ 878\ \ldots \end{aligned}$$

d
61.3 meters
237.8 meters

The calculator answer suggests an accuracy that is not justified. What accuracy is justified? To answer this question, we introduce the idea of **significant digits**.

The measurement 61.3 meters indicates that the measurement was made to the nearest tenth of a meter; that is, the actual width is between 61.25 and 61.35 meters. The number 61.3 has three significant digits. If we had written, instead, 61.30 meters as the width, then the actual width would be between 61.295 and 61.305 meters, and our measurement, 61.30 meters, would have four significant digits.

In all cases except one, the number of significant digits in a number is found by counting the digits from left to right, starting with the first nonzero digit and ending with the last digit present.

The significant digits in the following numbers are underlined:

$\underline{719.37}$ $\quad$ $\underline{82{,}395}$ $\quad$ $\underline{5.600}$ $\quad$ $0.000\ \underline{830}$ $\quad$ $0.000\ 0\underline{8}$

As we said, this definition takes care of all cases except one. Consider, for example, the number 7,800. It is not clear whether the number has been rounded to the hundreds place, the tens place, or the units place. This ambiguity can be resolved by writing this type of number in scientific notation:

7.8×10^3 has two significant digits

7.80×10^3 has three significant digits

7.800×10^3 has four significant digits

All three are equal to 7,800 when written without powers of 10.

In calculations involving multiplication, division, powers, and roots, we adopt the following convention:

We will round off the answer to match the number of significant digits in the number with the least number of significant digits used in the calculation.

In computing the length of the diagonal of the field, we would write the answer to three significant digits because the width, the least accurate of the two numbers involved, has three significant digits:

$d = 246$ meters   Three significant digits

One final note: in rounding a number that is exactly halfway between a larger and a smaller number, we will use the convention of making the final result even.

Example 1 Round each number to three significant digits:

(A) 43.0690 **(B)** 48.05 **(C)** 48.15 **(D)** $8.017\,632 \times 10^{-3}$

Solution

(A) 43.1

(B) 48.0

(C) 48.2

(B) and (C): Use the convention of making the digit before the 5 even if it is odd or leaving it alone if it is even.

(D) 8.02×10^{-3}

Problem 1 Round each number to three significant digits:

(A) 3.1495 **(B)** 0.004 135 **(C)** 32,450
(D) $4.314\,764\,09 \times 10^{12}$

ANSWERS TO MATCHED PROBLEM

1. **(A)** 3.15 **(B)** 0.004 14 **(C)** 32,400 **(D)** 4.31×10^{12}

APPENDIX C

DETERMINANTS AND CRAMER'S RULE

- Second-Order Determinants
- Third-Order Determinants
- Cramer's Rule

Determinants arise quite naturally in many areas in mathematics, including the solving of linear systems, vector analysis, calculus, and so on. We will consider a few of their uses in this appendix.

SECOND-ORDER DETERMINANTS

A square array of four real numbers, such as

$$\begin{vmatrix} 2 & -3 \\ 5 & 1 \end{vmatrix}$$

is called a determinant of order 2. (It is important to note that the array of numbers is between parallel lines and not square brackets. If square brackets are used, then the symbol has another meaning, namely, a matrix—see Section 8-4). The determinant shown here has two **rows** and two **columns**—rows are across and columns are up and down. Each number in the determinant is called an **element** of the determinant.

In general, we can symbolize a **second-order determinant** as follows:

$$\begin{vmatrix} a_{11} & a_{12} \\ a_{21} & a_{22} \end{vmatrix}$$

where we use a single letter with a **double subscript** to facilitate generalization to higher-order determinants. The first subscript number indicates the row in which the element lies, and the second subscript number indicates the column. Thus, a_{21} is the element in the second row and first column and a_{12} is the element in the first row and second column. Each second-order determinant represents a real number given by this formula:

$$\begin{vmatrix} a_{11} & a_{12} \\ a_{21} & a_{22} \end{vmatrix} = a_{11}a_{22} - a_{21}a_{12}$$

Example 1 $\begin{vmatrix} -1 & 2 \\ -3 & -4 \end{vmatrix} = (-1)(-4) - (-3)(2) = 4 - (-6) = 10$

Problem 1 Find: $\begin{vmatrix} 3 & -5 \\ 4 & -2 \end{vmatrix}$

THIRD-ORDER DETERMINANTS

A determinant of order 3 is a square array of nine elements and represents a real number given by the formula

$$\begin{vmatrix} a_{11} & a_{12} & a_{13} \\ a_{21} & a_{22} & a_{23} \\ a_{31} & a_{32} & a_{33} \end{vmatrix} = a_{11}a_{22}a_{33} - a_{11}a_{32}a_{23} + a_{21}a_{32}a_{13} - a_{21}a_{12}a_{33} + a_{31}a_{12}a_{23} - a_{31}a_{22}a_{13} \quad (1)$$

Note that each term in the expansion on the right of (1) contains exactly one element from each row and each column. Don't panic! You do not need to memorize Formula (1). After we introduce the ideas of "minor" and "cofactor," we will state a theorem that can be used to obtain the same result with much less memory strain.

The **minor of an element** in a third-order determinant is a second-order determinant obtained by deleting the row and column that contains the element. For example, in the determinant in Formula (1),

$$\text{Minor of } a_{23} = \begin{vmatrix} a_{11} & a_{12} & a_{13} \\ a_{21} & a_{22} & a_{23} \\ a_{31} & a_{32} & a_{33} \end{vmatrix} = \begin{vmatrix} a_{11} & a_{12} \\ a_{31} & a_{32} \end{vmatrix}$$

$$\text{Minor of } a_{32} = \begin{vmatrix} a_{11} & a_{12} & a_{13} \\ a_{21} & a_{22} & a_{23} \\ a_{31} & a_{32} & a_{33} \end{vmatrix} = \begin{vmatrix} a_{11} & a_{13} \\ a_{21} & a_{23} \end{vmatrix}$$

A quantity closely associated with the minor of an element is the cofactor of an element. The **cofactor of an element** a_{ij} (from the ith row and jth column) is the product of the minor of a_{ij} and $(-1)^{i+j}$. That is,

$$\text{Cofactor of } a_{ij} = (-1)^{i+j}(\text{Minor of } a_{ij})$$

Thus, a cofactor of an element is nothing more than a signed minor. The sign is determined by raising -1 to a power that is the sum of the numbers indicating the row and column in which the element lies. Note that $(-1)^{i+j}$ is -1 if $i+j$ is odd and 1 if $i+j$ is even. Referring again to the determinant in Formula (1),

$$\text{Cofactor of } a_{23} = (-1)^{2+3}\begin{vmatrix} a_{11} & a_{12} \\ a_{31} & a_{32} \end{vmatrix} = -\begin{vmatrix} a_{11} & a_{12} \\ a_{31} & a_{32} \end{vmatrix}$$

$$\text{Cofactor of } a_{11} = (-1)^{1+1}\begin{vmatrix} a_{22} & a_{23} \\ a_{32} & a_{33} \end{vmatrix} = \begin{vmatrix} a_{22} & a_{23} \\ a_{32} & a_{33} \end{vmatrix}$$

Example 2 Find the cofactor of -2 and 5 in the determinant

$$\begin{vmatrix} -2 & 0 & 3 \\ 1 & -6 & 5 \\ -1 & 2 & 0 \end{vmatrix}$$

Solution

$$\text{Cofactor of } -2 = (-1)^{1+1}\begin{vmatrix} -6 & 5 \\ 2 & 0 \end{vmatrix} = \begin{vmatrix} -6 & 5 \\ 2 & 0 \end{vmatrix}$$

$$= (-6)(0) - (2)(5) = -10$$

$$\text{Cofactor of } 5 = (-1)^{2+3}\begin{vmatrix} -2 & 0 \\ -1 & 2 \end{vmatrix} = -\begin{vmatrix} -2 & 0 \\ -1 & 2 \end{vmatrix}$$

$$= -[(-2)(2) - (-1)(0)] = 4$$

Problem 2 Find the cofactors of 2 and 3 in the determinant in Example 2.

Note: The sign in front of the minor, $(-1)^{i+j}$, can be determined mechanically by using a checkerboard pattern of plus and minus signs over the determinant, starting with $+$ in the upper left-hand corner:

$$\begin{matrix} + & - & + \\ - & + & - \\ + & - & + \end{matrix}$$

Use either the checkerboard or the exponent method, whichever is easier for you, to determine the sign in front of the minor.

Now we are ready for the central theorem of this section. It will provide us with an efficient means of evaluating third-order determinants. Moreover, it is worth noting that the theorem generalizes completely to include determinants of arbitrary order.

Theorem 1 The value of a determinant of order 3 is the sum of three products obtained by multiplying each element of any one row (or each element of any one column) by its cofactor.

To prove this theorem we must show that the expansions indicated by the theorem for any row or any column (six cases) produce the expression on the right of Formula (1). Proofs of special cases of this result are left to the C-level problems of Exercise C.

Example 3 Evaluate by expanding **(A)** by the first row and **(B)** by the second column:

$$\begin{vmatrix} 2 & -2 & 0 \\ -3 & 1 & 2 \\ 1 & -3 & -1 \end{vmatrix}$$

Solution **(A)**

$$\begin{vmatrix} 2 & -2 & 0 \\ -3 & 1 & 2 \\ 1 & -3 & -1 \end{vmatrix} \quad \text{Expand by the first row.}$$

$$= a_{11}\begin{pmatrix}\text{Cofactor}\\ \text{of } a_{11}\end{pmatrix} + a_{12}\begin{pmatrix}\text{Cofactor}\\ \text{of } a_{12}\end{pmatrix} + a_{13}\begin{pmatrix}\text{Cofactor}\\ \text{of } a_{13}\end{pmatrix}$$

$$= 2\left((-1)^{1+1}\begin{vmatrix} 1 & 2 \\ -3 & -1 \end{vmatrix}\right) + (-2)\left((-1)^{1+2}\begin{vmatrix} -3 & 2 \\ 1 & -1 \end{vmatrix}\right) + 0$$

$$= (2)(1)[(1)(-1) - (-3)(2)] + (-2)(-1)[(-3)(-1) - (1)(2)]$$

$$= (2)(5) + (2)(1) = 12$$

(B)

$$\begin{vmatrix} 2 & -2 & 0 \\ -3 & 1 & 2 \\ 1 & -3 & -1 \end{vmatrix} \quad \text{Expand by the second column.}$$

$$= a_{12}\begin{pmatrix}\text{Cofactor}\\ \text{of } a_{12}\end{pmatrix} + a_{22}\begin{pmatrix}\text{Cofactor}\\ \text{of } a_{22}\end{pmatrix} + a_{32}\begin{pmatrix}\text{Cofactor}\\ \text{of } a_{32}\end{pmatrix}$$

$$= (-2)\left((-1)^{1+2}\begin{vmatrix} -3 & 2 \\ 1 & -1 \end{vmatrix}\right) + (1)\left((-1)^{2+2}\begin{vmatrix} 2 & 0 \\ 1 & -1 \end{vmatrix}\right)$$

$$+ (-3)\left((-1)^{3+2}\begin{vmatrix} 2 & 0 \\ -3 & 2 \end{vmatrix}\right)$$

$$= (-2)(-1)[(-3)(-1) - (1)(2)] + (1)(1)[(2)(-1) - (1)(0)]$$

$$+ (-3)(-1)[(2)(2) - (-3)(0)]$$

$$= (2)(1) + (1)(-2) + (3)(4)$$

$$= 12$$

Problem 3 Evaluate by expanding **(A)** by the first row and **(B)** by the third column:

$$\begin{vmatrix} 2 & 1 & -1 \\ -1 & -3 & 0 \\ -1 & 2 & 1 \end{vmatrix}$$

It should now be clear that we can greatly reduce the work involved in evaluating a determinant by choosing to expand by a row or column with the greatest number of 0's.

Where are determinants used? Many equations and formulas have particularly simple and compact representations in determinant form that are easily remembered. Cramer's rule for solving a system of linear equations is one example. Problems 55 to 58 in Exercise C provide others.

CRAMER'S RULE

Now let us see how determinants arise rather naturally in the process of solving systems of linear equations. We will start by investigating two equations and two variables and then extend any results to three equations and three variables.

Instead of thinking of each system of linear equations in two variables as a different problem, let us see what happens when we attempt to solve the general system

$$a_{11}x + a_{12}y = k_1 \qquad (2A)$$
$$a_{21}x + a_{22}y = k_2 \qquad (2B)$$

once and for all in terms of the unspecified real constants a_{11}, a_{12}, a_{21}, a_{22}, k_1, and k_2.

We proceed by multiplying Equations (2A) and (2B) by suitable constants so that when the resulting equations are added, left side to left side and right side to right side, one of the variables drops out. Suppose we choose to eliminate y. What constants should we use to make the coefficients of y the same except for the signs? Multiply Equation (2A) by a_{22} and Equation (2B) by $-a_{12}$. Then add:

$$\begin{array}{rcll} a_{11}a_{22}x + a_{12}a_{22}y &=& k_1a_{22} & a_{22}(2A) \\ -a_{21}a_{12}x - a_{12}a_{22}y &=& -k_2a_{12} & -a_{12}(2B) \\ \hline a_{11}a_{22}x - a_{21}a_{12}x + 0y &=& k_1a_{22} - k_2a_{12} & \\ (a_{11}a_{22} - a_{21}a_{12})x &=& k_1a_{22} - k_2a_{12} & \end{array}$$

$$x = \frac{k_1a_{22} - k_2a_{12}}{a_{11}a_{22} - a_{21}a_{12}} \qquad a_{11}a_{22} - a_{21}a_{12} \neq 0$$

What do the numerator and denominator remind you of? From your experience with determinants you can see these expressions as

$$x = \frac{\begin{vmatrix} k_1 & a_{12} \\ k_2 & a_{22} \end{vmatrix}}{\begin{vmatrix} a_{11} & a_{12} \\ a_{21} & a_{22} \end{vmatrix}}$$

Similarly, starting with the same system and eliminating x (this is left as an exercise), we obtain

$$y = \frac{\begin{vmatrix} a_{11} & k_1 \\ a_{21} & k_2 \end{vmatrix}}{\begin{vmatrix} a_{11} & a_{12} \\ a_{21} & a_{22} \end{vmatrix}}$$

These results are summarized in the following rule, which is named after the Swiss mathematician G. Cramer (1704–1752):

Cramer's Rule for Two Equations and Two Variables

Given the system

$$\begin{aligned} a_{11}x + a_{12}y &= k_1 \\ a_{21}x + a_{22}y &= k_2 \end{aligned} \qquad \text{with} \qquad D = \begin{vmatrix} a_{11} & a_{12} \\ a_{21} & a_{22} \end{vmatrix} \neq 0$$

then

$$x = \frac{\begin{vmatrix} k_1 & a_{12} \\ k_2 & a_{22} \end{vmatrix}}{D} \qquad \text{and} \qquad y = \frac{\begin{vmatrix} a_{11} & k_1 \\ a_{21} & k_2 \end{vmatrix}}{D}$$

The determinant D is called the **coefficient determinant**. If $D \neq 0$, then the system has exactly one solution, which is given by Cramer's rule. If, on the other hand, $D = 0$, then it can be shown that the system is either inconsistent or dependent; that is, the system either has no solutions or has an infinite number of solutions.

Example 4 Solve using Cramer's rule:

$$\begin{aligned} 2x - 3y &= 7 \\ -3x + y &= -7 \end{aligned}$$

Solution $D = \begin{vmatrix} 2 & -3 \\ -3 & 1 \end{vmatrix} = -7$

$$x = \frac{\begin{vmatrix} 7 & -3 \\ -7 & 1 \end{vmatrix}}{-7} = \frac{-14}{-7} = 2 \qquad y = \frac{\begin{vmatrix} 2 & 7 \\ -3 & -7 \end{vmatrix}}{-7} = \frac{7}{-7} = -1$$

Problem 4 Solve using Cramer's rule:
$$\begin{aligned} 3x + 2y &= -3 \\ -4x + 3y &= -13 \end{aligned}$$

Cramer's rule generalizes completely for any size of linear system that has the same number of variables as equations. We state without proof the rule for three equations and three variables.

Cramer's Rule for Three Equations and Three Variables

Given the system

$$\begin{aligned} a_{11}x + a_{12}y + a_{13}z &= k_1 \\ a_{21}x + a_{22}y + a_{23}z &= k_2 \\ a_{31}x + a_{32}y + a_{33}z &= k_3 \end{aligned} \qquad \text{with} \qquad D = \begin{vmatrix} a_{11} & a_{12} & a_{13} \\ a_{21} & a_{22} & a_{23} \\ a_{31} & a_{32} & a_{33} \end{vmatrix} \neq 0$$

then

$$x = \frac{\begin{vmatrix} k_1 & a_{12} & a_{13} \\ k_2 & a_{22} & a_{23} \\ k_3 & a_{32} & a_{33} \end{vmatrix}}{D} \qquad y = \frac{\begin{vmatrix} a_{11} & k_1 & a_{13} \\ a_{21} & k_2 & a_{23} \\ a_{31} & k_3 & a_{33} \end{vmatrix}}{D}$$

$$z = \frac{\begin{vmatrix} a_{11} & a_{12} & k_1 \\ a_{21} & a_{22} & k_2 \\ a_{31} & a_{32} & k_3 \end{vmatrix}}{D}$$

It is easy to remember these determinant formulas for x, y, and z if you observe the following:

1. Determinant D is formed from the coefficients of x, y, and z, keeping the same relative position in the determinant as found in the system.
2. Determinant D appears in the denominators for x, y, and z.
3. The numerator for x can be obtained from D by replacing the coefficients of x—that is, a_{11}, a_{21}, and a_{31}—with the constants k_1, k_2, and k_3, respectively. Similar statements can be made for the numerators for y and z.

Example 5 Use Cramer's rule to solve:

$$\begin{aligned} x + y &= 1 \\ 3y - z &= -4 \\ x + z &= 3 \end{aligned}$$

Solution

$$D = \begin{vmatrix} 1 & 1 & 0 \\ 0 & 3 & -1 \\ 1 & 0 & 1 \end{vmatrix} = 2 \quad \text{Missing variables have 0 coefficients.}$$

$$x = \frac{\begin{vmatrix} 1 & 1 & 0 \\ -4 & 3 & -1 \\ 3 & 0 & 1 \end{vmatrix}}{2} = \frac{4}{2} = 2 \qquad y = \frac{\begin{vmatrix} 1 & 1 & 0 \\ 0 & -4 & -1 \\ 1 & 3 & 1 \end{vmatrix}}{2} = \frac{-2}{2} = -1$$

$$z = \frac{\begin{vmatrix} 1 & 1 & 1 \\ 0 & 3 & -4 \\ 1 & 0 & 3 \end{vmatrix}}{2} = \frac{2}{2} = 1$$

Problem 5 Use Cramer's rule to solve:

$$\begin{aligned} 3x - z &= 5 \\ x - y + z &= 0 \\ x + y &= 0 \end{aligned}$$

In practice, Cramer's rule is rarely used to solve systems of order higher than 2 or 3; more efficient methods are available. Cramer's rule is, however, a valuable tool in theoretical mathematics.

We now summarize the results of the various methods of solving two equations and two unknowns in Table 1.

TABLE 1 SOLVING SECOND-ORDER SYSTEMS OF LINEAR EQUATIONS

	METHOD		
SOLUTIONS	GRAPHING	ELIMINATION OR SUBSTITUTION	CRAMER'S RULE
Exactly one	Lines intersect in exactly one point	One unique pair of numbers	$D \neq 0$
None	Lines are parallel	Contradiction occurs—such as $0 = 5$	$D = 0$
Infinite	Lines coincide	$0 = 0$	$D = 0$

ANSWERS TO MATCHED PROBLEMS

1. 14 **2.** 13, -4 **3.** **(A)** 3 **(B)** 3
4. $x = 1, y = -3$ **5.** $x = 1, y = -1, z = -2$

EXERCISE C

A *Evaluate each second-order determinant.*

1. $\begin{vmatrix} 2 & 4 \\ 3 & -1 \end{vmatrix}$ **2.** $\begin{vmatrix} 2 & 2 \\ -3 & 1 \end{vmatrix}$ **3.** $\begin{vmatrix} 5 & -4 \\ -2 & 2 \end{vmatrix}$

4. $\begin{vmatrix} 6 & -2 \\ -1 & -3 \end{vmatrix}$ **5.** $\begin{vmatrix} 3 & -3.1 \\ -2 & 1.2 \end{vmatrix}$ **6.** $\begin{vmatrix} -1.4 & 3 \\ -0.5 & -2 \end{vmatrix}$

Given the determinant

$$\begin{vmatrix} a_{11} & a_{12} & a_{13} \\ a_{21} & a_{22} & a_{23} \\ a_{31} & a_{32} & a_{33} \end{vmatrix}$$

write the minor of each of the following elements:

7. a_{11} **8.** a_{33} **9.** a_{23} **10.** a_{22}

Write the cofactor of each of the following elements:

11. a_{11} **12.** a_{33} **13.** a_{23} **14.** a_{22}

Given the determinant

$$\begin{vmatrix} -2 & 3 & 0 \\ 5 & 1 & -2 \\ 7 & -4 & 8 \end{vmatrix}$$

write the minor of each of the following elements. (Leave your answer in determinant form.)

15. a_{11} **16.** a_{22} **17.** a_{32} **18.** a_{21}

Write the cofactor of each of the following elements and evaluate each:

19. a_{11} **20.** a_{22} **21.** a_{32} **22.** a_{21}

Solve using Cramer's rule.

23. $x + 2y = 1$, $x + 3y = -1$ **24.** $x + 2y = 3$, $x + 3y = 5$ **25.** $2x + y = 1$, $5x + 3y = 2$

26. $x + 3y = 1$, $2x + 8y = 0$ **27.** $2x - y = -3$, $-x + 3y = 4$ **28.** $2x + y = 1$, $5x + 3y = 2$

B *Evaluate each of the following determinants using cofactors:*

29. $\begin{vmatrix} 1 & 0 & 0 \\ -2 & 4 & 3 \\ 5 & -2 & 1 \end{vmatrix}$

30. $\begin{vmatrix} 2 & -3 & 5 \\ 0 & -3 & 1 \\ 0 & 6 & 2 \end{vmatrix}$

31. $\begin{vmatrix} 0 & 1 & 5 \\ 3 & -7 & 6 \\ 0 & -2 & -3 \end{vmatrix}$

32. $\begin{vmatrix} 4 & -2 & 0 \\ 9 & 5 & 4 \\ 1 & 2 & 0 \end{vmatrix}$

33. $\begin{vmatrix} 4 & -4 & 6 \\ 2 & 8 & -3 \\ 0 & -5 & 0 \end{vmatrix}$

34. $\begin{vmatrix} 3 & -2 & -8 \\ -2 & 0 & -3 \\ 1 & 0 & -4 \end{vmatrix}$

35. $\begin{vmatrix} -1 & 2 & -3 \\ -2 & 0 & -6 \\ 4 & -3 & 2 \end{vmatrix}$

36. $\begin{vmatrix} 0 & 2 & -1 \\ -6 & 3 & 1 \\ 7 & -9 & -2 \end{vmatrix}$

37. $\begin{vmatrix} 1 & 4 & 1 \\ 1 & 1 & -2 \\ 2 & 1 & -1 \end{vmatrix}$

38. $\begin{vmatrix} 3 & 2 & 1 \\ -1 & 5 & 1 \\ 2 & 3 & 1 \end{vmatrix}$

39. $\begin{vmatrix} 1 & 4 & 3 \\ 2 & 1 & 6 \\ 3 & -2 & 9 \end{vmatrix}$

40. $\begin{vmatrix} 4 & -6 & 3 \\ -1 & 4 & 1 \\ 5 & -6 & 3 \end{vmatrix}$

Solve using Cramer's rule.

41. $\begin{aligned} x + y &= 0 \\ 2y + z &= -5 \\ -x + z &= -3 \end{aligned}$

42. $\begin{aligned} x + y &= -4 \\ 2y + z &= 0 \\ -x + z &= 5 \end{aligned}$

43. $\begin{aligned} x + y &= 1 \\ 2y + z &= 0 \\ -x + z &= 0 \end{aligned}$

44. $\begin{aligned} x + y &= -4 \\ 2y + z &= 3 \\ -x + z &= 7 \end{aligned}$

45. $\begin{aligned} y + z &= -4 \\ x + 2z &= 0 \\ x - y &= 5 \end{aligned}$

46. $\begin{aligned} x - z &= 2 \\ 2x - y &= 8 \\ x + y + z &= 2 \end{aligned}$

47. $\begin{aligned} 2y - z &= -4 \\ x - y - z &= 0 \\ x - y + 2z &= 6 \end{aligned}$

48. $\begin{aligned} 2x + y &= 2 \\ x - y + z &= -1 \\ x + y + z &= -1 \end{aligned}$

49. $\begin{aligned} 2a + 4b + 3c &= 6 \\ a - 3b + 2c &= -7 \\ -a + 2b - c &= 5 \end{aligned}$

50. $\begin{aligned} 3u - 2v + 3w &= 11 \\ 2u + 3v - 2w &= -5 \\ u + 4v - w &= -5 \end{aligned}$

C *Assuming that Theorem 1 applies to determinants of arbitrary order, use it to evaluate the following fourth- and fifth-order determinants.*

51. $\begin{vmatrix} 0 & 1 & 0 & 1 \\ 2 & 4 & 7 & 6 \\ 0 & 3 & 0 & 1 \\ 0 & 6 & 2 & 5 \end{vmatrix}$

52. $\begin{vmatrix} 2 & 6 & 1 & 7 \\ 0 & 3 & 0 & 0 \\ 3 & 4 & 2 & 5 \\ 0 & 9 & 0 & 2 \end{vmatrix}$

53. $\begin{vmatrix} 2 & 0 & 0 & 0 & 0 \\ 0 & 3 & 0 & 0 & 0 \\ 0 & 0 & 2 & 0 & 0 \\ 0 & 0 & 0 & 1 & 0 \\ 0 & 0 & 0 & 0 & 4 \end{vmatrix}$

54. $\begin{vmatrix} -2 & 0 & 0 & 0 & 0 \\ 9 & -1 & 0 & 0 & 0 \\ 2 & 1 & 3 & 0 & 0 \\ -1 & 4 & 2 & 2 & 0 \\ 7 & -2 & 3 & 5 & 5 \end{vmatrix}$

55. Show that

$$\begin{vmatrix} x & y & 1 \\ 2 & 3 & 1 \\ -1 & 2 & 1 \end{vmatrix} = 0$$

is the equation of a line that passes through (2, 3) and (−1, 2).

56. Show that

$$\begin{vmatrix} x & y & 1 \\ x_1 & y_1 & 1 \\ x_2 & y_2 & 1 \end{vmatrix} = 0$$

is the equation of a line that passes through (x_1, y_1) and (x_2, y_2).

57. In analytic geometry it is shown that the area of a triangle with vertices (x_1, y_1), (x_2, y_2), and (x_3, y_3) is the absolute value of

$$\frac{1}{2}\begin{vmatrix} x_1 & y_1 & 1 \\ x_2 & y_2 & 1 \\ x_3 & y_3 & 1 \end{vmatrix}$$

Use this result to find the area of a triangle with given vertices (−1, 4), (4, 8), (1, 1).

58. Find the area of a triangle with given vertices (−1, 2), (2, 5), and (6, −3). (See Problem 57.)

59. Prove one case of Theorem 1 by expanding the left side of Formula (1) using the first row and cofactors to obtain the right side.

60. Prove one case of Theorem 1 by expanding the left side of Formula (1) using the second column and cofactors to obtain the right side.

It is clear that $x = 0$, $y = 0$, $z = 0$ is a solution to each of the following systems. Use Cramer's rule to determine if this solution is unique. [Hint: If $D \neq 0$, what can you conclude? If $D = 0$, what can you conclude?]

61.
$$\begin{aligned} x - 4y + 9z &= 0 \\ 4x - y + 6z &= 0 \\ x - y + 3z &= 0 \end{aligned}$$

62.
$$\begin{aligned} 3x - y + 3z &= 0 \\ 5x + 5y - 9z &= 0 \\ -2x + y - 3z &= 0 \end{aligned}$$

TABLES

TABLE I VALUES OF e^x AND e^{-x} (0.00 TO 3.00)

x	e^x	e^{-x}	x	e^x	e^{-x}	x	e^x	e^{-x}
0.00	1.0000	1.000 00	0.50	1.6487	0.606 53	1.00	2.7183	0.367 88
0.01	1.0101	0.990 05	0.51	1.6653	0.600 50	1.01	2.7456	0.364 22
0.02	1.0202	0.980 20	0.52	1.6820	0.594 52	1.02	2.7732	0.360 59
0.03	1.0305	0.970 45	0.53	1.6989	0.588 60	1.03	2.8011	0.357 01
0.04	1.0408	0.960 79	0.54	1.7160	0.582 75	1.04	2.8292	0.353 45
0.05	1.0513	0.951 23	0.55	1.7333	0.576 95	1.05	2.8577	0.349 94
0.06	1.0618	0.941 76	0.56	1.7507	0.571 21	1.06	2.8864	0.346 46
0.07	1.0725	0.932 39	0.57	1.7683	0.565 53	1.07	2.9154	0.343 01
0.08	1.0833	0.923 12	0.58	1.7860	0.559 90	1.08	2.9447	0.339 60
0.09	1.0942	0.913 93	0.59	1.8040	0.554 33	1.09	2.9743	0.336 22
0.10	1.1052	0.904 84	0.60	1.8221	0.548 81	1.10	3.0042	0.332 87
0.11	1.1163	0.895 83	0.61	1.8404	0.543 35	1.11	3.0344	0.329 56
0.12	1.1275	0.886 92	0.62	1.8589	0.537 94	1.12	3.0649	0.326 28
0.13	1.1388	0.878 10	0.63	1.8776	0.532 59	1.13	3.0957	0.323 03
0.14	1.1503	0.869 36	0.64	1.8965	0.527 29	1.14	3.1268	0.319 82
0.15	1.1618	0.860 71	0.65	1.9155	0.522 05	1.15	3.1582	0.316 64
0.16	1.1735	0.852 14	0.66	1.9348	0.516 85	1.16	3.1899	0.313 49
0.17	1.1853	0.843 66	0.67	1.9542	0.511 71	1.17	3.2220	0.310 37
0.18	1.1972	0.835 27	0.68	1.9739	0.506 62	1.18	3.2544	0.307 28
0.19	1.2092	0.826 96	0.69	1.9937	0.501 58	1.19	3.2871	0.304 22
0.20	1.2214	0.818 73	0.70	2.0138	0.496 59	1.20	3.3201	0.301 19
0.21	1.2337	0.810 58	0.71	2.0340	0.491 64	1.21	3.3535	0.298 20
0.22	1.2461	0.802 52	0.72	2.0544	0.486 75	1.22	3.3872	0.295 23
0.23	1.2586	0.794 53	0.73	2.0751	0.481 91	1.23	3.4212	0.292 29
0.24	1.2712	0.786 63	0.74	2.0959	0.477 11	1.24	3.4556	0.289 38
0.25	1.2840	0.778 80	0.75	2.1170	0.472 37	1.25	3.4903	0.286 50
0.26	1.2969	0.771 05	0.76	2.1383	0.467 67	1.26	3.5254	0.283 65
0.27	1.3100	0.763 38	0.77	2.1598	0.463 01	1.27	3.5609	0.280 83
0.28	1.3231	0.755 78	0.78	2.1815	0.458 41	1.28	3.5966	0.278 04
0.29	1.3364	0.748 26	0.79	2.2034	0.453 84	1.29	3.6328	0.275 27
0.30	1.3499	0.740 82	0.80	2.2255	0.449 33	1.30	3.6693	0.272 53
0.31	1.3634	0.733 45	0.81	2.2479	0.444 86	1.31	3.7062	0.269 82
0.32	1.3771	0.726 15	0.82	2.2705	0.440 43	1.32	3.7434	0.267 14
0.33	1.3910	0.718 92	0.83	2.2933	0.436 05	1.33	3.7810	0.264 48
0.34	1.4049	0.711 77	0.84	2.3164	0.431 71	1.34	3.8190	0.261 85
0.35	1.4191	0.704 69	0.85	2.3396	0.427 41	1.35	3.8574	0.259 24
0.36	1.4333	0.697 68	0.86	2.3632	0.423 16	1.36	3.8962	0.256 66
0.37	1.4477	0.690 73	0.87	2.3869	0.418 95	1.37	3.9354	0.254 11
0.38	1.4623	0.683 86	0.88	2.4109	0.414 78	1.38	3.9749	0.251 58
0.39	1.4770	0.677 06	0.89	2.4351	0.410 66	1.39	4.0149	0.249 08
0.40	1.4918	0.670 32	0.90	2.4596	0.406 57	1.40	4.0552	0.246 60
0.41	1.5068	0.663 65	0.91	2.4843	0.402 52	1.41	4.0960	0.244 14
0.42	1.5220	0.657 05	0.92	2.5093	0.398 52	1.42	4.1371	0.241 71
0.43	1.5373	0.650 51	0.93	2.5345	0.394 55	1.43	4.1787	0.239 31
0.44	1.5527	0.644 04	0.94	2.5600	0.390 63	1.44	4.2207	0.236 93
0.45	1.5683	0.637 63	0.95	2.5857	0.386 74	1.45	4.2631	0.234 57
0.46	1.5841	0.631 28	0.96	2.6117	0.382 89	1.46	4.3060	0.232 24
0.47	1.6000	0.625 00	0.97	2.6379	0.379 08	1.47	4.3492	0.229 93
0.48	1.6161	0.618 78	0.98	2.6645	0.375 31	1.48	4.3939	0.227 64
0.49	1.6323	0.612 63	0.99	2.6912	0.371 58	1.49	4.4371	0.225 37
0.50	1.6487	0.606 53	1.00	2.7183	0.367 88	1.50	4.4817	0.223 13

TABLE I *(continued)*

x	e^x	e^{-x}	x	e^x	e^{-x}	x	e^x	e^{-x}
1.50	4.4817	0.223 13	2.00	7.3891	0.135 34	2.50	12.182	0.082 085
1.51	4.5267	0.220 91	2.01	7.4633	0.133 99	2.51	12.305	0.081 268
1.52	4.5722	0.218 71	2.02	7.5383	0.132 66	2.52	12.429	0.080 460
1.53	4.6182	0.216 54	2.03	7.6141	0.131 34	2.53	12.554	0.079 659
1.54	4.6646	0.214 38	2.04	7.6906	0.130 03	2.54	12.680	0.078 866
1.55	4.7115	0.212 25	2.05	7.7679	0.128 73	2.55	12.807	0.078 082
1.56	4.7588	0.210 14	2.06	7.8460	0.127 45	2.56	12.936	0.077 305
1.57	4.8066	0.208 05	2.07	7.9248	0.126 19	2.57	13.066	0.076 536
1.58	4.8550	0.205 98	2.08	8.0045	0.124 93	2.58	13.197	0.075 774
1.59	4.9037	0.203 93	2.09	8.0849	0.123 69	2.59	13.330	0.075 020
1.60	4.9530	0.201 90	2.10	8.1662	0.122 46	2.60	13.464	0.074 274
1.61	5.0028	0.199 89	2.11	8.2482	0.121 24	2.61	13.599	0.073 535
1.62	5.0531	0.197 90	2.12	8.3311	0.120 03	2.62	13.736	0.072 803
1.63	5.1039	0.195 93	2.13	8.4149	0.118 84	2.63	13.874	0.072 078
1.64	5.1552	0.193 98	2.14	8.4994	0.117 65	2.64	14.013	0.071 361
1.65	5.2070	0.192 05	2.15	8.5849	0.116 48	2.65	14.154	0.070 651
1.66	5.2593	0.190 14	2.16	8.6711	0.115 33	2.66	14.296	0.069 948
1.67	5.3122	0.188 25	2.17	8.7583	0.114 18	2.67	14.440	0.069 252
1.68	5.3656	0.186 37	2.18	8.8463	0.113 04	2.68	14.585	0.068 563
1.69	5.4195	0.184 52	2.19	8.9352	0.111 92	2.69	14.732	0.067 881
1.70	5.4739	0.182 68	2.20	9.0250	0.110 80	2.70	14.880	0.067 206
1.71	5.5290	0.180 87	2.21	9.1157	0.109 70	2.71	15.029	0.066 537
1.72	5.5845	0.179 07	2.22	9.2073	0.108 61	2.72	15.180	0.065 875
1.73	5.6407	0.177 28	2.23	9.2999	0.107 53	2.73	15.333	0.065 219
1.74	5.6973	0.175 52	2.24	9.3933	0.106 46	2.74	15.487	0.064 570
1.75	5.7546	0.173 77	2.25	9.4877	0.105 40	2.75	15.643	0.063 928
1.76	5.8124	0.172 04	2.26	9.5831	0.104 35	2.76	15.800	0.063 292
1.77	5.8709	0.170 33	2.27	9.6794	0.103 31	2.77	15.959	0.062 662
1.78	5.9299	0.168 64	2.28	9.7767	0.102 28	2.78	16.119	0.062 039
1.79	5.9895	0.166 96	2.29	9.8749	0.101 27	2.79	16.281	0.061 421
1.80	6.0496	0.165 30	2.30	9.9742	0.100 26	2.80	16.445	0.060 810
1.81	6.1104	0.163 65	2.31	10.074	0.099 261	2.81	16.610	0.060 205
1.82	6.1719	0.162 03	2.32	10.176	0.098 274	2.82	16.777	0.059 606
1.83	6.2339	0.160 41	2.33	10.278	0.097 296	2.83	16.945	0.059 013
1.84	6.2965	0.158 82	2.34	10.381	0.096 328	2.84	17.116	0.058 426
1.85	6.3598	0.157 24	2.35	10.486	0.095 369	2.85	17.288	0.057 844
1.86	6.4237	0.155 67	2.36	10.591	0.094 420	2.86	17.462	0.057 269
1.87	6.4883	0.154 12	2.37	10.697	0.093 481	2.87	17.637	0.056 699
1.88	6.5535	0.152 59	2.38	10.805	0.092 551	2.88	17.814	0.056 135
1.89	6.6194	0.151 07	2.39	10.913	0.091 630	2.89	17.993	0.055 576
1.90	6.6859	0.149 57	2.40	11.023	0.090 718	2.90	18.174	0.055 023
1.91	6.7531	0.148 08	2.41	11.134	0.089 815	2.91	18.357	0.054 476
1.92	6.8210	0.146 61	2.42	11.246	0.088 922	2.92	18.541	0.053 934
1.93	6.8895	0.145 15	2.43	11.359	0.088 037	2.93	18.728	0.053 397
1.94	6.9588	0.143 70	2.44	11.473	0.087 161	2.94	18.916	0.052 866
1.95	7.0287	0.142 27	2.45	11.588	0.086 294	2.95	19.106	0.052 340
1.96	7.0993	0.140 86	2.46	11.705	0.085 435	2.96	19.298	0.051 819
1.97	7.1707	0.139 46	2.47	11.822	0.084 585	2.97	19.492	0.051 303
1.98	7.2427	0.138 07	2.48	11.941	0.083 743	2.98	19.688	0.050 793
1.99	7.3155	0.136 70	2.49	12.061	0.082 910	2.99	19.886	0.050 287
2.00	7.3891	0.135 34	2.50	12.182	0.082 085	3.00	20.086	0.049 787

TABLE II COMMON LOGARITHMS

x	0	1	2	3	4	5	6	7	8	9
1.0	0.0000	0.004321	0.008600	0.01284	0.01703	0.02119	0.02531	0.02938	0.03342	0.03743
1.1	0.04139	0.04532	0.04922	0.05308	0.05690	0.06070	0.06446	0.06819	0.07188	0.07555
1.2	0.07918	0.08279	0.08636	0.08991	0.09342	0.09691	0.1004	0.1038	0.1072	0.1106
1.3	0.1139	0.1173	0.1206	0.1239	0.1271	0.1303	0.1335	0.1367	0.1399	0.1430
1.4	0.1461	0.1492	0.1523	0.1553	0.1584	0.1614	0.1644	0.1673	0.1703	0.1732
1.5	0.1761	0.1790	0.1818	0.1847	0.1875	0.1903	0.1931	0.1959	0.1987	0.2014
1.6	0.2041	0.2068	0.2095	0.2122	0.2148	0.2175	0.2201	0.2227	0.2253	0.2279
1.7	0.2304	0.2330	0.2355	0.2380	0.2405	0.2430	0.2455	0.2480	0.2504	0.2529
1.8	0.2553	0.2577	0.2601	0.2625	0.2648	0.2673	0.2695	0.2718	0.2742	0.2765
1.9	0.2788	0.2810	0.2833	0.2856	0.2878	0.2900	0.2923	0.2945	0.2967	0.2989
2.0	0.3010	0.3032	0.3054	0.3075	0.3096	0.3118	0.3139	0.3160	0.3181	0.3201
2.1	0.3222	0.3243	0.3263	0.3284	0.3304	0.3324	0.3345	0.3365	0.3385	0.3404
2.2	0.3424	0.3444	0.3464	0.3483	0.3502	0.3522	0.3541	0.3560	0.3579	0.3598
2.3	0.3617	0.3636	0.3655	0.3674	0.3692	0.3711	0.3729	0.3747	0.3766	0.3784
2.4	0.3802	0.3820	0.3838	0.3856	0.3874	0.3892	0.3909	0.3927	0.3945	0.3962
2.5	0.3979	0.3997	0.4014	0.4031	0.4048	0.4065	0.4082	0.4099	0.4116	0.4133
2.6	0.4150	0.4166	0.4183	0.4200	0.4216	0.4232	0.4249	0.4265	0.4281	0.4298
2.7	0.4314	0.4330	0.4346	0.4362	0.4378	0.4393	0.4409	0.4425	0.4440	0.4456
2.8	0.4472	0.4487	0.4502	0.4518	0.4533	0.4548	0.4564	0.4579	0.4594	0.4609
2.9	0.4624	0.4639	0.4654	0.4669	0.4683	0.4698	0.4713	0.4728	0.4742	0.4757
3.0	0.4771	0.4786	0.4800	0.4814	0.4829	0.4843	0.4857	0.4871	0.4886	0.4900
3.1	0.4914	0.4928	0.4942	0.4955	0.4969	0.4983	0.4997	0.5011	0.5024	0.5038
3.2	0.5051	0.5065	0.5079	0.5092	0.5105	0.5119	0.5132	0.5145	0.5159	0.5172
3.3	0.5185	0.5198	0.5211	0.5224	0.5237	0.5250	0.5263	0.5276	0.5289	0.5302
3.4	0.5315	0.5328	0.5340	0.5353	0.5366	0.5378	0.5391	0.5403	0.5416	0.5428
3.5	0.5441	0.5453	0.5465	0.5478	0.5490	0.5502	0.5514	0.5527	0.5539	0.5551
3.6	0.5563	0.5575	0.5587	0.5599	0.5611	0.5623	0.5635	0.5647	0.5658	0.5670
3.7	0.5682	0.5694	0.5705	0.5717	0.5729	0.5740	0.5752	0.5763	0.5775	0.5786
3.8	0.5798	0.5809	0.5821	0.5832	0.5843	0.5855	0.5866	0.5877	0.5888	0.5899
3.9	0.5911	0.5922	0.5933	0.5944	0.5955	0.5966	0.5977	0.5988	0.5999	0.6010
4.0	0.6021	0.6031	0.6042	0.6053	0.6064	0.6075	0.6085	0.6096	0.6107	0.6117
4.1	0.6128	0.6138	0.6149	0.6160	0.6170	0.6180	0.6191	0.6201	0.6212	0.6222
4.2	0.6232	0.6243	0.6253	0.6263	0.6274	0.6284	0.6294	0.6304	0.6314	0.6325
4.3	0.6335	0.6345	0.6355	0.6365	0.6375	0.6385	0.6395	0.6405	0.6415	0.6425
4.4	0.6435	0.6444	0.6454	0.6464	0.6474	0.6484	0.6493	0.6503	0.6513	0.6522
4.5	0.6532	0.6542	0.6551	0.6561	0.6571	0.6580	0.6590	0.6599	0.6609	0.6618
4.6	0.6628	0.6637	0.6646	0.6656	0.6665	0.6675	0.6684	0.6693	0.6702	0.6712
4.7	0.6721	0.6730	0.6739	0.6749	0.6758	0.6767	0.6776	0.6785	0.6794	0.6803
4.8	0.6812	0.6821	0.6830	0.6839	0.6848	0.6857	0.6866	0.6875	0.6884	0.6893
4.9	0.6902	0.6911	0.6920	0.6928	0.6937	0.6946	0.6955	0.6964	0.6972	0.6981
5.0	0.6990	0.6998	0.7007	0.7016	0.7024	0.7033	0.7042	0.7050	0.7059	0.7067
5.1	0.7076	0.7084	0.7093	0.7101	0.7110	0.7118	0.7126	0.7135	0.7143	0.7152
5.2	0.7160	0.7168	0.7177	0.7185	0.7193	0.7202	0.7210	0.7218	0.7226	0.7235
5.3	0.7243	0.7251	0.7259	0.7267	0.7275	0.7284	0.7292	0.7300	0.7308	0.7316
5.4	0.7324	0.7332	0.7340	0.7348	0.7356	0.7364	0.7372	0.7380	0.7388	0.7396

TABLE II *(continued)*

x	0	1	2	3	4	5	6	7	8	9
5.5	0.7404	0.7412	0.7419	0.7427	0.7435	0.7443	0.7451	0.7459	0.7466	0.7474
5.6	0.7482	0.7490	0.7497	0.7505	0.7513	0.7520	0.7528	0.7536	0.7543	0.7551
5.7	0.7559	0.7566	0.7574	0.7582	0.7589	0.7597	0.7604	0.7612	0.7619	0.7627
5.8	0.7634	0.7642	0.7649	0.7657	0.7664	0.7672	0.7679	0.7686	0.7694	0.7701
5.9	0.7709	0.7716	0.7723	0.7731	0.7738	0.7745	0.7752	0.7760	0.7767	0.7774
6.0	0.7782	0.7789	0.7796	0.7803	0.7810	0.7818	0.7825	0.7832	0.7839	0.7846
6.1	0.7853	0.7860	0.7868	0.7875	0.7882	0.7889	0.7896	0.7903	0.7910	0.7917
6.2	0.7924	0.7931	0.7938	0.7945	0.7952	0.7959	0.7966	0.7973	0.7980	0.7987
6.3	0.7993	0.8000	0.8007	0.8014	0.8021	0.8028	0.8035	0.8041	0.8048	0.8055
6.4	0.8062	0.8069	0.8075	0.8082	0.8089	0.8096	0.8102	0.8109	0.8116	0.8122
6.5	0.8129	0.8136	0.8142	0.8149	0.8156	0.8162	0.8169	0.8176	0.8182	0.8189
6.6	0.8195	0.8202	0.8209	0.8215	0.8222	0.8228	0.8235	0.8241	0.8248	0.8254
6.7	0.8261	0.8267	0.8274	0.8280	0.8287	0.8293	0.8299	0.8306	0.8312	0.8319
6.8	0.8325	0.8331	0.8338	0.8344	0.8351	0.8357	0.8363	0.8370	0.8376	0.8382
6.9	0.8388	0.8395	0.8401	0.8407	0.8414	0.8420	0.8426	0.8432	0.8439	0.8445
7.0	0.8451	0.8457	0.8463	0.8470	0.8476	0.8482	0.8488	0.8494	0.8500	0.8506
7.1	0.8513	0.8519	0.8525	0.8531	0.8537	0.8543	0.8549	0.8555	0.8561	0.8567
7.2	0.8573	0.8579	0.8585	0.8591	0.8597	0.8603	0.8609	0.8615	0.8621	0.8627
7.3	0.8633	0.8639	0.8645	0.8651	0.8657	0.8663	0.8669	0.8675	0.8681	0.8686
7.4	0.8692	0.8698	0.8704	0.8710	0.8716	0.8722	0.8727	0.8733	0.8739	0.8745
7.5	0.8751	0.8756	0.8762	0.8768	0.8774	0.8779	0.8785	0.8791	0.8797	0.8802
7.6	0.8808	0.8814	0.8820	0.8825	0.8831	0.8837	0.8842	0.8848	0.8854	0.8859
7.7	0.8865	0.8871	0.8876	0.8882	0.8887	0.8893	0.8899	0.8904	0.8910	0.8915
7.8	0.8921	0.8927	0.8932	0.8938	0.8943	0.8949	0.8954	0.8960	0.8965	0.8971
7.9	0.8976	0.8982	0.8987	0.8993	0.8998	0.9004	0.9009	0.9015	0.9020	0.9025
8.0	0.9031	0.9036	0.9042	0.9047	0.9053	0.9058	0.9063	0.9069	0.9074	0.9079
8.1	0.9085	0.9090	0.9096	0.9101	0.9106	0.9112	0.9117	0.9122	0.9128	0.9133
8.2	0.9138	0.9143	0.9149	0.9154	0.9159	0.9165	0.9170	0.9175	0.9180	0.9186
8.3	0.9191	0.9196	0.9201	0.9206	0.9212	0.9217	0.9222	0.9227	0.9232	0.9238
8.4	0.9243	0.9248	0.9253	0.9258	0.9263	0.9269	0.9274	0.9279	0.9284	0.9289
8.5	0.9294	0.9299	0.9304	0.9309	0.9315	0.9320	0.9325	0.9330	0.9335	0.9340
8.6	0.9345	0.9350	0.9355	0.9360	0.9365	0.9370	0.9375	0.9380	0.9385	0.9390
8.7	0.9395	0.9400	0.9405	0.9410	0.9415	0.9420	0.9425	0.9430	0.9435	0.9440
8.8	0.9445	0.9450	0.9455	0.9460	0.9465	0.9469	0.9474	0.9479	0.9484	0.9489
8.9	0.9494	0.9499	0.9504	0.9509	0.9513	0.9518	0.9523	0.9528	0.9533	0.9538
9.0	0.9542	0.9547	0.9552	0.9557	0.9562	0.9566	0.9571	0.9576	0.9581	0.9586
9.1	0.9590	0.9595	0.9600	0.9605	0.9609	0.9614	0.9619	0.9624	0.9628	0.9633
9.2	0.9638	0.9643	0.9647	0.9652	0.9657	0.9661	0.9666	0.9671	0.9675	0.9680
9.3	0.9685	0.9689	0.9694	0.9699	0.9703	0.9708	0.9713	0.9717	0.9722	0.9727
9.4	0.9731	0.9736	0.9741	0.9745	0.9750	0.9754	0.9759	0.9763	0.9768	0.9773
9.5	0.9777	0.9782	0.9786	0.9791	0.9795	0.9800	0.9805	0.9809	0.9814	0.9818
9.6	0.9823	0.9827	0.9832	0.9836	0.9841	0.9845	0.9850	0.9854	0.9859	0.9863
9.7	0.9868	0.9872	0.9877	0.9881	0.9886	0.9890	0.9894	0.9899	0.9903	0.9908
9.8	0.9912	0.9917	0.9921	0.9926	0.9930	0.9934	0.9939	0.9943	0.9948	0.9952
9.9	0.9956	0.9961	0.9965	0.9969	0.9974	0.9978	0.9983	0.9987	0.9991	0.9996

TABLE III NATURAL LOGARITHMS ($\ln x = \log_e x$)

ln 10 = 2.3026 5 ln 10 = 11.5130 9 ln 10 = 20.7233
2 ln 10 = 4.6052 6 ln 10 = 13.8155 10 ln 10 = 23.0259
3 ln 10 = 6.9078 7 ln 10 = 16.1181
4 ln 10 = 9.2103 8 ln 10 = 18.4207

x	.00	.01	.02	.03	.04	.05	.06	.07	.08	.09
1.0	0.0000	0.0100	0.0198	0.0296	0.0392	0.0488	0.0583	0.0677	0.0770	0.0862
1.1	0.0953	0.1044	0.1133	0.1222	0.1310	0.1398	0.1484	0.1570	0.1655	0.1740
1.2	0.1823	0.1906	0.1989	0.2070	0.2151	0.2231	0.2311	0.2390	0.2469	0.2546
1.3	0.2624	0.2700	0.2776	0.2852	0.2927	0.3001	0.3075	0.3148	0.3221	0.3293
1.4	0.3365	0.3436	0.3507	0.3577	0.3646	0.3716	0.3784	0.3853	0.3920	0.3988
1.5	0.4055	0.4121	0.4187	0.4253	0.4318	0.4383	0.4447	0.4511	0.4574	0.4637
1.6	0.4700	0.4762	0.4824	0.4886	0.4947	0.5008	0.5068	0.5128	0.5188	0.5247
1.7	0.5306	0.5365	0.5423	0.5481	0.5539	0.5596	0.5653	0.5710	0.5766	0.5822
1.8	0.5878	0.5933	0.5988	0.6043	0.6098	0.6152	0.6206	0.6259	0.6313	0.6366
1.9	0.6419	0.6471	0.6523	0.6575	0.6627	0.6678	0.6729	0.6780	0.6831	0.6881
2.0	0.6931	0.6981	0.7031	0.7080	0.7129	0.7178	0.7227	0.7275	0.7324	0.7372
2.1	0.7419	0.7467	0.7514	0.7561	0.7608	0.7655	0.7701	0.7747	0.7793	0.7839
2.2	0.7885	0.7930	0.7975	0.8020	0.8065	0.8109	0.8154	0.8198	0.8242	0.8286
2.3	0.8329	0.8372	0.8416	0.8459	0.8502	0.8544	0.8587	0.8629	0.8671	0.8713
2.4	0.8755	0.8796	0.8838	0.8879	0.8920	0.8961	0.9002	0.9042	0.9083	0.9123
2.5	0.9163	0.9203	0.9243	0.9282	0.9322	0.9361	0.9400	0.9439	0.9478	0.9517
2.6	0.9555	0.9594	0.9632	0.9670	0.9708	0.9746	0.9783	0.9821	0.9858	0.9895
2.7	0.9933	0.9969	1.0006	1.0043	1.0080	1.0116	1.0152	1.0188	1.0225	1.0260
2.8	1.0296	1.0332	1.0367	1.0403	1.0438	1.0473	1.0508	1.0543	1.0578	1.0613
2.9	1.0647	1.0682	1.0716	1.0750	1.0784	1.0818	1.0852	1.0886	1.0919	1.0953
3.0	1.0986	1.1019	1.1053	1.1086	1.1119	1.1151	1.1184	1.1217	1.1249	1.1282
3.1	1.1314	1.1346	1.1378	1.1410	1.1442	1.1474	1.1506	1.1537	1.1569	1.1600
3.2	1.1632	1.1663	1.1694	1.1725	1.1756	1.1787	1.1817	1.1848	1.1878	1.1909
3.3	1.1939	1.1969	1.2000	1.2030	1.2060	1.2090	1.2119	1.2149	1.2179	1.2208
3.4	1.2238	1.2267	1.2296	1.2326	1.2355	1.2384	1.2413	1.2442	1.2470	1.2499
3.5	1.2528	1.2556	1.2585	1.2613	1.2641	1.2669	1.2698	1.2726	1.2754	1.2782
3.6	1.2809	1.2837	1.2865	1.2892	1.2920	1.2947	1.2975	1.3002	1.3029	1.3056
3.7	1.3083	1.3110	1.3137	1.3164	1.3191	1.3218	1.3244	1.3271	1.3297	1.3324
3.8	1.3350	1.3376	1.3403	1.3429	1.3455	1.3481	1.3507	1.3533	1.3558	1.3584
3.9	1.3610	1.3635	1.3661	1.3686	1.3712	1.3737	1.3762	1.3788	1.3813	1.3838
4.0	1.3863	1.3888	1.3913	1.3938	1.3962	1.3987	1.4012	1.4036	1.4061	1.4085
4.1	1.4110	1.4134	1.4159	1.4183	1.4207	1.4231	1.4255	1.4279	1.4303	1.4327
4.2	1.4351	1.4375	1.4398	1.4422	1.4446	1.4469	1.4493	1.4516	1.4540	1.4563
4.3	1.4586	1.4609	1.4633	1.4656	1.4679	1.4702	1.4725	1.4748	1.4770	1.4793
4.4	1.4816	1.4839	1.4861	1.4884	1.4907	1.4929	1.4951	1.4974	1.4996	1.5019
4.5	1.5041	1.5063	1.5085	1.5107	1.5129	1.5151	1.5173	1.5195	1.5217	1.5239
4.6	1.5261	1.5282	1.5304	1.5326	1.5347	1.5369	1.5390	1.5412	1.5433	1.5454
4.7	1.5476	1.5497	1.5518	1.5539	1.5560	1.5581	1.5602	1.5623	1.5644	1.5665
4.8	1.5686	1.5707	1.5728	1.5748	1.5769	1.5790	1.5810	1.5831	1.5851	1.5872
4.9	1.5892	1.5913	1.5933	1.5953	1.5974	1.5994	1.6014	1.6034	1.6054	1.6074
5.0	1.6094	1.6114	1.6134	1.6154	1.6174	1.6194	1.6214	1.6233	1.6253	1.6273
5.1	1.6292	1.6312	1.6332	1.6351	1.6371	1.6390	1.6409	1.6429	1.6448	1.6467
5.2	1.6487	1.6506	1.6525	1.6544	1.6563	1.6582	1.6601	1.6620	1.6639	1.6658
5.3	1.6677	1.6696	1.6715	1.6734	1.6752	1.6771	1.6790	1.6808	1.6827	1.6845
5.4	1.6864	1.6882	1.6901	1.6919	1.6938	1.6956	1.6974	1.6993	1.7011	1.7029

Note: $\ln 35{,}200 = \ln (3.52 \times 10^4) = \ln 3.52 + 4 \ln 10$
$\ln 0.008\,64 = \ln (8.64 \times 10^{-3}) = \ln 8.64 - 3 \ln 10$

TABLE III *(continued)*

x	.00	.01	.02	.03	.04	.05	.06	.07	.08	.09
5.5	1.7047	1.7066	1.7084	1.7102	1.7120	1.7138	1.7156	1.7174	1.7192	1.7210
5.6	1.7228	1.7246	1.7263	1.7281	1.7299	1.7317	1.7334	1.7352	1.7370	1.7387
5.7	1.7405	1.7422	1.7440	1.7457	1.7475	1.7492	1.7509	1.7527	1.7544	1.7561
5.8	1.7579	1.7596	1.7613	1.7630	1.7647	1.7664	1.7681	1.7699	1.7716	1.7733
5.9	1.7750	1.7766	1.7783	1.7800	1.7817	1.7834	1.7851	1.7867	1.7884	1.7901
6.0	1.7918	1.7934	1.7951	1.7967	1.7984	1.8001	1.8017	1.8034	1.8050	1.8066
6.1	1.8083	1.8099	1.8116	1.8132	1.8148	1.8165	1.8181	1.8197	1.8213	1.8229
6.2	1.8245	1.8262	1.8278	1.8294	1.8310	1.8326	1.8342	1.8358	1.8374	1.8390
6.3	1.8405	1.8421	1.8437	1.8453	1.8469	1.8485	1.8500	1.8516	1.8532	1.8547
6.4	1.8563	1.8579	1.8594	1.8610	1.8625	1.8641	1.8656	1.8672	1.8687	1.8703
6.5	1.8718	1.8733	1.8749	1.8764	1.8779	1.8795	1.8810	1.8825	1.8840	1.8856
6.6	1.8871	1.8886	1.8901	1.8916	1.8931	1.8946	1.8961	1.8976	1.8991	1.9006
6.7	1.9021	1.9036	1.9051	1.9066	1.9081	1.9095	1.9110	1.9125	1.9140	1.9155
6.8	1.9169	1.9184	1.9199	1.9213	1.9228	1.9242	1.9257	1.9272	1.9286	1.9301
6.9	1.9315	1.9330	1.9344	1.9359	1.9373	1.9387	1.9402	1.9416	1.9430	1.9445
7.0	1.9459	1.9473	1.9488	1.9502	1.9516	1.9530	1.9544	1.9559	1.9573	1.9587
7.1	1.9601	1.9615	1.9629	1.9643	1.9657	1.9671	1.9685	1.9699	1.9713	1.9727
7.2	1.9741	1.9755	1.9769	1.9782	1.9796	1.9810	1.9824	1.9838	1.9851	1.9865
7.3	1.9879	1.9892	1.9906	1.9920	1.9933	1.9947	1.9961	1.9974	1.9988	2.0001
7.4	2.0015	2.0028	2.0042	2.0055	2.0069	2.0082	2.0096	2.0109	2.0122	2.0136
7.5	2.0149	2.0162	2.0176	2.0189	2.0202	2.0215	2.0229	2.0242	2.0255	2.0268
7.6	2.0281	2.0295	2.0308	2.0321	2.0334	2.0347	2.0360	2.0373	2.0386	2.0399
7.7	2.0412	2.0425	2.0438	2.0451	2.0464	2.0477	2.0490	2.0503	2.0516	2.0528
7.8	2.0541	2.0554	2.0567	2.0580	2.0592	2.0605	2.0618	2.0631	2.0643	2.0656
7.9	2.0669	2.0681	2.0694	2.0707	2.0719	2.0732	2.0744	2.0757	2.0769	2.0782
8.0	2.0794	2.0807	2.0819	2.0832	2.0844	2.0857	2.0869	2.0882	2.0894	2.0906
8.1	2.0919	2.0931	2.0943	2.0956	2.0968	2.0980	2.0992	2.1005	2.1017	2.1029
8.2	2.1041	2.1054	2.1066	2.1078	2.1090	2.1102	2.1114	2.1126	2.1138	2.1150
8.3	2.1163	2.1175	2.1187	2.1199	2.1211	2.1223	2.1235	2.1247	2.1258	2.1270
8.4	2.1282	2.1294	2.1306	2.1318	2.1330	2.1342	2.1353	2.1365	2.1377	2.1389
8.5	2.1401	2.1412	2.1424	2.1436	2.1448	2.1459	2.1471	2.1483	2.1494	2.1506
8.6	2.1518	2.1529	2.1541	2.1552	2.1564	2.1576	2.1587	2.1599	2.1610	2.1622
8.7	2.1633	2.1645	2.1656	2.1668	2.1679	2.1691	2.1702	2.1713	2.1725	2.1736
8.8	2.1748	2.1759	2.1770	2.1782	2.1793	2.1804	2.1815	2.1827	2.1838	2.1849
8.9	2.1861	2.1872	2.1883	2.1894	2.1905	2.1917	2.1928	2.1939	2.1950	2.1961
9.0	2.1972	2.1983	2.1994	2.2006	2.2017	2.2028	2.2039	2.2050	2.2061	2.2072
9.1	2.2083	2.2094	2.2105	2.2116	2.2127	2.2138	2.2148	2.2159	2.2170	2.2181
9.2	2.2192	2.2203	2.2214	2.2225	2.2235	2.2246	2.2257	2.2268	2.2279	2.2289
9.3	2.2300	2.2311	2.2322	2.2332	2.2343	2.2354	2.2364	2.2375	2.2386	2.2396
9.4	2.2407	2.2418	2.2428	2.2439	2.2450	2.2460	2.2471	2.2481	2.2492	2.2502
9.5	2.2513	2.2523	2.2534	2.2544	2.2555	2.2565	2.2576	2.2586	2.2597	2.2607
9.6	2.2618	2.2628	2.2638	2.2649	2.2659	2.2670	2.2680	2.2690	2.2701	2.2711
9.7	2.2721	2.2732	2.2742	2.2752	2.2762	2.2773	2.2783	2.2793	2.2803	2.2814
9.8	2.2824	2.2834	2.2844	2.2854	2.2865	2.2875	2.2885	2.2895	2.2905	2.2915
9.9	2.2925	2.2935	2.2946	2.2956	2.2966	2.2976	2.2986	2.2996	2.3006	2.3016

ANSWERS

TO SELECTED PROBLEMS

CHAPTER 1 EXERCISE 1-1

1. T **3.** T **5.** T **7.** T **9.** T **11.** F
13. $-3, 0, 5$ (Infinitely many more answers are possible, except for 0.)
15. $\frac{2}{3}, -\frac{7}{8}$, 2.65 are three of infinitely many **17.** $\{6, 7, 8, 9\}$ **19.** $\{a, s, t, u\}$ **21.** $\varnothing$ **23.** $\{5\}$
25. $\varnothing$ **27.** $\{-2, 2\}$ **29.** **(A)** T **(B)** F **(C)** T
31. **(A)** 3 and 4 **(B)** -2 and -1 **(C)** -3 and -2 **33.** **(A)** $\{1, 2, 3, 4, 6\}$ **(B)** $\{2, 4\}$
35. **(A)** J, Q, R **(B)** Q, R **(C)** R **(D)** Q, R **37.** $\frac{1}{11}$
39. **(A)** 0.88888888 . . . **(B)** 0.27272727 . . . **(C)** 2.23606797 . . . **(D)** 1.37500000 . . . **41.** 6

EXERCISE 1-2

1. $11 - 5 = \frac{12}{2}$ **3.** $4 > -18$ **5.** $-12 < -3$ **7.** $x \geq -8$ **9.** $-2 < x < 2$ **11.** Is equal to
13. Is greater than **15.** Is less than **17.** Is less than or equal to **19.** Is not equal to
21. Is greater than or equal to **23.** $<$ **25.** $>$ **27.** $<$ **29.** $=$ **31.** $<$ **33.** $>$
35. $<, <$ **37.** $>$ **39.** $>$ **41.** $<$ **43.** Greater than **45.** −5 0 5 x

47. −5 0 5 x **49.** −5 0 5 x **51.** −5 0 5 x

53. $x - 8 > 0$ **55.** $x + 4 \geq 0$ **57.** $80 = 3 + 2x$ **59.** $-3 \leq x < 4$ **61.** $26 = x - 12$
63. $x < 2x - 6$ **65.** $6x = 4 + 3x$ **67.** $x - 6 = 5(7 + x)$ **69.** $63 \leq \frac{9}{5}C + 32 \leq 72$
71. $x + (x + 1) + (x + 2) = 186$ **73.** $t = -5$ **75.** $5x + 7x = 12x$ **77.** $3 - x$
79. "Is" does not translate into "equal" in this case. (The number 8 is actually an element in the set of even numbers.) The properties of equality do not apply.
81. $90 = x(2x - 3)$ **83.** $2x + 2(3x - 10) = 210$ **85.** **(A)** $A = x(200 - x)$ **(B)** $0 \leq x \leq 200$

EXERCISE 1-3

1. $3 + x$ **3.** $(5 \cdot 7)z$ **5.** mn **7.** $(9 + 11) + M$ **9.** $3x + 3$ **11.** $2x + x^2$ **13.** $7x$
15. $x + y$ **17.** Commutative property for addition **19.** Associative property for addition
21. Commutative property for multiplication **23.** Distributive property
25. Associative property for multiplication **27.** Identity property for addition
29. Identity property for multiplication **31.** Distributive property **33.** $x + 9$ **35.** $20y$
37. $u + 15$ **39.** $21x$ **41.** x **43.** Commutative property for addition
45. Commutative property for multiplication **47.** Commutative property for addition
49. Associative property for multiplication **51.** $x + y + z + 12$ **53.** $3x + 4y + 11$ **55.** $x + y + 5$
57. $36mnp$ **59.** $x^2 + 5x$ **61.** $2(x + 9)$ **63.** $8x + 16$ **65.** $6(x + 2)$
67. (B) and (D) are false, since $12 - 4 \neq 4 - 12$ and $12 \div 4 \neq 4 \div 12$.
69. *1.* Commutative property for addition *2.* Associative property for addition
3. Associative property for addition *4.* Substitution principle for equality
5. Commutative property for addition *6.* Associative property for addition

EXERCISE 1-4

1. -7 **3.** 6 **5.** 2 **7.** 27 **9.** 0 **11.** -10 **13.** 4 **15.** -6 **17.** 12
19. Sometimes **21.** -3 **23.** -2 **25.** -6 **27.** -5 **29.** -5 **31.** -5 **33.** -1
35. -6 **37.** 28 **39.** -29.191 **41.** 76.025 **43.** -16.179 **45.** -5 **47.** 7 or -7
49. -5 **51.** 5 **53.** -5 **55.** -11 **57.** -3 **59.** 6 **61.** -5 **63.** True
65. False; $(+7) - (-3) = +10$, $(-3) - (+7) = -10$ **67.** True
69. False: $|(+9) + (-3)| = +6$, $|+9| + |-3| = +12$
71. Commutative property for addition, associative property for addition, additive inverse property, identity property for addition **73.** \$23.75

EXERCISE 1-5

1. 15 **3.** 3 **5.** -18 **7.** -3 **9.** 0 **11.** 0 **13.** Not defined **15.** Not defined
17. -7 **19.** -1 **21.** 11 **23.** 9 **25.** Both are 8 **27.** -10 **29.** 3 **31.** -14
33. -5 **35.** -70 **37.** 0 **39.** 10 **41.** 40 **43.** 12 **45.** -18 **47.** -2 **49.** 6
51. 5 **53.** 1 **55.** -12 **57.** Not defined **59.** 56 **61.** 0 **63.** -6 **65.** 8 **67.** 0
69. -50 **71.** 0 **73.** Not defined (cannot divide by 0) **75.** 100 **77.** Never **79.** Always
81. When x and y are of opposite signs
83. *1.* Identity property for addition *2.* Distributive property *3.* Addition property for equality
4. Inverse and associative properties for addition *5.* Inverse property for addition
6. Identity property for addition *7.* Symmetric property for equality

EXERCISE 1-6

1. $17x$ **3.** x **5.** $8x$ **7.** $-13t$ **9.** $3x + 5y$ **11.** $4 + 4x$ **13.** $4m - 6n$ **15.** $9u - 4v$
17. $-2m - 24n$ **19.** $5u - 6v$ **21.** $4x - 3$ **23.** $x + 11$ **25.** 18 **27.** All real numbers
29. 9 **31.** No solution **33.** 10 **35.** 1 **37.** All real numbers **39.** 4 **41.** No solution

43. $15 - \frac{1}{2}y$ **45.** $y - 1$ **47.** $8 - x$ **49.** $6 - 2y$ **51.** $2x$ **53.** $x - 3$ **55.** $2x + 2$
57. $3x - 4$ **59.** $-2x + 1$ **61.** $7x = 4x - 12$, $x = -4$ **63.** $2x + 3 = 3x - 12$, $x = 15$
65. $x + (x + 1) + (x + 2) = 96$; 31, 32, 33 **67.** $x + (x + 2) + (x + 4) = 42$; 12, 14, 16

REVIEW EXERCISE 1-7

1. **(A)** T **(B)** T **(C)** F **(D)** F **(E)** T **(F)** T **(G)** T **(H)** F *(1-1)* **2.** Rational *(1-1)*
3. 17 *(1-5)* **4.** 13 *(1-5)* **5.** -5 *(1-4)* **6.** -13 *(1-4)* **7.** 6 *(1-4)* **8.** -3 *(1-4)*
9. 3 *(1-4)* **10.** -12 *(1-4)* **11.** 28 *(1-5)* **12.** -18 *(1-5)* **13.** -4 *(1-5)*
14. 6 *(1-5)* **15.** Not defined *(1-5)* **16.** 0 *(1-5)* **17.** 4 *(1-5)* **18.** -14 *(1-5)*
19. -5 *(1-5)* **20.** -12 *(1-5)* **21.** 8 *(1-4)* **22.** 5 *(1-4)* **23.** -3 *(1-4)*
24. -2 *(1-4)* **25.** -5 *(1-4)* **26.** -5 *(1-4)* **27.** $x + 10$ *(1-3)* **28.** $15x$ *(1-3)*
29. $8xy$ *(1-3)* **30.** $x + y + z + 12$ *(1-3)* **31.** $x + 2$ *(1-3)* **32.** x *(1-3)*
33. $3x - 15$ *(1-3)* **34.** $7a + 7b$ *(1-3)* **35.** $3x + 14$ *(1-3)* **36.** $2x + 1$ *(1-3)* **37.** $>$ *(1-2)*
38. $<$ *(1-2)* **39.** $<$ *(1-2)* **40.** $<$ *(1-3)* **41.** $>$ *(1-3)* **42.** $<$ *(1-3)*
43. **(A)** F **(B)** T **(C)** T **(D)** F **(E)** T **(F)** T **(G)** F **(H)** T *(1-1)*

44. **(A)** 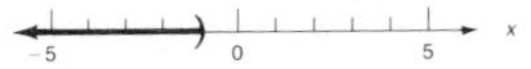 −5 0 5 x **(B)**  −5 0 5 x *(1-2)* **45.** 6 *(1-5)*

46. 2 *(1-4)* **47.** 4 *(1-4)* **48.** 1 *(1-4)* **49.** 4 *(1-4)* **50.** -26 *(1-5)* **51.** 10 *(1-5)*
52. 35 *(1-5)* **53.** -2 *(1-5)* **54.** Not defined *(1-5)* **55.** 6 *(1-4)* **56.** -2 *(1-4)*
57. 8 *(1-4)* **58.** -6 *(1-4)* **59.** -9 *(1-5)* **60.** -7 *(1-5)* **61.** -1 *(1-5)*
62. -4 *(1-6)* **63.** $-4y - 4$ *(1-6)* **64.** $x - 1 > 0$ *(1-2)* **65.** $2x + 3 \geq 0$ *(1-2)*
66. $50 = 2x - 10$ *(1-2)* **67.** $x < 2x - 12$ *(1-2)* **68.** $-5 \leq x < 5$ *(1-2)*
69. $8 + x = 5(x - 6)$ *(1-2)* **70.** $x(x - 10) = 1{,}200$ *(1-2)* **71.** $2x + 2(x + 5) = 43$ *(1-2)*
72. a *(1-3)* **73.** 0 *(1-3)* **74.** 0 *(1-3)* **75.** a *(1-3)* **76.** 1 *(1-3)* **77.** $1/a$ *(1-3)*
78. Less than *(1-2)* **79.** Greater than *(1-2)* **80.** $A = P + Prt$ *(1-2)* **81.** $x + y < z$ *(1-2)*
82. $Q + P$ *(1-3)* **83.** $A + 1$ *(1-3)* **84.** $-x$ *(1-3)* **85.** $u + 4$ *(1-2)* **86.** 1 *(1-3)*
87. $x + (3 + 5)$ *(1-3)* **88.** $y < 5$ *(1-2)* **89.** Commutative property for addition *(1-3)*
90. Associative property for addition *(1-3)* **91.** Commutative property for multiplication *(1-3)*
92. Associative property for multiplication *(1-3)* **93.** Additive identity *(1-3)*
94. Additive inverse *(1-3)* **95.** Distributive property *(1-3)* **96.** -42 *(1-5)* **97.** 6 *(1-5)*
98. 3, $<$ *(1-1, 1-2)* **99.** **(A)** $\{-1, 0, 1, 2\}$ **(B)** $\varnothing$ *(1-1, 1-2)*
100. **(A)** $\{3, 4, 5, 6, 7\}$ **(B)** $\{4, 5\}$ *(1-1)*
101. **(A)** All nonnegative real numbers **(B)** All nonpositive real numbers *(1-4)* **102.** 6 *(1-6)*
103. 15, 16, 17 *(1-6)* **104.** 7, 8 *(1-6)*

CHAPTER 2 EXERCISE 2-1

1. 9 **3.** 5 **5.** 5 **7.** 4 **9.** 6 **11.** Binomial, 2 **13.** Trinomial, 6 **15.** Binomial, 3
17. Monomial, 8 **19.** -3 **21.** 3 **23.** 1 **25.** $10x^{11}$ **27.** $12y^{10}$ **29.** 35×10^{17}
31. $9x - 3$ **33.** $-2x - 4$ **35.** $7x^2 - x - 12$ **37.** $-x + 1$ **39.** $-y^2 - 2$ **41.** $-3x^2y$
43. $3y^3 + 4y^2 - y - 3$ **45.** $3a^2 - b^2$ **47.** $-7x + 9y$ **49.** $-5x + 3y$ **51.** $4x - 6$
53. $-8x + 12$ **55.** $10t - 18$ **57.** $x - 14$ **59.** $-m + 2n$ **61.** $y + z$
63. $2x^4 + 3x^3 + 7x^2 - x - 8$ **65.** $-3x^3 + x^2 + 3x - 2$ **67.** $-2m^3 - 5$ **69.** $-t + 27$
71. $2x - w$ **73.** -3 **75.** $3x - 2$ **77.** $P = 2x + 2(x - 5) = 4x - 10$
79. Value in cents $= 5x + 10(x - 5) + 25[(x - 5) + 2] = 40x - 125$ **81.** $x + 2x + 2x + 3x = 8x$
83. $8t + 12 \cdot 2t = 32t$

EXERCISE 2-2

1. y^5 **3.** $10y^5$ **5.** $-24x^{20}$ **7.** $6u^{16}$ **9.** c^3d^4 **11.** $15x^2y^3z^5$ **13.** $y^2 + 7y$
15. $10y^2 - 35y$ **17.** $3a^5 + 6a^4$ **19.** $2y^3 + 4y^2 - 6y$ **21.** $7m^6 - 14m^5 - 7m^4 + 28m^3$
23. $10u^4v^3 - 15u^2v^4$ **25.** $2c^3d^4 - 4c^2d^4 + 8c^4d^5$ **27.** $6y^3 + 19y^2 + y - 6$
29. $m^3 - 2m^2n - 9mn^2 - 2n^3$ **31.** $6m^4 + 2m^3 - 5m^2 + 4m - 1$ **33.** $a^3 + b^3$
35. $2x^4 + x^3y - 7x^2y^2 + 5xy^3 - y^4$ **37.** $x^2 + 5x + 6$ **39.** $a^2 + 4a - 32$ **41.** $t^2 - 16$
43. $m^2 - n^2$ **45.** $4t^2 - 11t + 6$ **47.** $3x^2 - 7xy - 6y^2$ **49.** $4m^2 - 49$ **51.** $30x^2 - 2xy - 12y^2$
53. $6s^2 - 11st + 3t^2$ **55.** $x^4 + x^3y + xy^3 + y^4$ **57.** $2x^3 + 6xy - x^2y^2 - 3y^3$ **59.** $9x^2 + 12x + 4$
61. $4x^2 - 20xy + 25y^2$ **63.** $36u^2 + 60uv + 25v^2$ **65.** $4m^2 - 20mn + 25n^2$
67. $x^3 + 6x^2y + 12xy^2 + 8y^3$ **69.** $-x^2 + 17x - 11$ **71.** $2x^3 - 13x^2 + 25x - 18$
73. $9x^3 - 9x^2 - 18x$ **75.** $12x^5 - 19x^3 + 12x^2 + 4x - 3$ **77.** $9x^2 + 6xy + y^2 - 12x - 4y + 4$
79. $x^2 + 2xy + y^2 - x - y - 2$ **81.** $m + n$ **83.** Area $= y(y - 8) = y^2 - 8y$

EXERCISE 2-3

1. $3x(z - 2)$ **3.** $2(4x + y)$ **5.** $3x(2x + 3)$ **7.** $7xy(2x - y)$ **9.** $(x - 3)(2x + z)$
11. $(a + b)(c - d)$ **13.** $(x - y)^2$ **15.** $(c + d)(ab - 1)$ **17.** $x^3(x^2 + x + 1)$
19. $a^2b(1 - ab - a^2)$ **21.** $c(ab + bd + de)$ **23.** $xyz(yz^2 - xy^2 + x^2z)$ **25.** $(a - 2)(a - 3)$
27. $(x - 1)(x + 1)$ **29.** $6(x - 2)$ **31.** $-3(x - 1)$ **33.** $(x^2 + 3)(2x + 1)$ **35.** $(ab - c^2)(c - ab)$
37. $(3x - y)(y + 2)$ **39.** $(x^2 + 3)(2x + 1)$ **41.** $(ab - c^2)(c - ab)$ **43.** $(3x - y)(y + 2)$
45. $(x^2 + 3)(2x^2 - 3)$ **47.** $(x^2y^2 + 3)(x - y)$ **49.** $(b - a)(4a - b)$ **51.** $(xyz - x)(y - z)$
53. $(2 + 3z)(x - 3y)$ **55.** $(a + c)(3a + b)$ **57.** $(4 - y^2)(3 - x)$ **59.** $(a + d)(b + c)$
61. $(z - y)(y - 2x)$

EXERCISE 2-4

1. $(x + 1)(x + 2)$ **3.** $(x + 2)(x + 5)$ **5.** $(x - 3)(x + 1)$ **7.** $(x + 4)(x - 1)$ **9.** $(x - 4)(x - 5)$
11. $(x - 1)(x - 5)$ **13.** $(x - 3)^2$ **15.** $(x + 7)^2$ **17.** $(x - 3y)^2$ **19.** $(2x + 1)(x - 1)$
21. $(2x + 5)(x + 1)$ **23.** Not factorable **25.** Not factorable **27.** $(2x - 5y)(x - 2y)$
29. $(2x + 3)(2x - 1)$ **31.** $(4x - 1)(x + 1)$ **33.** Not factorable **35.** $(2x + 3)^2$
37. $(x - 2)(x + 1)$ **39.** $(3x + 2)^2$ **41.** $(3x - 2)(2x - 1)$ **43.** $(3x + y)(2x - y)$ **45.** $(3x + y)^2$
47. Not factorable **49.** Not factorable **51.** $(4x + 3)(2x - 5)$ **53.** Not factorable
55. $3(2x - y)(x + 4y)$ **57.** $(x^2 + 1)(x^2 + 3)$ **59.** $(2x^2 - 3)(x^2 + 1)$ **61.** $(x^3 - 2)(x^3 + 3)$

EXERCISE 2-5

1. $(4x - 1)(x + 3)$ **3.** $(2x + 3)(2x - 1)$ **5.** Not factorable **7.** $(4x - 1)(x - 6)$
9. $(6x - 1)(x - 4)$ **11.** $(2x + 5)(2x - 3)$ **13.** Not factorable **15.** $(6x - 5)(x - 4)$
17. $(2x + 3)(3x - 1)$ **19.** $(2x + 5)(3x - 4)$ **21.** $(3x + 5)(2x - 3)$ **23.** Not factorable
25. $(x - 2y)(4x - 5y)$ **27.** Not factorable **29.** $2(3x - 2y)(x + 3y)$ **31.** $(4x + 5y)(3x + y)$
33. Not factorable **35.** $(4x + 3y)(x - 5y)$ **37.** Not factorable **39.** $(9x - 4)(2x + 5)$
41. $(3x + 4y)(6x - 5y)$ **43.** $-17, -7, -3, 3, 7, 17$ **45.** $0, 4, 6$ **47.** $(2x^2 + 1)(3x^2 + 2)$
49. $(x^2 + y^2)(2x^2 + 3y^2)$ **51.** $(2x^2 - 3)(x^3 + 1)$

EXERCISE 2-6

1. $(v-5)(v+5)$ **3.** $(3x-2)(3x+2)$ **5.** Not factorable **7.** $(3x-4y)(3x+4y)$
9. $(x+1)(x^2-x+1)$ **11.** $(m-n)(m^2+mn+n^2)$ **13.** $(2x+3)(4x^2-6x+9)$ **15.** $3uv^2(2u-v)$
17. $2(x-2)(x+2)$ **19.** $2x(x^2+4)$ **21.** $3x(2x-y)(2x+y)$ **23.** $2x(x+1)(x^2-x+1)$
25. $6(x+2)(x+4)$ **27.** $3x(x^2-2x+5)$ **29.** $(xy-4)(xy+4)$ **31.** $(ab+2)(a^2b^2-2ab+4)$
33. $2xy(2x+y)(x+3y)$ **35.** $4(u+2v)(u^2-2uv+4v^2)$ **37.** $5y^2(6x+y)(2x-7y)$
39. $(y+2)(x+y)$ **41.** $(x-5)(x+y)$ **43.** $(a-2b)(x-y)$ **45.** $(3c-4d)(5a+b)$
47. $(x-2)(x+1)(x-1)$ **49.** $(y-x)[(y-x)-1]=(y-x)(y-x-1)$ **51.** $(xy+2)(xy-3)$
53. $(z^2-3)(z^2+2)$ **55.** $(x^4+2)(x^4-2)$ **57.** $(r^2+s^2)(r-s)(r+s)$
59. $(x^2-4)(x^2+1)=(x-2)(x+2)(x^2+1)$ **61.** $[(x-3)-4y][(x-3)+4y]=(x-3-4y)(x-3+4y)$
63. $[(a-b)-2(c-d)][(a-b)+2(c-d)]$ **65.** $[5(2x-3y)-3ab][5(2x-3y)+3ab]$
67. $(x-1)(x^2+x+1)(x+1)(x^2-x+1)$ **69.** $(2x-1)(x-2)(x+2)$ **71.** $[5-(a+b)][5+(a+b)]$
73. $[4x^2-(x-3y)][4x^2+(x-3y)]$ **75.** $(x-2)(x^2+3)$ **77.** $(x^4+1)(x-1)$
79. $(3x-1)(x^2+4)$ **81.** $(x+2+y)(x+2-y)$

EXERCISE 2-7

1. $0, -5$ **3.** $0, -4$ **5.** $12, -1$ **7.** $1, -5$ **9.** $-\frac{2}{3}, 4$ **11.** $-1, 3$ **13.** $-\frac{2}{3}, 1$
15. Not factorable in the integers **17.** $-\frac{1}{2}, 3$ **19.** $-2, 2$ **21.** $-6, 2$ **23.** $-\frac{1}{2}, 2$ **25.** $\frac{1}{2}, 2$
27. $2, -3$ **29.** 11×3 in. **31.** $3, -3$ **33.** $1, 2, -2$ **35.** $3, -3$ **37.** $-2, 1$
39. $-1, 1$ **41.** $-1, 1$

REVIEW EXERCISE 2-8

1. x^{11} *(2-1)* **2.** $10x^7$ *(2-1)* **3.** **(A)** 5 **(B)** 3 *(2-1)* **4.** **(A)** 5 **(B)** 11 *(2-2)*
5. x^2+2x+1 *(2-1)* **6.** $-x^2+2x+9$ *(2-1)* **7.** $2x^3+5x^2-8x-20$ *(2-2)*
8. x^2+5x-3 *(2-1)* **9.** $5x^2-x+5$ *(2-2)* **10.** $6x^4-5x^3+8x^2+5x+4$ *(2-2)*
11. $2x^2-2$ *(2-1)* **12.** $2x^3-4x^2+12x$ *(2-1)* **13.** $3x^3(x+3)$ *(2-3)* **14.** $x^2y(2y^2+x)$ *(2-3)*
15. $(x-5)(2x+1)$ *(2-3)* **16.** $(x+3)(4x-1)$ *(2-4)* **17.** $(2x+5)(2x-5)$ *(2-6)*
18. $(x+y)(x+2y)$ *(2-4)* **19.** Not factorable *(2-6)* **20.** $(x+2)(x+2)$ or $(x+2)^2$ *(2-3)*
21. $(2x+1)(x-3)$ *(2-4)* **22.** $(x-3)(x-3)$ or $(x-3)^2$ *(2-3)* **23.** $(3x+2)(x-1)$ *(2-3)*
24. $(a-2)(a^2+2a+4)$ *(2-6)* **25.** $(x+2)(x+4)$ *(2-4)* **26.** $(x-2y)(2x-y)$ *(2-4)*
27. $(x+4)(x^2-4x+16)$ *(2-6)* **28.** $(a-3b)(a+3b)$ *(2-6)* **29.** $-4, -3$ *(2-7)*
30. $-2, +2$ *(2-7)* **31.** $(x^2+1)(x^4-x^2+1)$ *(2-6)* **32.** $x^2(2x+1)(x+3)$ *(2-6)*
33. $a^3(x-2)^2$ *(2-6)* **34.** $3xy(x+y)(x-y)$ *(2-6)* **35.** $(x^2+1)(x-3)$ *(2-6)*
36. $x(x^4+1)(x+1)$ *(2-6)* **37.** $2x(x+1)(x-2)$ *(2-6)* **38.** $(x+y+1)(x+y-1)$ *(2-6)*
39. $(x^2+1)(x-1)(x+1)$ *(2-6)* **40.** $(2x-5)(4x^2+10x+25)$ *(2-6)* **41.** $(3x+2)(x^2-5)$ *(2-6)*
42. $x^2(x+3)(x-1)$ *(2-6)* **43.** $-x^3y(xy+1)(xy+1)$ *(2-6)* **44.** $2a(a+1)(a^2-a+1)$ *(2-6)*
45. $-1, \frac{3}{2}$ *(2-7)* **46.** $-4, 5$ *(2-7)*

CHAPTER 3 EXERCISE 3-1

1. $\dfrac{1}{2x^2}$ **3.** $\dfrac{2x^2}{3y}$ **5.** $\dfrac{3(x-9)}{y}$ **7.** $\dfrac{2x-1}{3x}$ **9.** $\dfrac{x}{2}$ **11.** $\dfrac{1}{n}$ **13.** $12xy$ **15.** $14x^3y$

17. $\frac{x+2}{3x}$ **19.** $\frac{x-3}{x+3}$ **21.** $\frac{2x-3y}{2xy}$ **23.** $\frac{x+2}{x+y}$ **25.** $\frac{x+5}{2x}$ **27.** $\frac{x^2+2x+4}{x+2}$
29. $\frac{x+3}{x-2}$ **31.** $\frac{2x}{y}$ **33.** $\frac{x^2+2}{x-2}$ **35.** $3x^2+3xy$ **37.** x^2-y^2 **39.** $\frac{x-y}{3x}$ **41.** $\frac{x-y}{2x+y}$
43. $\frac{x^2+y^2}{(x+y)^2}$ **45.** $x+1$ **47.** $\frac{x^2+y^2}{x^3+y^3}$ **49.** $\frac{x+y}{u+v}$

EXERCISE 3-2

1. $\frac{8}{9}$ **3.** $\frac{6}{b}$ **5.** $\frac{y}{x}$ **7.** $\frac{3}{2}$ **9.** $\frac{3c}{a}$ **11.** $\frac{x}{9y^2}$ **13.** $\frac{16xy}{3}$ **15.** $\frac{9xy}{8c}$ **17.** $\frac{-45u^2}{16v^2}$
19. $\frac{c^3d^2}{a^6b^6}$ **21.** $\frac{x}{2}$ **23.** $\frac{x}{x-3}$ **25.** $\frac{3y}{x+3}$ **27.** $\frac{1}{2y}$ **29.** $t(t-4)$ **31.** $\frac{1}{m}$
33. $-x(x-2)$ or $2x-x^2$ **35.** $\frac{a^2}{2}$ **37.** 2 **39.** -1 **41.** $\frac{(x-y)^2}{y^2(x+y)}$ **43.** $x=3$
45. $\frac{R}{S}\cdot\left(\frac{P}{Q}\cdot\frac{S}{R}\right)=\frac{\not{R}P\not{S}}{\not{S}Q\not{R}}=\frac{P}{Q}$

EXERCISE 3-3

1. $3x$ **3.** x **5.** v^3 **7.** $12x^2$ **9.** $(x+1)(x-2)$ **11.** $3y(y+3)$ **13.** $\frac{7x+2}{5x^2}$ **15.** 2
17. $\frac{1}{y+3}$ **19.** $\frac{3-2x}{k}$ **21.** $\frac{12x+y}{4y}$ **23.** $\frac{2+y}{y}$ **25.** $\frac{u^3+uv-v^2}{v^3}$ **27.** $\frac{9x^2+8x-2}{12x^2}$
29. $\frac{5x-1}{(x+1)(x-2)}$ **31.** $\frac{7y-6}{3y(y+3)}$ **33.** $24x^3y^2$ **35.** $75x^2y^2$ **37.** $18(x-1)^2$
39. $24(x-7)(x+7)^2$ **41.** $(x-2)(x+2)^2$ **43.** $12x^2(x+1)^2$ **45.** $\frac{8v-6u^2v^2+3u^3}{36u^3v^3}$
47. $\frac{15t^2+14t-6}{36t^3}$ **49.** $\frac{2}{t-1}$ **51.** $\frac{5a^2-2a-5}{(a+1)(a-1)}$ **53.** $\frac{5x+55}{12(x-5)^2(x+5)}$ **55.** $\frac{15x-11}{18(x-1)^2}$
57. $\frac{-4}{(x-1)(x+3)}$ **59.** $\frac{2s^2+s-2}{2s(s-2)(s+2)}$ **61.** $\frac{2(x+4)}{(x-2)(x+2)^2}$ **63.** $\frac{3}{x+3}$ **65.** $\frac{7}{y-3}$
67. -1 **69.** $\frac{-17}{15(x-1)}$ **71.** $\frac{x+3}{(x-2)(x+7)}$ **73.** $\frac{(3x+1)(x+3)}{12x^2(x+1)^2}$ **75.** $\frac{xy^2-xy+y^2}{x^3-y^3}$

EXERCISE 3-4

1. $3x+1$ **3.** $2y^2+y-3$ **5.** $3x+1, R=3$ **7.** $4x-1$ **9.** $3x-4, R=-1$ **11.** $x+2$
13. $x^2+4x+11, R=26$ **15.** $x^2+x+2, R=2$ **17.** $2x^2+5x+16, R=46$
19. $2x^2-3x+4, R=-6$ **21.** $x^3+5x^2+23x+95, R=385$ **23.** $x^3-x^2+5x-7, R=19$
25. $4x+1, R=-4$ **27.** $4x+6, R=25$ **29.** $x-4, R=3$ **31.** x^2+x+1
33. $x^3+3x^2+9x+27$ **35.** $4a+5, R=-7$ **37.** x^2+3x-5 **39.** $x^2+3x+8, R=27$
41. $x^2+2x+2, R=8$ **43.** $x^2-x-1, R=5$ **45.** $x^3+4x^2+16x+61, R=249$

47. $x^3 - 2x^2 + 4x - 11, R = 27$ **49.** $3x^3 + x^2 - 2, R = -4$ **51.** $4x^2 - 2x - 1, R = -2$
53. $2x^3 + 6x^2 + 32x + 84, R = 186x - 170$ **55.** $Q = x - 3, R = x + 2$ **57.** 170 **59.** −2
61. −5 **63.** −0.389 000 **65.** −1.234 625 **67.** −43.817 000 **69.** 5.297 889

EXERCISE 3-5

1. $\frac{3}{4}$ **3.** $\frac{9}{10}$ **5.** $\frac{8}{13}$ **7.** $\frac{22}{51}$ **9.** xy **11.** $\frac{3xy}{2}$ **13.** $\frac{1}{x-3}$ **15.** $\frac{x+y}{x}$
17. $\frac{1}{y-x}$ **19.** $\frac{x-y}{x+y}$ **21.** 1 **23.** $-\frac{1}{2}$ **25.** $\frac{1}{1-x}$ **27.** $-x$ **29.** $\frac{3x+5}{2x+3}$
31. $r = \frac{2r_R r_G}{r_R + r_G}$

EXERCISE 3-6

1. 13 **3.** 8 **5.** −6 **7.** $-\frac{1}{12}$ **9.** 30 **11.** 20 **13.** 10 **15.** 3 **17.** $-\frac{7}{4}$
19. 3 **21.** 10 **23.** −9 **25.** 4 **27.** No solution **29.** 5 **31.** $\frac{53}{11}$ **33.** No solution
35. 1 **37.** $\frac{31}{24}$ **39.** −4 **41.** No solution **43.** $\frac{2}{3}$ **45.** −2 **47.** 4, −3 **49.** No solution

REVIEW EXERCISE 3-7

1. $\frac{3x^2}{2(z+3)^2}$ *(3-1)* **2.** $\frac{x+1}{x-1}$ *(3-1)* **3.** $\frac{3x+2}{3x}$ *(3-3)* **4.** $\frac{2x+11}{6x}$ *(3-3)*
5. $\frac{2-9x^2-8x^3}{12x^3}$ *(3-3)* **6.** $\frac{2xy}{ab}$ *(3-2)* **7.** $2(x+1)$ *(3-3)* **8.** $\frac{2}{m+1}$ *(3-3)*
9. $\frac{x+7}{(x-2)(x+1)}$ *(3-3)* **10.** $\frac{(d-2)^2}{d+2}$ *(3-2)* **11.** $\frac{-1}{(x+2)(x+3)}$ *(3-3)* **12.** $\frac{3}{8}$ *(3-5)*
13. $\frac{11}{6}$ *(3-5)* **14.** $\frac{y-2}{y+1}$ *(3-5)* **15.** 9 *(3-6)* **16.** $-\frac{10}{9}$ *(3-6)* **17.** 60 *(3-6)*
18. −12 *(3-6)* **19.** 5 *(3-6)* **20.** No solution *(3-6)* **21.** 11 *(3-6)* **22.** −2 *(3-6)*
23. −5 *(3-6)* **24.** $-\frac{3}{5}$ *(3-6)* **25.** No solution *(3-6)* **26.** $\frac{3}{4}$ *(3-6)*
27. $x^2 - 2x - 1, R = -4$ *(3-4)* **28.** $x, R = 0$ *(3-4)* **29.** $x^2 + 2x + 1, R = 3$ *(3-4)*
30. $x^3 - 2x^2 + 5x - 10, R = 19$ *(3-4)* **31.** $x^3 + x^2 + x + 1, R = 0$ *(3-4)*
32. $x^2 - 1, R = 2x + 1$ *(3-4)* **33.** $\frac{12a^2b^2 - 40a^2 - 5b}{30a^2b^3}$ *(3-3)* **34.** $\frac{5-2x}{2x-3}$ *(3-3)*
35. $\frac{2y^4}{9a^4}$ *(3-2)* **36.** $\frac{5x-12}{3(x-4)(x+4)}$ *(3-2)* **37.** $\frac{x}{x+1}$ *(3-2)* **38.** $\frac{x-y}{x}$ *(3-5)*
39. $\frac{y}{x^3-y^3}$ *(3-3)* **40.** $\frac{x}{y(x+y)}$ *(3-5)* **41.** $x+1$ *(3-2)* **42.** $\frac{x^2+24x-9}{12x(x-3)(x+3)^2}$ *(3-3)*
43. $\frac{-1}{s+2}$ *(3-3)* **44.** −1 *(3-2)* **45.** $\frac{y^2}{x}$ *(3-2)* **46.** $\frac{x-y}{x+y}$ *(3-5)*

47. $\frac{(x+1)(x-2)}{2x}$ *(3-5)* **48.** $-\frac{13}{5}$ *(3-6)* **49.** -15 *(3-6)* **50.** No solution *(3-6)*
51. All real numbers except 0 and -1 *(3-6)*

CHAPTER 4 EXERCISE 4-1

1. $\frac{3}{2}$ **3.** $\frac{5}{2}$ **5.** $\frac{25}{2}$ **7.** 20 **9.** 7 **11.** $(2x-3)+x=12$; 5 ft and 7 ft
13. $2x+2(x-6)=36$; 12×6 ft **15.** 60 quarters **17.** 40 m **19.** 162 km
21. $2(2x+3)+2x=66$; 23×10 cm **23.** $\frac{x}{6}-2=\frac{x}{4}+1$; -36 **25.** $\frac{x}{52}=\frac{9}{46}$; 10.17 ml
27. $x+(x+2)=(x+4)+5$; 7, 9, 11 **29.** $\frac{x}{4}=\frac{1}{1.06}$; 3.77 liters **31.** $\frac{x}{1}=\frac{1}{0.62}$; 1.61 km
33. $2x+2\cdot\frac{x}{6}=84$; 36×6 m **35.** $\frac{x}{500}=\frac{240}{200}$; \$600 **37.** $\frac{x}{23}=\frac{10}{35}$; 6.57 in.
39. $\frac{3x}{5}-4=\frac{x}{3}+8$; 45 **41.** $\frac{x}{5}=\frac{1}{0.26}$; 19.23 liters **43.** $\frac{2P}{5}+70+\frac{P}{4}=P$; 200 cm
45. $\frac{x}{300}=\frac{250}{25}$; 3,000 trout **47.** $\frac{D}{3}+6+\frac{D}{2}=D$; 36 km **49.** 3 kg

EXERCISE 4-2

1. 5 hr **3.** 6 hr **5.** 65 min **7.** 5 hr **9.** 71.25 min **11.** 3.43 hr **13.** 7.5 days
15. 670 m **17.** 180 cm **19.** 85.8 ft **21.** Solve $\frac{N}{600}=\frac{500}{60}$; $N=5{,}000$ chipmunks
23. 205 points **9.** \$6,400 **25.** **(A)** 216 mi **(B)** 225 mi
27. **(A)** 15 in. **(B)** 20 in. **(C)** 22.5 in. **(D)** 24 in. **(E)** 25 in. **(F)** 18 in. **(G)** 18.75 in.
29. Solve $x-0.2x=160$; $x=\$200$ **31.** 5,300 copies **33.** 8 sec **35.** 7 cm^2
37. Solve $20x=50(180)$; $x=450$ kg **39.** 200,000 mi/sec **41.** Solve $\frac{x}{5}+4+\frac{2x}{3}=x$; $x=30$ m
43. 15 min or $\frac{1}{4}$ hr

EXERCISE 4-3

1. $h=\frac{2A}{b}$ **3.** $a=\frac{2A-bh}{h}$ **5.** $x=\frac{y-7}{3}$ **7.** $\theta=\frac{180l}{\pi r}$ **9.** $t=\frac{S-P}{dS}$ **11.** $a=\frac{3V}{4\pi b^2}$
13. $p=\frac{Pq}{1-P}$ **15.** $I=\frac{ЯR}{ERA}$ **17.** $v=\frac{Ft+mv_0}{m}$ **19.** $V=\frac{QP-y-F}{Q}$ **21.** $f=\frac{ab}{a+b}$
23. $D=\frac{Q}{C_0}\left(C-\frac{QC_h}{2}\right)$ **25.** $T_2=\frac{T_1P_2V_2}{P_1V_1}$ **27.** $x=\frac{5y-3}{3y-2}$

EXERCISE 4-4

1. $-8 \le x \le 7$ 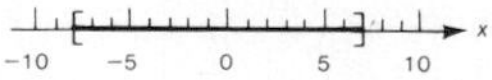 **3.** $-6 \le x < 6$
5. $x \ge -6$ 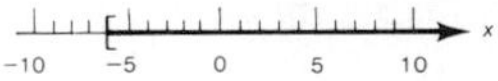 **7.** $(-2, 6]$ 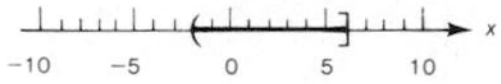
9. $(-7, 8)$ 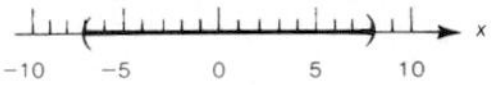 **11.** $(-\infty, -2]$ 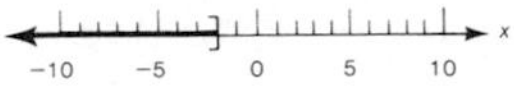
13. $[-7, 2)$; $-7 \le x < 2$ **15.** $(-\infty, 0]$; $x \le 0$
17. $x < 5$ or $(-\infty, 5)$ **19.** $x \ge 3$ or $[3, \infty)$
21. $N < -8$ or $(-\infty, -8)$ **23.** $t > 2$ or $(2, \infty)$ 
25. $m > 3$ or $(3, \infty)$ 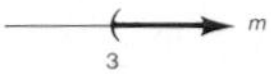 **27.** $B \ge -4$ or $[-4, \infty)$
29. $-2 < t \le 3$ or $(-2, 3]$ **31.** $2x - 3 \ge -6$; $x \ge \frac{-3}{2}$
33. $15 - 3x < 6$; $x > 3$ **35.** $q < -14$ or $(-\infty, -14)$ **37.** $x \ge 4.5$ or $[4.5, \infty)$
39. $-20 \le x \le 20$ or $[-20, 20]$ 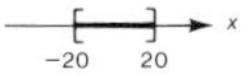 **41.** $-30 \le x < 18$ or $[-30, 18)$
43. $-8 \le x < -3$ or $[-8, -3)$ **45.** $-14 < x \le 11$ or $(-14, 11]$

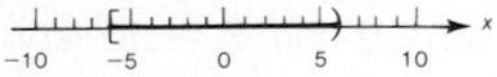

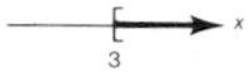

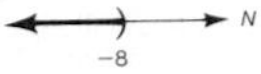

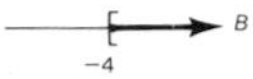

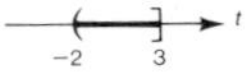

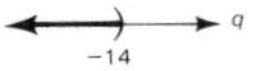

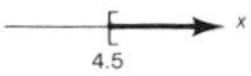

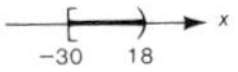

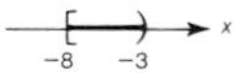

47. Positive **49.** **(A)** F **(B)** T **(C)** T **51.** $8{,}000 \le h \le 20{,}000$ or $[8{,}000, 20{,}000]$ **53.** $x > 600$
55. If r is the worker's maximum running rate and R is the train's rate, then he will escape running toward the train if $r > R/3 = 7R/21$, and he will escape running away from the train if $r > 3R/7 = 9R/21$. Thus, his chances are better if he runs toward the train!

EXERCISE 4-5

1. $\sqrt{5}$ **3.** 4 **5.** $5 - \sqrt{5}$ **7.** $5 - \sqrt{5}$ **9.** 12 **11.** 12 **13.** 9 **15.** 4 **17.** 4
19. 9 **21.** $x = \pm 7$  **23.** $-7 \le x \le 7$ 
25. $x \le -7$ or $x \ge 7$  **27.** $y = 2$ or 8 **29.** $2 < y < 8$
31. $y < 2$ or $y > 8$ 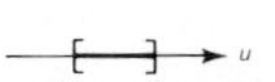 **33.** $u = -11$ or -5 
35. $-11 \le u \le -5$ 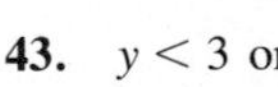 **37.** $u \le -11$ or $u \ge -5$ 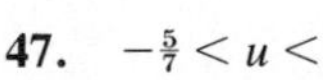 **39.** $x = -4, \frac{4}{3}$
41. $-\frac{9}{5} \le x \le 3$ **43.** $y < 3$ or $y > 5$ **45.** $t = -\frac{4}{5}, \frac{18}{5}$ **47.** $-\frac{5}{7} < u < \frac{23}{7}$
49. $x \le -6$ or $x \ge 9$ **51.** $-35 < C < -\frac{5}{9}$ **53.** $x \ge 5$ **55.** $x \le -8$ **57.** $x \ge -\frac{3}{4}$ **59.** $x \le \frac{2}{5}$
61. Case 1: $a = b$; $|b - a| = |0| = 0$; $|a - b| = |0| = 0$
Case 2: $a > b$; $|b - a| = -(b - a) = a - b$
$|a - b| = a - b$
Case 3: $b > a$; $|b - a| = b - a$
$|a - b| = -(a - b) = b - a$

REVIEW EXERCISE 4-6

1. $x = -2$ *(4-1)* **2.** $x = 2$ *(4-1)* **3.** $x < -2$ *(4-4)* **4.** $1 < x < 6$ *(4-4)*
5. $x = \pm 6$ *(4-5)* **6.** $-6 < x < 6$ *(4-5)* **7.** $x < -6$ or $x > 6$ *(4-5)* **8.** $-14, -4$ *(4-5)*
9. $-14 < y < -4$ *(4-5)* **10.** $y < -14$ or $y > -4$ *(4-5)* **11.** $x \le -12$ *(4-4)*
12. $b = \frac{2A}{h}$ *(4-3)* **13.** **(A)** 6 **(B)** 6 *(4-5)* **14.** $-4 \le x < 3$ *(4-4)*

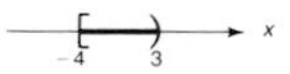

15. $1 < x \le 4$ *(4-4)* **16.** $x \ge 1$ *(4-4)* **17.** 41 *(4-1)*

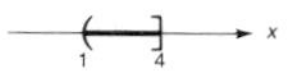

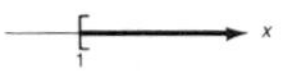

18. $\frac{1}{2}, 3$  *(4-5)* **19.** $\frac{1}{2} \le x \le 3$ 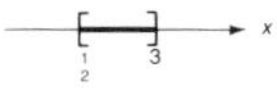 *(4-5)*

20. $x < \frac{1}{2}$ or $x > 3$ 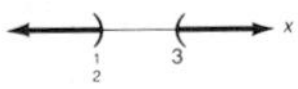 *(4-5)* **21.** $x \ge -19$  *(4-4)*

22. $-6 < x \le -1$ *(4-4)* **23.** $L = \frac{2S}{n} - a$ or $L = \frac{2S - an}{n}$ *(4-3)*

24. $M = \frac{P}{1 - dt}$ *(4-3)* **25.** **(A)** T **(B)** T *(4-4)*
26. $-1 \le x \le 4$ or $[-1, 4]$  *(4-5)* **27.** $x \ge \frac{25}{7}$ 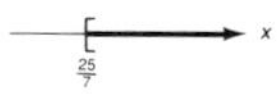 *(4-4)*

28. $-3 \le x \le 6$ 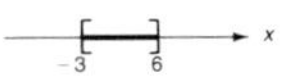 *(4-4)* **29.** No solution *(4-5)* **30.** $x = \frac{5y + 3}{2y - 4}$ *(4-3)*

31. $f_1 = \frac{ff_2}{f_2 - f}$ *(4-3)* **32.** $x \ge \frac{3}{2}$ *(4-3)* **33.** $x \le \frac{3}{2}$ *(4-3)*

34. Solve $2x + 2\left(\frac{3x}{5} - 2\right) = 76$; 25 by 13 cm *(4-2)* **35.** Solve $56(x + 1.5) = 76x$; $x = 4.2$ hr *(4-2)*

36. Solve $x - 0.3x = 210$; $x = \$300$ *(4-2)* **37.** Solve $45x + 55(x - 10) = 3{,}000$; $x = 35.5$ min *(4-2)*

38. Solve $\frac{x}{127} = \frac{1}{2.54}$; $x = 50$ in. *(4-1)* **39.** Solve $\frac{x}{70} = \frac{18}{50}$; $x = 25.2$ ml of alcohol *(4-1)*

40. Solve $\frac{65 + 80 + x}{3} \ge 75$; $x \ge 80$ *(4-2)* **41.** Solve $\frac{x}{360} = \frac{55}{6}$; $x = 3{,}300$ squirrels *(4-1)*

42. Solve $10 \le \frac{5}{9}(F - 32) \le 15$; $50° \le F \le 59°$ *(4-4)*

CHAPTER 5 — EXERCISE 5-1

1. 10 **3.** 6 **5.** 12 **7.** 2 **9.** u^7v^7 **11.** 4 **13.** $\frac{a^8}{b^8}$ **15.** 3 **17.** 6 **19.** 2

21. 7 **23.** 12 **25.** $8x^{12}$ **27.** $3x^2$ **29.** $\frac{3}{4m^2}$ **31.** $x^{10}y^{10}$ **33.** $\frac{m^5}{n^5}$ **35.** $20y^{11}$

37. 48×10^{17} **39.** 10^{14} **41.** x^6 **43.** m^6n^{15} **45.** $\frac{c^6}{d^{15}}$ **47.** $\frac{3u^4}{v^2}$ **49.** $2^4s^8t^{16}$ or $16s^8t^{16}$

51. $6x^5y^{15}$ **53.** $\frac{m^4n^{12}}{p^8q^4}$ **55.** $\frac{u^3}{v^9}$ **57.** $9x^4$ **59.** -1 **61.** $\frac{-1}{x^5}$ **63.** $-\frac{wy}{x^3}$

65. $\frac{(x - y)^2}{2(x + y)^2}$ **67.** x^8 **69.** x^n **71.** x^{2n+2} **73.** $\frac{u^2}{v^4}$

EXERCISE 5-2

1. 1 **3.** 1 **5.** $\frac{1}{3^3}$ **7.** $\frac{1}{m^7}$ **9.** 4^3 **11.** y^5 **13.** 10^2 **15.** y **17.** 1 **19.** 10^{10}
21. x^{11} **23.** $\frac{1}{z^5}$ **25.** $\frac{1}{10^7}$ **27.** 10^{12} **29.** y^8 **31.** $u^{10}v^6$ **33.** $\frac{x^4}{y^6}$ **35.** $\frac{x^2}{y^3}$ **37.** 1
39. 10^2 **41.** y **43.** 10 **45.** 3×10^{16} **47.** y^9 **49.** $3^2m^2n^2$ **51.** $\frac{2^3m^3}{n^9}$ **53.** $\frac{n^{15}}{m^{12}}$
55. $\frac{3^3}{2^2}$ **57.** 1 **59.** $\frac{4y^3}{3x^5}$ **61.** $\frac{a^9}{8b^4}$ **63.** $\frac{1}{x^7}$ **65.** $\frac{n^8}{m^{12}}$ **67.** $\frac{m^3n^3}{8}$ **69.** $\frac{t^2}{x^2y^{10}}$
71. 4 **73.** $\frac{1}{a^2 - b^2}$ **75.** $\frac{1}{xy}$ **77.** $-cd$ **79.** $\frac{xy}{x + y}$ **81.** $\frac{(y - x)^2}{x^2y^2}$ **83.** $\frac{y - x}{y}$

EXERCISE 5-3

1. 7×10 **3.** 8×10^2 **5.** 8×10^4 **7.** 8×10^{-3} **9.** 8×10^{-8} **11.** 5.2×10
13. 6.3×10^{-1} **15.** 3.4×10^2 **17.** 8.5×10^{-2} **19.** 6.3×10^3 **21.** 6.8×10^{-6} **23.** 800
25. 0.04 **27.** 300,000 **29.** 0.0009 **31.** 56,000 **33.** 0.0097 **35.** 430,000
37. 0.000 000 38 **39.** 5.46×10^9 **41.** 7.29×10^{-8} **43.** 10^{13} **45.** 10^{-5}
47. 83,500,000,000 **49.** 0.000 000 000 006 14 **51.** 865,000
53. 0.000 000 000 000 000 000 000 001 7 **55.** 9×10^4 **57.** 6×10^{-4} **59.** 3×10^5
61. 5×10^4 **63.** 3×10 or 30 **65.** 3×10^{-4} or 0.0003 **67.** 6.6×10^{21} tons **69.** 10^7; 6×10^8
71. 562

EXERCISE 5-4

1. 5 **3.** Not a real number **5.** 2 **7.** −2 **9.** −2 **11.** 64 **13.** 4 **15.** x
17. $\frac{1}{x^{1/5}}$ **19.** x^2 **21.** ab^3 **23.** $\frac{x^3}{y^4}$ **25.** x^2y^3 **27.** $\frac{2}{5}$ **29.** $\frac{8}{125}$ **31.** $\frac{1}{4}$ **33.** $\frac{1}{6}$
35. $\frac{1}{125}$ **37.** 25 **39.** $\frac{1}{9}$ **41.** $\frac{1}{x^{1/2}}$ **43.** $n^{1/12}$ **45.** x^4 **47.** $\frac{2v^2}{u}$ **49.** $\frac{1}{x^2y^3}$ **51.** $\frac{x^4}{y^3}$
53. $\frac{5}{4}x^4y^2$ **55.** $64y^{1/3}$ **57.** $12m - 6m^{35/4}$ **59.** $2x + 3x^{1/2}y^{1/2} + y$ **61.** $x + 2x^{1/2}y^{1/2} + y$
63. Not defined **65.** $\frac{2}{a} + \frac{5}{a^{1/2}b^{1/2}} - \frac{3}{b}$ **67.** $a^{1/2}b^{1/3}$ **69.** x **71.** $\frac{1}{x^m}$
73. **(A)** Any negative number **(B)** n even and x any negative number

EXERCISE 5-5

1. $\sqrt{11}$ **3.** $\sqrt[3]{5}$ **5.** $\sqrt[5]{u^3}$ **7.** $4\sqrt[7]{y^3}$ **9.** $\sqrt[7]{(4y)^3}$ **11.** $\sqrt[5]{(4ab^3)^2}$ **13.** $\sqrt{a + b}$
15. $6^{1/2}$ **17.** $m^{1/4}$ **19.** $y^{3/5}$ **21.** $(xy)^{3/4}$ **23.** $(x^2 - y^2)^{1/2}$ **25.** $-5\sqrt[5]{y^2}$
27. $\sqrt[7]{(1 + m^2n^2)^3}$ **29.** $\frac{1}{\sqrt[3]{w^2}}$ **31.** $\frac{1}{\sqrt[5]{(3m^2n^3)^3}}$ **33.** $\sqrt{a} + \sqrt{b}$ **35.** $\sqrt[3]{(a^3 + b^3)^2}$

37. $(a + b)^{2/3}$ **39.** $-3x(a^3b)^{1/4}$ **41.** $(-2x^3y^7)^{1/9}$ **43.** $\frac{3}{y^{1/3}}$ or $3y^{-1/3}$

45. $\frac{-2x}{(x^2 + y^2)^{1/2}}$ or $-2x(x^2 + y^2)^{-1/2}$ **47.** $m^{2/3} - n^{1/2}$ **49.** $\sqrt{2^2 + 3^2} = \sqrt{13} \neq 2 + 3 = 5$

51. **(A)** Any negative number **(B)** n even and x any negative number

EXERCISE 5-6

1. y **3.** $2u$ **5.** $7x^2y$ **7.** $3\sqrt{2}$ **9.** $m\sqrt{m}$ **11.** $2x\sqrt{2x}$ **13.** $\frac{1}{3}$ **15.** $\frac{1}{y}$ **17.** $\frac{\sqrt{5}}{5}$

19. $\frac{\sqrt{5}}{5}$ **21.** $\frac{\sqrt{y}}{y}$ **23.** $\frac{\sqrt{y}}{y}$ **25.** $3xy^2\sqrt{xy}$ **27.** $3x^4y^2\sqrt{2y}$ **29.** $\frac{\sqrt{2x}}{2x}$ **31.** $2x\sqrt{3x}$

33. $\frac{3\sqrt{2ab}}{2b}$ **35.** $\frac{\sqrt{42xy}}{7y}$ **37.** $\frac{3m^2\sqrt{2mn}}{2n}$ **39.** $2x^2y$ **41.** $2xy^2\sqrt[3]{2xy}$ **43.** $\sqrt{x}$ **45.** 4

47. $6m^3n^3$ **49.** $2\sqrt[3]{9}$ **51.** $\frac{2a\sqrt{3ab}}{3b}$ **53.** Is in the simplest radical form **55.** $\frac{2x}{3y^2}$

57. $-3m^2n^2\sqrt[5]{3m^2n}$ **59.** $\sqrt[3]{x^2(x - y)}$ **61.** $x^2y\sqrt[3]{6xy}$ **63.** $2x^2y\sqrt[3]{4x^2y}$ **65.** $-\sqrt[3]{6x^2y^2}$

67. $\sqrt[3]{(x - y)^2}$ **69.** $\frac{\sqrt[4]{12x^3y^3}}{2x}$ **71.** $-x\sqrt{x^2 + 2}$ **73.** $4x^9y\sqrt[3]{2y}$ **75.** $mn\sqrt[12]{3^7m^5n^2}$

77. $x^n(x + y)^{n+2}$ **79.** **(A)** $6x$ **(B)** $-2x$ **81.** **(A)** $2x$ **(B)** 0 **83.** **(A)** x (B) $-5x$

EXERCISE 5-7

1. $9\sqrt{3}$ **3.** $-5\sqrt{a}$ **5.** $-5\sqrt{n}$ **7.** $4\sqrt{5} - 2\sqrt{3}$ **9.** $\sqrt{m} - 3\sqrt{n}$ **11.** $4\sqrt{2}$

13. $-6\sqrt{2}$ **15.** $7 - 2\sqrt{7}$ **17.** $3\sqrt{2} - 2$ **19.** $y - 8\sqrt{y}$ **21.** $4\sqrt{n} - n$ **23.** $3 + 3\sqrt{2}$

25. $3 - \sqrt{3}$ **27.** $9 + 4\sqrt{5}$ **29.** $m - 7\sqrt{m} + 12$ **31.** $\sqrt{5} - 2$ **33.** $\frac{\sqrt{5} - 1}{2}$

35. $\frac{\sqrt{5} + \sqrt{2}}{3}$ **37.** $\frac{y - 3\sqrt{y}}{y - 9}$ **39.** $8\sqrt{2mn}$ **41.** $6\sqrt{2} - 2\sqrt{5}$ **43.** $-\sqrt[5]{a}$ **45.** $5\sqrt[3]{x} - \sqrt{x}$

47. $\frac{9\sqrt{2}}{4}$ **49.** $\frac{-3\sqrt{6uv}}{2}$ **51.** $38 - 11\sqrt{3}$ **53.** $x - y$ **55.** $10m - 11\sqrt{m} - 6$

57. $5 + \sqrt[3]{18} + \sqrt[3]{12}$ **59.** $(3 - \sqrt{2})^2 - 6(3 - \sqrt{2}) + 7 = 9 - 6\sqrt{2} + 2 - 18 + 6\sqrt{2} + 7 = 0$

61. $-7 - 4\sqrt{3}$ **63.** $5 + 2\sqrt{6}$ **65.** $\frac{x + 5\sqrt{x} + 6}{x - 9}$ **67.** $\frac{6x + 9\sqrt{x}}{4x - 9}$ **69.** $3\sqrt{3}$ **71.** $\frac{10}{3}\sqrt[3]{9}$

73. $x + 2\sqrt[3]{xy} - \sqrt[3]{x^2y^2} - 2y$ **75.** $x + y$ **77.** $\frac{8x + 2\sqrt{xy} - 15y}{16x - 25y}$ **79.** $\frac{\sqrt[3]{x^2} - \sqrt[3]{x}\sqrt[3]{y} + \sqrt[3]{y^2}}{x + y}$

81. $\frac{(\sqrt{x} + \sqrt{y} + \sqrt{z})[(x + y - z) - 2\sqrt{xy}]}{(x + y - z)^2 - 4xy}$

EXERCISE 5-8

1. $8 + 3i$ **3.** $-5 + 3i$ **5.** $5 + 3i$ **7.** $6 + 13i$ **9.** $3 - 2i$ **11.** -15 or $-15 + 0i$
13. $-6 - 10i$ **15.** $15 - 3i$ **17.** $-4 - 33i$ **19.** 65 or $65 + 0i$ **21.** $\frac{2}{5} - \frac{1}{5}i$ **23.** $\frac{3}{13} + \frac{11}{13}i$
25. $5 + 3i$ **27.** $7 - 5i$ **29.** $-3 + 2i$ **31.** $8 + 25i$ **33.** $\frac{5}{3} - \frac{2}{3}i$ **35.** $\frac{2}{13} + \frac{3}{13}i$
37. $-\frac{2}{5}i$ or $0 - \frac{2}{5}i$ **39.** $\frac{3}{2} - \frac{1}{2}i$ **41.** $4 - 7i$ **43.** 0 or $0 + 0i$ **45.** 0
47. $-1, -i, 1, i, -1, -i, 1$ **49.** $(a + c) + (b + d)i$ **51.** $a^2 + b^2$ or $(a^2 + b^2) + 0i$
53. $(ab - bd) + (ad + bc)i$ **55.** 1 **57.** $\pm 6i$ **59.** $9 \pm 3i$ **61.** $-1, i, 1, -i, -1, i, 1$
63. For $x \geq 10$ **65.** 1

REVIEW EXERCISE 5-9

1. x^5 *(5-1)* **2.** x^3y^3 *(5-1)* **3.** $\dfrac{x^3}{y^3}$ *(5-1)* **4.** $\dfrac{1}{x^5}$ *(5-1)* **5.** x^{24} *(5-1)* **6.** 1 *(5-1)*
7. x^{11} *(5-1)* **8.** $-8x^3$ *(5-1)* **9.** $-6x^{11}$ *(5-1)* **10.** 1 *(5-2)* **11.** $\frac{1}{9}$ *(5-2)* **12.** 8 *(5-2)*
13. $\frac{1}{2}$ *(5-4)* **14.** Not a real number *(5-4)* **15.** 4 *(5-4)*
16. **(A)** 4.28×10^9 **(B)** 3.18×10^{-5} *(5-3)* **17.** **(A)** 729,000 **(B)** 0.000 603 *(5-3)*
18. $6x^4y^7$ *(5-2)* **19.** $\dfrac{3u^4}{v^2}$ *(5-2)* **20.** $6x^5y^{15}$ *(5-2)* **21.** $\dfrac{c^6}{d^{15}}$ *(5-2)* **22.** $\dfrac{4x^4}{9y^6}$ *(5-2)*
23. x^{12} *(5-2)* **24.** y^2 *(5-2)* **25.** $\dfrac{y^3}{x^2}$ *(5-2)* **26.** x^3 *(5-4)* **27.** $\dfrac{1}{x^2}$ *(5-4)*
28. $\dfrac{1}{x^{1/3}}$ *(5-4)* **29.** u *(5-4)* **30.** **(A)** $\sqrt{3m}$ **(B)** $3\sqrt{m}$ *(5-5)*
31. **(A)** $(2x)^{1/2}$ **(B)** $(a + b)^{1/2}$ *(5-5)* **32.** $2xy^2$ *(5-6)* **33.** $\dfrac{5}{y}$ *(5-6)* **34.** $6x^2y^3\sqrt{y}$ *(5-6)*
35. $\dfrac{\sqrt{2y}}{2y}$ *(5-6)* **36.** $2b\sqrt{3a}$ *(5-6)* **37.** $6x^2y^3\sqrt{xy}$ *(5-6)* **38.** $\dfrac{\sqrt{2xy}}{2x}$ *(5-6)*
39. $-3\sqrt{x}$ *(5-7)* **40.** $\sqrt{7} - 2\sqrt{3}$ *(5-7)* **41.** $5 + 2\sqrt{5}$ *(5-7)* **42.** $1 + \sqrt{3}$ *(5-7)*
43. $\dfrac{5 + 3\sqrt{5}}{4}$ *(5-7)* **44.** $3 - 6i$ *(5-8)* **45.** $15 + 3i$ *(5-8)* **46.** $2 + i$ *(5-8)*
47. $-\frac{1}{2} - i$ *(5-8)* **48.** $\dfrac{4x^6}{y^{16}}$ *(5-1)* **49.** $-x^7y^8$ *(5-1)* **50.** $-x^4y^3$ *(5-1)* **51.** $9x^8y^9$ *(5-1)*
52. $\dfrac{-8x^3}{y^6}$ *(5-1)* **53.** $\dfrac{9x^2}{4y^2}$ *(5-1)* **54.** 2×10^{-3} or 0.002 *(5-3)* **55.** $\dfrac{m^2}{2n^5}$ *(5-2)*
56. $\dfrac{x^6}{y^4}$ *(5-2)* **57.** $\dfrac{4x^4}{y^6}$ *(5-2)* **58.** $\dfrac{c}{a^2b^4}$ *(5-2)* **59.** $\frac{1}{4}$ *(5-2)* **60.** $\dfrac{n^{10}}{9m^{10}}$ *(5-2)*
61. $\dfrac{1}{(x - y)^2}$ *(5-2)* **62.** $\dfrac{3a^2}{b}$ *(5-4)* **63.** $\dfrac{3x^2}{2y^2}$ *(5-4)* **64.** $\dfrac{1}{m}$ *(5-4)* **65.** $6x^{1/6}$ *(5-4)*
66. $\dfrac{x^{1/12}}{2}$ *(5-4)* **67.** $\frac{5}{9}$ *(5-2)* **68.** $x + 2x^{1/2}y^{1/2} + y$ *(5-4)* **69.** $a^2 = b$ *(5-4)*
70. **(A)** $\sqrt[3]{4m^2n^2}$ **(B)** $3\sqrt[5]{x^2}$ *(5-5)* **71.** **(A)** $x^{5/7}$ **(B)** $-3(xy)^{2/3}$ *(5-5)* **72.** $2x^2y$ *(5-6)*
73. $3x^2y\sqrt[3]{x^2y}$ *(5-6)* **74.** $\dfrac{n^2\sqrt{6m}}{3}$ *(5-6)* **75.** $\sqrt[4]{y^3}$ *(5-6)* **76.** $-6x^2y^2\sqrt[5]{3x^2y}$ *(5-6)*
77. $x\sqrt[3]{2x^2}$ *(5-6)* **78.** $\dfrac{\sqrt[5]{12x^3y^2}}{2x}$ *(5-7)* **79.** $2x - 3\sqrt{xy} - 5y$ *(5-7)* **80.** $\dfrac{x - 4\sqrt{x} + 4}{x - 4}$ *(5-7)*

81. $\frac{6x + 3\sqrt{xy}}{4x - y}$ *(5-7)* **82.** $\frac{5\sqrt{6}}{6}$ *(5-7)* **83.** $-1 - i$ *(5-8)* **84.** $\frac{4}{13} - \frac{7}{13}i$ *(5-8)*

85. $5 + 4i$ *(5-8)* **86.** $\frac{xy}{x + y}$ *(5-2)* **87.** $\frac{a^2b^2}{a^3 + b^3}$ *(5-2)* **88.** $y\sqrt[3]{2x^2y}$ *(5-6)* **89.** 0 *(5-7)*

90. **(A)** x **(B)** $5x$ *(5-6)*

CHAPTER 6 EXERCISE 6-1

1. $-6, 1$ **3.** $2, 5$ **5.** $\frac{1}{2}, -4$ **7.** ± 4 **9.** $\pm 4i$ **11.** $\pm 3\sqrt{5}$ **13.** $\pm \frac{3}{2}$ **15.** $\pm \frac{3}{4}$
17. $x^2 + 4x + 4 = (x + 2)^2$ **19.** $x^2 - 6x + 9 = (x - 3)^2$ **21.** $x^2 + 12x + 36 = (x + 6)^2$
23. $-2 \pm \sqrt{2}$ **25.** $3 \pm 2\sqrt{3}$ **27.** $-2, 2$ **29.** $3, -4$ **31.** $-\frac{1}{2}, 2$ **33.** $\frac{1}{2}, 2$ **35.** $2, -3$

37. $\pm\sqrt{2}$ **39.** $\pm\frac{3}{4}i$ **41.** $\pm\sqrt{\frac{7}{9}}$ or $\pm\frac{\sqrt{7}}{3}$ **43.** $8, -2$ **45.** $-1 \pm 3i$ **47.** $-\frac{1}{3}, 1$

49. $x^2 + 3x + \frac{9}{4} = (x + \frac{3}{2})^2$ **51.** $u^2 - 5u + \frac{25}{4} = (u - \frac{5}{2})^2$ **53.** $\frac{-1 \pm \sqrt{5}}{2}$ **55.** $\frac{5 \pm \sqrt{17}}{2}$

57. $2 \pm 2i$ **59.** $\frac{2 \pm \sqrt{2}}{2}$ **61.** $\frac{-3 \pm \sqrt{17}}{4}$ **63.** $-1, 1$ **65.** No solution **67.** $\frac{-5 \pm \sqrt{10}}{2}$

69. $2 \pm i$ **71.** $\frac{3 \pm i\sqrt{7}}{4}$ **73.** $\frac{-1 \pm i\sqrt{3}}{2}$ **75.** $-\sqrt{2} \pm 2$ **77.** $2\sqrt{3} \pm i$ **79.** $i \pm \sqrt{3}$

81. $x = \frac{-m \pm \sqrt{m^2 - 4n}}{2}$ **83.** $a = \sqrt{c^2 - b^2}$ **85.** 90¢/gal

EXERCISE 6-2

1. $a = 2, b = -5, c = 3$ **3.** $a = 3, b = 1, c = -1$ **5.** $a = 3, b = 0, c = -5$ **7.** $-4 \pm \sqrt{13}$

9. $5 \pm 2\sqrt{7}$ **11.** $\frac{-3 \pm \sqrt{13}}{2}$ **13.** $1 \pm i\sqrt{2}$ **15.** $\frac{3 \pm \sqrt{3}}{2}$ **17.** $\frac{-1 \pm \sqrt{13}}{6}$

19. Two real roots **21.** One real root **23.** Two nonreal complex roots **25.** $5 \pm \sqrt{7}$

27. $-1 \pm \sqrt{3}$ **29.** $0, -\frac{3}{2}$ **31.** $2 \pm 3i$ **33.** $5 \pm 2\sqrt{7}$ **35.** $\frac{2 \pm \sqrt{2}}{2}$ **37.** $1 \pm i\sqrt{2}$

39. $\frac{2}{5}, 3$ **41.** $t = \sqrt{2d/g}$ **43.** $r = -1 + \sqrt{A/P}$ **45.** $\frac{\sqrt{7} \pm i}{2}$ **47.** $-\sqrt{3}, \frac{-\sqrt{3}}{3}$

49. $\frac{1}{2}i, -2i$ **51.** $\frac{\pm\sqrt{3} - i}{2}$ **53.** $1.35, 0.48$ **55.** $-1.05, 0.63$ **57.** $\frac{y \pm \sqrt{5y^2}}{2}$

59. $y \pm \sqrt{2y^2}$ **61.** Has real solutions, since discriminant is positive
63. Has no real solutions, since discriminant is negative **65.** $\frac{9}{8}$

67. $\left(\frac{-b + \sqrt{b^2 - 4ac}}{2a}\right)\left(\frac{-b - \sqrt{b^2 - 4ac}}{2a}\right) = \frac{b^2 - (b^2 - 4ac)}{4a^2} = \frac{c}{a}$

EXERCISE 6-3

1. 12, 14 **3.** 0, 2 **5.** 127 mi **7.** 5.12 by 3.12 cm **9.** 1 ft **11.** 5.66 ft/sec
13. 50 mi/hr **15.** 1.41 min **17.** 2 hr; 3 hr **19.** 2 km/hr **21.** \$6

EXERCISE 6-4

1. 4 **3.** 18 **5.** $\pm 1, \pm 3$ **7.** $\pm 3, \pm i\sqrt{2}$ **9.** $-1, 4$ **11.** 9, 16 **13.** 4
15. No solution **17.** No solution **19.** No solution **21.** 5, 13 **23.** $-1, 2$
25. $\pm 1, \pm 2, \pm 2i, \pm i$ **27.** $-8, 125$ **29.** 1, 16 **31.** $\frac{2}{3}, -\frac{3}{2}$ **33.** $\pm 2, \pm \frac{1}{2}$
35. $-2, 3, \frac{1}{2} \pm \frac{\sqrt{7}}{2}i$ **37.** $3 \pm 2i, 4, 2$ **39.** 4 **41.** 11 **43.** 1, 5 **45.** 4, 1.4

EXERCISE 6-5

1. $(-4, 3)$  **3.** $(-\infty, -4] \cup [3, \infty)$ **5.** $(-4, 3)$

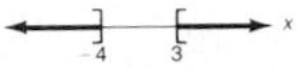

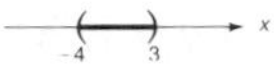

7. $(-\infty, 3) \cup (7, \infty)$  **9.** $\{1, 2, 3, 4, 5\}$ **11.** $\{3\}$ **13.** $\varnothing$ **15.** $[-2, 0]$

17. $(-1, \infty)$ **19.** $(-\infty, 5]$ **21.** $(-\infty, -6] \cup [0, \infty)$

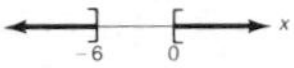

23. $(-\infty, -3] \cup [3, \infty)$ **25.** $(-2, 5]$

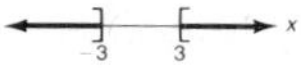

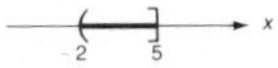

27. $(-\infty, -2) \cup (5, \infty)$  **29.** $(-\infty, -2) \cup (0, 4]$

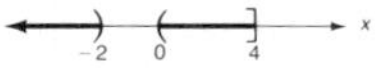

31. $(-\infty, 0) \cup (\frac{1}{4}, \infty)$  **33.** All real numbers; graph is the whole real line

35. No solution **37.** $(-\infty, -\sqrt{3}] \cup [\sqrt{3}, \infty)$ **39.** $[2, 3)$

41. $(-2, 3)$  **43.** $(-3, 0) \cup (1, \infty)$

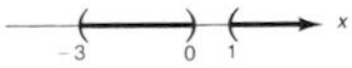

45. $(-1, \frac{1}{2}] \cup (2, \infty)$ 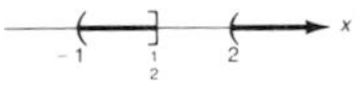 **47.** $(-3, 1)$ 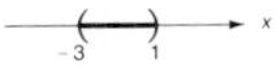 **49.** $x \geq 2$ or $x \leq 1$

51. $2 \leq t \leq 8$ or $[2, 8]$

53.

$ax + b = 0$	$x < -b/a$	$x > -b/a$
$ax = -b$	$ax < -b$	$ax > -b$
$x = -b/a$ (critical point)	$ax + b < 0$	$ax + b > 0$

REVIEW EXERCISE 6-6

1. 0, 3 *(6-1)* **2.** ± 5 *(6-1)* **3.** 2, 3 *(6-1)* **4.** $-3, 5$ *(6-1)* **5.** $\pm\sqrt{7}$ *(6-1)*
6. $a = 3, b = 4, c = -2$ *(6-2)* **7.** $x = \frac{-b \pm \sqrt{b^2 - 4ac}}{2a}$ *(6-2)* **8.** $\frac{-3 \pm \sqrt{5}}{2}$ *(6-2)*
9. 3, 9 *(6-3)* **10.** -19, two nonreal complex roots *(6-2)* **11.** $\{1, 2, 3, 4, 5, 7, 9\}$ *(6-6)*

12. $\{1, 3, 5\}$ *(6-4)* **13.** $(-1, 5)$ *(6-6)* **14.** $[0, 3]$ *(6-6)* **15.** $(-5, 4)$ 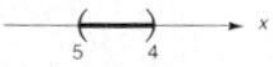 *(6-5)*

16. $(-\infty, -5] \cup [4, \infty)$ *(6-5)* **17.** 0, 2 *(6-1)* **18.** $\pm 2\sqrt{3}$ *(6-1)*
19. $\pm 3i$ *(6-1)* **20.** $-2, 6$ *(6-1)* **21.** $\frac{-1}{3}, 3$ *(6-1)* **22.** $\frac{1}{2}, -3$ *(6-1)* **23.** $\frac{1 \pm \sqrt{7}}{3}$ *(6-2)*

24. $\frac{1 \pm \sqrt{7}}{2}$ *(6-2)* **25.** $-4, 5$ *(6-1)* **26.** $\frac{3}{4}, \frac{5}{2}$ *(6-1)* **27.** $\frac{-1 \pm \sqrt{5}}{2}$ *(6-2)*
28. $1 \pm i\sqrt{2}$ *(6-2)* **29.** 2, 3 *(6-4)* **30.** 9, 25 *(6-4)* **31.** $\pm 2, \pm 3i$ *(6-4)*
32. $64, \frac{-27}{8}$ *(6-4)* **33.** $(-\infty, -3] \cup [7, \infty)$ *(6-5)*

34. $(-\infty, 0) \cup (\frac{1}{2}, \infty)$ *(6-5)* **35.** No solution *(6-5)*

36. All real numbers; graph is the real line *(6-5)* **37.** 6×5 in. *(6-3)* **38.** $3 \pm 2\sqrt{3}$ *(6-1)*
39. $\frac{3 \pm i\sqrt{6}}{2}$ or $\frac{3}{2} \pm \frac{\sqrt{6}}{2}i$ *(6-4)* **40.** $\frac{-3 \pm \sqrt{57}}{6}$ *(6-2)* **41.** $\pm 1, \pm 2, \pm 2i, \pm i$ *(6-4)*

42. $3, \frac{9}{4}$ *(6-4)* **43.** $(-\infty, 1] \cup (3, 4)$ *(6-5)* **44.** 9 cm and 12 cm *(6-3)*
45. **(A)** 2,000 and 8,000 **(B)** 5,000 *(6-3)*

CHAPTER 7 EXERCISE 7-1

1. $A(-10, 10)$, $B(10, -10)$, $C(16, 14)$, $D(16, 0)$, $E(-14, -16)$, $F(0, 4)$, $G(6, -16)$, $H(-14, 0)$

3.

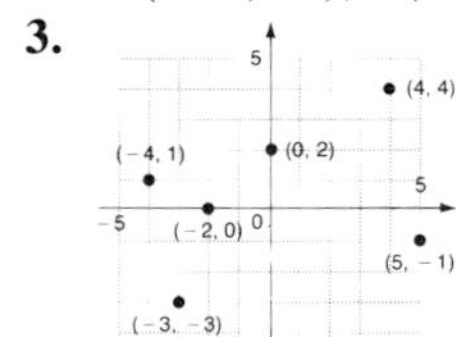

5.

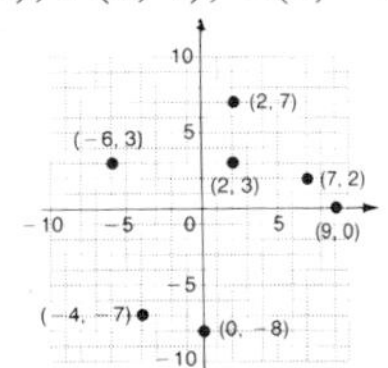

7. $A(-3\frac{1}{2}, 2)$, $B(-2, -4\frac{1}{2})$, $C(0, -2\frac{1}{2})$, $D(2\frac{3}{4}, 0)$, $E(3\frac{3}{4}, 2\frac{3}{4})$, $F(4\frac{1}{4}, -3\frac{1}{2})$

9.

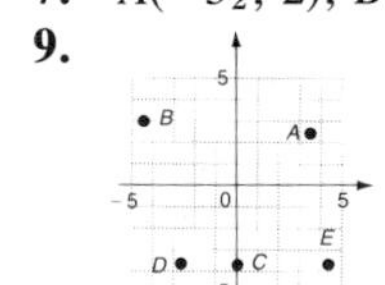

11.

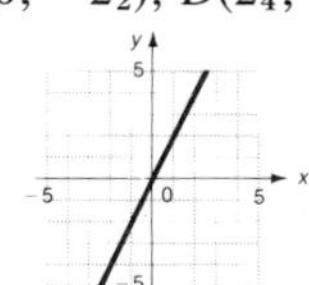

13.

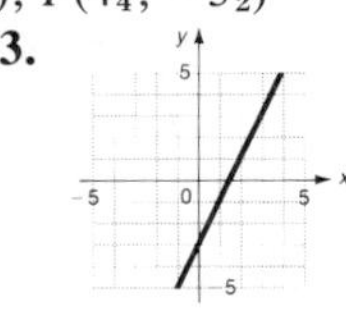

15.

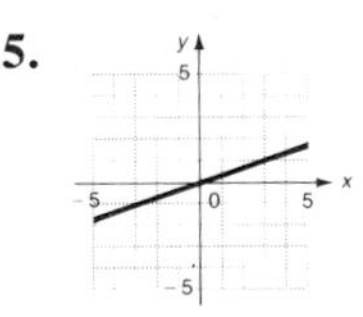

17.

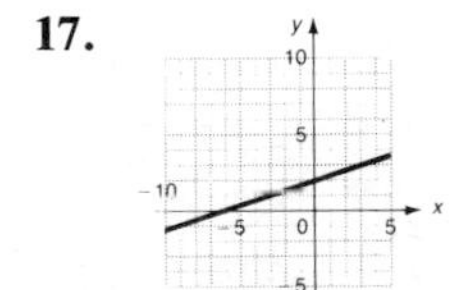

19.

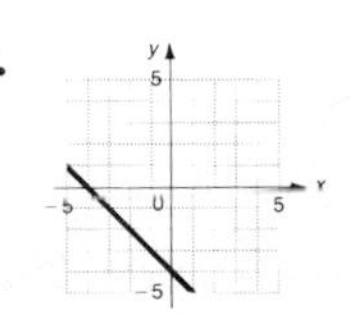

21.

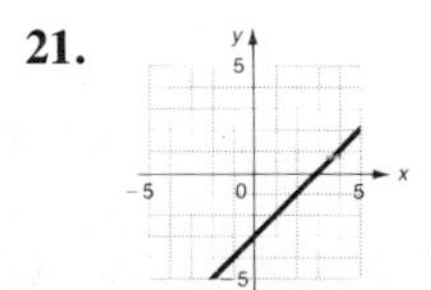

23.

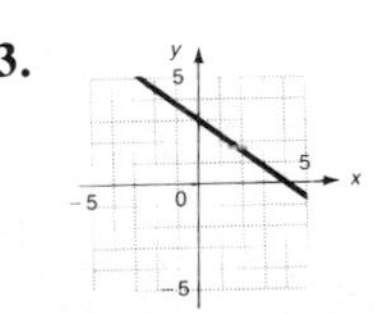

25.

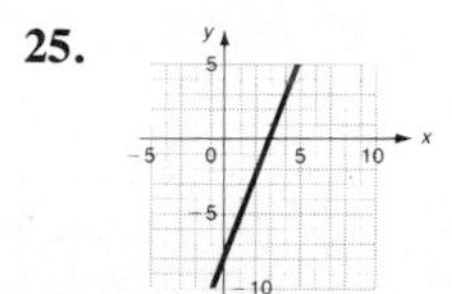

27.

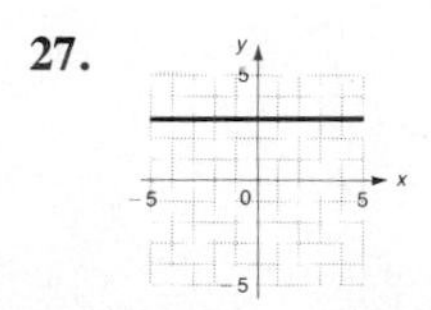

29.

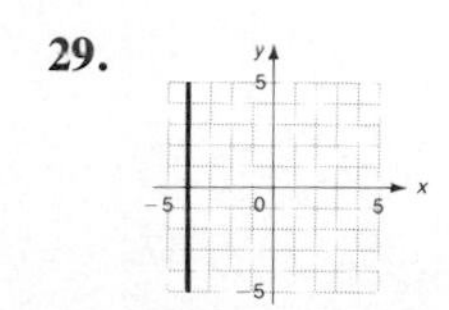

31.

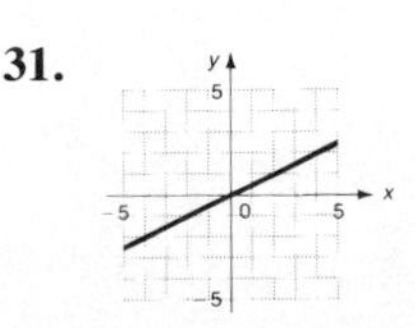

33.

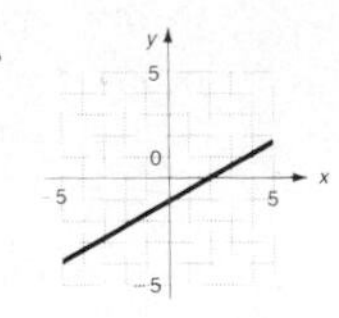

35.

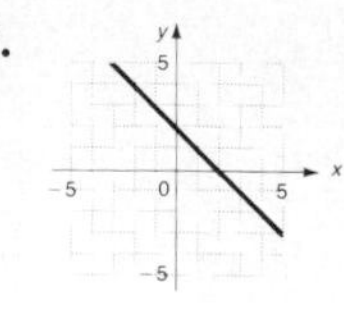

37.

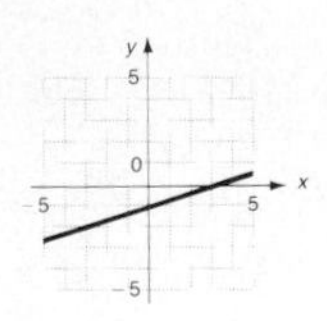

39.

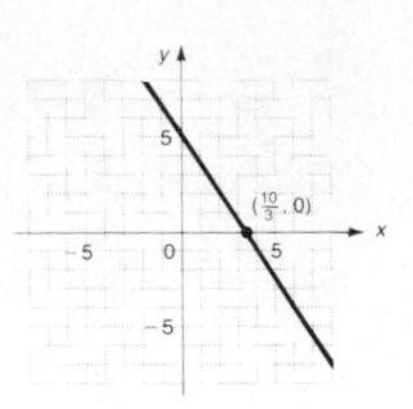

41.

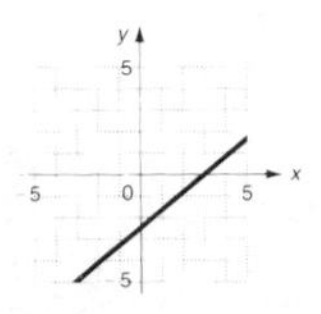

43.

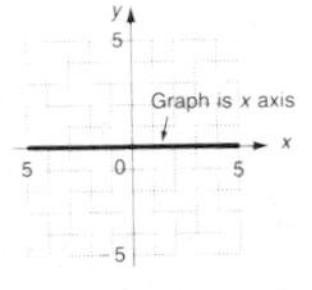

45. $y = 2x - 4$

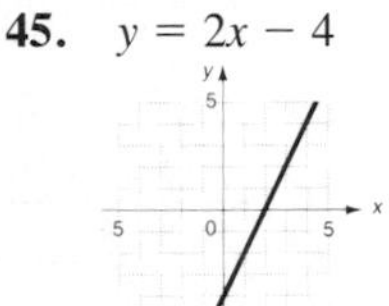

47. $x + 2y = -6$

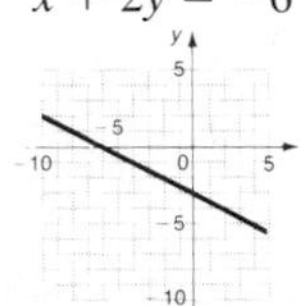

49.

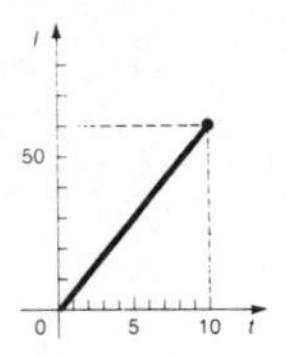

51.

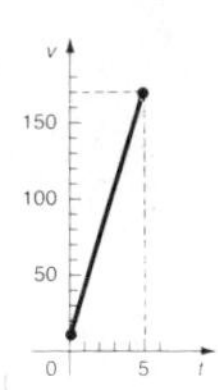

53.

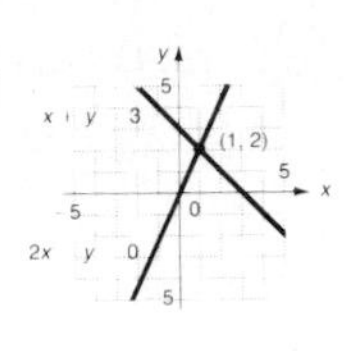

55.

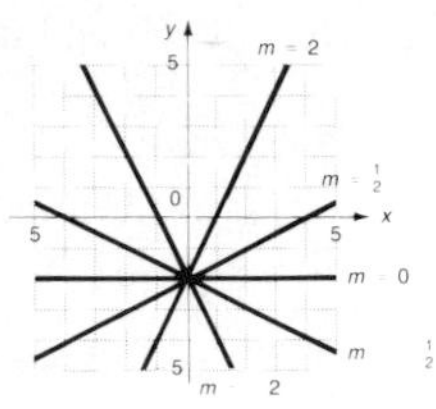

57.

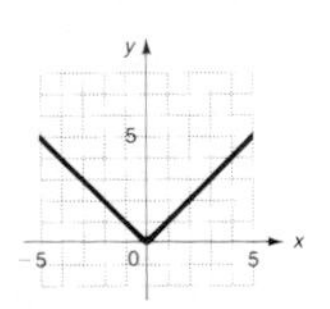

59.

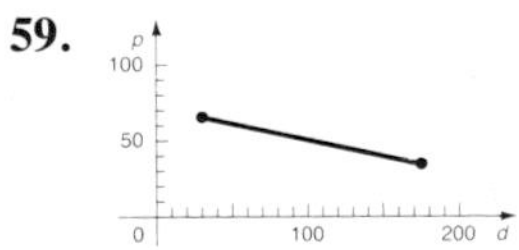

61.

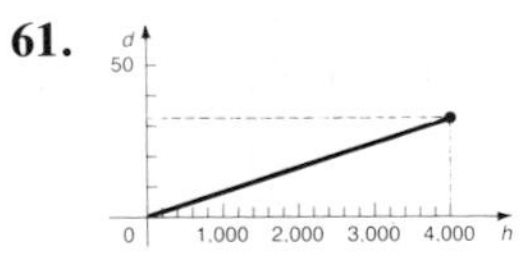

EXERCISE 7-2

1. Slope: 2
y intercept: -3

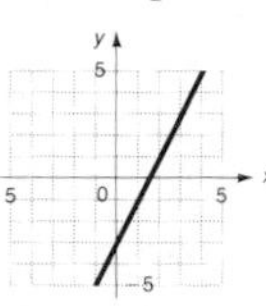

3. Slope: -1
y intercept: 2

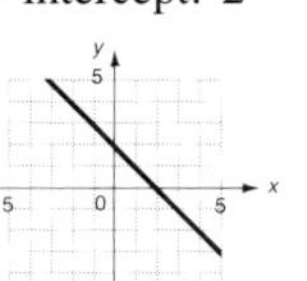

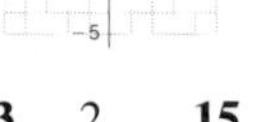

5. $y = 5x - 2$ **7.** $y = -2x + 4$ **9.** $y - 4 = 2(x - 5)$

11. $y - 1 = -2(x - 2)$ **13.** 2 **15.** $\frac{1}{2}$ **17.** $y - 6 = 2(x - 5)$ or $y - 2 = 2(x - 3)$

19. $y - 5 = \frac{1}{2}(x - 10)$ or $y - 1 = \frac{1}{2}(x - 2)$

21. Slope: $-\frac{1}{3}$
y intercept: 2

23. Slope: $-\frac{1}{2}$
y intercept: 2

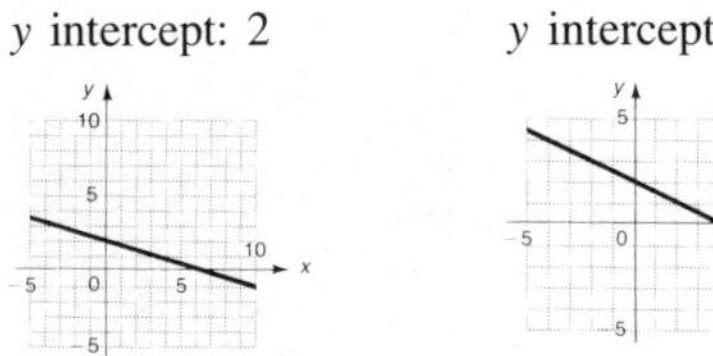

25. Slope: $-\frac{2}{3}$
y intercept: 2

27. $y = -\dfrac{x}{2} - 2$ **29.** $y = \frac{2}{3}x + \frac{3}{2}$ **31.** $y - 2 = -2(x + 3)$, $y = -2x - 4$

33. $y - 3 = \frac{1}{2}(x + 4)$, $y = \dfrac{x}{2} + 5$ **35.** $\frac{1}{3}$ **37.** $-\frac{1}{4}$

39. $y - 4 = \frac{1}{3}(x + 6)$ or $y - 7 = \frac{1}{3}(x - 3)$, $y = \dfrac{x}{3} + 6$

41. $y = -\frac{1}{4}(x + 4)$ or $y + 2 = -\frac{1}{4}(x - 4)$, $y = -\dfrac{x}{4} - 1$ **43.** $x = -3, y = 5$ **45.** $x = -1, y = 22$

47. **(A)** $y = \frac{3}{5}x + \frac{17}{5}$ **(B)** $y = -\frac{5}{3}x + \frac{17}{3}$ **49.** **(A)** $y = -3x + 2$ **(B)** $y = \frac{1}{3}x + 2$

51. **(A)** $y = -x + 2$ **(B)** $y = x$ **53.** **(A)** $y = 5$ **(B)** $x = -2$

55. **(A)** $R = \frac{3}{2}C + 3$ **(B)** **(C)** \$158 **57.** $F = \frac{9}{5}C + 32$

59. **(A)** $V = -1{,}800t + 20{,}000$ **(B)** \$12,800, \$5,600 **(C)** $-1{,}800$ **(D)**

EXERCISE 7-3

1.

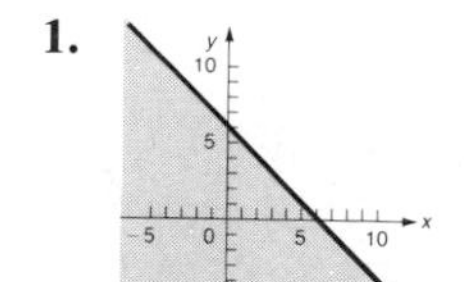

3.

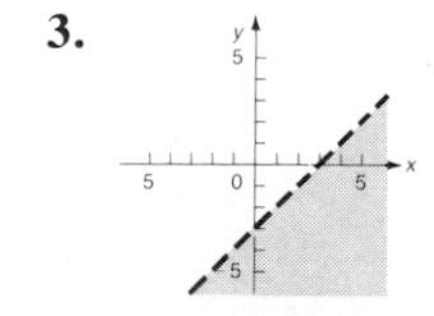

5.

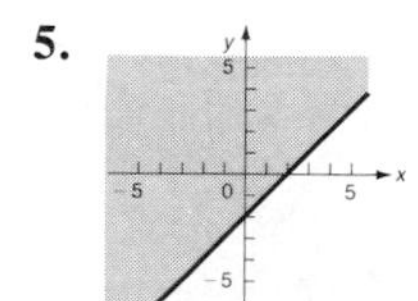

7.

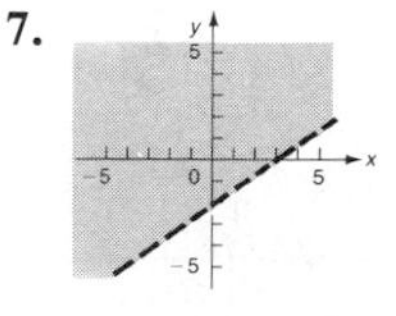

9.

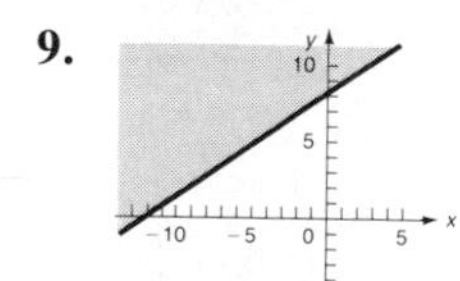

11.

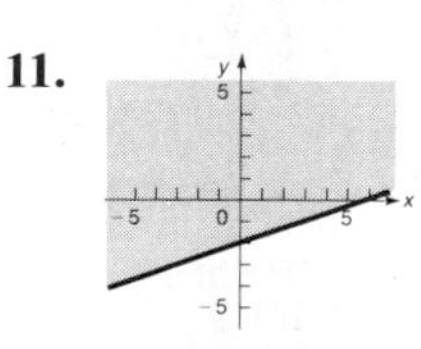

13.

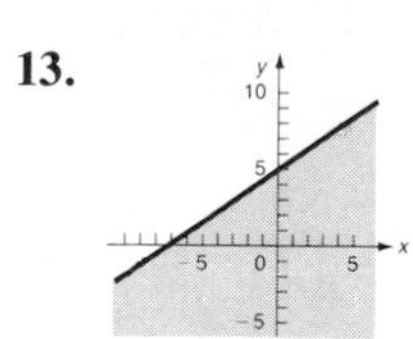

15.

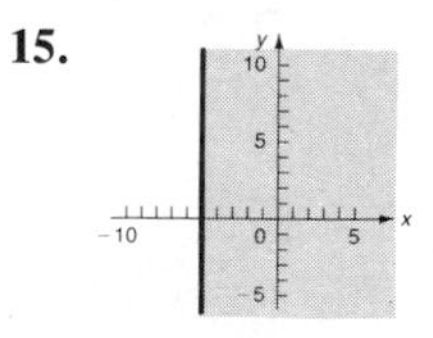

17.

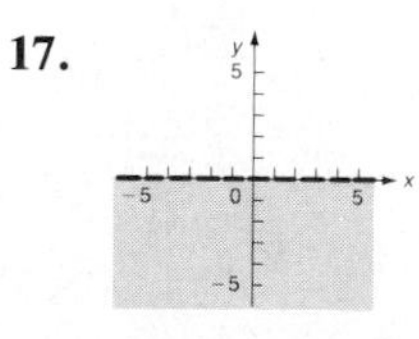

19.

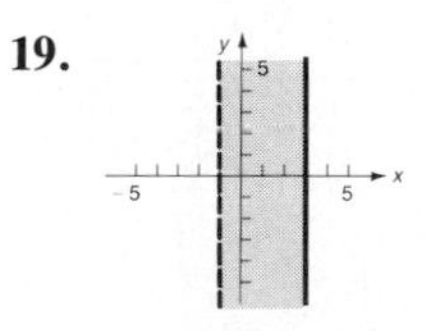

21.

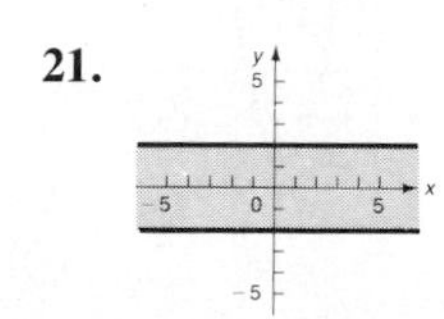

23.

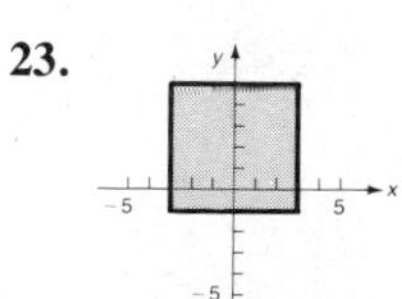

25.

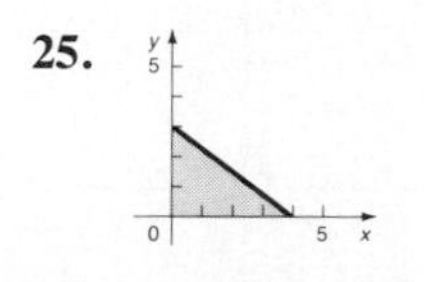

27.
$$6x + 8y \le 120$$
$$x + 3y \le 30$$
$$x \ge 0$$
$$y \ge 0$$

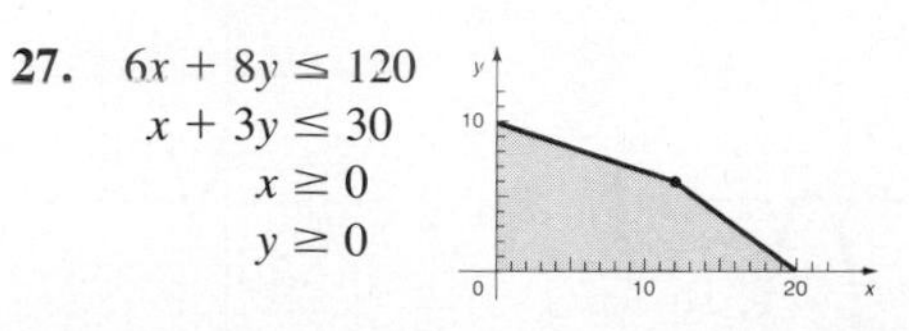

EXERCISE 7-4

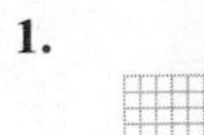

1. 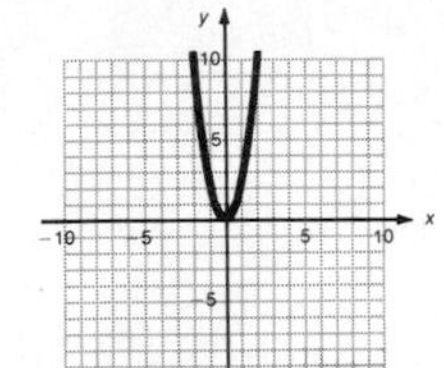

3. 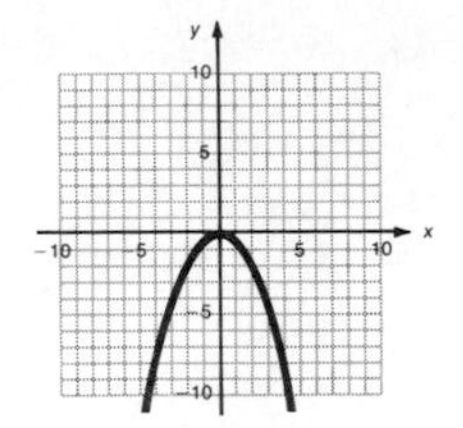

5. 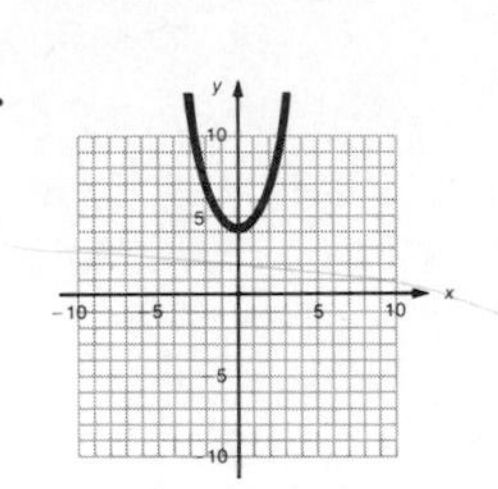

7. 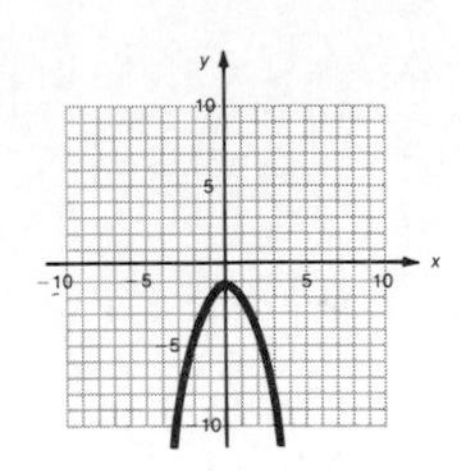

9. 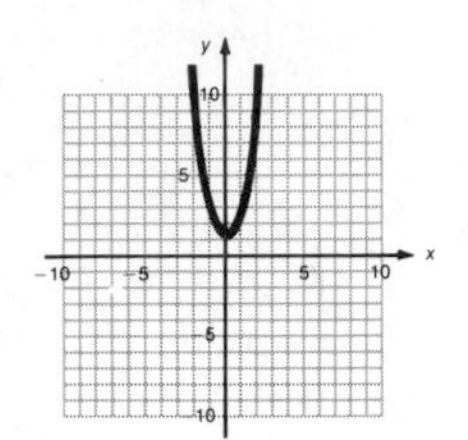

11. 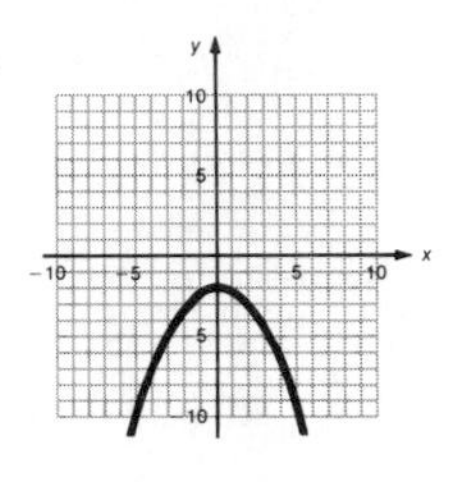

13. 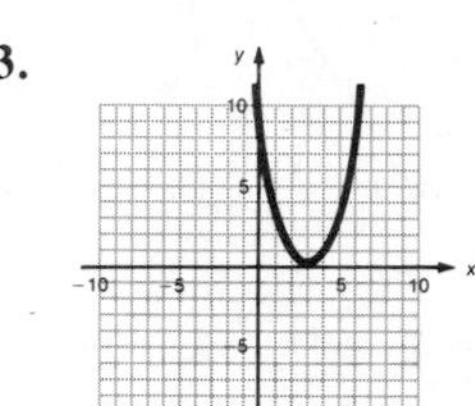

15. 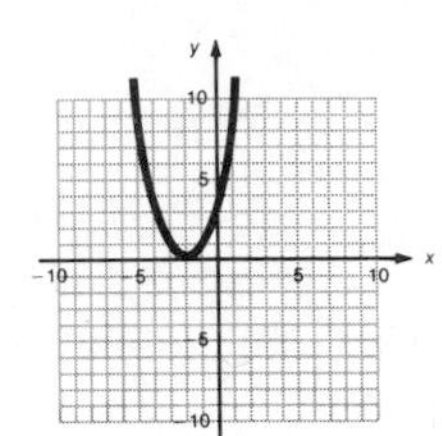

17. 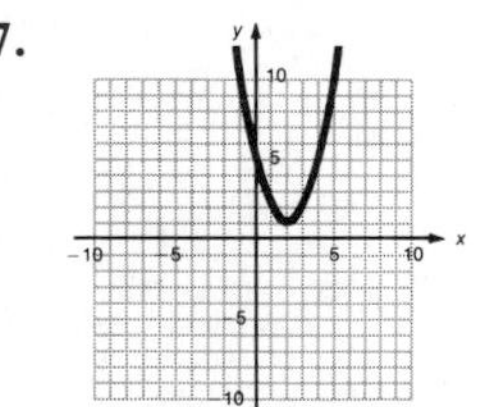

19. 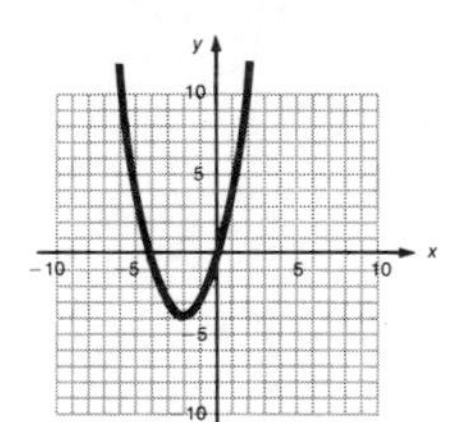

21. 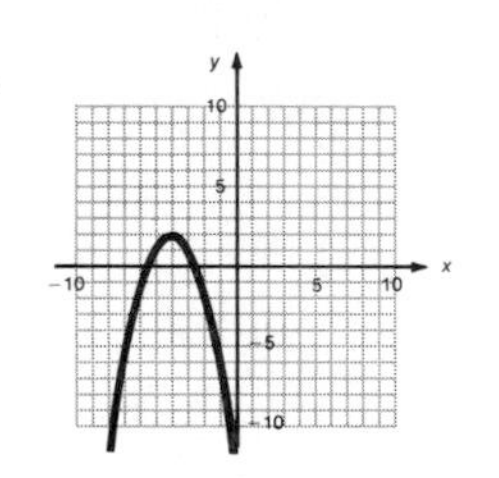

23. 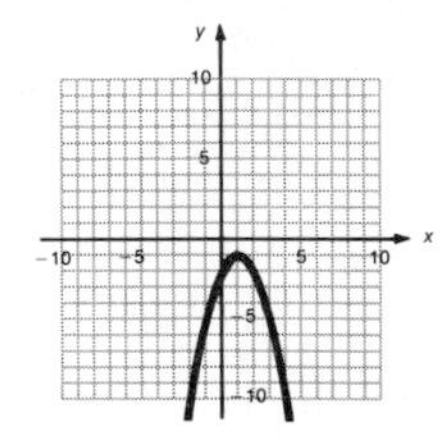

25. 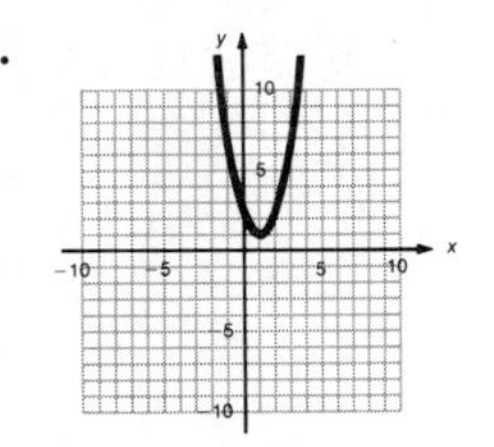

27. 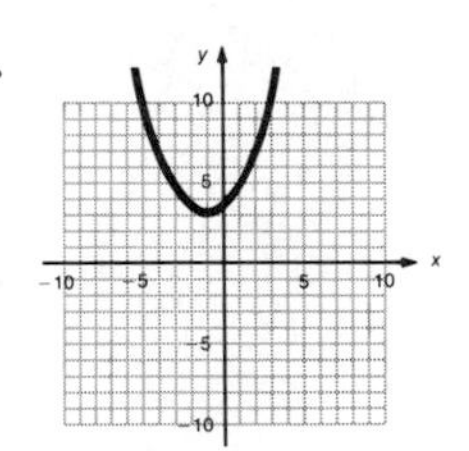

29. 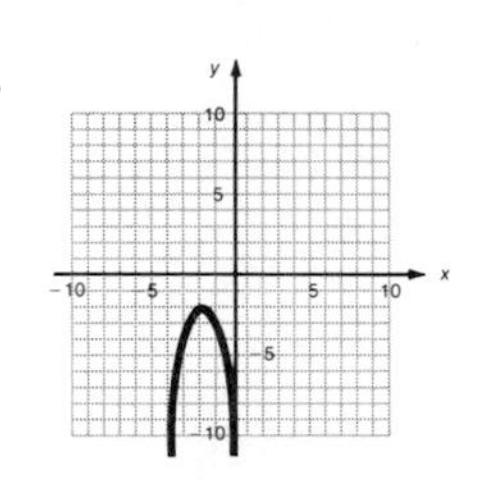

31. 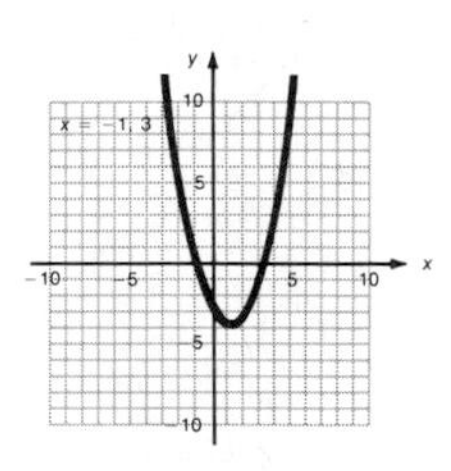

33. 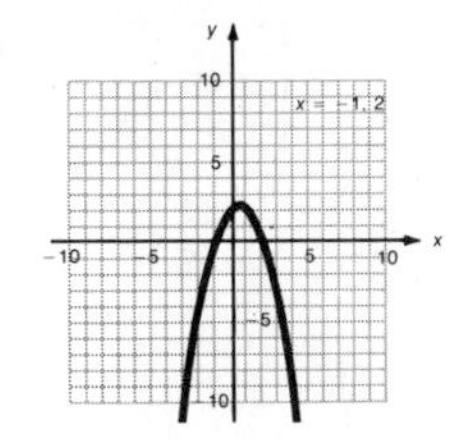

35. 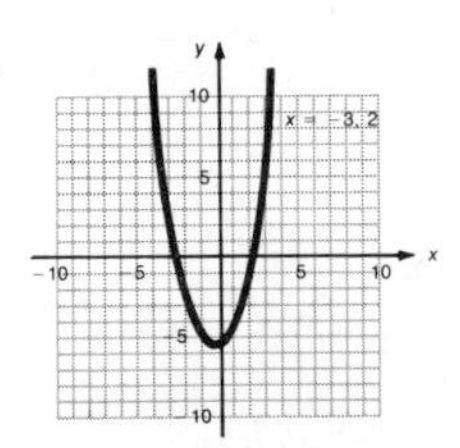

37. 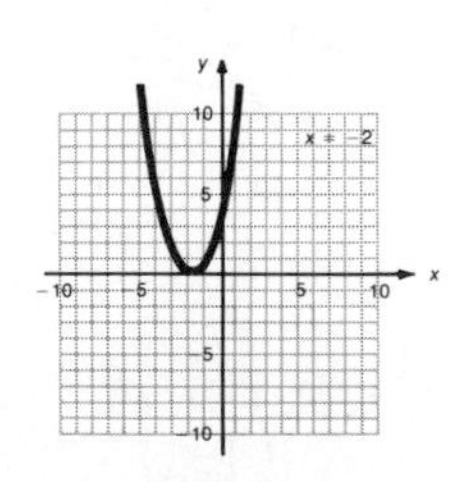

39. 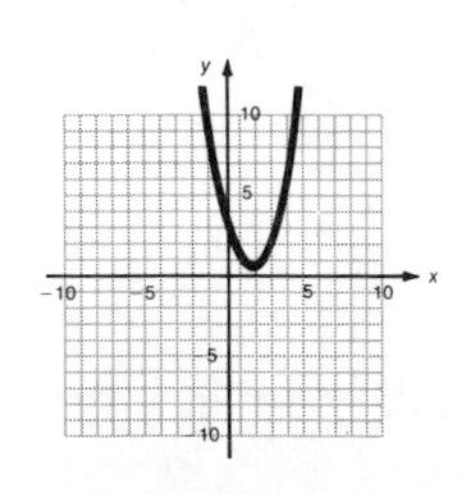

41. 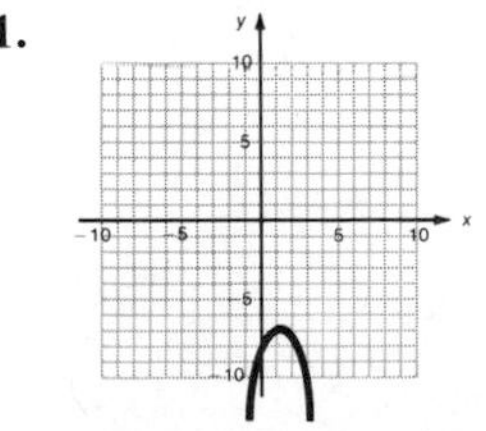

43. 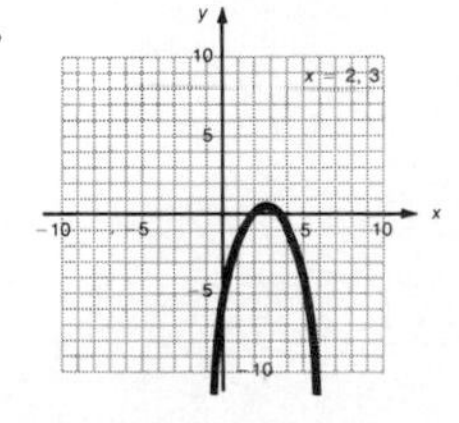

45. 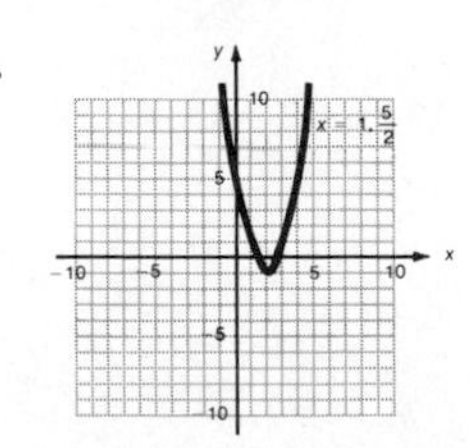

47. 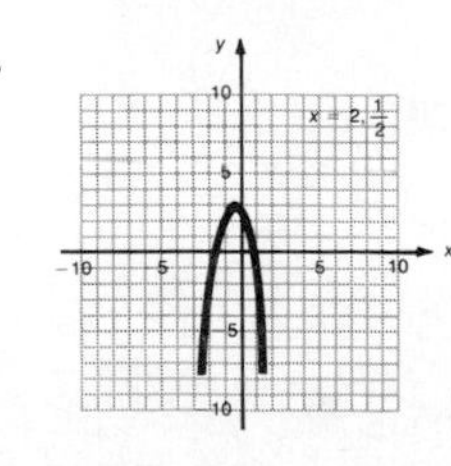

EXERCISE 7-5

1. $\sqrt{5}$ **3.** $\sqrt{89}$ **5.** $\sqrt{18}$ or $3\sqrt{2}$  **7.**

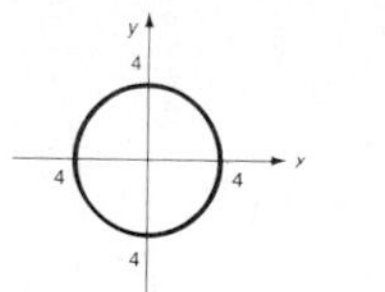

9.

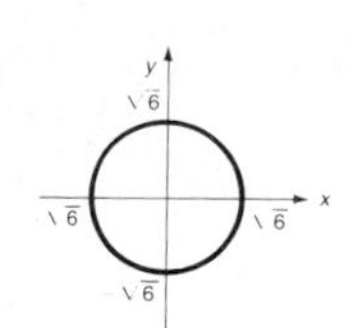

11. $x^2 + y^2 = 49$

13. $x^2 + y^2 = 5$ **15.**

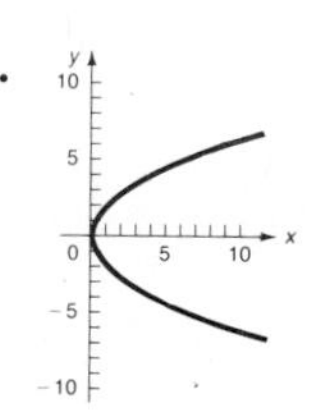

17.

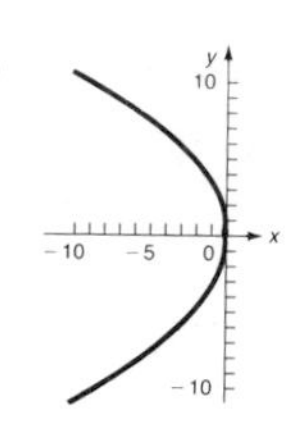

19.

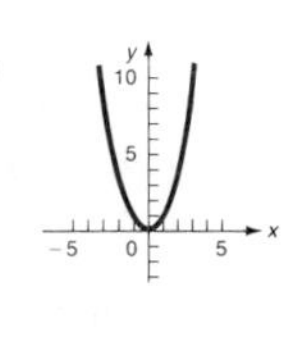

21.

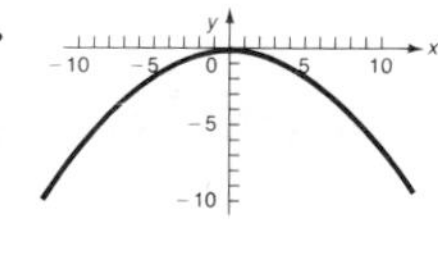

23. Yes, since two sides have length $\sqrt{17}$. **25.**

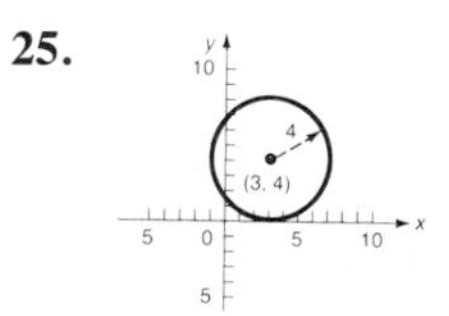

27.

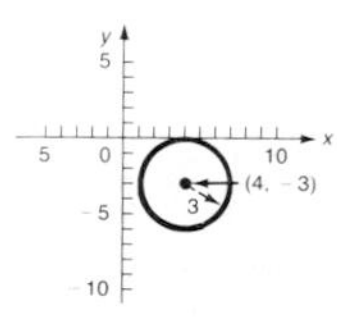

29.

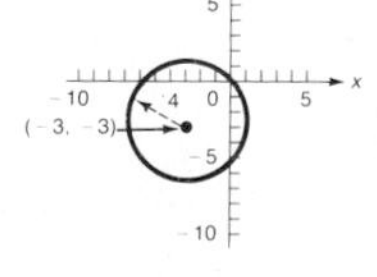

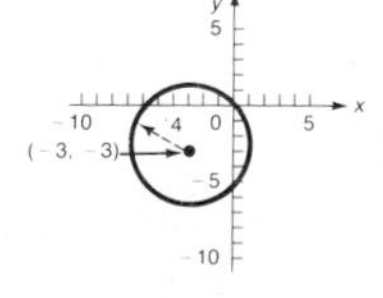

31. $(x - 3)^2 + (y - 5)^2 = 49$ **33.** $(x + 3)^2 + (y - 3)^2 = 64$

35. $(x + 4)^2 + (y + 1)^2 = 3$ **37.** $x = \frac{1}{2}y^2$

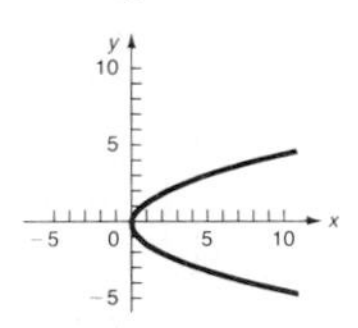

39. $(x - 2)^2 + (y - 3)^2 = 9$

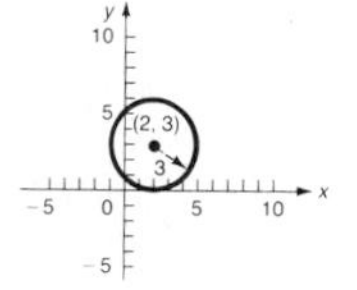

41. $(x - 3)^2 + (y + 3)^2 = 16$ **43.** $(x + 3)^2 + (y + 2)^2 = 9$

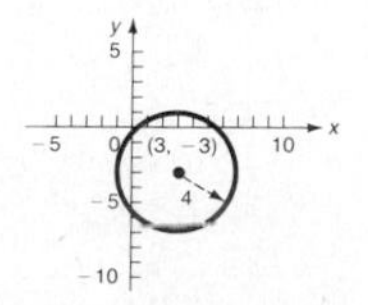

45. $(1, 0)$, $x = -1$ **47.** $x = -3, 7$ **49.** $x^2 + y^2 = 25$
51. $(x - 2)^2 = -4(y - 3)$ or $x^2 - 4x + 4y - 8 = 0$ **53.** $x^2 + y^2 = 50^2$ **55.** $x^2 = -100y$

EXERCISE 7-6

1.

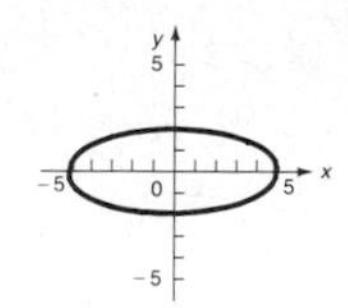

3.

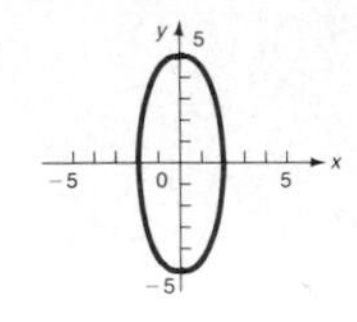

5.

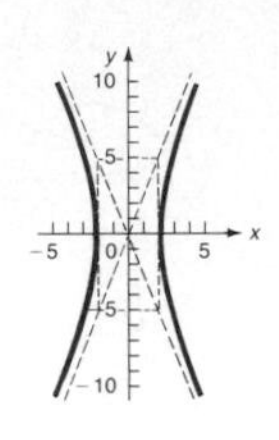

7.

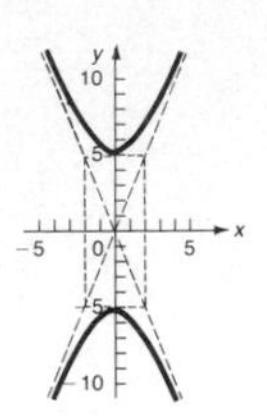

9. $\dfrac{x^2}{9} + \dfrac{y^2}{4} = 1$

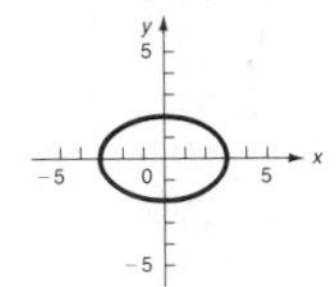

11. $\dfrac{x^2}{9} - \dfrac{y^2}{4} = 1$

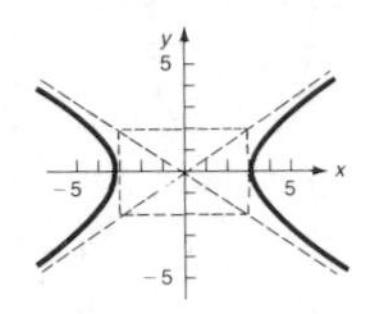

13. $(0, -\sqrt{21}), (0, \sqrt{21})$ **15.** $(5, 0), (-5, 0)$

REVIEW EXERCISE 7-7

1. *(7-1)*

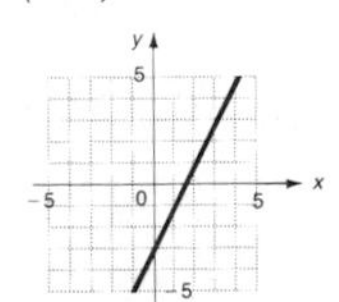

2. *(7-1)*

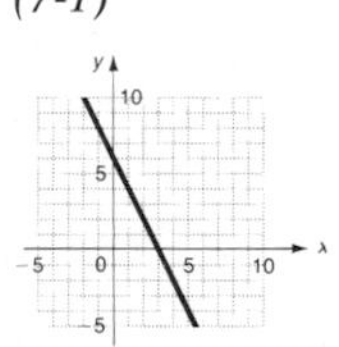

3. *(7-3)*

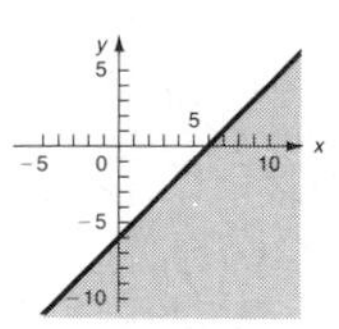

4. *(7-3)*

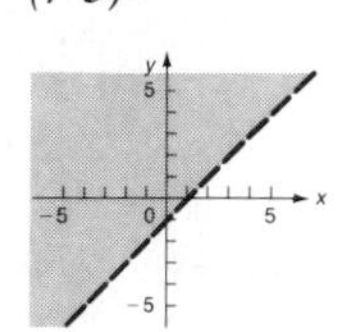

5. *(7-5)*

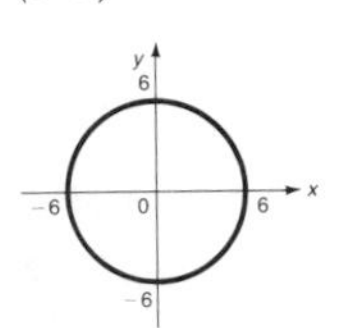

6. *(7-4)*

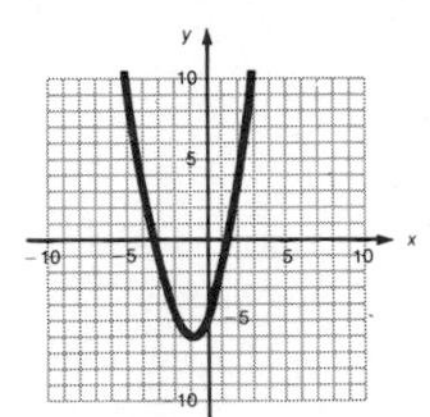

7. Slope $= -2$, y intercept $= -3$ *(7-2)* **8.** $2x + y = 8$ *(7-2)* **9.** 2 *(7-2)*
10. $y = 2x + 1$ *(7-2)* **11.** $\sqrt{20}$ or $2\sqrt{5}$ *(7-5)* **12.** $x^2 + y^2 = 25$ *(7-5)*

13. *(7-1)*

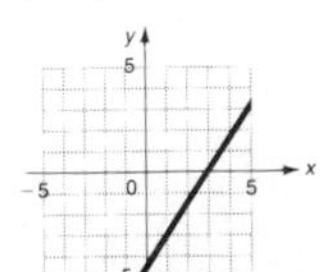

14. *(7-1)*

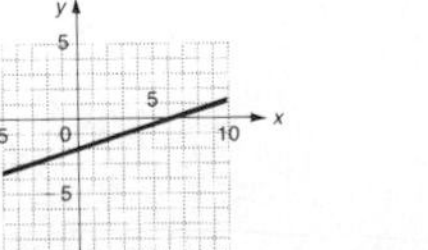

15. *(7-1)*

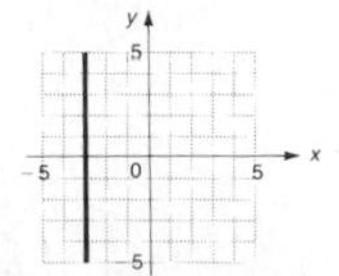

16. *(7-3)*

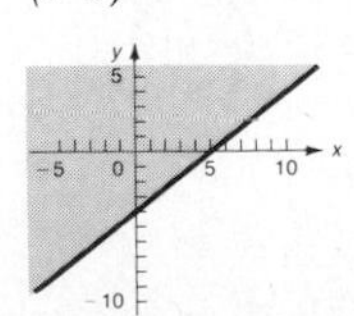

17. *(7-4)*

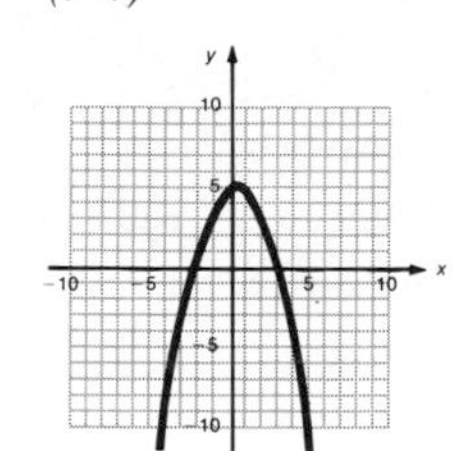

18. *(7-3)*

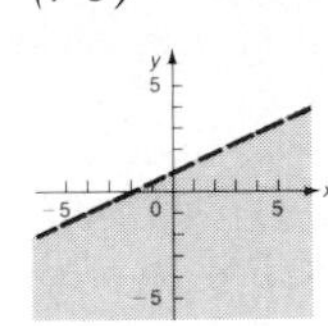

19. *(7-3)*

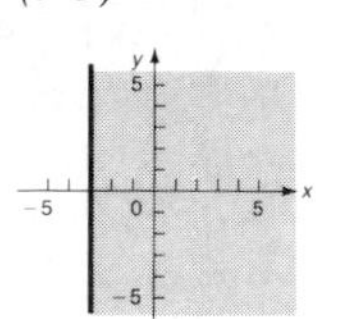

20. *(7-3)*

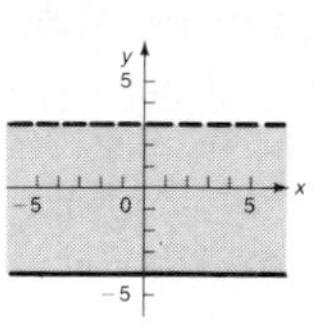

21. *(7-3)*

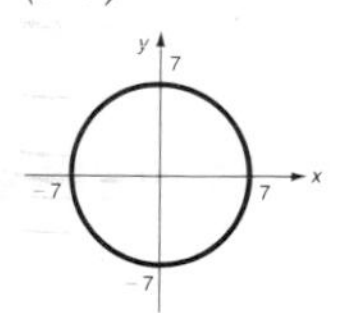

22. *(7-5)*

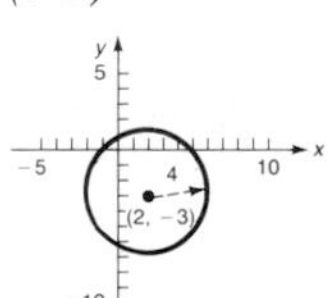

23. *(7-5)*

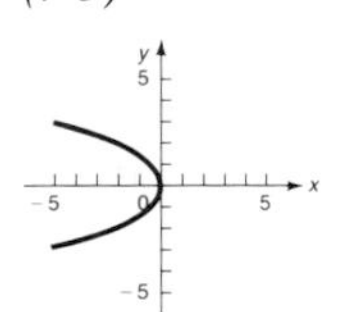

24. *(7-6)*

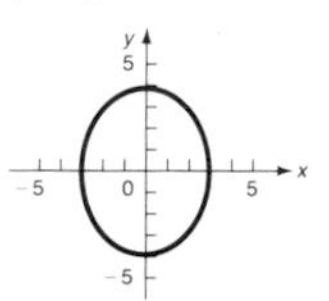

25. *(7-6)*

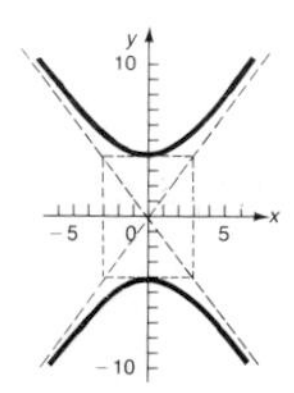

26. Slope $= -\frac{1}{2}$, y intercept $= -3$ *(7-2)* **27.** $y = -\frac{1}{3}x + 1$ *(7-2)*

28. $2x + 3y = 0$ *(7-2)* **29.** $y = 2x - 10$ *(7-2)* **30.** Vertical: $x = 5$; horizontal: $y = -2$ *(7-2)*
31. $(x + 3)^2 + (y - 4)^2 = 49$ *(7-5)* **32.** $x^2 + y^2 = 169$ *(7-5)* **33.** $y = \frac{3}{2}x + 11$ *(7-2)*

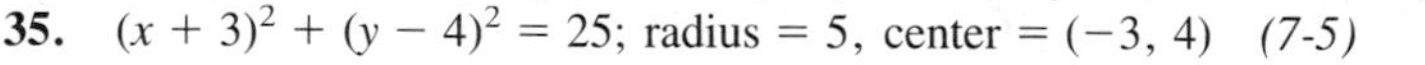

34. *(7-4)* **35.** $(x + 3)^2 + (y - 4)^2 = 25$; radius $= 5$, center $= (-3, 4)$ *(7-5)* **36.** *(7-3)*

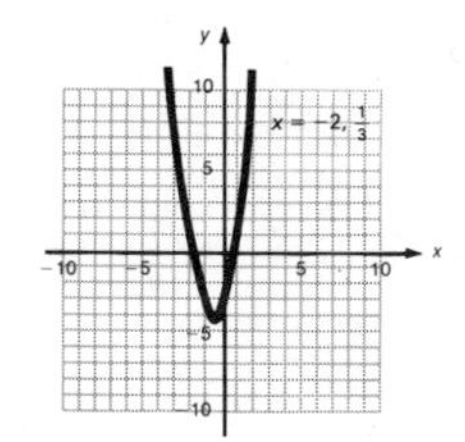

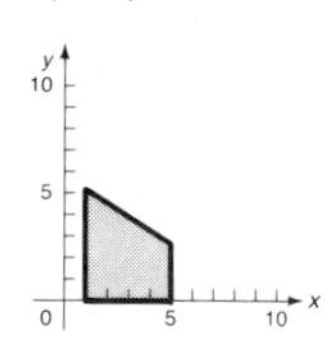

CHAPTER 8 EXERCISE 8-1

1. $(2, -3)$ **3.** No solution **5.** $(1, 4)$ **7.** $(-2, -3)$ **9.** $(-\frac{4}{3}, 1)$
11. Infinite number of solutions, $(x, 3x - 9)$ for any real number x **13.** $(-2, 3)$ **15.** $(1, -5)$
17. $(\frac{1}{3}, -2)$ **19.** Infinite number of solutions, $(x, \frac{1}{2}x + 3)$ for any real number x **21.** $(-2, 2)$
23. $(1.1, 0.3)$ **25.** $(8, 6)$

EXERCISE 8-2

1. 30 quarters and 70 dimes **3.** 700 \$2 tickets; 2800 \$4 tickets **5.** 20 dl
7. 14 nickels and 8 dimes **9.** 52°; 38° **11.** 20 liters

13. 60 dl of 20% solution; 30 dl of 50% solution **15.** 25 kg of \$5/kg tea and 50 kg of \$6.50/kg tea
17. \$8,000 at 10% and \$12,000 at 15% **19.** 84 $\frac{1}{4}$-lb packages; 60 $\frac{1}{2}$-lb packages
21. 60 ml of 80% solution and 40 ml of 50% solution
23. **(A)** $1\frac{9}{13}$ sec, $7\frac{9}{13}$ sec **(B)** Approximately 8,462 ft **25.** 3.6 liters
27. Both companies pay \$136 on sales of \$1,700. The straight-commission company pays better to the right of this point; the other company pays better to the left. **29.** At \$2 per record, $d = s = 3{,}000$.

EXERCISE 8-3

1. $x = -1, y = -1, z = -2$ **3.** $x = 0, y = -2, z = 5$ **5.** $x = 2, y = 0, z = -1$
7. $a = -1, b = 2, c = 0$ **9.** $x = 0, y = 2, z = -3$ **11.** $x = 1, y = -2, z = 1$ **13.** No solution
15. $w = 1, x = -1, y = 0, z = 2$ **17.** $D = -4, E = -4, F = -17$
19. 1,200 style A; 800 style B; 2,000 style C
21. 60 grams of mix A, 50 grams of mix B, 40 grams of mix C
23. Oldest press: 12 hr; middle press: 6 hr; newest press: 4 hr

EXERCISE 8-4

1. $\left[\begin{array}{rr|r} 4 & -6 & -8 \\ 1 & -3 & 2 \end{array}\right]$ **3.** $\left[\begin{array}{rr|r} -4 & 12 & -8 \\ 4 & -6 & -8 \end{array}\right]$ **5.** $\left[\begin{array}{rr|r} 1 & -3 & 2 \\ 8 & -12 & -16 \end{array}\right]$ **7.** $\left[\begin{array}{rr|r} 1 & -3 & 2 \\ 0 & 6 & -16 \end{array}\right]$

9. $\left[\begin{array}{rr|r} 1 & -3 & 2 \\ 2 & 0 & -12 \end{array}\right]$ **11.** $\left[\begin{array}{rr|r} 1 & -3 & 2 \\ 3 & -3 & -10 \end{array}\right]$ **13.** $x = 3, y = 2$ **15.** $x = 3, y = 1$
17. $x = 2, y = 1$ **19.** $x = 2, y = 4$ **21.** No solution **23.** $x = 1, y = 4$
25. Infinitely many solutions: $y = s$, $x = 2s - 3$ for any real number s
27. Infinitely many solutions: $y = s$, $x = \frac{1}{2}s + \frac{1}{2}$ for any real number s **29.** $x = 2, y = -1$
31. $x = 2, y = -1$ **33.** $x = 1.1, y = 0.3$ **35.** $x = -2, y = 3, z = 1$ **37.** $x = 0, y = -2, z = 2$
39. $x = 1, y = 0, z = -1, w = 0$

EXERCISE 8-5

1.
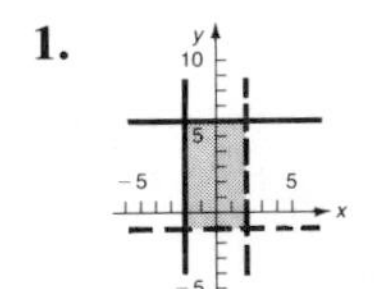

3.
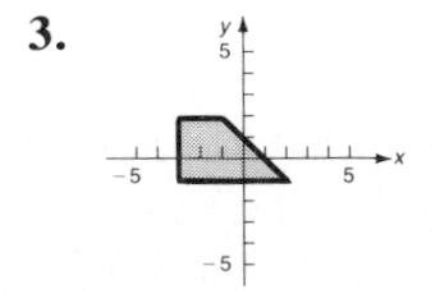

5.
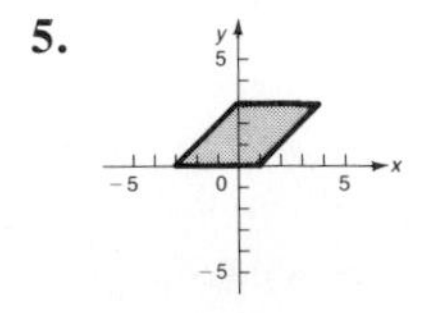

7.
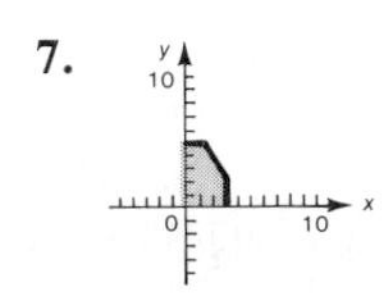

9.
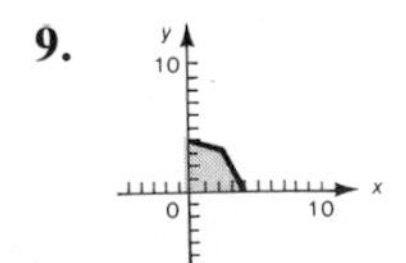

11.
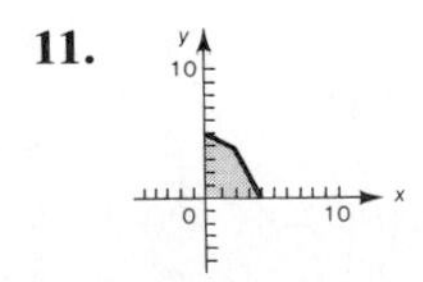

13.
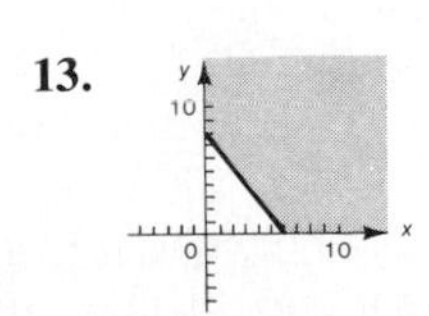

15.
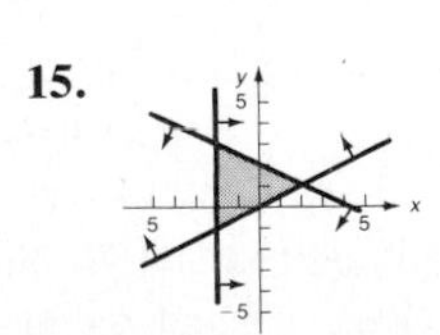

17.

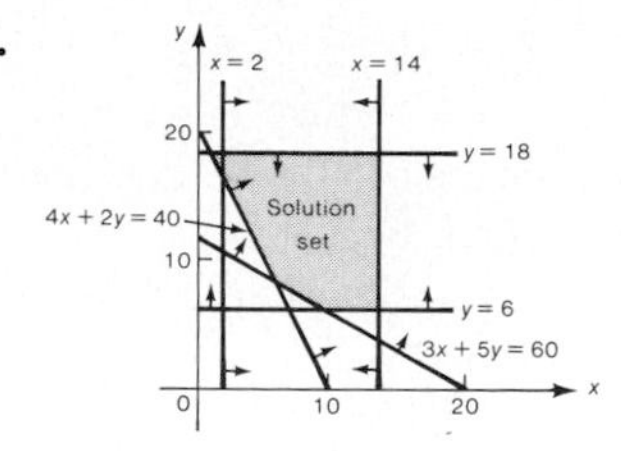

19. $6x + 4y \leq 108$
$x + \ y \leq 24$
$x \geq 0$
$y \geq 0$

EXERCISE 8-6

1. $(-3, -4), (3, -4)$ **3.** $(\frac{1}{2}, 1)$ **5.** $(4, -2), (-4, 2)$ **7.** $(3 - i, 4 - 3i), (3 + i, 4 + 3i)$
9. $(2, 1), (2, -1), (-2, 1), (-2, -1)$ **11.** $(-3, -2), (-3, 2), (3, -2), (3, 2)$
13. $(2 + \sqrt{10}, -2 + \sqrt{10}), (2 - \sqrt{10}, -2 - \sqrt{10})$ **15.** $(-1, 2), (4, 7)$
17. $(i, 2i), (i, -2i), (-i, 2i), (-i, -2i)$ **19.** $(2, 4), (-2, 4), (i\sqrt{5}, -5), (-i\sqrt{5}, -5)$
21. $(4, 0), (-3, \sqrt{7}), (-3, -\sqrt{7})$ **23.** 2 by 16 ft **25.** $(2\sqrt{2}, \sqrt{2}), (-2\sqrt{2}, -\sqrt{2}), (1, 4), (-1, -4)$
27. $(3, 3), (-3, -3), (0, 6), (0, -6)$ **29.** $(4, 4), (-4, -4), (\frac{4}{5}\sqrt{5}, -\frac{4}{5}\sqrt{5}), (-\frac{4}{5}\sqrt{5}, \frac{4}{5}\sqrt{5})$

REVIEW EXERCISE 8-7

1. $x = 6, y = 1$ *(8-1)* 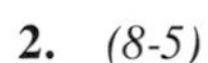 **2.** *(8-5)*

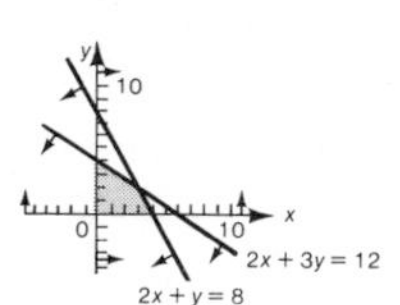

3. $x = 2, y = 1$ *(8-1)*

4. $x = 2, y = 1$ *(8-1)* **5.** $x = -1, y = 2, z = 1$ *(8-3)* **6.** $(1, -1), (\frac{7}{5}, -\frac{1}{5})$ *(8-6)*
7. $(4, 3), (-4, 3), (4, -3), (-4, -3)$ *(8-6)* **8.** $(2, 1, -1)$ *(8-3)* 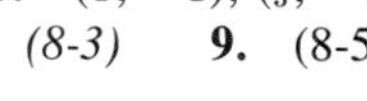 **9.** (8-5)

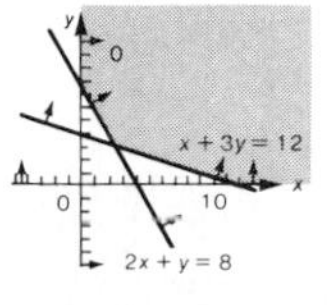

10. No solution *(8-1)* **11.** $(2, -1, 2)$ *(8-3)* **12.** $(1, 3), (1, -3), (-1,3), (-1, -3)$ *(8-6)*
13. $(1 + i, 2i), (1 - i, -2i)$ *(8-6)*
14. Lines coincide—infinitely many solutions; $(x, \frac{1}{3}x + \frac{1}{2})$ for all real x *(8-1)*
15. $(0, 2), (0, -2), (i\sqrt{2}, i\sqrt{2}), (-i\sqrt{2}, -i\sqrt{2})$ *(8-6)* **16.** $x = -1, y = 3$ *(8-4)*
17. $x = -1, y = 2, z = 1$ *(8-4)* **18.** $x = 2, y = 1, z = -1$ *(8-4)* **19.** $x = -12, y = 7, z = 0$ *(8-4)*
20. 16 dimes and 14 nickels *(8-2)* **21.** \$4,000 at 6% and \$2,000 at 10% *(8-2)*
22. 6 by 5 cm *(8-1)* **23.** 30 grams of 70% solution and 70 grams of 40% solution *(8-2)*
24. 48 $\frac{1}{2}$-lb packages, 72 $\frac{1}{3}$-lb packages *(8-2)* **25.** 40 grams mix *A*, 60 grams mix *B*, 30 grams mix *C* *(8-2)*

CHAPTER 9 EXERCISE 9-1

1. Function **3.** Not a function **5.** Function **7.** Function **9.** Not a function **11.** Function
13. Function **15.** Function **17.** Not a function **19.** Not a function **21.** Function
23. Function
25. Not a function **27.** A function

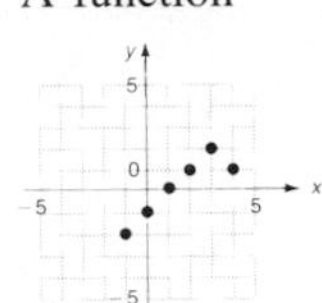

29. A function

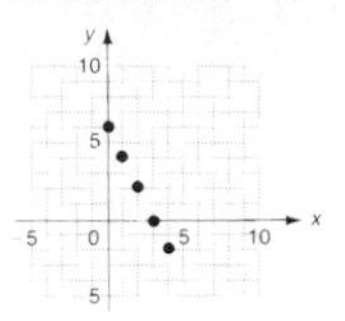

31. Not a function

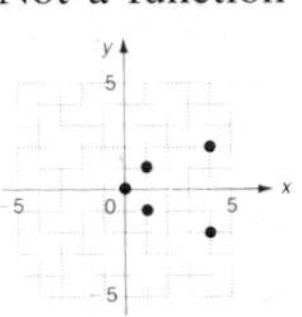

33. Not a function

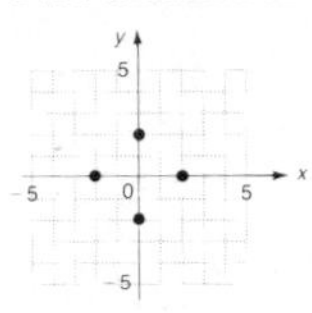

35. A function

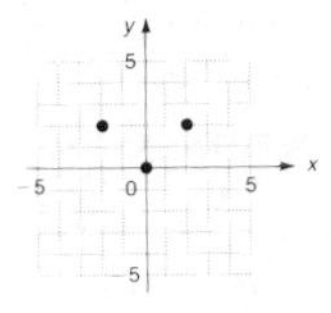

37. R **39.** R **41.** All real numbers except $x = 0$ **43.** All real numbers except $x = -2$ and $x = 3$
45. All real numbers except $x = -4$ and $x = 3$ **47.** $x \leq 4$ or $(-\infty, 4]$
49. $x < -3$ or $x \geq 1$, or $(-\infty, -3) \cup [1, \infty)$
51. A function **53.** Not a function **55.** A function

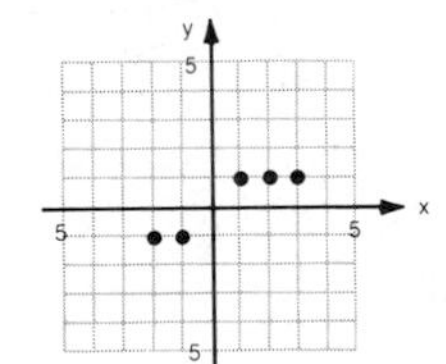

57. **(A)** PARTIAL TABLE

t	d
0	0
1	144
2	256
3	336
4	384
5	400
6	384
7	336
8	256
9	144
10	0

(B)

DOMAIN Set of all real t	RANGE Set of all real d
$0 \leq t \leq 10$	$0 \leq d \leq 400$

(C) The relation is a function.

EXERCISE 9-2

1. 4 **3.** -8 **5.** -2 **7.** -2 **9.** -12 **11.** -6 **13.** -27 **15.** 2 **17.** 6
19. 25 **21.** 22 **23.** -91 **25.** $10 - 2u$ **27.** $3a - 9a^2$ **29.** $2 - 2h$ **31.** -2
33. 10 **35.** 48 **37.** 0 **39.** -7 **41.** $\frac{1}{5}, -\frac{3}{5}$, not defined **43.** 10
45. **(A)** Yes **(B)** Yes **(C)** No **47.** $C(x) = 8.6x$ **49.** $C(F) = \frac{5}{9}(F - 32)$
51. **(A)** 30 mi, 300 mi **(B)** 30

EXERCISE 9-3

1. Slope: 2
y intercept: -4

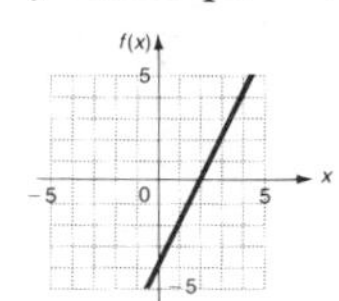

3. Slope: -2
y intercept: 4

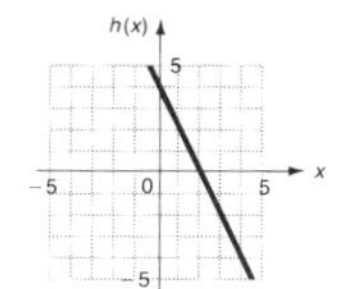

5. Slope: $-\frac{2}{3}$
y intercept: 4

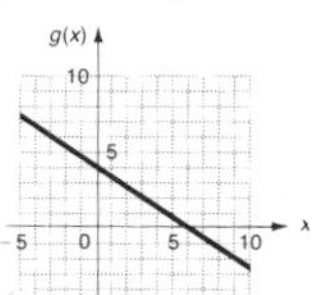

7. Vertex: $(-4, 0)$
Intercepts: $x = -4$
$y = 16$

9. Vertex: $(1, 3)$
Intercepts: x none
$y = 4$

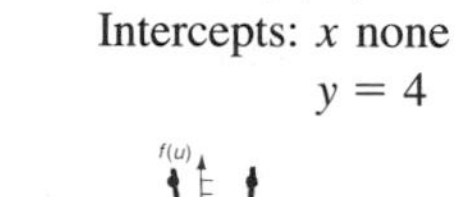

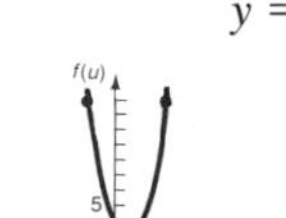

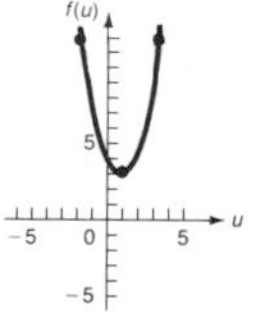

11. Vertex: $(2, 6)$
Intercepts: $x = 2 \pm \sqrt{6}$
$y = 2$

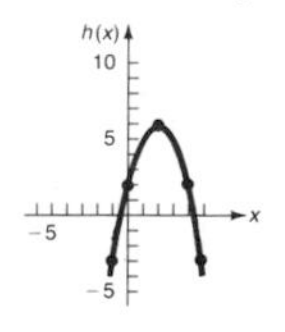

13. Vertex: $(3, 9)$
Intercepts: $x = 0, 6$
$y = 0$

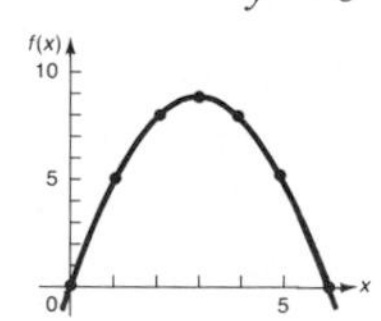

15. Vertex: $(0, -4)$
Intercepts: $x = 2, -2$
$y = -4$

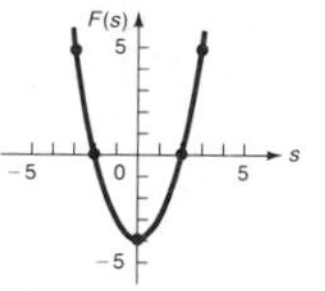

17. Vertex: $(0, 4)$
Intercepts: $x = 2, -2$
$y = 4$

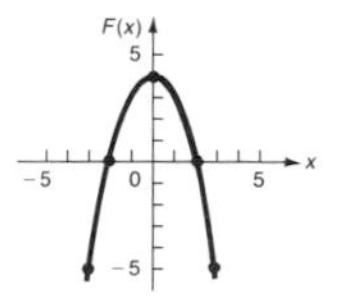

19. Vertex: $(3.5, -2.25)$
Intercepts: $x = 2, 5$
$y = 10$

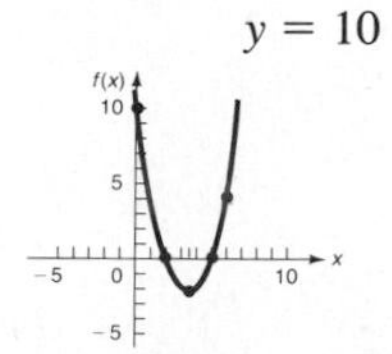

21. Vertex: $(1.5, 6.25)$
Intercepts: $t = -1, 4$
$y = 4$

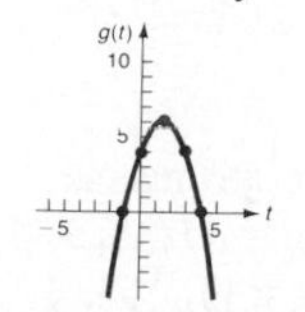

23. Vertex: $(-2, -2)$
Intercepts: $x = 0, -4$
$y = 0$

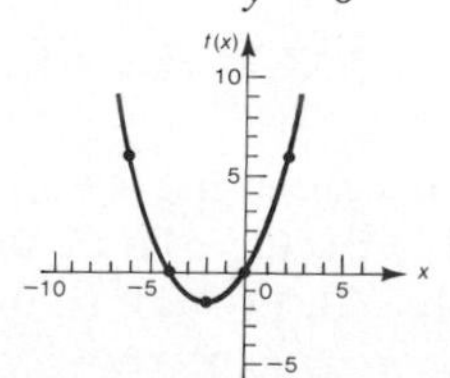

25. Vertex: $(-2, 6)$
Intercepts: $x = -2 \pm \sqrt{3}$
$y = 0$

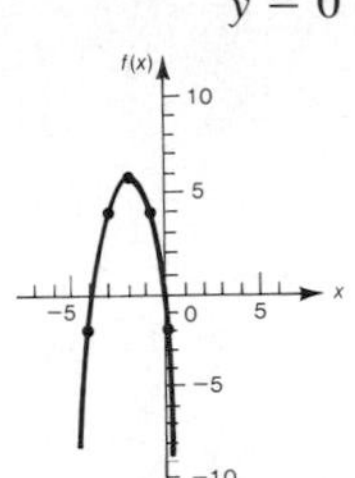

27.

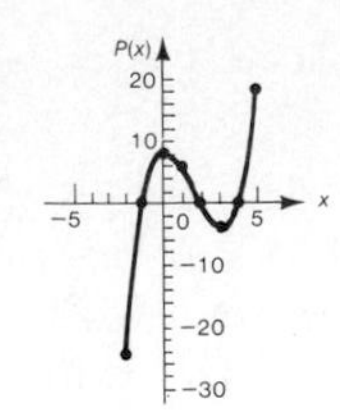

29.

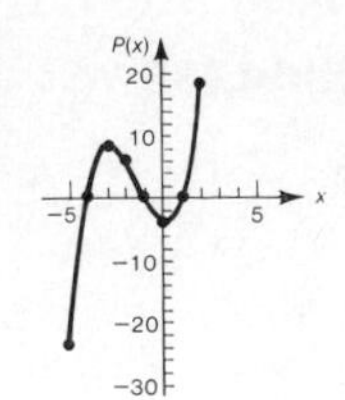

31.

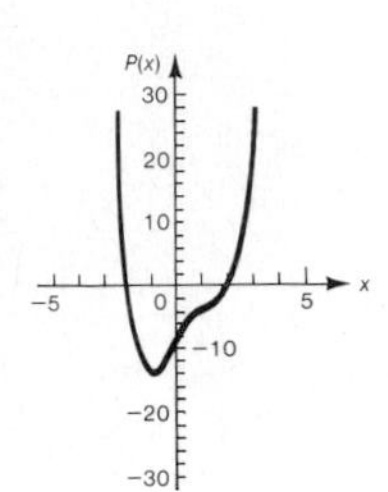

33.

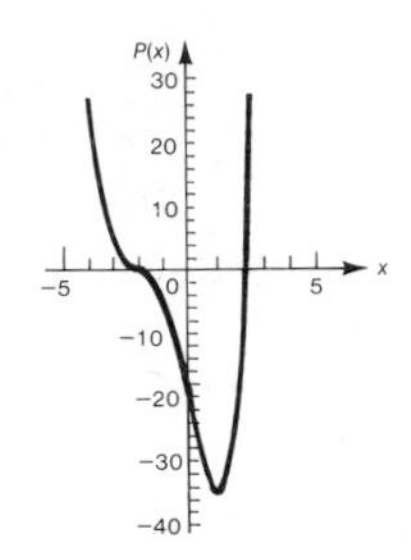

35.

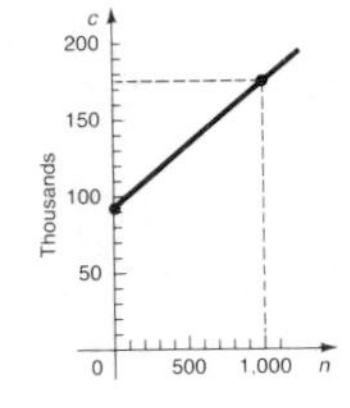

37. **(A)** $A(x) = x(50 - x)$
(B) Domain: $0 \leq x \leq 50$
(C)

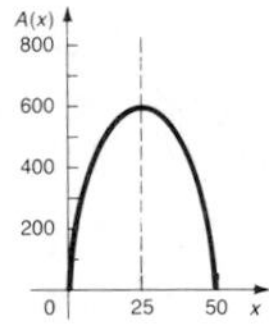

(D) Largest area: $A(25) = 625$ m^2

39. **(A)** $V(x) = (12 - 2x)(8 - 2x)x$
$= 4x^3 - 40x^2 + 96x$
(B) Domain: $0 < x < 4$. [*Note:* At $x = 0$ and $x = 4$, we have zero volume.]
(C)

(D) Largest volume: $V(x) \approx V(1.5) \approx 67.5$ in.^{3}
A 1.5-in. square should be cut from each corner.
Dimensions of box are $5 \times 9 \times 1.5$ inches.

EXERCISE 9-4

1. Function has an inverse **3.** Function does not have an inverse
5. $f^{-1} = \{(1, -2), (2, -1), (3, 0), (4, 1), (5. 2)\}$; domain $= \{1, 2, 3, 4, 5\}$
7. $F^{-1} = \{(\frac{1}{5}, -2), (\frac{1}{4}, -1), (\frac{1}{3}, 0), (\frac{1}{2}, 1), (1, 2)\}$; domain $= \{\frac{1}{5}, \frac{1}{4}, \frac{1}{3}, \frac{1}{2}, 1\}$

9.

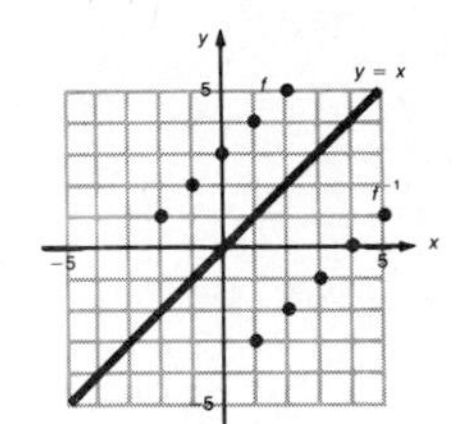

11.

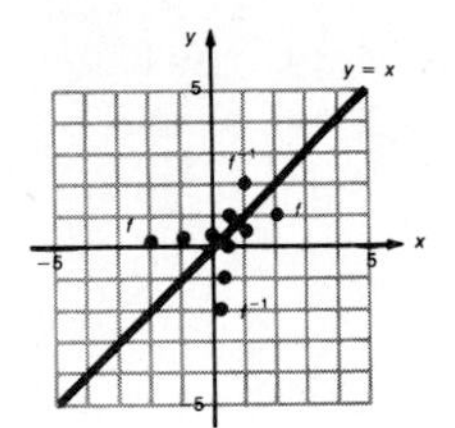

13. f^{-1}: $x = 3y - 2$ or $y = (x + 2)/3$ **15.** F^{-1}: $x = y/3 - 2$ or $y = 3(x + 2)$ **17.** h^{-1}: $x = y^2/2$ or $y = \sqrt{2x}$

19.

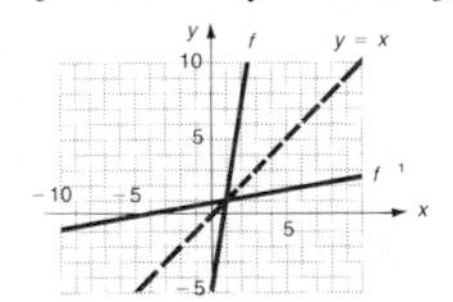

21.

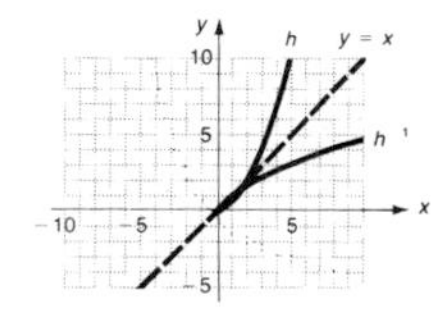

23. **(A)** $f^{-1}(x) = \dfrac{x + 2}{3}$ **(B)** $\frac{4}{3}$ **(C)** 3 **25.** **(A)** $F^{-1}(x) = 3(x + 2)$ **(B)** 3 **(C)** 4

27. **(A)** $f^{-1}(x) = 3(x - 2)$ **(B)** a **29.** x

31. **(A)** $G^{-1} = \{(\frac{1}{9}, -2), (\frac{1}{3}, -1), (1, 0), (3, 1), (9, 2)\}$

(B)

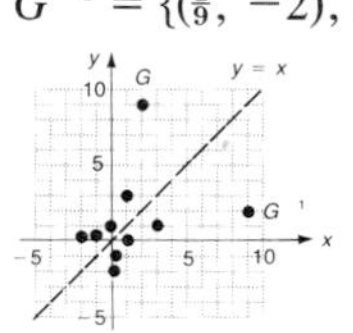

EXERCISE 9-5

1. $F = kv^2$ **3.** $f = k\sqrt{T}$ **5.** $y = k/\sqrt{x}$ **7.** $t = k/T$ **9.** $R = kSTV$ **11.** $V = khr^2$ **13.** 4

15. $9\sqrt{3}$ **17.** $U = k(ab/c^3)$ **19.** $L = k(wh^2/l)$ **21.** -12 **23.** 83 lb **25.** 20 amp

27. The new horsepower must be eight times the old. **29.** No effect **31.** $t^2 = kd^3$

33. 1.47 hr (approx.) **35.** 20 days **37.** Quadrupled **39.** 540 lb

41. **(A)** $\Delta S = kS$ **(B)** 10 oz **(C)** 8 candlepower **43.** 32 times/sec **45.** $N = k(F/d)$

47. 1.2 mi/sec **49.** 20 days **51.** The volume is increased by a factor of 8.

REVIEW EXERCISE 9-6

1. Not a function *(9-1)* **2.** A function *(9-1)* **3.** A function *(9-1)* **4.** A function *(9-1)*

5. Not a function *(9-1)* **6.** A function *(9-1)* **7.** A function *(9-1)* **8.** Not a function *(9-1)*

9. Not a fundtion *(9-1)* **10.** Not a function *(9-1)* **11.** A function *(9-1)*

12. A function *(9-1)* **13.** Domain = {1, 3, 5} *(9-1)* **14.** Domain: $x \geq 1$ *(9-1)*

15. Domain: all real numbers *(9-1)* **16.** **(A)** 0 **(B)** 6 **(C)** 9 **(D)** $6 - m$ *(9-2)*

17. **(A)** -6 **(B)** 0 **(C)** -3 **(D)** $c - 2c^2$ *(9-2)*

18. Slope = 2
y intercept: -4 *(9-3)*

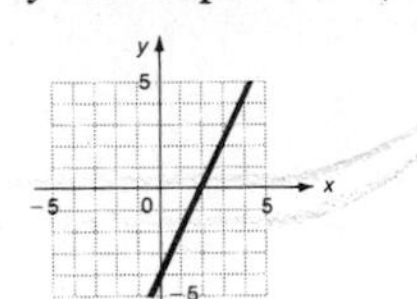

19. *(9-3)*

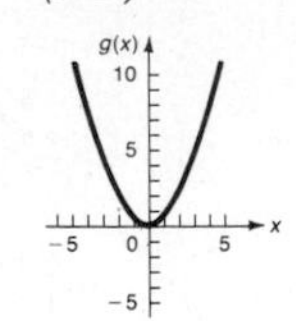

20. (9-4)

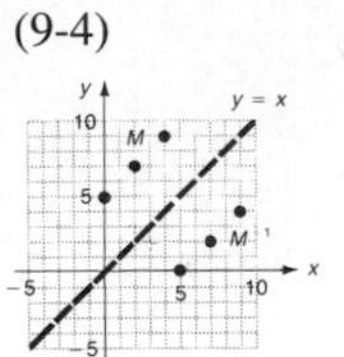

21. Domain = {5, 7, 9}, Range = {0, 2, 4} *(9-4)* **22.** $m = kn^2$ *(9-5)* **23.** $P = \dfrac{k}{Q^3}$ *(9-5)*

24. $A = kab$ *(9-5)* **25.** $y = k\dfrac{x^3}{\sqrt{z}}$ *(9-5)* **26.** No. A function has an inverse only when it is one-to-one, that is, only when every element in the range corresponds to exactly one element in the domain. *(9-4)*

27. **(A)** 7 **(B)** 1 **(C)** $x^2 + x - 7$ **(D)** 3 *(9-2)* **28.** **(A)** $3 + 2h$ **(B)** 2 *(9-2)*

29. Slope: $-\frac{3}{2}$
y intercept: 6 *(9-3)*

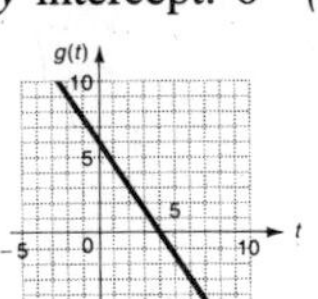

30. Vertex: (2, 1)
Intercepts: x none
$y = 5$ *(9-3)*

31. *(9-3)*

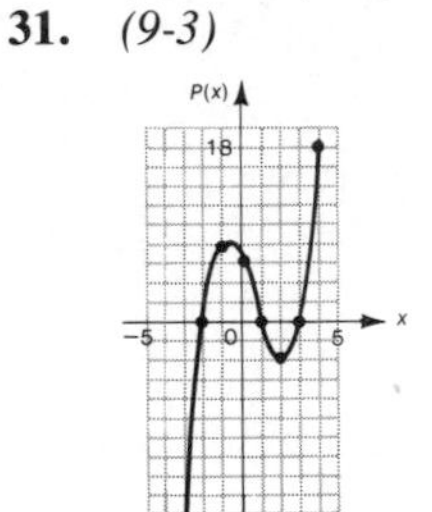

32. Problem 3 *(9-4)* **33.** Problem 6 *(9-4)* **34.** None *(9-4)* **35.** Problem 11 *(9-4)*
36. Problem 6 *(9-4)* **37.** Problem 3 *(9-4)* **38.** $(-\infty, -2] \cup (5, \infty)$ *(9-1)*

39. **(A)** $M^{-1}(x) = 2x - 3$ **(B)** 1 **(C)** 3 *(9-4)* **40.** **(A)** $y = k\dfrac{x}{z}$ **(B)** $y = \frac{4}{3}$ *(9-5)*

41. **(A)** $-3 - 4h - h^2$ **(B)** $-4 - h$ *(9-2)*

42. Vertex: (3,144)
Intercepts: $t = 0, 6$
$y = 0$ *(9-3)*

43. **(A)** $f^{-1}(x) = \sqrt{x}, x \geq 0$ *(9-4)*
(B)

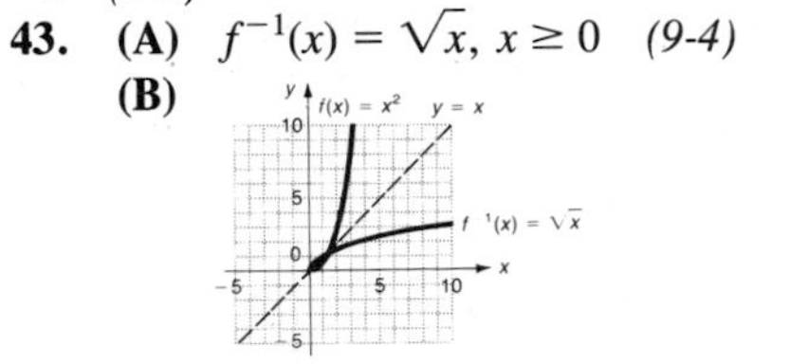

(C) $f^{-1}(9) = 3, f^{-1}[f(x)] = x, x \geq 0$

44. *(9-4)*

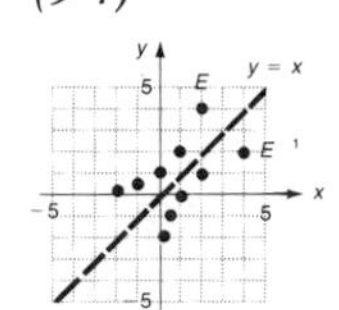

45. *(9-3)*

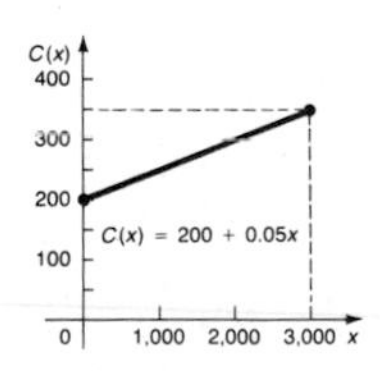

46. **(A)**

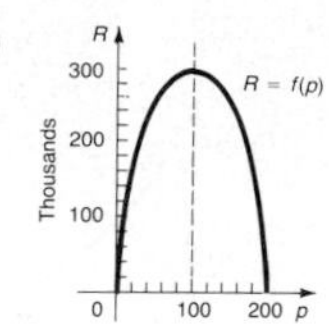

(B) $p = \$100$, largest revenue $= f(100) = \$300{,}000$ *(9-3)*

47. $t = k\dfrac{wd}{p}$, $t = 24$ sec *(9-5)* **48.** The total force is doubled. *(9-5)*

CHAPTER 10 EXERCISE 10-1

1. $y = 3^x$

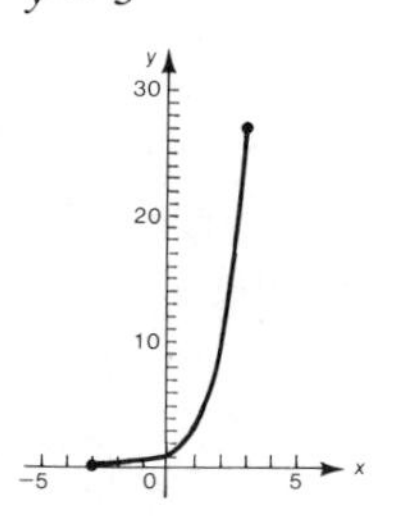

3. $y = 3^{-x}$

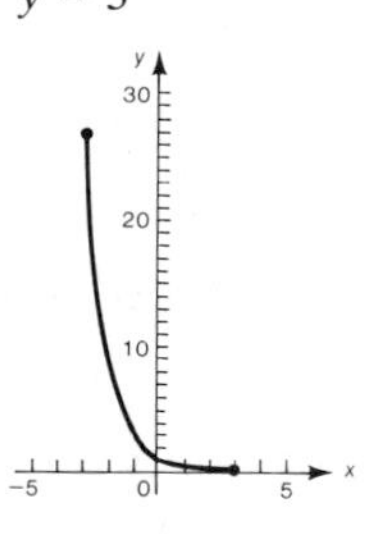

5. $y = 4 \cdot 3^x$

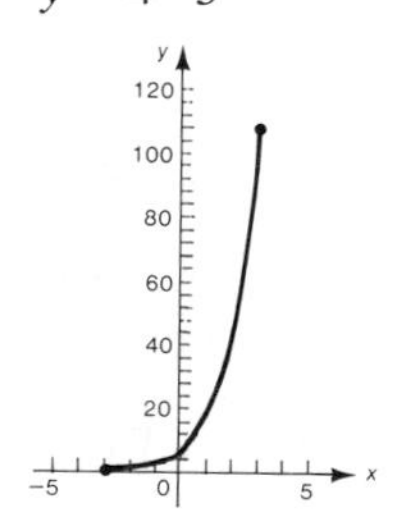

7. $y = 2^{x+3}$

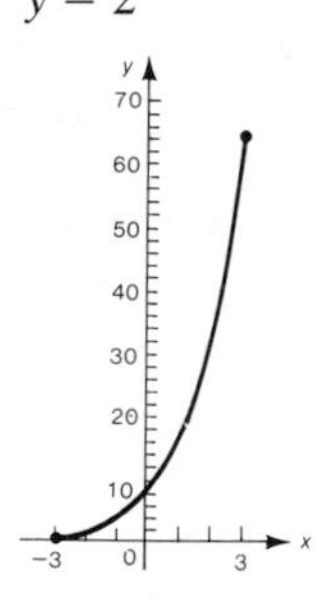

9. $y = 7 \cdot 2^{-2x}$

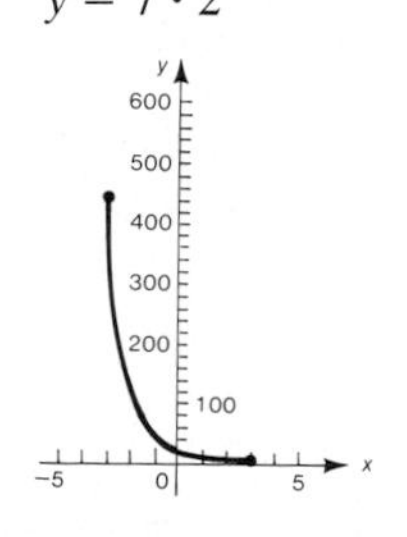

11. $y = e^x$

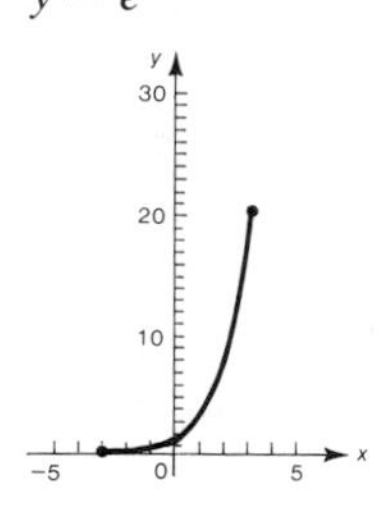

13. $y = 10e^{-0.12x}$

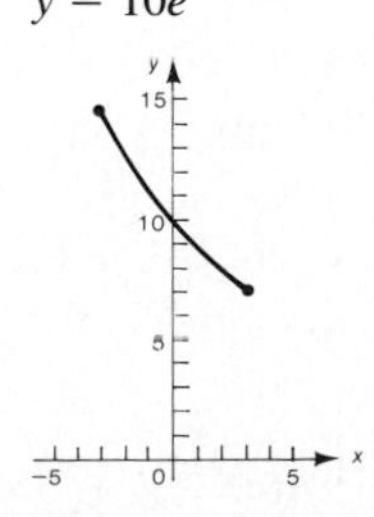

15. $y = 10e^{-x^2}$

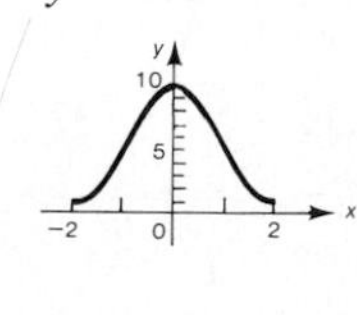

17. $y = y_0 2^x$

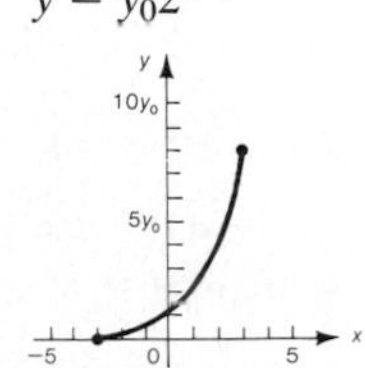

19.

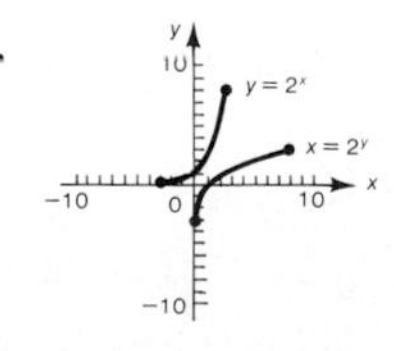

21. $f(n) = 2^n$

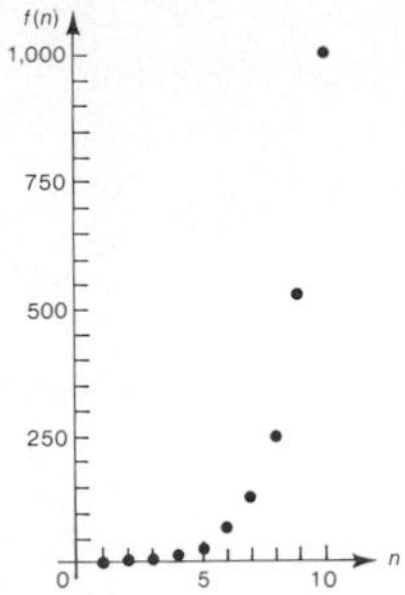

23. $P = 14.7e^{-0.21x}$

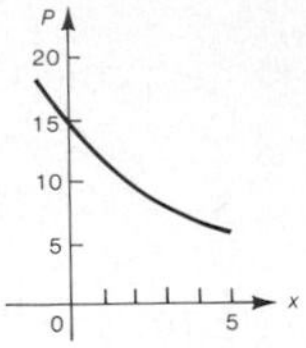

25. $A = 100(\frac{1}{2})^{t/28}$

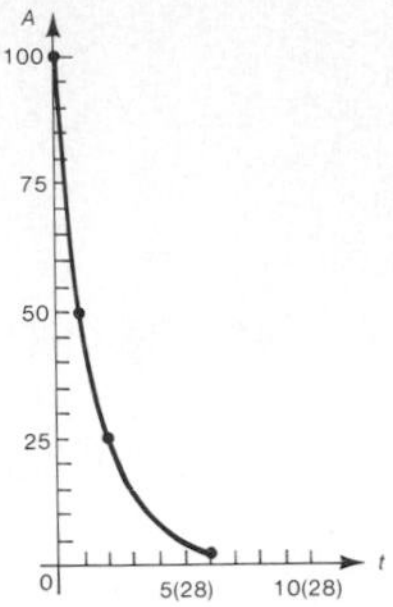

27. $N(i) = 100e^{-0.11(i-1)}$

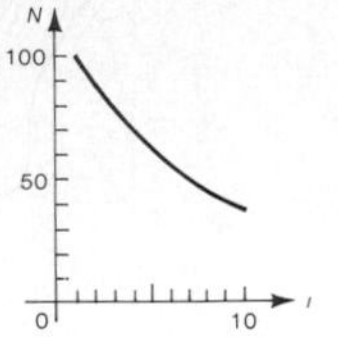

EXERCISE 10-2

1. $9 = 3^2$ **3.** $81 = 3^4$ **5.** $1{,}000 = 10^3$ **7.** $1 = e^0$ **9.** $\log_8 64 = 2$ **11.** $\log_{10} 10{,}000 = 4$
13. $\log_v u = x$ **15.** $\log_{27} 9 = \frac{2}{3}$ **17.** 5 **19.** -4 **21.** 2 **23.** 3 **25.** $x = 4$
27. $y = 2$ **29.** $b = 4$ **31.** $0.001 = 10^{-3}$ **33.** $3 = 81^{1/4}$ **35.** $16 = (\frac{1}{2})^{-4}$ **37.** $N = a^e$
39. $\log_{10} 0.01 = -2$ **41.** $\log_e 1 = 0$ **43.** $\log_2(\frac{1}{8}) = -3$ **45.** $\log_{81}(\frac{1}{3}) = -\frac{1}{4}$ **47.** $\log_{49} 7 = \frac{1}{2}$
49. u **51.** $\frac{1}{2}$ **53.** $\frac{3}{2}$ **55.** 0 **57.** $\frac{3}{2}$ **59.** $x = 2$ **61.** $y = -2$ **63.** $b = 100$
65. Any positive real number except 1
67. Domain of f is the set of all real numbers; range of f is 1. No, f does not have an inverse function.
69. **(A)**

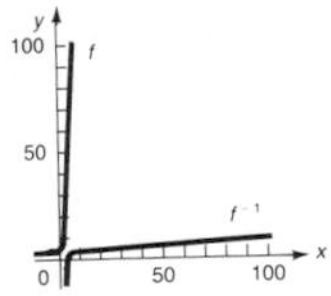

(B) Domain of f is the set of real numbers; range of f is the set of positive real numbers. Domain of f is the range of f^{-1}; range of f is the domain of f^{-1}.
(C) f^{-1} is called the logarithmic function with base 10.

71. -3

EXERCISE 10-3

1. $\log_b u + \log_b v$ **3.** $\log_b A - \log_b B$ **5.** $5 \log_b u$ **7.** $\frac{3}{5} \log_b N$ **9.** $\frac{1}{2} \log_b Q$
11. $\log_b u + \log_b v + \log_b w$ **13.** $\log_b AB$ **15.** $\log_b \dfrac{X}{Y}$ **17.** $\log_b \dfrac{wx}{y}$ **19.** 3.40 **21.** -0.92
23. 3.30 **25.** $2 \log_b u + 7 \log_b v$ **27.** $-\log_b a$ **29.** $\frac{1}{3} \log_b N - 2 \log_b p - 3 \log_b q$
31. $\frac{1}{4}(2 \log_b x + 3 \log_b y - \frac{1}{2} \log_b z)$ **33.** $\log_b \dfrac{x^2}{y}$ **35.** $\log_b \dfrac{x^3y^2}{z^4}$ **37.** $\log_b \sqrt[5]{x^2y^3}$ **39.** 2.02
41. 0.23 **43.** -0.05 **45.** 8 **47.** $y = cb^{-kt}$
49. Let $u = \log_b M$ and $v = \log_b N$; then $M = b^u$ and $N = b^v$.

Thus, $\log_b \dfrac{M}{N} = \log_b \dfrac{b^u}{b^v} = \log_b b^{u-v} = u - v = \log_b M - \log_b N.$

51. $MN = b^{\log_b M} b^{\log_b N} = b^{\log_b M + \log_b N}$; hence, by definition of logarithm, $\log_b MN = \log_b M + \log_b N$.

EXERCISE 10-4

1. 4.9177 **3.** −2.8419 **5.** 3.7623 **7.** −2.5128 **9.** 200,800 **11.** 0.000 664 8
13. 47.73 **15.** 0.6760 **17.** 4.959 **19.** 7.861 **21.** 3.301 **23.** 3.6776 **25.** −1.6094
27. −1.7372
29.

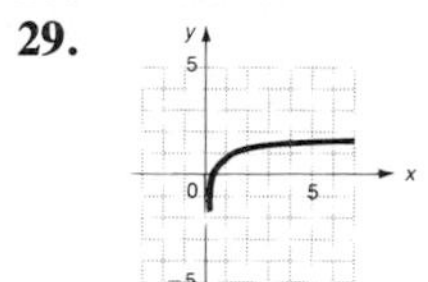

31. **33.**

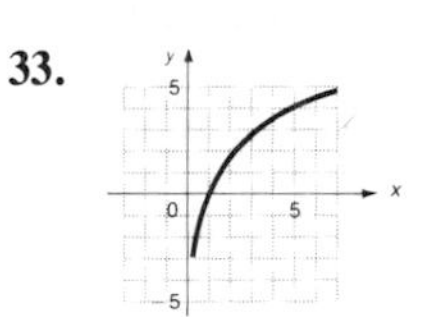

35. 12.725 **37.** −25.715 **39.** 1.1709×10^{32} **41.** 4.2672×10^{-7}

EXERCISE 10-5

1. 1.46 **3.** 0.321 **5.** 1.29 **7.** 3.50 **9.** 1.80 **11.** 2.07 **13.** 20 **15.** 5
17. 14.2 **19.** −1.83 **21.** 11.7 **23.** 5 **25.** 1, e^2, e^{-2} **27.** $x = e^e$ **29.** 100, 0.1
31. $x = -(1/k)\ln(I/I_0)$ **33.** $I = I_0 10^{N/10}$ **35.** $t = (-L/R)\ln(1 - (RI/E))$
37. Inequality sign should have been reversed when both sides were multiplied by $\log \frac{1}{2}$, a negative quantity.
39. 5 years to the nearest year **41.** Approx. 3.8 hr **43.** Approx. 35 years **45.** Approx. 28 years
47. Divide both sides by I_0, take logs of both sides, and then multiply both sides by 10. **49.** 95 ft, 489 ft

REVIEW EXERCISE 10-6

1. $n = \log_{10} m$ *(10-2)* **2.** $x = 10^y$ *(10-2)* **3.** 8 *(10-2)* **4.** 5 *(10-2)* **5.** 3 *(10-2)*
6. 1.24 *(10-5)* **7.** 11.9 *(10-5)* **8.** 900 *(10-3, 10-5)* **9.** 5 *(10-3, 10-5)*
10. $y = e^x$ *(10-2)* **11.** $y = \ln x$ *(10-2)* **12.** −2 *(10-2)* **13.** $\frac{1}{3}$ *(10-2)* **14.** 64 *(10-2)*
15. e *(10-2)* **16.** 33 *(10-2)* **17.** 1 *(10-2)* **18.** 2.32 *(10-5)* **19.** 3.92 *(10-5)*
20. 92.1 *(10-5)* **21.** 300 *(10-3, 10-5)* **22.** 2 *(10-3, 10-5)* **23.** 1, 10^3, 10^{-3} *(10-3, 10-5)*
24. 10^e *(10-3, 10-5)* **25.** 1.95 *(10-4)* **26.** *(10-1)* **27.** *(10-1)*

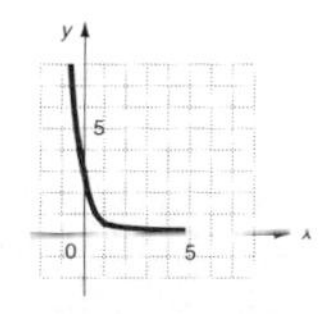

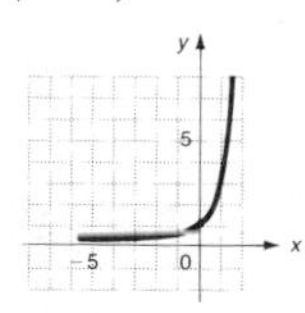

28. *(10-4)* **29.** *(10-4)* **30.** $y = ce^{-5t}$ *(10-3, 10-5)*

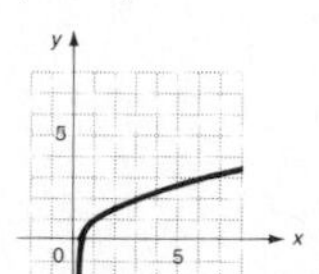

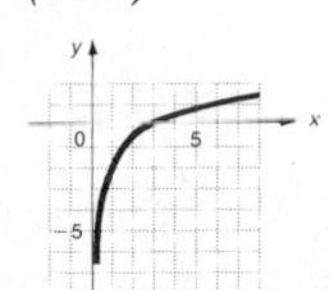

31. Domain $f = (0, \infty)$ = Range f^{-1}; Range $f = R$ = Domain f^{-1} *(10-2)*

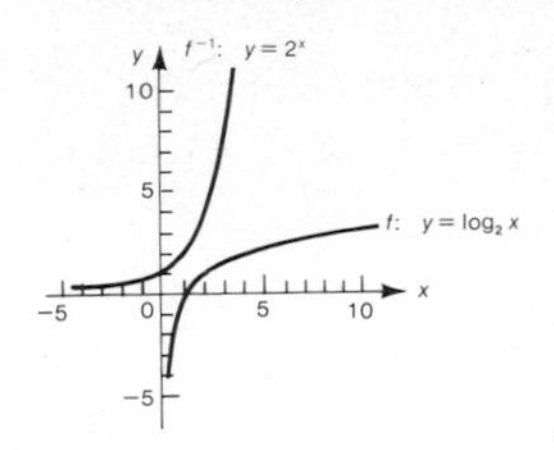

32. If $\log_1 x = y$, then we would have to have $1^y = x$; that is, $1 = x$ for arbitrary positive x, which is impossible. *(10-2)*
33. Let $u = \log_b M$ and $v = \log_b N$; then $M = b^u$ and $N = b^v$.
Thus, $\log_b(M/N) = \log_b(b^u/b^v) = \log_b b^{u-v} = u - v = \log_b M - \log_b N$. *(10-3)*
34. 23.4 years *(10-5)* **35.** 23.1 years *(10-5)* **36.** 37,100 years *(10-5)*
37. $I = I_0 e^{-kx}$ *(10-3, 10-5)* **38.** $n = -\log[1 - (Pi/R)]/\log(1 + i)$ *(10-3, 10-5)*

CHAPTER 11 EXERCISE 11-1

1. $-1, 0, 1, 2$ **3.** $0, \frac{1}{3}, \frac{1}{2}, \frac{3}{5}$ **5.** $4, -8, 16, -32$ **7.** 6 **9.** $\frac{99}{101}$
11. $S_5 = 1 + 2 + 3 + 4 + 5$ **13.** $S_3 = \frac{1}{10} + \frac{1}{100} + \frac{1}{1000}$ **15.** $S_4 = -1 + 1 - 1 + 1$
17. $1, -4, 9, -16, 25$ **19.** 0.3, 0.33, 0.333, 0.3333, 0.33333 **21.** $7, 3, -1, -5, -9$
23. $4, 1, \frac{1}{4}, \frac{1}{16}, \frac{1}{64}$ **25.** $a_n = n + 3$ **27.** $a_n = 3n$ **29.** $a_n = \dfrac{n}{n+1}$ **31.** $a_n = (-1)^{n+1}$
33. $a_n = (-2)^n$ **35.** $a_n = \dfrac{x^n}{n}$ **37.** $\frac{4}{1} - \frac{8}{2} + \frac{16}{3} - \frac{32}{4}$ **39.** $S_3 = x^2 + \dfrac{x^3}{2} + \dfrac{x^4}{3}$
41. $x - \dfrac{x^2}{2} + \dfrac{x^3}{3} - \dfrac{x^4}{4} + \dfrac{x^5}{5}$ **43.** $S_4 = \sum_{k=1}^{4} k^2$ **45.** $S_5 = \sum_{k=1}^{5} 1/2^k$ **47.** $S_n = \sum_{k=1}^{n} 1/k^2$
49. $S_n = \sum_{k=1}^{n} (-1)^{k+1}k^2$
53. **(A)** 3, 1.833, 1.462, 1.415 **(B)** Table, $\sqrt{2} = 1.414$ **(C)** 1, 1.5, 1.417, 1.414

EXERCISE 11-2

1. **(B)** $d = -0.5$; 5.5, 5 **(C)** $d = -5$; $-26, -31$ **3.** $a_2 = -1, a_3 = 3, a_4 = 7$
5. $a_{15} = 67$; $S_{11} = 242$ **7.** $S_{21} = 861$ **9.** $a_{15} = -21$ **11.** $d = 6$; $a_{101} = 603$ **13.** $S_{40} = 200$
15. $a_{11} = 2$, $S_{11} = \frac{77}{6}$ **17.** $a_1 = 1$ **19.** $S_{51} = 4{,}131$ **21.** $-1{,}071$ **23.** 4,446
27. $a_n = -3 + (n - 1)3$ **29.** **(A)** 336 ft **(B)** 1,936 ft **(C)** $16t^2$

EXERCISE 11-3

1. **(A)** $r = -2$; $-16, 32$ **(D)** $r = \frac{1}{3}$; $\frac{1}{54}, \frac{1}{162}$ **3.** $a_2 = 3, a_3 = -\frac{3}{2}, a_4 = \frac{3}{4}$ **5.** $a_{10} = \frac{1}{243}$
7. $S_7 = 3{,}279$ **9.** $r = 0.398$ **11.** $S_{10} = -1{,}705$ **13.** $a_2 = 6, a_3 = 4$ **15.** $S_7 = 547$

17. $\frac{1023}{1024}$ **19.** $x = 2\sqrt{3}$ **21.** $S_\infty = \frac{9}{2}$ **23.** No sum **25.** $S_\infty = \frac{8}{5}$ **27.** $\frac{7}{9}$ **29.** $\frac{6}{11}$
31. $3\frac{8}{37}$ or $\frac{119}{37}$ **33.** $a = P(1 + r)^n$; approx. 12 years **35.** 900 **37.** $4,000,000

EXERCISE 11-4

1. 720 **3.** 20 **5.** 720 **7.** 15 **9.** 1 **11.** 28 **13.** 9!/8! **15.** 8!/5! **17.** 126
19. 6 **21.** 1 **23.** 2,380 **25.** $u^5 + 5u^4v + 10u^3v^2 + 10u^2v^3 + 5uv^4 + v^5$
27. $y^4 - 4y^3 + 6y^2 - 4y + 1$ **29.** $32x^5 - 80x^4y + 80x^3y^2 - 40x^2y^3 + 10xy^4 - y^5$ **31.** $5{,}005u^9v^6$
33. $264m^2n^{10}$ **35.** $924w^6$ **37.** 1.1046 **39.** $\binom{n}{r} = \dfrac{n!}{r!(n-r)!} = \dfrac{n!}{(n-r)![n-(n-r)]!} = \binom{n}{n-r}$

REVIEW EXERCISE 11-5

1. Arithmetic: (B) and (C); Geometric: (A) and (E) *(11-2, 11-3)*
2. **(A)** 5, 7, 9, 11 **(B)** $a_{10} = 23$ **(C)** $S_{10} = 140$ *(11-2)*
3. **(A)** 16, 8, 4, 2 **(B)** $a_{10} = \frac{1}{32}$ **(C)** $S_{10} = 31\frac{31}{32}$ *(11-3)*
4. **(A)** $-8, -5, -2, 1$ **(B)** $a_{10} = 19$ **(C)** $S_{10} = 55$ *(11-2)*
5. **(A)** $-1, 2, -4, 8$ **(B)** $a_{10} = 512$ **(C)** $S_{10} = 341$ *(11-3)* **6.** $S_\infty = 32$ *(11-3)*
7. 720 *(11-4)* **8.** $20 \cdot 21 \cdot 22 = 9{,}240$ *(11-4)* **9.** 21 *(11-4)*
10. $S_{10} = -6 - 4 - 2 + 0 + 2 + 4 + 6 + 8 + 10 + 12 = 30$ *(11-2)*
11. $S_7 = 8 + 4 + 2 + 1 + \frac{1}{2} + \frac{1}{4} + \frac{1}{8} = 15\frac{7}{8}$ *(11-3)* **12.** $S_\infty = \frac{81}{5}$ *(11-3)*
13. $S_n = \sum_{k=1}^{n} \dfrac{(-1)^{k+1}}{3^k}$, $S_\infty = \frac{1}{4}$ *(11-3)* **14.** $d = 3$, $a_5 = 25$ *(11-2)* **15.** 190 *(11-4)*
15. 190 *(11-5)* **16.** 1,820 *(11-4)* **17.** 1 *(11-4)*
18. $x^5 - 5x^4y + 10x^3y^2 - 10x^2y^3 + 5xy^4 - y^5$ *(11-4)* **19.** $-1{,}760x^3y^9$ *(11-4)* **20.** $\frac{8}{11}$ *(11-3)*
21. $49g/2$ ft; $625g/2$ ft *(11-2)* **22.** $x^6 + 6ix^5 - 15x^4 - 20ix^3 + 15x^2 + 6ix - 1$ *(11-4)*

APPENDIX EXERCISE C

1. -14 **3.** 2 **5.** -2.6 **7.** $\begin{vmatrix} a_{22} & a_{23} \\ a_{32} & a_{33} \end{vmatrix}$ **9.** $\begin{vmatrix} a_{11} & a_{12} \\ a_{31} & a_{32} \end{vmatrix}$ **11.** $(-1)^{1+1}\begin{vmatrix} a_{22} & a_{23} \\ a_{32} & a_{33} \end{vmatrix}$
13. $(-1)^{2+3}\begin{vmatrix} a_{11} & a_{12} \\ a_{31} & a_{32} \end{vmatrix}$ **15.** $\begin{vmatrix} 1 & -2 \\ -4 & 8 \end{vmatrix}$ **17.** $\begin{vmatrix} -2 & 0 \\ 5 & -2 \end{vmatrix}$ **19.** $(-1)^{1+1}\begin{vmatrix} 1 & -2 \\ -4 & 8 \end{vmatrix} = 0$
21. $(-1)^{3+2}\begin{vmatrix} -2 & 0 \\ 5 & -2 \end{vmatrix} = -4$ **23.** $x = 5, y = -2$ **25.** $x = 1, y = -1$ **27.** $x = -1, y = 1$
29. 10 **31.** -21 **33.** -120 **35.** -40 **37.** -12 **39.** 0 **41.** $x = 2, y = -2, z = -1$
43. $x = 2, y = -1, z = 2$ **45.** $x = 2, y = -3, z = -1$ **47.** $x = 1, y = -1, z = 2$
49. $a = -1, b = 2, c = 0$ **51.** -8 **53.** 48
55. Expand the determinant about the first row to obtain $x - 3y + 7 = 0$; then show that the two points satisfy this linear equation.
57. $\frac{23}{2}$ **61.** $D = 0$; infinitely many solutions

INDEX

SEQUENCES AND SERIES (11-1)

$a_1, a_2, \ldots, a_n, \ldots$	Infinite sequence
$a_n = f(n), \quad n \in N$	
$S_n = a_1 + a_2 + \cdots + a_n$	Finite series
$S_n = \sum_{k=1}^{n} a_k$	Summation notation

ARITHMETIC SEQUENCES (11-2)

$a_1, a_2, \ldots, a_n, \ldots$

$a_n - a_{n-1} = d$ — Common difference

$a_n = a_1 + (n-1)d$ — nth-term formula

$$S_n = a_1 + \cdots + a_n = \frac{n}{2}[2a_1 + (n-1)d] \qquad \text{Sum of } n \text{ terms}$$

$$S_n = \frac{n}{2}(a_1 + a_n)$$

FACTORIAL (11-4)

$n! = 1 \cdot 2 \cdot 3 \cdot \cdots \cdot n$ — n factorial ($n \in N$)

$0! = 1$

$1! = 1$

$$\binom{n}{r} = \frac{n!}{r!(n-r)!}, \quad 0 \le r \le n,\ n \in N$$

GEOMETRIC SEQUENCE (11-3)

$a_1, a_2, \ldots, a_n, \ldots$

$\frac{a_n}{a_{n-1}} = r$ — Common ratio

$a_n = a_1 r^{n-1}$ — nth-term formula

$$S_n = a_1 + \cdots + a_n = \frac{a_1 - a_1 r^n}{1 - r}, \quad r \ne 1 \qquad \text{Sum of } n \text{ terms}$$

$$S_n = \frac{a_1 - r a_n}{1 - r}, \quad r \ne 1$$

$$S_\infty = a_1 + a_2 + \cdots = \frac{a_1}{1 - r}, \quad |r| < 1$$

BINOMIAL FORMULA (11-4)

$$(a + b)^n = \sum_{k=0}^{n} \binom{n}{k} a^{n-k} b^k$$